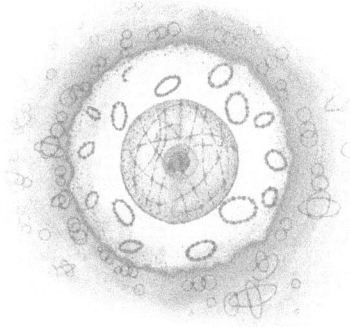

The Thermodynamics of Quantum Yang-Mills Theory

Theory and Applications
Second Edition

The Thermodynamics of
Quantum Yang-Mills Theory

Theory and Applications
Second Edition

Ralf Hofmann
Heidelberg University, Germany

World Scientific

NEW JERSEY · LONDON · SINGAPORE · BEIJING · SHANGHAI · HONG KONG · TAIPEI · CHENNAI · TOKYO

Published by

World Scientific Publishing Co. Pte. Ltd.

5 Toh Tuck Link, Singapore 596224

USA office: 27 Warren Street, Suite 401-402, Hackensack, NJ 07601

UK office: 57 Shelton Street, Covent Garden, London WC2H 9HE

Library of Congress Cataloging-in-Publication Data
Names: Hofmann, Ralf, author.
Title: The thermodynamics of quantum Yang-Mills theory : theory and applications /
 Ralf Hofmann, Heidelberg University, Germany.
Description: Second edition. | Singapore ; Hackensack, NJ : World Scientific
 Publishing Co. Pte. Ltd., [2016] | 2016 | Includes bibliographical references and index.
Identifiers: LCCN 2016004241| ISBN 9789813100473 (hardcover ; alk. paper) |
 ISBN 9813100478 (hardcover ; alk. paper) | ISBN 9789813100480 (pbk.) |
 ISBN 9813100486 (pbk.)
Subjects: LCSH: Quantum field theory. | Yang-Mills theory. | Thermodynamics.
Classification: LCC QC174.45 .H64 2016 | DDC 530.14/35--dc23
LC record available at http://lccn.loc.gov/2016004241

British Library Cataloguing-in-Publication Data
A catalogue record for this book is available from the British Library.

To Karin, Cattleya, and Emmy Eva-Lina

Preface of Second Edition

Work on the second edition of this book was motivated by three considerations. First, during research on radiatively induced scattering processes it became apparent over the last four years that the first edition contained a few incorrect statements concerning the formulation of kinematic constraints on four-momenta in effective loop expansions within the deconfining phase. These statements and some of their implications are now rectified in Chapter 5. Second, a number of deeper insights into the role of Harrington-Shepard (anti)calorons, being the basic constituents of the deconfining thermal ground state, has emerged over the last years. In particular, it appears that the action of those (anti)calorons of radius $\rho \sim |\phi|^{-1}$, which dominate the thermal ground state, is given by Planck's quantum of action $\hbar$. As a consequence, the effective quantum field theory of the deconfining phase is induced by classical, Euclidean field configurations of the fundamental theory. That is, the inert and adjoint scalar field ϕ of the effective theory is now conceived as describing spatially densely packed (anti)caloron centers each of whose unit of winding ($S_3 \to S_3$) implies the action $\hbar$. Moreover, an effective, point-like vertex is regarded as induced by an isolated, just-not-resolved (anti)caloron whose Euclidean time dependence is incommensurate with determinism in a Minkowskian spacetime. On the other hand, static peripheries of (anti)calorons, now discussed explicitly in Chapter 4, give rise to the effective pure-gauge configuration a_μ^{gs} by overlap, in turn implying a finite energy density of the thermal ground state. Peripheries are associated with electric and magnetic dipole densities which can be interpreted in terms of an electric permittivity and a magnetic permeability of the thermal ground state. This is the objective of a new section in Chapter 9. Also, work on radiatively induced one-loop scattering of massless modes, which is insightful concerning the convergence properties of loop expansions in general, was added to Chapter 5. Finally, we now offer an investigation in Chapter 10 of the redshift-temperature relation of the cosmic microwave background (CMB), based on the equation of state of the SU(2) Yang-Mills theory postulated to describe all thermal photon gases, and implications thereof for (i) resolving a discrepancy between the redshift of instantaneous re-ionization of the Universe as extracted from quasar spectra and from the angular power spectrum of the CMB and (ii) the physics of cosmic neutrinos.

For historical reasons we denote vectors in three-dimensional space sometimes by boldfaced symbols and sometimes in normal font but with an arrow or hat (in case of dimensionless vectors) on top. We apologize for imposing this extra demand on flexibility in reading this book.

We hope that this second edition will enjoy a similarly large readership as the first edition did.

December 2015
Bad Bergzabern

Preface of First Edition

This book's goal is to provide a certain amount of insight into, computational tools for, quantitative results on, and physics applications of Quantum Yang–Mills theory defined on a four-dimensional flat spacetime continuum. Subjecting the theory to thermodynamics, formulations of effective theories are feasible. Their derivations and evaluations, though partially dependent on numerical methods, appear to yield a definite picture of the involved physics at a high level of quantitative accuracy. As a consequence, applications to the real world suggest themselves.

Part I of the book intends to equip advanced students and researchers, interested in expediting the development of nonperturbatively accessed Quantum Yang–Mills theory (possibly even beyond the realm of thermodynamics), with well established and recent results. This theoretical part contains a selection of problems at the end of each chapter whose contemplation may stimulate the reader to further investigation.

Certainly there is much room for the elaboration of the material presented in Part I towards its physical implications. To point out potential applications, as judged from the vantage point of Yang–Mills thermodynamics, we have added Part II. The focus is on low-temperature photon physics in astrophysical systems and in cosmology. Some of the methods and results of Part II are indicative of the infancy of the subject and, with future re-investigation, may be in need of amendment.

For this book to be of optimal benefit to the reader a certain familiarity with the perturbative method in four-dimensional gauge theory, including its perturbative renormalizability, is presumed. Moreover, the new developments presented rely on milestones like the Euclidean formulation of thermal quantum field theory, the (adjoint) Anderson–Englert–Brout–Higgs–Guralnik–Hagen–Kibble mechanism, the small-mass expansion in thermal field theory, closed-form expressions for particular topological gauge-field configurations at finite temperature, and knowledge on how fundamental radiative corrections affect these configurations (semiclassical approximation). The presentation of this classic material in Chapters 2–4 is sketchy. To guide further study, reference to the literature is, however, provided at the end of each of these chapters.

In presenting recent results on Yang–Mills thermodynamics and its potential applications to low-temperature photon physics, we have, in Chapters 5–10, resorted largely to research papers written on the subject throughout the period 2004–2010. A few stretches of original work are, however, interspersed in between the exposition of material already published. In detail, Chapters 5–7 discuss the theoretical foundations of nonperturbative thermodynamics in the three phases of SU(2) and SU(3) Yang–Mills theory, Chapter 8 is devoted to a comparison with results obtained by the method of lattice gauge theory, and Chapters 9 and 10 are concerned with pre- and post-dicted experimental and observational consequences of SU(2) Yang–Mills theory for low-temperature photon physics.

Thermalization implies spacetime symmetries that are exploited in this book to identify and process fundamental, Bogomoln'yi–Prasad–Sommerfield saturated field configurations in their role to imply a useful *a priori* estimate for the deconfining thermal ground state. The notion of the latter is the starting point for a treatment of the Yang–Mills partition function in effective variables. The reader may object that thermalization is too narrow a framework to address many phenomena of the real world. Such an objection is, of course, justified. For example, man-made or astrophysically driven isolated particle collisions, successfully described by the present Standard Model (SM), are hard if not impossible to describe in terms of thermodynamics. The framework of extended thermalization, however, may well be capable of de-camouflaging the apparent point-like nature of the SM's degrees of freedom, of revealing the true nature of particle physics' ground state and electroweak symmetry breaking and of providing a framework for the description of collective phenomena in particle collisions at the energy scale of Tera-electron volts. Also, a promising new look at the cosmological vacuum-energy problem emerges as a result of thermal Yang–Mills theory, and a realistic theoretical access to low-temperature photon gases of cosmological or astrophysical extents could be provided. With the impressive advancement of observational and experimental machinery, precise tests of particular predictions will become available in the near future. Moreover, it is conceivable that controlled deformations of pure Yang–Mills thermodynamics (local thermalization, simultaneous existence of several, weakly interacting global thermal equilibria of the same theory) are accessible quantitatively. This could yield successful descriptions for a wider range of natural phenomena than those that are obviously subject to thermalization.

If not stated otherwise, we work in supernatural units $\hbar = c = k_B = 1$ throughout this book where $h = 2\pi\hbar$ is Planck's quantum of action, c denotes the velocity of light in vacuum, k_B refers to Boltzmann's constant, and Einstein's summation convention for repeated indices is used.

July 2011
Heidelberg

Contents

PART 1

Theory

Chapter 1
The Classical Yang–Mills Action

1.1 Historical Remarks

The extraction of the specific gauge principle of electromagnetism from accumulated and theoretically distilled experience, the discovery that generalizations of this principle provide in a fundamental way for dynamical physical law, and subsequent successful exploitations in the realm of particle physics belong to the greatest triumphs of the human mind.

The evolution of gauge theory started to take its course when, during the first half of the nineteenth century, M. Faraday and others undertook their insightful experiments on static electric and magnetic phenomena, on electromagnetic induction, and on light propagation in vacuum and media. These results constituted the experimental basis for the paper "A dynamical theory of the electromagnetic field" authored by J. C. Maxwell in 1864. An important result of Maxwell's theory was the prediction that light is nothing but the wavelike propagation of the electromagnetic field. Twenty years later H. Hertz could confirm this consequence of the theoretical unification of electricity and magnetism experimentally by verifying that periodically accelerated electric charges emit electromagnetic waves.

It was A. Einstein who in 1905 understood that the covariance of Maxwell's equations is reconcilable with G. Galilei's principle of relativity only for those coordinate changes between uniformly moving frames that leave the speed of light unchanged. The respective

3

Lorentz-covariant form

$$\partial^\mu F_{\mu\nu} = j_\nu \tag{1.1}$$

of Maxwell's equations, j_ν the electromagnetic current four-vector and $F_{\mu\nu} \equiv \partial_\mu A_\nu - \partial_\nu A_\mu$ the antisymmetric, rank-two, electromagnetic field-strength tensor, makes an invariance of these equations under the transformation

$$A_\mu \to A_\mu + \partial_\mu \varphi \tag{1.2}$$

most explicit (φ a Lorentz-invariant and real function of space and time). The quantity A_μ ($\mu = 0, 1, 2, 3$) is known as the gauge potential.

That the symmetry of Eq. (1.2) is actually associated with the local action of the unitary group U(1) was understood in the framework of quantum mechanics, where it became clear that the *dynamical* four-current j_ν can be written as a bilinear of a quantum mechanical wave function ψ (principle bundle). Demanding the covariance of the equation obeyed by ψ under local phase changes (U(1) rotations) makes the gauge potential A_μ a connection[1] on the manifold of the gauge group. Notice that the transformation law of Eq. (1.2) can be rewritten in the form

$$A_\mu \xrightarrow{\Omega} \Omega A_\mu \Omega^\dagger + i\Omega \partial_\mu \Omega^\dagger, \tag{1.3}$$

where Ω is an element of the group U(1): $\Omega \equiv e^{i\varphi}$.

The concept of covariance or invariance of dynamical equations under gauge transformations is generalizable to non-Abelian groups. It was W. Pauli who first formulated such a generalization

---

[1]H. Weyl in his 1929 article 'Elektron und Gravitation' did not consider a multiplication of the wave function by a pure phase but introduced a conformal rescaling suggesting a connection with gravitation. The concept of the covariance of a quantum mechanical wave equation (or invariance of the associated action functional) under the gauge group U(1) was first discussed by E. Schrödinger in his 1932 article "Sur la theorie relativiste de l'electron et l'interpretation de la mecanique quantique".

and together with his collaborators Barker and Gulmanelli tried to investigate its physical consequences. When, during a Princeton seminar in early 1954, C.-N. Yang reported about his work with R. L. Mills on a local four-dimensional field theory with SU(2) gauge invariance Pauli was in the audience and very explicitly expressed his dislike. According to A. Pais [von Meyenn (1999)] Yang described the situation as follows: "Pauli asked, 'What is the mass of this field B_μ'. I said we did not know. Then I resumed my presentation, but soon Pauli asked the same question again. I said something to the effect that that was a very complicated problem, we had worked on it and come to no definite conclusions. I still remember his repartee: 'That is not sufficient excuse!'." In a note to Yang issued on the same day Pauli pointed out that the theory of Yang and Mills was just a special case of something he already had discussed, and he came back to his seminar questions [von Meyenn (1999)]: "But I was and still am disgusted and discouraged of the vector field corresponding to particles with zero rest-mass (I do not take your excuse for it with 'complications' seriously), and the difficulty with the group due to the distinction of the electromagnetic field remains".

In hindsight it is plausible that Yang and Mills sensed that the "complications" of a non-Abelian gauge theory could actually help in overcoming precipitating a conclusion about all gauge modes being massless,[2] and it is probably for this reason that we use the

---

[2]For the semisimple non-Abelian Lie group SU(2) there are three such gauge modes. Without a mechanism at hand to single out one of them to remain massless while giving mass to the other two, this gauge group would have to be discarded as a valid basis for the description of the electromagnetic field. We shall discuss in Chapters 4 and 5 that Pauli's view on the masslessness of the propagating Yang–Mills field ignores the existence of nonpropagating field configurations in SU(2) Yang–Mills theory that possess winding number. Due to their extended nature these "instantons" or "calorons" support a finite correlation length. For the gauge group SU(2) this implies mass for two out of the three propagating vector modes

term Yang–Mills theory today when we refer to a field theory based on a non-Abelian gauge principle [Yang and Mills (1954)]. For an overview on milestones in the development of Quantum Yang–Mills theory the reader is referred to the book "50 Years of Yang–Mills Theory" ['t Hooft (2005)].

1.2   The Semisimple Lie Group SU(N) (N ≥ 2)

Let us now discuss non-Abelian gauge groups. We only consider the group SU(N) which in its fundamental representation is defined by the group under multiplication of unitary $N \times N$ matrices with strictly positive determinant. Thus these matrices are associated with linear, special unitary transformations of the vector space $\mathbf{C}^N$:

$$\Omega \in SU(N) \iff \Omega\Omega^\dagger = \Omega^\dagger\Omega = 1 \quad \text{and} \quad \det \Omega = 1. \quad (1.4)$$

by the emergence of a scalar field of finite ground-state expectation. Phenomena of this type are known under the name *dynamical* gauge-symmetry breaking. Assuming an external mechanism to induce a *spontaneous* gauge-symmetry breaking, F. Englert and R. Brout (1964), P. W. Higgs (1964), and G. S. Guralnik *et al.* (1964) clarified how vector particles do acquire mass thanks to the (partial) breaking of gauge symmetry induced by a prescribed ground state within a relativistic gauge-field theory. For the nonrelativistic case the existence of such a mechanism was already pointed out one year earlier by P. W. Anderson (1963).

To write down an explicit model for spontaneous gauge-symmetry breaking, a gauge-invariant sector for one or more scalar fields is postulated in the action in addition to the pure Yang–Mills sector. These scalar fields, which transform under certain representations of the gauge group, then are to acquire ground state expectations by virtue of their potentials. For vector particles to acquire mass by a nontrivial ground state it is inessential whether the latter emerges in Yang–Mills theory itself (dynamical breaking) or by an external mechanism (spontaneous breaking). To acquire a quantitative understanding of dynamical breaking, however, requires an essential break with the perturbative approach to Quantum Yang–Mills theory. In 1954, that is, ten years prior to the Anderson–Englert–Brout–Higgs–Guralnik–Hagen–Kibble discovery essentially only perturbation theory or perturbation-theory inspired methods were known to address gauge theories. Thus Pauli's reaction to Yang's talk was quite natural.

The group SU(N) is a continuous Lie group. A Lie group can be regarded a differentiable manifold. For example, the group U(1), whose elements are parameterized as $\Omega = e^{2\pi i s}(0 \leq s \leq 1)$, is the unit circle S_1 centered at the origin of the complex plane $\mathbf{C}$. Elements of the group SU(2) can be parameterized as

$$\Omega = \frac{i\tau_\mu x_\mu}{\sqrt{x \cdot x}}$$

where the four 2×2 matrices τ_μ are the Pauli matrices for $\mu = 1, 2, 3$ and $\tau_4 = -i\mathbf{1}_2$. The matrices τ_μ form the basis of a four-dimensional vector space with scalar product $v \cdot w = \frac{1}{2} \operatorname{tr} v^\dagger w = \sum_{\mu=1}^{4} v_\mu w_\mu$. Since $x_\mu/\sqrt{x \cdot x}$ is the set of all unit vectors in $\mathbf{R}^4$ we conclude that the group manifold of SU(2) is the unit three-sphere S_3. For N > 2 the global structure of the group manifold of SU(N) is complicated. For example, one can only say that *locally* the eight-dimensional group manifold of SU(3) is given by the product of a unit three- and a unit five-sphere: $S_3 \times S_5$.

Each element $\Omega(\xi)$ of a Lie group G can be expanded about the unit element as

$$\Omega(\xi) = 1 + i\xi_a t^a + O(\xi^2). \tag{1.5}$$

The matrices t^a are called generators. Group elements a finite distance $|\xi|$ away from unity are generated by the exponential map $e^{i\xi_a t^a}$ which is the same as $\lim_{K \to \infty} (1 + i\xi_a t^a/K)^K$. That is, an infinite succession of applications of the infinitesimal transformation $1 + i\xi_a t^a/K, (K \to \infty)$ is applied to generate a group element finitely away from unity. From Eqs. (1.4) and (1.5) it follows that for SU(N) the generator matrices t^a are traceless, $\operatorname{tr} t^a = 0$, and hermitian, $(t^a)^\dagger = t^a$. The t_a span a vector space, and for SU(N) this space is of dimension $N^2 - 1$. For the expansion in Eq. (1.5) to hold also for the product $\Omega(\xi)\Omega(\xi')$ of two group elements $\Omega(\xi)$ and $\Omega(\xi')$, the commutator of two generators t^a and t^b must be a linear combination of the t^c:

$$[t^a, t^b] = i f^{abc} t^c. \tag{1.6}$$

The numbers f^{abc} are called structure constants. The vector space spanned by the t^a supplemented by the relation (1.6) is called a Lie algebra. For the Lie group G one refers to its Lie algebra as g. The semisimplictity of the group SU(N) refers to the fact that su(N) is a direct sum of simple Lie algebras. For a simple Lie algebra g, in turn, the only ideals[3] are g itself and zero. The rank of a Lie group is defined to be the maximal number of mutually commuting, linearly independent generators. For SU(N) one has rank $SU(N) = N - 1$.

From the following identity for the commutator $[,]$

$$[t^a, [t^b, t^c]] + [t^b, [t^c, t^a]] + [t^c, [t^a, t^b]] = 0 \qquad (1.7)$$

the Jacobi identity

$$f^{ade}f^{bcd} + f^{bde}f^{cad} + f^{cde}f^{abd} = 0 \qquad (1.8)$$

for the structure constants follows from Eq. (1.6). In addition to the *fundamental* representation, that is used to derive the Lie algebra with its structure constants f^{abc}, the action of the group SU(N) can be represented on vector spaces other than $\mathbf{C}^N$. The dimension $d(r)$ of a representation is defined to be the dimension of the vector space the group G is represented on. For example, for the group SU(2) the dimension $d(r)$ of its irreducible representations is determined by their total "spin" J: $d(J) = 2J + 1$ where J is a half-integer ($J = 0, 1/2, 1, 3/2, 2, 5/2, 3, \dots$). Due to a theorem by H. Weyl the semisimplicity of the Lie group SU(N) implies that any finite-dimensional representation, for example built from an M-fold product of the fundamental representation ($M = 2, 3, 4, 5, \dots$), is completely reducible. That is, whatever the subspace of the representation space identified to be invariant under the action of the group is, its complement is also invariant.

---

[3]The subset i of g is called an ideal iff $[x, y] \in i$ for all $x \in g$ and $y \in i$.

For each irreducible representation r of SU(N) it can be shown that the generators t_r^a can be normalized as

$$\mathrm{tr}\, t_r^a t_r^b = C(r)\, \delta^{ab}, \tag{1.9}$$

where $C(r)$ is a number chosen such that for a given set of structure constants f^{abc} the Lie bracket relation (1.6) is satisfied.

From Eqs. (1.6) and (1.9) it follows that

$$f^{abc} = -\frac{i}{C(r)} \mathrm{tr}\, ([t_r^a, t_r^b] t_r^c), \tag{1.10}$$

and, as a consequence, we observe total antisymmetry of f^{abc}:

$$f^{abc} = -f^{bac} = -f^{acb} = -f^{cba}. \tag{1.11}$$

In Yang–Mills theory an important representation of a Lie group G takes place on the vector space spanned by the generators, that is, on its own Lie algebra. In this *adjoint* representation the generator matrices are given by the structure constants as

$$(t_{\mathrm{ad}}^b)^{ac} = i f^{abc}. \tag{1.12}$$

The dimension $d(\mathrm{ad})$ of the adjoint representation is given as $d(\mathrm{ad}) = N^2 - 1$ for SU(N). For an arbitrary representation r of any Lie group G one computes

$$[t_r^a, t_r^b t_r^b] = 0, \tag{1.13}$$

where the index b now is summed over. Equation (1.13) is the statement that the operator $t_r^b t_r^b$ is left invariant under any infinitesimal and thus also finite action of the Lie group G: $t_r^b t_r^b \to \Omega t_r^b t_r^b \Omega^\dagger$, $(\Omega \in G)$. The only operator for which this is the case must be a multiple of unity. Therefore

$$t_r^b t_r^b = C_2(r)\mathbf{1}, \tag{1.14}$$

where $C_2(r)$ is a constant which depends on the representation r. From Eqs. (1.14) and (1.9) is follows that

$$d(r)C_2(r) = d(\mathrm{ad})C(r).\qquad(1.15)$$

For SU(2) we may normalize the generators t^a of the two-dimensional fundamental representations as $t^a = \frac{1}{2}\sigma^a$, where σ^a, $(a = 1, 2, 3)$, are the Pauli matrices. With this normalization one has $f^{abc} = \epsilon^{abc}$, $(\epsilon^{123} = 1)$, and $\mathrm{tr}\, t^a t^b = \frac{1}{2}\delta^{ab}$. For the fundamental representation of SU(N) the SU(2) generators t^1, t^2, and t^3 can be extended to $N \times N$ matrices by filling in zeros. These $N \times N$ matrices can be taken as the first three generators of SU(N). Therefore, one has $\mathrm{tr}\, t^a t^b = \frac{1}{2}\delta^{ab}$ also for the fundamental representation of SU(N). Invoking Eq. (1.15), we thus find for this representation that

$$C = \frac{1}{2}, \quad C_2 = \frac{N^2 - 1}{2N}.\qquad(1.16)$$

Reducing out the product of a fundamental SU(N) representation with its complex conjugate, one obtains (without derivation)

$$C(\mathrm{adj}) = C_2(\mathrm{adj}) = N.\qquad(1.17)$$

The center subgroup of a group G is defined as the subgroup whose elements commute with all other elements of G. For example, the center of the Lie group G_2 is trivial, and the Lie group SU(N) has a nontrivial center group which consists of all N roots of unity:

$$\text{center of SU(N)} = \mathbf{Z}_N \equiv \{e^{\frac{2\pi i k}{N}}\mathbf{1} \mid k = 0, 1, \dots, N-1\}.\qquad(1.18)$$

The fact that the center of SU(N) is $\mathbf{Z}_N$ has consequences for the nature of the corresponding gauge theory's phase transitions and for its low-temperature physics.

1.3   Gauge Connection A_μ and its Field Strength $F_{\mu\nu}$

We consider the fundamental representation of SU(N) generators normalized as $\mathrm{tr}\, t^a t^b = \frac{1}{2}\delta^{ab}$. The Lie-algebra valued gauge connection (or gauge field) $A_\mu(x)$ is defined as

$$A_\mu(x) = A_\mu^a(x)\, t^a \quad (a = 1, \ldots, N^2 - 1), \tag{1.19}$$

where the $A_\mu^a(x)$ are real functions of the coordinates x_ν of four-dimensional, flat spacetime. Generalizing the transformation law of Eq. (1.2) to the non-Abelian gauge group SU(N), we demand that $A_\mu(x)$ inhomogeneously gauge-transforms as in Eq. (1.3), where now $\Omega \equiv \exp(i\omega^a(x)t^a) \in$ SU(N), and the functions $\omega^a(x)$ are real. Since A_μ lives in the Lie algebra we expect its transformation under gauge rotations to be associated with the adjoint representation. Indeed, the homogeneous part of the transformation (1.3) reads in components as

$$A_\mu^a \xrightarrow{\ \Omega\ } \Omega_{ad}^{ab} A_\mu^b + \text{inhomogeneous term}, \tag{1.20}$$

where $\Omega_{ad}^{ab} \equiv 2\,\text{tr}\, t^a \Omega t^b \Omega^\dagger$ is the matrix of Ω in the adjoint representation.

To construct local densities which are functionals of A_μ and which are invariant under the transformation (1.3), we are looking for a quantity A that transforms homogeneously in the fundamental representation

$$A \xrightarrow{\ \Omega\ } \Omega A \Omega^\dagger. \tag{1.21}$$

Gauge-invariant quantities then simply are formed by raising A to integer powers M (including unity) and by subsequently performing the trace operation within the Lie algebra:

$$\text{tr}\, A^M \xrightarrow{\ \Omega\ } \text{tr}\, \Omega A \Omega^\dagger \Omega A \Omega^\dagger \cdots \Omega A \Omega^\dagger = \text{tr}\, \Omega^\dagger \Omega A^M = \text{tr}\, A^M. \tag{1.22}$$

Keeping the Abelian form of the field-strength tensor, $F_{\mu\nu} \equiv \partial_\mu A_\nu - \partial_\nu A_\mu$, the gauge group SU(N) does not yield a homogeneously transforming object because the inhomogeneous part of the gauge transformation in Eq. (1.3) does not cancel as in the Abelian U(1) case. This is due to

$$\partial_\mu(\Omega\partial_\nu\Omega^\dagger) \neq \partial_\nu(\Omega\partial_\mu\Omega^\dagger). \tag{1.23}$$

Additional antisymmetric terms need to be added to the analog of the Abelian field strength to cancel the difference between the left-hand side and the right-hand side of inequality (1.23). The following form works [Yang and Mills (1954)]:

$$F_{\mu\nu} \equiv \partial_\mu A_\nu - \partial_\nu A_\mu - i[A_\mu, A_\nu] \xrightarrow{\Omega} \Omega F_{\mu\nu}\Omega^\dagger. \tag{1.24}$$

1.4   Gauge-Invariant Objects

1.4.1   *Action density and energy–momentum tensor*

We have just seen in Sec. 1.3 that a Lie algebra trace over local polynomials of the SU(N) field strength tensor $F_{\mu\nu}$, as defined in Eq. (1.24), is guaranteed to be gauge-invariant. In general, for an adjoint object $B(x)$, which transforms homogeneously,

$$B \equiv\xrightarrow{\Omega} \Omega B\Omega^\dagger$$

one defines its covariant derivative D_μ to transform in the same way:

$$D_\mu B \equiv \partial_\mu B - i[A_\mu, B] \xrightarrow{\Omega} \Omega\left(D_\mu B\right)\Omega^\dagger. \tag{1.25}$$

It is easy to check that

$$[D_\mu, D_\nu]B = [F_{\mu\nu}, B], \tag{1.26}$$

and thus in particular $[D^\mu, D^\nu]F_{\mu\nu} = [F^{\mu\nu}, F_{\mu\nu}]$. To define a *scale-invariant*, fundamental Yang–Mills action that is Poincaré (also parity) and gauge-invariant only the bilinear

$$\text{const.} \times \text{tr}\, F_{\mu\nu}F^{\mu\nu} \tag{1.27}$$

qualifies for the local action density. With the normalization $\text{tr}\, t^a t^b = \frac{1}{2}\delta^{ab}$ for the generators t^a it is convenient to adopt const. $= \frac{1}{2g^2}$ in (1.27), where g denotes the fundamental gauge coupling constant.

This is because a redefinition of the gauge field A_μ according to

$$A_\mu \to g A_\mu \tag{1.28}$$

cancels the factor $1/g^2$ in (1.27). The rescaled field-strength tensor then reads as follows in components

$$F^a_{\mu\nu} = \partial_\mu A^a_\nu - \partial_\nu A^a_\mu + g f^{abc} A^b_\mu A^c_\nu, \tag{1.29}$$

and the fundamental Yang–Mills action S_{YM} takes the form

$$S_{\mathrm{YM}} = -\int d^4x \frac{1}{4} F^a_{\mu\nu} F^{\mu\nu,a} = \int d^4x \mathcal{L}_{\mathrm{YM}}. \tag{1.30}$$

The term proportional to g in Eq. (1.29) vanishes for gauge field components residing in the Lie algebra of the maximally Abelian subgroup $U(1)^{N-1}$. Restricting to this part of su(N) would yield a theory of $N-1$ species of noninteracting gauge fields (photons) of Bose–Einstein quantum statistics. For $g \to 0$ this situation extends to the entire su(N): All propagating gauge modes become "photons" in this limit. Switching on the coupling g, one would expect the situation of free "photons" to be a good *a priori* estimate for the expansion of physical quantities in powers of g provided that g is kept sufficiently small. This is the essence of the perturbative philosophy discussed in Chapter 2, and it is the basis for W. Pauli's rejection of the physical relevance of Yang–Mills theories since the masslessness of "photons" does persist order for order in perturbation theory. It turns out, however, that the Yang–Mills action of Eq. (1.30) provides for a much richer ensemble of relevant field configurations than the perturbative approach presumes. Some of these gauge-field configurations are associated with topologically nontrivial mappings of the boundary S_3 of (Euclidean) four-dimensional spacetime $\mathbf{R}^4$ into the group manifold S_3 of SU(2) which is a subgroup of SU(N) for $N \geq 2$. The lower bound on the (Euclidean) action of such configurations is given by $S_k = \frac{8\pi^2|k|}{g^2}$ where k refers to the winding number or topological charge. Taylor-expanding the weight $\exp(-S_k)$ in the (Euclidean) partition function in powers of g yields identically zero.

Thus perturbation theory *ignores* topologically nontrivial field configurations from the start. Indeed, a mechanism for the generation of mass for the $k = 0$ sector of propagating gauge fields, which would invalidate Pauli's concerns, cannot occur in perturbation theory.

A variation of the action of Eq. (1.30) with respect to the gauge fields A_μ yields:

$$\frac{\delta S_{YM}}{\delta A_\mu} \overset{!}{=} 0 \quad \Rightarrow \quad D_\mu F^{\mu\nu} = 0. \tag{1.31}$$

From the Bianchi identity for the field strength $F_{\mu\nu}$

$$D_\mu F^{\kappa\lambda} + D_\kappa F^{\lambda\mu} + D_\lambda F^{\mu\kappa} = 0 \tag{1.32}$$

it follows that a solution to the first-order equations

$$F_{\mu\nu} = \pm\frac{1}{2}\epsilon_{\mu\nu\kappa\lambda}F^{\kappa\lambda} \equiv \pm\tilde{F}_{\mu\nu} \tag{1.33}$$

also solves the second-order Yang–Mills equations in Eq. (1.31). Here $\epsilon_{\mu\nu\kappa\lambda}$ is the totally antisymmetric tensor in four dimensions with $\epsilon_{0123} = 1$. The duality operation $\frac{1}{2}\epsilon_{\mu\nu\kappa\lambda}F^{\kappa\lambda}$ in Eq. (1.33) generates the *dual* field-strength tensor $\tilde{F}_{\mu\nu}$ whose components have the electric field $E_i \equiv F^{0i} = -F^{i0}$, $(i = 1,2,3)$ interchanged with the magnetic field $B_i \equiv \frac{1}{2}\epsilon_{ijk}F^{jk}$, $(i,j,k = 1,2,3)$. In a four-dimensional Minkowski spacetime no analytic solutions to the (anti) selfduality equation (1.33) are known. After a transition to four-dimensional Euclidean spacetime by the following 90-degree rotation in the complex plane,

$$x^0 \to ix_4 \quad (x^0, x_4 \text{ real}), \quad A_0^a \to -iA_4^a \quad (A_0^a, A_4^a \text{ real}), \tag{1.34}$$

the (anti)selfduality equation (1.33) possesses, however, stable solutions owing to the aforementioned topologically nontrivial mappings of $\partial \mathbf{R}^4 = S_3$ onto the SU(2) group manifold S_3.

Since the Yang–Mills action is invariant under a translation of spacetime there is according to Noether's theorem a set of four conserved currents. A symmetric and gauge-invariant form of these conservation laws, which is achieved by adding a harmless term to the Noether currents (not changing the conservation equations), invokes the following definition

$$\Theta^{\mu\nu} \equiv 2\,\mathrm{tr}\left(-F^{\mu\lambda}F_{\lambda}^{\nu} + \frac{1}{4}g^{\mu\nu}F^{\kappa\lambda}F_{\kappa\lambda} \right) = \Theta^{\nu\mu}. \tag{1.35}$$

One then has that

$$\partial_{\mu}\Theta^{\mu\nu} = \partial_{\nu}\Theta^{\mu\nu} = 0. \tag{1.36}$$

Equation (1.36) expresses the local conservation of the quantity $\Theta^{\mu\nu}$ which is interpreted as the symmetric energy–momentum tensor. From Eq. (1.35) it follows that

$$\Theta^{\mu}_{\mu} = 0 = \partial_{\mu}S^{\mu}, \tag{1.37}$$

where $S^{\mu} \equiv \Theta^{\mu\nu}x_{\nu}$ is the conserved current associated with the dilation symmetry of the classical Yang–Mills action (1.30). As we shall see in Sec. 5.2.4, this symmetry is violated by quantum effects which introduce a mass scale Λ to the theory.

After the Euclidean rotation of Eq. (1.34) we have $g^{\mu\nu} \to -\delta_{\mu\nu}$ in Eq. (1.35). For (anti)selfdual field configurations on a Euclidean spacetime [obeying Eq. (1.33) or $F_{4i} = E_i = \pm\frac{1}{2}\epsilon_{ijk}F_{jk} = \pm B_i$, $(i, j, k = 1, 2, 3)$] we have

$$\begin{aligned}
\Theta_{\mu\nu} = -2\,\mathrm{tr}\bigg\{ &\delta_{\mu\nu}\left(\mp\vec{E}\cdot\vec{B} \pm \frac{1}{4}(2\vec{E}\cdot\vec{B} + 2\vec{B}\cdot\vec{E}) \right) \\
&\mp (\delta_{\mu4}\delta_{\nu i} + \delta_{\mu i}\delta_{\nu4})(\vec{E}\times\vec{E})_i \pm \delta_{\mu i}\delta_{\nu(j\neq i)}(E_iB_j - E_iB_j) \\
&\pm \delta_{\mu(j\neq i)}\delta_{\nu i}(E_jB_i - E_jB_i) \bigg\} \equiv 0.
\end{aligned} \tag{1.38}$$

Equation (1.38) is important: it guarantees that an (anti)selfdual Euclidean gauge field configuration of finite action possesses no energy–momentum and thus does not propagate. This is true regardless of whether one chooses to stay in Euclidean signature or to rotate back to Minkowski spacetime where such a gauge-field configuration no longer solves the field equations. Notice that any quantity obtained by performing a spatial average over noninteracting (anti)selfdual Euclidean gauge field configurations inherits the property of vanishing energy–momentum.

1.4.2 *Nonlocal objects*

In Sec. 1.4.1 we have discussed local gauge-invariant bilinears of $F_{\mu\nu}$: the Yang–Mills action density and the energy–momentum tensor. In searching gauge-invariant or gauge-covariant objects one may work beyond locality.

Consider a smooth curve C either in four-dimensional Minkowski or Euclidean spacetime as a function of the parameter $0 \le s \le 1$:

$$x^\mu = x^\mu(s), \quad x_0^\mu \equiv x^\mu(s=0), \quad x_1^\mu \equiv x^\mu(s=1). \qquad (1.39)$$

The Wilson line along C is defined as the following exponential

$$\{x_0, x_1\}_C = \mathcal{P}\exp\left(i\int_C dx^\mu\, A_\mu\right) = \mathcal{P}\exp\left(i\int_0^1 ds\, \frac{dx^\mu}{ds}\, A_\mu\right). \qquad (1.40)$$

In Eq. (1.40) the symbol $\mathcal{P}$ demands path-ordering of the exponential as explained in Fig. 1. In contrast to the gauge field A_μ, which lives in the Lie algebra su(N), the Wilson line is an element of the group SU(N). One can show that under the gauge transformation (1.3)

$$\{x_0, x_1\}_C \xrightarrow{\Omega} \Omega(x_0)\{x_0, x_1\}_C\Omega^\dagger(x_1). \qquad (1.41)$$

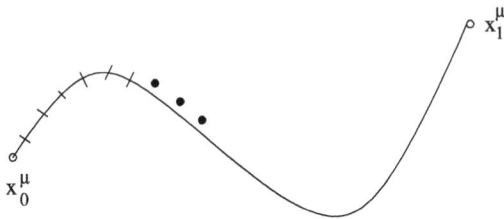

$$\ldots \exp[-i\tfrac{dx^\mu}{ds}\,(2ds)\, A_\mu(2ds)ds]\ \exp[-i\tfrac{dx^\mu}{ds}\,(ds)\, A_\mu(ds)\,ds]\ \exp[-i\tfrac{dx^\mu}{ds}\,(0)\, A_\mu(0)\,ds]$$

Figure 1.1. Path-ordering of the exponential of a line integral along a given smooth curve C is defined by multiplying the exponentials of the integrals, which belong to infinitesimal curve sections, consecutively onto each other starting at x_0^μ and terminating at x_1^μ.

If the curve C forms a closed loop, that is $x_0 = x_1$, then the nonlocal functional $\{x_0, x_0\}_C$ of the gauge field A_μ locally transforms under the adjoint representation:

$$\{x_0, x_0\}_C \xrightarrow{\Omega} \Omega(x_0)\{x_0, x_0\}_C \Omega^\dagger(x_0). \qquad (1.42)$$

The Wilson line for a closed curve is called Wilson loop. Shifting the base-point of the curve from x_0 to $x_{s>0}$, the group element Ω mediating the gauge transformation in Eq. (1.42) is shifted accordingly. The quantity

$$W_C \equiv \text{tr} \{x_0, x_0\}_C \qquad (1.43)$$

is gauge-invariant. In Euclidean spacetime and for a rectangular contour extending along x_4 and a spatial direction, the quantity $\langle W_C \rangle$ was introduced by K. Wilson (1973) to define a criterion for confinement in lattice formulations of Quantum Yang–Mills Theory. Here the average $\langle \, \rangle$ is over all gauge-inequivalent configurations A_μ contributing to the partition function of the theory. According to Wilson (1973), the theory exhibits confinement of infinitely heavy, fundamentally transforming test-charges if $\langle W_C \rangle$ grows with the area of the contour (area law).

1.5 Spontaneous Gauge-Symmetry Breaking

We have not yet considered the possibility of a spontaneous breaking of gauge symmetry introduced by a noninvariance of the theory's ground state under gauge transformations. Such a situation can be induced by a scalar sector introduced ad hoc in addition to the pure Yang–Mills action of Eq. (1.30). Alternatively, we show in Chapter 5 that a scalar field ϕ emerges thermodynamically by *partially* executing the average over field configurations in the partition function in combination with a spatial coarse-graining. In this section we just assume the existence of the field ϕ without asking about its origin.

Insisting on the extended theory's ground state estimate to honor Poincaré invariance (or the Euclidean version of it), the spontaneous breaking of gauge invariance is mediated by a nonpropagating field configuration ϕ_0 of finite modulus. This can be achieved by a gauge-invariant potential $V(\phi)$ or by a boundary condition. Depending on the degree of completeness of the spontaneous breaking of gauge symmetry, the scalar field ϕ transforms (is charged) under the appropriate representation of the gauge group SU(N). We only consider the case that ϕ is charged under the adjoint representation.

A minimally extended Yang–Mills actions reads

$$S_{\text{YM,ext}} = \text{tr} \int d^4x \left(-\frac{1}{2} F_{\mu\nu} F^{\mu\nu} + D_\mu \phi D^\mu \phi \right) = \int d^4x \mathcal{L}_{\text{YM,ext}},$$

(1.44)

where D_μ refers to the adjoint covariant derivative defined in Eq. (1.25). We assume here that by some external mechanism the field modulus $|\phi|$ a priori acquires a constant ground state expectation $|\phi_0|$. Therefore, $\phi = \phi_0 + \delta\phi$ with $\delta\phi$ referring to possible propagating fluctuations of the field ϕ. Notice that $\phi_0 = \text{const.}$ and $A_\mu^a \equiv 0$ (or a gauge transformation thereof) solve the classical Euler–Lagrange equations of motion

$$D_\mu D^\mu \phi = 0, \quad D_\mu F^{\mu\nu} = ig[\phi, D^\nu \phi],$$

(1.45)

which follow from action $S_{\text{YM,ext}}$ in Eq. (1.44). Notice further that the energy density ρ associated with this solution is constant (zero in this case). Therefore, $\phi_0 = \text{const.}$ and $A_\mu^a \equiv 0$ represent a valid ground state estimate. According to Eq. (1.21), an infinitesimal gauge transformation $\Omega(x) = 1 + it^a \omega^a(x)$ acts on ϕ_0 as

$$\phi_0 \xrightarrow{\Omega} \phi_0 + i[t^a \omega^a, \phi_0].$$

(1.46)

The quantity $i[t^a \omega^a, \phi_0]$ may not be vanishing. In this case ϕ_0 breaks (part of the) gauge symmetry.

Without constraining generality, the direction of ϕ_0 in the Lie algebra su(N) can be chosen to lie pararallel with that of a particular generator t^b. What is then the effect of ϕ_0 on the propagation of the gauge-field mode A_μ^a? Setting $\phi = \phi_0$ in Eq. (1.44), that is, neglecting fluctuation of the field ϕ, the Lagrangian in Eq. (1.44) takes the following form

$$\mathcal{L}_{\text{YM,ext}} = -\frac{1}{4} F_{\mu\nu}^a F^{\mu\nu,a} - g^2 \, \text{tr} \, [A_\mu^a t^a, \phi_0][A_\mu^c t^c, \phi_0]. \qquad (1.47)$$

From Eq. (1.47) one reads off that all those modes A_μ^a, whose associated generators t^a commute with the generator t^b, propagate as massless gauge-field excitations with two polarization states. Since rank SU(N) = N − 1 there are N − 1 such modes in an SU(N) gauge theory. Therefore, spontaneous gauge-symmetery breaking by a ground state expectation of an adjointly transforming scalar field ϕ is incomplete:

$$\text{SU(N)} \xrightarrow{\phi_0} \text{U(1)}^{N-1}. \qquad (1.48)$$

All other modes, associated with "broken generators" t^a, acquire a mass m^a. They propogate with three polarization states. One has

$$(m^a)^2 = -2g^2 \text{tr} \, [t^a, \phi_0][t^a, \phi_0]. \qquad (1.49)$$

This is, exemplarily, the essence of the celebrated Anderson–Englert–Brout–Higgs–Guralnik–Hagen–Kibble mechanism [Anderson (1963); Englert and Brout (1964); Higgs (1964); Guralnik et al. (1964)]. If the field ϕ were assumed to be charged under the fundamental representation of SU(N), $\phi \to \Omega\phi$, gauge-symmetry breaking would be complete.

1.6   Homotopy Groups: Concept and Use

The situation of a *constant* field-configuration ϕ_0, as described in the previous section, is generalizable towards field configurations of *finite* (i) action, (ii) static energy, or (iii) static energy per spatial line element. In four-dimensional Euclidean spacetime case (i) actually assumes the absence of the field ϕ, but cases (ii) and (iii) require a nontrivial dependence of ϕ_0 on spacetime.

Let us first sketch possible breaking patterns for a general non-Abelian gauge group G. For a much more detailed exposition see Weinberg (1996). We assume G to be broken down to its subgroup H by a ground-state estimate $\phi_0 \neq 0$:

$$G \xrightarrow{\phi_0} H. \qquad (1.50)$$

Since ϕ_0 is invariant under gauge transformations associated with elements of H we consider the coset G/H defined as the set of group elements g_1 and g_2 that are related as $g_1 = g_2 h$ where $h \in H$. For conditions (i) or (ii) or (iii) to be satisfied the four-dimensional action density or the three-dimensional or the two-dimensional energy density, respectively, must vanish at infinity. Recall that boundaries of the according spaces are given as: $\partial \mathbf{R}^4 = S^3, \partial \mathbf{R}^3 = S^2$, and $\partial \mathbf{R}^2 = S^1$. This suggests the consideration of four-dimensional, three-dimensional, or two-dimensional unit vectors $\hat{x}$ associated with the spheres S_3, S_2, or S_1, respectively. The field A_μ must approach a pure-gauge configuration

$$A_\mu(x \to \infty) = i\Omega \partial_\mu \Omega^\dagger \qquad (1.51)$$

so that the action density $\propto \frac{1}{2} \mathrm{tr}\, F_{\mu\nu} F_{\mu\nu}$ vanishes at infinity. Let us first discuss case (i). The group elements Ω associated with Eq. (1.51) may cover the entire group manifold S_3 of SU(2), which we assume to be a subgroup of G, and the investigation of maps

$$S_3 \to G \qquad (1.52)$$

is natural. The set of all classes of topologically equivalent maps from S_3 to G is known as the third homotopy group of G and is denoted as $\Pi_3(G)$. For example, consider the case of SU(2) $= S_3$. Then the homotopy group $\Pi_3(S_3)$ is given by the set of whole numbers $\mathbf{Z}$ since the maps from S_3 to S_3 are topologically distinguished by different winding numbers ($n \in \mathbf{Z}$ fold coverage of S_3 when moving through the domain S_3). Since a given map associated with an element of $\Pi_3(S_3)$ cannot smoothly be deformed onto a map associated with another element of $\Pi_3(S_3)$ we are guaranteed the stability of the minimal-action field configurations whose asymptotic behavior is associated with distinct elements of $\Pi_3(S_3)$. Necessarily, such configurations must be solutions to the field equations. As we shall see in Chapter 4, for SU(N) these are the (anti)selfdual configurations solving Eq. (1.33). Because of essential contributions to the action being localized within compact regions of four-dimensional Euclidean spacetime these configurations are called (anti)instantons [Belavin *et al.* (1975)].

For cases (ii) or (iii) the points at infinity, for which field ϕ_0 needs to induce vanishing static energy density or static energy density per line element,[4] are parameterized by three-dimensional or two-dimensional unit vectors, respectively. To identify topologically stabilized field configurations, one thus is lead to the investigation of the second or first homotopy groups, respectively: $\Pi_2(G/H)$ or $\Pi_1(G/H)$. Nontriviality of the latter means that closed curves $\{S_1\}$ are not contractible to a point on G/H. The so-called fundamental group Π_1 was orginally introduced by H. Poincaré in his search for a set of promising topological invariants uniquely determining the topology of a given manifold. An example of $\Pi_2(G/H)$ being nontrivial is SU(2) $\rightarrow$ U(1) where $\Pi_2(G/H = S_2) = \mathbf{Z}$. The associated topologically stabilized field configurations are finite-mass magnetic monopoles ['t Hooft (1974); Polyakov (1974)] or electric-magnetically charged dyons [Julia and Zee (1974); Prasad and Sommerfield (1975)]. Here the term "charge" refers to the three-dimensional charge "seen" by the unbroken gauge group U(1). This charge is a topological invariant associated with a member of $\Pi_2(G/H = S_2) = \mathbf{Z}$. An example of $\Pi_1(G/H)$ being nontrivial is U(1) $\rightarrow$ 1 since $\Pi_1(G/H = S_1) = \mathbf{Z}$. Here the associated topologically stabilized field configurations are infinitely long, magnetic vortex lines with a finite line density of mass. Here the two-dimensional

---

[4]This can, for example, be achieved by the vanishing of a gauge-invariant potential $V(\phi)$ whose zeros correspond to discrete values of the modulus of ϕ_0.

charge is a topological invariant [Abrikosov (1957);Nielsen and Olesen (1973)]. Notice that the case of nonselfintersecting *closed* vortex lines is topologically not stabilized against decay into nothingness since it corresponds to this two-dimensional charge being zero. (A closed vortex line intersects a given two-dimensional plane an even number of times, and the projections of the vortex tangentials onto a fixed normal to the plane alternate in sign.)

Problems

1.1. Show that the dim su(N) = $N^2 - 1$.

1.2. Find arguments for why Eq. (1.9) holds. *Hint*: Show that from $(t_r^a)^\dagger = t_r^a$ the positivity of tr $t_r^a t_r^a$ (no sum over a) follows and that the matrix tr $t_r^a t_r^b$ is Hermitian and thus can be diagonalized.

1.3. Why does Eq. (1.11) hold? *Hint*: Use the cyclicity under the trace-symbol and the antisymmetry of the commutator [,].

1.4. Verify that the generators as defined in Eq. (1.12) satisfy the Lie bracket of Eq. (1.6). *Hint*: Use the Jacobi identity (1.8).

1.5. Show that the definition of Eq. (1.24) for the non-Abelian field strength $F_{\mu\nu}$ indeed yields a homogeneous transformation law under gauge transformations. *Hint*: Use $\partial_\mu(\Omega\Omega^\dagger) = 0$.

1.6. Verify Eq. (1.26) and show that the covariant derivative of Eq. (1.25) satisfies the Leibnitz rule: $D_\mu(AB) = (D_\mu A)B + A D_\mu B$ for $A, B \in$ su(N).

1.7. Appealing to Eq. (1.8), show that the Bianchi identity (1.32) holds.

1.8. Perform the variational problem in Eq. (1.31) explicitly. *Hint*: The associated Euler–Lagrange equations read

$$\partial_\mu \frac{\partial \mathcal{L}_{YM}}{\partial(\partial_\mu A_\nu^a)} = \frac{\partial \mathcal{L}_{YM}}{\partial A_\nu^a}. \tag{1.53}$$

This equation for coordinates in the Lie algebra can be recast into the matrix equation (1.31).

1.9. Show that the transformation law (1.41) holds. *Hint*: First, consider an infinitesimal line element and show (1.41) for this case.

The Wilson line for a finite line section is obtained by multiplying the contributions of each infinitesimal element onto one another in an ordered way.

1.10. Verify Eq. (1.48).

1.11. Show that the ground state expectation ϕ_0 of a field ϕ charged under the *fundamental* representation of SU(N) spontaneously breaks the gauge symmetry SU(N) entirely: $\text{SU(N)} \xrightarrow{\phi_0} 1$.

1.12. Derive the equations of motion (1.45) by demanding stationarity of the action $S_{\text{YM,ext}}$ in Eq. (1.44).

References

Abrikosov, A. A. (1957). On the magnetic properties of superconductors of the second group, *Sov. Phys. JETP*, **5**, 1174.

Anderson, P. W. (1963). Plasmons, gauge invariance, and mass, *Phys. Rev.*, **130**, 439.

Belavin, A. A., *et al.* (1975). Pseudoparticle solution of the Yang–Mills equations, *Phys. Lett. B*, **59**, 85.

Englert, F. and Brout, R. (1964). Broken symmetry and the mass of gauge vector mesons, *Phys. Rev. Lett.*, **13**, 321.

Guralnik, G. S. *et al.* (1964) Global conservation laws and massless particles, *Phys. Rev. Lett.*, **13**, 585.

Higgs, P. W. (1964). Broken symmetries and the masses of gauge bosons, *Phys. Rev. Lett.*, **13**, 508.

Julia, B. and Zee, A. (1974). Poles with both magnetic and electric charges in nonabelian gauge theory, *Phys. Rev. D*, **11**, 2227.

von Meyenn, K., ed. (1999). *Scientific Correspondence*, Vol. IV, Part II, (Springer), pp. 494–496, 501–503.

Nielsen, H. B. and Olesen, P. (1973). Vortexline models for dual strings, *Nucl. Phys. B*, **61**, 45.

Polyakov, A. M. (1974). Particle spectrum in the quantum field theory, *JETP Lett.*, **20**, 194.

Prasad, M. K. and Sommerfield, C. M. (1975). An exact solution for the 't Hooft monopole and the Julia–Zee dyon, *Phys. Rev. Lett.*, **35**, 760.

't Hooft, G. (1974). Magnetic monopoles in unified gauge theories, *Nucl. Phys. B*, **79**, 276.

't Hooft, G. ed. (2005). *50 Years of Yang–Mills Theory* (World Scientific Publishing).

Weinberg, S. (1996). *The Quantum Theory of Fields*, Vol. 2 (Cambridge University Press).

Wilson, K. G. (1973). Confinement of quarks, *Phys. Rev. D*, **10**, 2445.

Yang, C. N. and Mills, R. L. (1954). Conservation of isotopic spin and isotopic gauge invariance, *Phys. Rev.*, **96**, 191.

Chapter 2

The Perturbative Approach at Zero Temperature

2.1 General Remarks

The quantity that is primarily addressed in perturbation theory, the scattering or S-matrix, is a concept driven by the set-up of collider experiments. In such experiments a well prepared particle state of given energy and momentum travels from spatial infinity (in-asymptote) into the interaction region transforming into some intermediate state, whose emergence is dictated by the nature of the interaction and the laws of quantum mechanics, to eventually releases a state of particles traveling again towards spatial infinity (out-asymptote) without any further interaction. By the well-known reduction formulae [Lehmann *et al.* (1955)] the elements of the S-matrix can be obtained from a theory's *n*-point functions which, in turn, are given by functional derivatives with respect to external sources of the vacuum-to-vacuum transition amplitude (or partition function in a Euclidean formulation).

In principle, the description of a process studied in collider experiments should be based on a solution to the classical equations of motion of a field theory, subject to appropriate boundary conditions and modified by quantum corrections. In most realistic four-dimensional field theories, however, even the classical problem is not resolvable in an exact way. One thus resorts to the perturbative approach where *both* the *classical* solution *and* quantum corrections about it are expanded into powers of a small parameter determining the strength of the interaction: the coupling constant. This expansion describes fluctuations about a free-particle state again in terms of free-particle propagation in between local vertices whose structure is fixed by (a gauge-fixed version of) the classical action. In the case of classically approached fluctuations the free-particle state in between vertices obeys a standard dispersion law (mass shell, tree-level approximation), in the case of quantum fluctuations the particle may be far off this mass shell. That is, its momentum no longer is fixed by the in- and out-asymptotes through conservation. In the language of Feynman diagrams this situation corresponds to a closed loop which in turn is at the heart of the problem of ultraviolet divergences (UV divergences) of four-dimensional perturbative quantum field theory. A great achievement in the development of a Quantum Theory of Yang–Mills fields was the demonstration that the so-called renormalization program, designed to absorb these UV divergences into re-defined parameters, actually only invokes structures already present in the classical (gauge-fixed) action ['t Hooft and Veltman (1972a,b); 't Hooft (1971); Lee and Zinn-Justin (1972); Slavnov and Faddeev (1970)]. That is, in computing corrections to higher and higher orders in the coupling constant g the UV divergences can always be absorbed into the parameters of the classical theory. This miracle is called perturbative renormalizability and arises in pure Yang–Mills theory because of the severe constraints imposed by gauge invariance onto the elements of the S-matrix. Historically, two independent ways of showing the perturbative renormalizability of Quantum Yang–Mills theory have been pursued: the cumbersome but insightful diagrammatic technique ['t Hooft and Veltman (1972a,b); 't Hooft (1971)] and the formal argument via the functional integral [Slavnov and Faddeev (1970); Lee and Zinn-Justin (1972)].

On one hand, the renormalizability of Quantum Yang–Mills theory implies that the theory does not loose its predictive power when accessed perturbatively. Varying the order of the expansion in powers of g, the number of renormalization conditions required to experimentally fix ambiguities in the physical S-matrix never changes. On the other hand, there is a deep implication in the context of the so-called renormalization group evolution [Wilson (1971a,b, 1975)]: Since no new local or nonlocal structures emerge in the action in pursuing the perturbative expansion up to *any* order one concludes that integrating out quantum fluctuations residing in the topologically trivial sector[1] the perturbative effective action remains form-invariant as the flow-parameter (or renormalization scale) changes. The entire flow-parameter dependence then is described by that of the classical parameters of the theory. Thus, if the construction of an effective theory of Quantum Yang–Mills theory, valid below a certain physical scale for resolution, somehow manages to separate off the perturbatively accessible sector from the sector containing topologically nontrivial field configurations then the structure in the effective action associated with integrated-out perturbative fluctuations is the same as in the fundamental Yang–Mills action.

In pure Yang–Mills theories a complication in the perturbative expansion is the occurrence of infrared divergences (IR divergences). As we shall see in Chapter 5, the occurence of this class of divergences in Yang–Mills theory is an artifact of the perturbative construction: Not ignoring field configurations of nontrivial topology, a part of the free-particle spectrum of the in-asymptotics and out-asymptotics acquires mass due to a nontrivial ground state, and IR divergences no longer take place.

2.2   Gauge Fixing in the Functional Integration

We consider the gauge group SU(N). An element of the S-matrix ultimately is determined by the functional integral

$$\langle \text{out}|\text{in}\rangle = \mathcal{N}^{-1} \int \prod_{x,\mu,a} dA_\mu^a(x) \exp\left(iS[A]\right), \qquad (2.1)$$

---

[1]Recall that perturbation theory considers small quantum flutuations about an approximate solution to the classical field equations. The latter's topological properties are not changed by adding small fluctuations of trivial topology to it.

where $S[A]$ is the Yang–Mills action functional

$$S[A] = -\frac{1}{2g^2}\mathrm{tr}\int d^4x\, F^{\mu\nu}(x)F_{\mu\nu}(x), \qquad (2.2)$$

$F_{\mu\nu} \equiv \partial_\mu A_\nu - \partial_\nu A_\mu - i[A_\mu, A_\nu]$, $A_\mu \equiv A_\mu^a t_a$, $\mathrm{tr}\, t_a t_b = \frac{1}{2}\delta_{ab}$, and $\mathcal{N}^{-1}$ represents an (inessential) normalization factor. The integration measure $\prod_{x,\mu,a} dA_\mu^a(x)$ is for an integration over a system of an infinite, uncountable number of degrees of freedom labeled by the continuous four-dimensional spacetime coordinate x, and the asymptotic configurations of the gauge field A_μ for $x_0 \to \pm\infty$ represent the out-asymptote and the in-asymptote, respectively.

Definition (2.1) as it stands is quite useless because a functional integration over all gauge copies of a given configuration $A_\mu(x)$ is implied. The integral over all gauge copies should be factored out before executing the functional integral. A corresponding procedure, valid in perturbation theory, was orginally proposed by L. Faddeev and A. A. Popov (1967), and we will sketch it in what follows.

In order to perform actual calculations in a Lorentz-covariant way one should only allow for those points in the gauge orbit

$$A_\mu^\Omega = \Omega^\dagger(x)A_\mu\Omega(x) + i\Omega^\dagger(x)\partial_\mu\Omega(x) \qquad (2.3)$$

of a given configuration A_μ which make Lorentz covariance explicit. In Eq. (2.3) the group element Ω is parameterized as

$$\Omega(x) \equiv \exp\left(i\omega^a(x)t_a\right). \qquad (2.4)$$

A gauge condition accomplishing this is

$$G[A_\mu] \equiv \partial^\mu A_\mu(x) - K(x) = 0, \qquad (2.5)$$

where $K(x)$ represents an arbitrary su(N)-valued but gauge-invariant Lorentz-scalar function.

If the group element $\Omega(x)$ is parametrized as in Eq. (2.4) then the following identity holds[2]

$$1 = \int \prod_{x,a} d\omega^a(x) \delta(G[A_\mu^\Omega]) \det \frac{\delta G[A_\mu^\Omega]}{\delta \omega}. \qquad (2.6)$$

This factor of unity is now multiplied onto the integrand of Eq. (2.1), and the order of the functional integration over ω and A_μ is interchanged. The gauge-rotated field A_μ^Ω reads in components

$$A_\mu^{\Omega,a} = R[\omega]_b^a A_\mu^b + f_\mu[\omega]^a, \qquad (2.7)$$

where $R[\omega]_b^a$ is the group element in the adjoint representation, and $f_\mu[\omega]^a$ is an ω-dependent shift. The gauge transformation (2.7) leaves the measure $\prod_{x,\mu,a} dA_\mu^a(x)$ in Eq. (2.1) invariant. Due to the gauge invariance of the Yang–Mills action the integrand in Eq. (2.1) is also invariant under $A_\mu \to A_\mu^\Omega$, and one may perform the functional integration over the variable $A_\mu' = A_\mu^\Omega$ instead of A_μ. In doing so one observes that to linear order in ω

$$A_\mu^{\Omega,a} = A_\mu^a + \left(D_\mu \omega\right)^a. \qquad (2.8)$$

Thus, to this order, the replacement $A_\mu \to A_\mu^\Omega$ in the covariant derivative D_μ does not change the right-hand side of Eq. (2.8).

The functional determinant of the operator $\partial^\mu D_\mu$, demanded by Eqs. (2.5), (2.6) and (2.8), can be expressed by a Gaussian functional integration over complex, su(N)-valued Grassmannian scalar fields $\bar{c}(x)$ and $c(x)$:

$$\det \partial^\mu D_\mu = \int \prod_{y,b} d\bar{c}^b(y) \prod_{z,d} c^d(z) \exp\left(-i\mathrm{tr} \int d^4x \, \partial^\mu \bar{c} D_\mu c\right). \qquad (2.9)$$

---

[2]First think of a system of finitely many degrees of freedom to check the validity of Eq. (2.6) and then perform the limit towards a system of infinite, uncountably many degrees of freedom.

Thus, after renaming A'_μ to A_μ Eq. (2.1) is re-cast as

$$\langle \text{out}|\text{in}\rangle = \mathcal{N}^{-1}\left(\int \prod_{u,a} d\omega^a(u)\right)\int \prod_{x,y,z,b,d,e} dA_\mu^b(x)d\bar{c}^d(y)\,dc^e(z)$$

$$\times\, \delta(\partial^\mu A_\mu(x) - K(x))\exp\left(iS[A] + iS_{\text{F.P.}}[\bar{c},c,A]\right), \quad (2.10)$$

where

$$S_{\text{F.P.}} \equiv -\text{tr}\int d^4x\,\partial^\mu\bar{c}D_\mu c. \tag{2.11}$$

Compared to an ordinary Klein–Gordon field the kinetic term of the Grassmannian field c comes with the "wrong" sign in the action $S_{\text{F.P.}}$ of Eq. (2.11). This and the fact that the field c obeys Fermi instead of Bose statistics prompt the name "ghost". In the practice of computing Feynman diagrams this implies that for each closed ghost loop a factor of -1 needs to be inserted.

Notice that the integration over gauge transformations merely appears as a factor in Eq. (2.10). Since (modulo a constant factor) this functional integral does not depend on the function $K(x)$ of the gauge condition (2.5), one may generate yet another constant factor by functionally integrating over $K(x)$, subject to some weight functional $F[K]$. Choosing $F[K] = \exp\left(-i\text{tr}\int d^4x\frac{K^2(x)}{2\alpha}\right)$ (α a real number) the result for the gauge fixed amplitude reads

$$\langle \text{out}|\text{in}\rangle = \frac{1}{\mathcal{N}'}\int \prod_{x,y,z,b,d,e} dA_\mu^b(x)\,d\bar{c}^d(y)\,dc^e(z)$$

$$\times\, \exp\left(iS[A] + iS_{\text{F.P.}}[\bar{c},c,A] + iS_{\text{g.f.}}[A,\alpha]\right), \quad (2.12)$$

where

$$S_{\text{g.f.}}[A,\alpha] \equiv -\frac{1}{\alpha}\text{tr}\int d^4x(\partial^\mu A_\mu)^2, \tag{2.13}$$

and $1/\mathcal{N}'$ is a modified normalization factor.

The usefulness of the modified action $S_{\text{mod}} \equiv S[A] + S_{\text{F.P.}}[\bar{c},c,A] + S_{\text{g.f.}}[A,\alpha]$ action in the exponent of Eq. (2.12) lies in the fact that certain contributions to cutting rules arising from unphysical polarizations of the gauge field A_μ are canceled.

The Faddeev–Popov procedure, as sketched above, does not amount to a *complete* fixing of the gauge. This is a consequence of the fact that the gauge condition of Eq. (2.5) is invariant under certain gauge transformations. Symptomatic for the incompleteness of the gauge-fixing associated with the functional integral of Eq. (2.12) is a residual global BRST symmetry (after their discoverers Becchi *et al.* (1976), Iofa and Tyutin (1976)) of the action S_{mod}. This symmetry demands in perturbation theory certain relations between divergences arising in Feynman diagrams — the Slavnov–Taylor identities [Slavnov (1972); Taylor (1971)] — which assure the complete cancelation of contributions due to unphysical gauge boson polarizations in S-matrix amplitudes. As pointed out by V. N. Gribov (1978), when working beyond perturbation theory, the Faddeev–Popov procedure is insufficient to fix the gauge. Namely, a given configuration A_μ satisfying Eq. (2.5) and nonperturbatively contributing to ⟨out|in⟩ exhibits a set of gauge copies whose cardinality depends on A_μ! To eliminate these gauge copies Gribov proposed to demand the positivity of det $\partial^\mu D_\mu$ in the functional integration of Eq. (2.12). However, it can be shown that gauge copies are not entirely avoided even under this additional constraint [Van Baal (1992); Dell'Antonio and Zwanziger (1991)].

The Slavnov–Taylor (or Ward) identities are instrumental in showing that in Quantum Yang–Mills theory for the expansion of amplitudes in powers of the coupling g to any order, all that is needed to make the theory finite is a renormalization of the parameters of the classical theory (perturbative renormalizability). For a pure Yang–Mills theory this statement essentially implies that whatever is perturbatively integrated out in the ultraviolet influences the physics at the sliding scale μ only insofar as to induce a running of the gauge coupling: g becomes a function of μ. We do not here address the technically-involved diagrammatic proof by induction of perturbative renormalizability ['t Hooft and Veltman (1972a,b);

't Hooft (1971)].[3] Rather, we explain below the one-loop result for the running coupling.

2.3   One-loop Running of the Gauge Coupling

2.3.1   *Feynman rules*

Let us address the running of g at the one-loop level and the rules to compute it. Since we would like to count powers of g in a perturbation theory about the free-field situation we need to let $A_\mu \to gA_\mu$ in Eq. (2.10). From the resulting Lagrangian one can read off the Feynman rules. We formulate them in momentum space. For the vertices, as depicted in Fig. 2.1, we have

$$
\begin{aligned}
\text{gauge-boson three-vertex}: \quad & gf^{abc}[g^{\mu\nu}(k-p)^\rho + g^{\nu\rho}(p-q)^\mu \\
& + g^{\mu\rho}(q-k)^\nu], \\
\text{gauge-boson four-vertex}: \quad & -ig^2[f^{abe}f^{cde}(g^{\mu\rho}g^{\nu\sigma} - g^{\mu\rho}g^{\nu\sigma}) \\
& + f^{ace}f^{bde}(g^{\mu\nu}g^{\rho\sigma} - g^{\mu\sigma}g^{\nu\rho}) \\
& + f^{ade}f^{bce}(g^{\mu\nu}g^{\rho\sigma} - g^{\mu\rho}g^{\nu\sigma})], \\
\text{gauge-boson-ghost three-vertex}: \quad & gf^{abc}p^\mu,
\end{aligned}
\tag{2.14}
$$

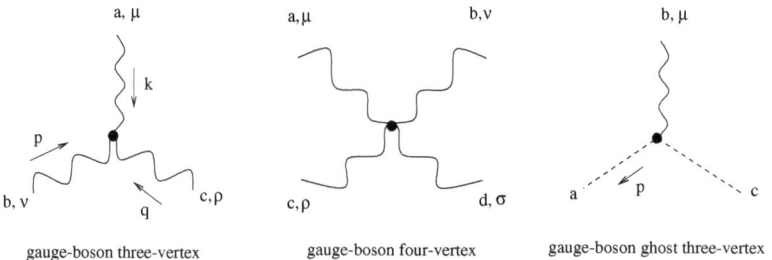

Figure 2.1.   Bare vertices in a gauge-fixed perturbative expansion of the S-matrix in Quantum Yang–Mills theory.

---

[3]The argument in ['t Hooft and Veltman (1972a,b); 't Hooft (1971)] is based on dimensional regularization which does not introduce a breaking of gauge symmetry.

where $g_{\mu\nu} = \mathrm{diag}(1, -1, -1, -1)$, f^{ace} are the structure constants of SU(N), and k, p, q are four momenta associated with the external lines. We work in Feynman gauge: $\alpha = 1$. In this gauge and going to momentum space (four-momentum k) the propagators read:

$$\text{gauge boson propagator}: \quad \delta^{ab} \frac{-i g_{\mu\nu}}{k^2},$$

$$\text{ghost propagator}: \quad \delta^{ab} \frac{i}{k^2}. \qquad (2.15)$$

Computing a specific diagram, an integration operation $\int \frac{d^4 k}{(2\pi)^4}$ for each loop momentum l and the symmetry factor, which is the inverse of the number of possible ways of interchanging components without changing the diagram, need to be inserted. In addition, a factor of -1 appears for each ghost loop.

2.3.2 Callan–Symanzik equation

The Callan–Symanzik equation is the implication of the fact that bare connected n-point functions $G^n(x_1, \ldots, x_n)$ do not know about a renormalization condition imposed at a momentum scale μ. Thus

$$\frac{d}{d\mu} G^{(n)}(x_1, \ldots, x_n) = 0. \qquad (2.16)$$

This means that the μ dependences of the *renormalized* parameters of the theory (coupling g and wave function A_μ in four-dimensional Yang–Mills theory), which do know about the renormalization point μ, must cancel any additional μ dependences introduced by subtractions. The renormalized n-point function $G^{(n)}(x_1, \ldots, x_n; \mu)$ depends on μ via the coupling-constant and the wave-function subtractions and also explicitly. Therefore

$$\left(\mu \frac{\partial}{\partial \mu} + \beta \frac{\partial}{\partial g} + n\gamma \right) G^{(n)}(x_1, \ldots, x_n; \mu) = 0, \qquad (2.17)$$

where $\beta \equiv \mu \, dg/d\mu$, $\gamma \equiv -\mu \, d(Z^{1/2})/d\mu$, and $Z^{1/2}$ represents the wave-function renormalization. Since the dimensionless quantities g

and $Z^{1/2}$ do not depend on any regulator scale after renormalization and since the only renormalized *parameter* is g it follows that β and γ are functions of g only: $\beta = \beta(g)$ and $\gamma = \gamma(g)$. The solution to the equation

$$\mu \frac{dg}{d\mu} = \beta(g) \tag{2.18}$$

subject to an experimentally imposed boundary condition thus determines the dependence of g on the renormalization scale μ.

To one-loop order and looking at the three-point function $G^{(3)}$ with the external legs being gauge fields it follows from Eq. (2.17) that the β function for Yang–Mills theory is given as

$$\beta(g) = \mu \frac{d}{d\mu} \left(-\delta_g + \frac{g}{2} \delta_{Z_A} \right), \tag{2.19}$$

where δ_g is the factor in the counterterm involving the bare three-vertex, and δ_{Z_A} is the factor in the counterterm for the self-energy subtraction including the local gauge-field loop (one four vertex), the nonlocal gauge-field loop (two three vertices of gauge fields only), and the nonlocal ghost loop (two three vertices of gauge field and ghost fields) (see Fig. 2.2). At one loop one has $\frac{d}{d\mu} \delta_g = 0$.

2.3.3 *Computation of the β function*

Let us now show how the β function for a pure SU(N) Yang–Mills theory is computed. We closely follow the presentation of the book by M. E. Peskin and D. V. Schroeder (1995).

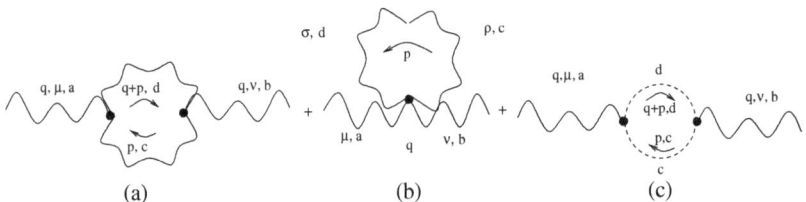

Figure 2.2. The sum of self-energy diagrams whose counterterm generates the running gauge coupling g in a *pure* Yang–Mills theory.

To cast δ_{Z_A} into the form divergence $\times$ μ-dependence, where the latter factor emerges from a renormalization condition imposed at the scale μ, we need to discuss a number of technical tools.

Integration over symmetric loop-momentum bilinears in numerators:

Consider the following four-dimensional integration over the loop four-momentum l

$$\int d^4 l \frac{l^\mu l^\nu}{D(l)}, \qquad (2.20)$$

where the denominator function $D(l) = \tilde{D}(l^2)$ ($l^2 \equiv l_\mu l^\mu$) is a Lorentz scalar. Notice that for $\mu \neq \nu$ the integration over the component l^μ exhibits an integrand which is antisymmetric under $l^\mu \to -l^\mu$, and thus the integral vanishes. Since no other Lorentz nonscalar is available to construct a second-rank tensor one concludes that the integral (2.20) needs to be proportional to the Lorentz-invariant metric tensor $g_{\mu\nu}$. This implies that

$$\int d^4 l \frac{l^\mu l^\nu}{D(l)} = \frac{1}{4} g^{\mu\nu} \int d^4 l \frac{l^2}{D(l)}. \qquad (2.21)$$

Rotation to Euclidean signature (Wick rotation):

To avoid the discussion of poles in momentum integrations involving denominators of the form $p^2 - \Delta^2$, where Δ is a Lorentz-invariant function of the external momenta only, one performs a Wick rotation to Euclidean signature $p^2 = p_1^2 + p_2^2 + p_3^2 + p_4^2$ as

$$p^0 \to i p_4, \qquad (2.22)$$

where both p^0 and p_4 are real.

Dimensional regularization:

In what follows we assume that the Wick rotation has been performed. To single out the UV divergences of loop integrals in a way that does not introduce any explicit breaking of gauge invariance

G. 't Hooft and M. Veltman proposed a regularization by continuation from the physical dimension $d = 4$ of spacetime to $d = 4 - \epsilon$ where $0 \le \epsilon < 1$. In typical expressions for loop integrals the angular integration in d dimensions is split off from the radial integration as

$$\int \frac{d^d l}{(2\pi)^d} \frac{1}{(l^2 + \Delta)^n} = \int \frac{d\Omega_d}{(2\pi)^d} \int_0^\infty dl \frac{l^{d-1}}{(l^2 + \Delta)^n}, \quad \text{or}$$

$$\int \frac{d^d l}{(2\pi)^d} \frac{l^2}{(l^2 + \Delta)^n} = \int \frac{d\Omega_d}{(2\pi)^d} \int_0^\infty dl \frac{l^{d+1}}{(l^2 + \Delta)^n}, \quad (n \ge 2). \qquad (2.23)$$

By observing that $\int d\Omega_d = \frac{2\pi^{d/2}}{\Gamma(d/2)}$, by performing the substitution $x = \frac{\Delta}{l^2 + \Delta}$ in radial integrations, and recalling that

$$\int_0^1 dx \, x^{\gamma-1}(1 - x)^{\delta-1} = \frac{\Gamma(\gamma)\Gamma(\delta)}{\Gamma(\gamma + \delta)} \qquad (2.24)$$

one arrives at

$$\int \frac{d^d l}{(2\pi)^d} \frac{1}{(l^2 + \Delta)^n} = \frac{\Delta^{\frac{d}{2}-n}}{2^d \pi^{d/2}} \frac{\Gamma(n - \frac{d}{2})}{\Gamma(n)}, \quad \text{or}$$

$$\int \frac{d^d l}{(2\pi)^d} \frac{l^2}{(l^2 + \Delta)^n} = \frac{\Delta^{\frac{d}{2}-n+1}}{2^d \pi^{d/2}} \frac{d}{2} \frac{\Gamma(n - \frac{d}{2} - 1)}{\Gamma(n)}, \quad (n \ge 2). \qquad (2.25)$$

Here the Gamma function $\Gamma(x)$ is defined as $\Gamma(x) = \int_0^\infty dy \, y^{x-1} e^{-y}$, one has $\Gamma(x + 1) = x\Gamma(x)$, $\Gamma(1) = 1$, and there are single poles of $\Gamma(x)$ at $x = -n$ of residues $(-1)^n/n!$ for $n = 0, 1, 2, \dots$. Writing $d = 4 - \epsilon$, setting $n = 2$ (the situation relevant for one-loop self-energies), and considering the limit $\epsilon \to 0$, the expressions on the right-hand sides of Eq. (2.25) diverge as

$$\int \frac{d^d l}{(2\pi)^d} \frac{1}{(l^2 + \Delta)^2} \longrightarrow \frac{\Delta^{-\frac{\epsilon}{2}}}{8\pi^2} \frac{1}{\epsilon}, \quad \text{or}$$

$$\int \frac{d^d l}{(2\pi)^d} \frac{l^2}{(l^2 + \Delta)^2} \longrightarrow -\frac{\Delta^{-\frac{\epsilon}{2}+1}}{4\pi^2} \frac{1}{\epsilon}. \qquad (2.26)$$

Notice that the Euclidean expression in d dimensions corresponding to Eq. (2.20) is given as

$$\int d^d l \frac{l^\mu l^\nu}{D(l)} = \frac{1}{d} \delta^{\mu\nu} \int d^d l \frac{l^2}{D(l)}. \tag{2.27}$$

Feynman parametrization:

Finally, working out the diagrams of Fig. 2.2, it is advantageous to express the product of two fractions in terms of an integral. Thereby, one fraction contains the Euclidean loop momentum p only, and the other fraction is a function of both external, Euclidean momentum q and Euclidean loop momentum p. To do this, we observe that for A and B real and positive such that $B > A$ one has

$$\frac{1}{A}\frac{1}{B} = \int_0^\infty dx \frac{1}{((1-x)A + xB)^2}. \tag{2.28}$$

Setting $A = p^2$ and $B = (p+q)^2$, one thus has

$$\frac{1}{p^2}\frac{1}{(p+q)^2} = \int_0^\infty dx \frac{1}{(P^2 + \Delta)^2}, \tag{2.29}$$

where $P = p + xq$ and $\Delta = x(1-x)q^2$. Due to $\partial_{p_\mu} P_\nu = \delta_\nu^\mu$ a loop integration over p can also be performed over variable P, and, setting $n = 2$, we reach the form of the expressions in (2.23) for one-loop self-energies where Δ is an invariant formed from the external momentum q.

We now have all tools at our disposal to compute the subtraction constant δ_{Z_A} as demanded by Eq. (2.19) for the determination of the β function. To compute the β function we are only interested in the coefficient of the pole in ϵ^{-1} of δ_{Z_A} which does not depend on details of the renormalization condition. Modulo a factor arising from the Lorentz-tensor structure of the free gauge field propagator this coefficient is related to minus the coefficient of ϵ^{-1} emerging from the sum of the diagrams in Fig. 2.2. Bearing in mind that $f^{acd} f^{bcd} = N \delta^{ab}$, appealing to the Feynman rules of Sec. 2.3.1 and to Eqs. (2.22), (2.26), (2.27) and (2.29), and asigning a symmetry factor of $1/2, 1/2$, and 1 to diagrams A, B and C in Fig. 2.2, respectively, we obtain for $\epsilon \to 0$ the following expressions:

diagram A: $\dfrac{ig^2N}{16\pi^2}\delta^{ab}\displaystyle\int_0^\infty dx\, \Delta^{-\frac{\epsilon}{2}} \times \epsilon^{-1}((5-11x(1-x))q^2g^{\mu\nu}$

$\qquad\qquad - (2+10x(1-x))q^\mu q^\nu),$

diagram B: $\dfrac{ig^2N}{16\pi^2}\delta^{ab}\displaystyle\int_0^\infty dx\, \Delta^{-\frac{\epsilon}{2}}$

$\qquad\qquad \times \epsilon^{-1}(-6+6x(4-3x))q^2\,g^{\mu\nu},$

diagram C: $\dfrac{ig^2N}{16\pi^2}\delta^{ab}\displaystyle\int_0^\infty dx\, \Delta^{-\frac{\epsilon}{2}}$

$\qquad\qquad \times \epsilon^{-1}(x(1-x)q^2\,g^{\mu\nu}+2x(1-x)q^\mu q^\nu).$ (2.30)

Performing the x integrations in Eq. (2.30) (by setting $\epsilon = 0$ in $\Delta^{-\frac{\epsilon}{2}}$) and summing up the contributions of all diagrams, yields

$$\text{diagram A} + \text{diagram B} + \text{diagram C}$$

$$= \frac{ig^2}{16\pi^2}\delta^{ab}(q^2\,g^{\mu\nu}-q^\mu q^\nu)\frac{10\,N}{3}\epsilon^{-1}.$$ (2.31)

Notice the transverse tensor structure in Eq. (2.31) being a consequence of gauge invariance $(q_\mu(q^2g^{\mu\nu}-q^\mu q^\nu)=0)$. Setting the typical momentum-transfer invariant Δ of the self-energies equal to the square of the renormalization scale μ, we read off from Eq. (2.30) the normalization of the counterterm relevant for the β function:

$$\delta Z_A = \frac{10N}{3}\frac{g^2}{16\pi^2}\epsilon^{-1}\mu^{-\epsilon}.$$ (2.32)

Thus Eq. (2.19) finally yields

$$\beta(g) = -\frac{5N}{3}\frac{g^3}{16\pi^2}.$$ (2.33)

The occurrence of a minus sign on the right-hand side of Eq. (2.33) was puzzling to first investigators: Theories with an Abelian gauge invariance always produce positive β functions.

Without derivation, we also give the result for the β function when allowing for the presence of n_f species of fermions in the fundamental representation of SU(N):

$$\beta(g) = -\frac{g^3}{16\pi^2} \left(\frac{11N}{3} - \frac{2}{3}n_f \right). \tag{2.34}$$

Notice that the result in Eq. (2.34) does not approach the result of Eq. (2.33) by simply letting $n_f \to 0$. This is because diagrams with two *external* fermion legs and one gauge-field leg contribute to the β function in addition to the gauge-field self-energy contribution coming from the fermion *loop*. It is the latter which gives rise to the term with factor n_f in Eq. (2.34). The result of Eq. (2.34) was discovered independently by I. B. Khriplovich (1969, 1970), G. 't Hooft (1973), H. D. Politzer (1973), and D. J. Gross and F. Wilczek (1973). Namely, the right-hand sides of Eq. (2.33) and, for sufficiently small n_f, Eq. (2.34) are *negative* implying that the coupling g becomes smaller as the scale μ of a scattering process increases. This phenomenon is known as "asymptotic freedom". On the other hand, the fact that the gauge coupling g grows towards the infrared is called "infrared slavery". Notice that in the realm of perturbation theory nothing definite can be said about the regime of "infrared slavery" since the presumed smallness of the expansion parameter is not selfconsistent.

Integration of

$$\mu \frac{d}{d\mu} g = -\frac{5N}{3} \frac{g^3}{16\pi^2} \tag{2.35}$$

yields

$$g^2(\mu) = \frac{g_0^2}{1 + \frac{5N}{3} \frac{g_0^3}{16\pi^2} \log \frac{\mu^2}{\mu_0^2}}, \tag{2.36}$$

where $g_0 = g(\mu_0)$ denotes the boundary condition prescribed to the evolution of g. The position of the pole μ_p on the right-hand side of Eq. (2.36),

$$\mu_p^2 = \mu_0^2 \exp\left(-\frac{48\pi^2}{5Ng_0^2}\right) \equiv \Lambda^2, \qquad (2.37)$$

is a free mass parameter of the quantum theory which is absent at the classical level. This phenomenon is called "dimensional transmutation".

In the nonperturbative derivation of the deconfining thermal Yang–Mills ground state, which we will perform in Chapter 5 for the cases of SU(2) and SU(3), the emergence of a mass parameter due to the process of integrating out (nonperturbative, topologically nontrivial) quantum fluctuations does not suffer from the abovementioned infrared inconsistency of the perturbative approach.

Problems

2.1. Derive the Faddeev–Popov action for the axial gauge given by the condition $G[A_\mu] \equiv A_0(x) - K(x) = 0$.
2.2. For SU(2) derive the invariant (Haar) measure for an integration over this gauge group in the Euler-angle parameterization of the group element U:

$$U(\alpha, \beta, \gamma) = \mathbf{1}\cos\alpha + i\vec{\tau} \cdot \hat{t}(\beta, \gamma)\sin\alpha,$$

where $\hat{t}$ is a three-dimensional unit vector parameterized by the two angles β and γ, and $\vec{\tau}$ is the vector of the three Pauli matrices.
2.3. Same as in Problem 2.2 but now for the parameterization

$$U^\pm = i\frac{\tau_\mu^\pm x_\mu}{\sqrt{x_\nu x_\nu}},$$

where the four-vector $\tau^\pm$ of 2×2-matrix valued entries is defined as $\tau^\pm = (\vec{\tau}, \mp i\mathbf{1}_2)$ (τ_1, τ_2, τ_3 the Pauli matrices) and the contraction of indices is understood in a Euclidean sense: $v_\mu w_\mu \equiv v_1 w_1 + v_2 w_2 + v_3 w_3 + v_4 w_4$.

2.4. Derive the expression for the one-loop β function of Eq. (2.19) from Eq. (2.17) applied to the three-point function involving external gauge-field legs only.

2.5. Show by direct calculation that $\frac{d}{d\mu} \delta_g = 0$ in Eq. (2.19). Without calculation, what is a reason which implies this result?

2.6. Derive the expressions on the right-hand sides of Eq. (2.23).

2.7. Derive Eq. (2.30).

References

Becchi, C. *et al.* (1976). Renormalization of gauge theories, *Ann. Phys.*, **98**, 287.

Dell'Antonio, G. and Zwanziger, D. (1991). Every gauge orbit passes inside the Gribov horizon, *Commun. Math. Phys.*, **138**, 291.

Faddeev, L. D. and Popov, V. N. (1967). Feynman diagrams for the Yang–Mills field, *Phys. Lett. B*, **25**, 697.

Gribov, V. N. (1978). Quantization of nonabelian gauge theories, *Nucl. Phys. B*, **139**, 1.

Gross, D. J. and Wilczek, F. (1973). Ultraviolet behavior of nonabelian gauge theories, *Phys. Rev. Lett.*, **30**, 1343.

Iofa, M. Z. and Tyutin, I. V. (1976). Gauge invariance of spontaneously broken nonabelian theories in the Bogolyubov–Parachuk–Hepp–Zimmermann method, *Theor. Math. Phys.*, **27**, 38.

Khriplovich, I. B. (1969). Green's functions in theories with non-Abelian gauge group, *Yad. Fiz.*, **10**, 409.

Khriplovich, I. B. (1970). Green's functions in theories with non-Abelian gauge group, *Sov. J. Nucl. Phys.*, **10**, 235.

Lee, B. W. and Zinn-Justin, J. (1972). Spontaneously broken gauge symmetries. I. Preliminaries, *Phys. Rev. D*, **5**, 3121.

Lehmann, H., Symanzik, K., and Zimmermann, W. (1955). On the formulation of quantized field theories, *Nuovo Cimento*, **1**, 1425.

Peskin, M. E. and Schroeder, D. V. (1995). *An Introduction to Quantum Field Theory* (Addison-Wesley).

Politzer, H. D. (1973). Reliable perturbative results for strong interactions?, *Phys. Rev. Lett.*, **30**, 1346.

Slavnov, A. A. and Faddeev, L. D. (1970). Massless and massive Yang–Mills field, *Theor. Math. Phys.*, **3**, 312.

Slavnov, A. A. (1972). Ward identities in gauge theories, *Theor. Math. Phys.*, **10**, 99.

Taylor, J. C. (1971). Ward identities and charge renormalization of the Yang–Mills field, *Nucl. Phys. B*, **33**, 436.

't Hooft, G. (1971). Renormalization of massless Yang–Mills fields, *Nucl. Phys. B*, **33**, 173.

't Hooft, G. and Veltman, M. (1972a). Regularization and renormalization of gauge fields, *Nucl. Phys. B*, **44**, 189.

't Hooft, G. and Veltman, M. (1972b). Combinatorics of gauge fields, *Nucl. Phys. B*, **50**, 318.

't Hooft, G. (1973) unpublished.

Van Baal, P. (1992). More (thoughts on) Gribov copies, *Nucl. Phys. B*, **369**, 259.

Wilson, K. G. (1971a). Renormalization group and critical phenomena I: Renormalization group and the Kadanoff scaling picture, *Phys. Rev. B*, **4**, 3174.

Wilson, K. G. (1971b). Renormalization group and critical phenomena II: Phase space cell analysis of critical behavior, *Phys. Rev. B*, **4**, 3174.

Wilson, K. G. (1975). The renormalization group: Critical phenomena and the Kondo problem, *Rev. Mod. Phys.*, **47**, 773.

Chapter 3
Aspects of Finite-Temperature Field Theory

3.1 Free-Particle Partition Function: Real Scalar Field

That the path integral or "sum over histories" for a transition amplitude of a particular asymptotic field configuration ϕ_{as} onto itself is
linked to thermodynamics was discovered by R. P. Feynman [Feyn-
man (1953); Feynman and Hibbs (1965)] under the influence of
J. Schwinger [Migdal (2006)]. More precisely, an analytic continuation of this transition amplitude to compactified imaginary time and
a subsequent "summation" over all ϕ_{as} provides for a representation of the thermodynamical partition function. To strip Feynman's
derivation of inessential gauge-theory complications we sketch it in
Sec. 3.1.1 for the simplest case of a four-dimensional field theory of a
real and free scalar field ϕ. The importance of the Euclidean formulation of thermal field theory cannot be overemphasized because in
Yang–Mills theory it enables the appeal to topologically nontrivial
solutions of the equations of motion. A subclass of the latter turns out
to provide for the building blocks in a nonperturbative construction
of the thermal ground state at high temperature.

In contrast to the thermal ground state the proper consideration
of effective excitations in Yang–Mills theory requires a momentum
space formulation reminiscent of the real-time formalism of finite-
temperature field theory: Only then is it possible to discern quantum
from thermal fluctuations. Because the thermal ground state sets a
scale of maximal resolution such a distinction is necessary. In this
chapter we mainly will discuss, for the example of a real and free

43

scalar field, how the connection to Minkowskian signature is made for the physics of excitations. Our exposition partially is based on the textbook by J. I. Kapusta (1989) and a pioneering article by L. Dolan and R. Jackiw (1974) on the small-mass expansion. Because diagrammatic techniques in interacting finite-temperature field theories are thoroughly exposed in the literature (see for example [Kapusta (1989);Le Bellac (1991)] and the review article [Landsman and Weert (1987)]), we will not repeat them here but rather refer to these publications when necessary.

3.1.1   Euclidean formulation of thermal field theory

In Minkowskian signature the path-integral representation for the transition of a field configuration $\phi_\alpha(\vec{x})$, after evolution for a time t by the dynamics of the theory, onto itself reads [Feynman and Hibbs (1965)]

$$
\langle \phi_\alpha | e^{-iHt} | \phi_\alpha \rangle = \int \prod_{\vec{x},t'} d\pi(\vec{x},t') \int_{\phi(\vec{x},0)=\phi_\alpha(\vec{x})}^{\phi(\vec{x},t)=\phi_\alpha(\vec{x})} \prod_{\vec{y},t''} d\phi(\vec{y},t'')
$$

$$
\times \exp\left[i \int_0^t dt''' \int d^3z \left(\pi \frac{\partial \phi}{\partial t'''} - \mathcal{H}(\pi,\phi) \right) \right], \qquad (3.1)
$$

where $\mathcal{H} = \Theta^{00}$ denotes the spatial Hamiltonian density belonging to the Hamiltonian function H, and $0 \geq t', t'' \geq t$. If the dependence of $\mathcal{H}$ on the conjugate momentum variable π is quadratic, $\mathcal{H}_\pi = \frac{1}{2}\pi^2$, then the π integration in Eq. (3.1) can be performed in terms of a Gaussian integral, and one obtains

$$
\langle \phi_\alpha | e^{-iHt} | \phi_\alpha \rangle = N \int_{\phi(\vec{x},0)=\phi_\alpha(\vec{x})}^{\phi(\vec{x},t)=\phi_\alpha(\vec{x})} \prod_{\vec{x},t'} d\phi(\vec{x},t')
$$

$$
\times \exp\left[i \int_0^t dt'' \int d^3y \left(\frac{1}{2}\partial_\mu \phi \partial^\mu \phi - V(\phi) \right) \right]. \qquad (3.2)
$$

Here N is an inessential normalization, $V(\phi)$ denotes ϕ's potential, and the exponent in Eq. (3.2) thus just equals i times the theory's action $S = \int_0^t dt'' d^3x\, \mathcal{L} \equiv \int_0^t dt'' d^3x \left(\frac{1}{2}\partial_\mu\phi\partial^\mu\phi - V(\phi)\right)$ for a given field configuration evolving from $t'' = 0$ to $t'' = t$. In the following, we always will assume that the Gaussian integration leading to Eq. (3.2) can be and has been performed.

The partition function Z for the canonical ensemble of the same theory reads

$$Z = \text{Tr}\, e^{-\beta H} = \int d\phi_\alpha \langle \phi_\alpha | e^{-\beta H} | \phi_\alpha \rangle, \qquad (3.3)$$

where the trace symbol Tr demands summation over all possible field configurations, labeled by ϕ_α. Moreover, $\beta \equiv 1/T$, and T denotes temperature. Rotating $t \to -i\beta$ in the exponent on the left-hand side of Eq. (3.2), $t', t'' \to -i\tau', -i\tau''$, where $0 \le \tau', \tau'' \le \beta$ are real, and subsequently integrating over all configurations ϕ_α, a representation of the partition function Z of Eq. (3.3) is obtained as

$$Z = \text{Tr}\, e^{-\beta H} = N \int_{\phi(\vec{x},0) = \phi_\alpha(\vec{x})}^{\phi(\vec{x},\beta) = \phi_\alpha(\vec{x})} \prod_{\vec{x},\tau'} d\phi(\vec{x},\tau')$$

$$\times \exp\left[-\int_0^\beta d\tau'' \int d^3y \left(\frac{1}{2}\partial_{\tau''}\phi\partial_{\tau''}\phi + \frac{1}{2}\nabla\phi \cdot \nabla\phi + V(\phi)\right)\right]$$

$$\equiv N \int_{\phi(\vec{x},0) = \phi_\alpha(\vec{x})}^{\phi(\vec{x},\beta) = \phi_\alpha(\vec{x})} \prod_{\vec{x},\tau'} d\phi(\vec{x},\tau') \exp\left[-\int_0^\beta d\tau'' \int d^3y\, \mathcal{L}_E\right]. \qquad (3.4)$$

Notice that only field configurations ϕ, which are periodic in "time" τ, contribute to the functional integration in Eq. (3.4).

Let us now investigate the implications of Eq. (3.4) for a free theory, that is, for the situation where $V(\phi) = \frac{1}{2}m^2\phi^2$. The Euler–Lagrange equation for the Euclidean theory reads

$$\partial_\mu \frac{\partial \mathcal{L}_E}{\partial_\mu\phi} - \frac{\partial \mathcal{L}_E}{\partial\phi} = 0 = (-\partial_\tau^2 - \nabla^2 + m^2)\phi. \qquad (3.5)$$

The imaginary-time propagator $\langle \phi(\tau,\vec{x})\phi(\bar{\tau},\vec{y}) \rangle \equiv D(\vec{x} - \vec{y}, \tau - \bar{\tau})$ solves Eq. (3.5) subject to modification by the inhomogeneity

$\delta^{(4)}(\vec{x} - \vec{y}, \tau - \bar{\tau})$. The solution to Eq. (3.5) must be periodic $\phi(\vec{x}, \tau) = \phi(\vec{x}, \tau + \beta)$. The equation

$$(-\partial_\tau^2 - \nabla^2 + m^2)D(\vec{x} - \vec{y}, \tau - \bar{\tau}) \equiv \mathcal{D} D(\vec{x} - \vec{y}, \tau - \bar{\tau})$$
$$= \delta^{(4)}(\vec{x} - \vec{y}, \tau - \bar{\tau}) \qquad (3.6)$$

is solved most efficiently in Fourier space (discrete variable conjugate to $\tau - \bar{\tau}$, continuous variables conjugate to $\vec{x} - \vec{y}$), and we have

$$\bar{D}(\vec{p}, \omega_n) = \frac{1}{\omega_n^2 + \vec{p}^2 + m^2}, \qquad (3.7)$$

where $\omega_n \equiv 2\pi nT$ ($n \in \mathbb{Z}$) is called the n-th Matsubara frequency, and $\bar{D}$ is the Fourier transform of D. The Euclidean action

$$S_E \equiv \int_0^\beta d\tau \int d^3x \mathcal{L}_E = \frac{1}{2} \int_0^\beta d\tau \int d^3x (\partial_\tau \phi \partial_\tau \phi + \nabla \phi \cdot \nabla \phi + m^2 \phi^2)$$
$$(3.8)$$

is re-cast after partial integrations as

$$S_E = \frac{1}{2} \int_0^\beta d\tau \int d^3x (-\phi \partial_\tau^2 \phi - \phi \nabla^2 \phi + m^2 \phi^2)$$
$$= \frac{1}{2} \int_0^\beta d\tau \int d^3x \, \phi \mathcal{D} \phi. \qquad (3.9)$$

Thus the partition function in Eq. (3.4) can be written as a Gaussian integral:

$$Z = N \int_{\phi(\vec{x},0) = \phi_\alpha(\vec{x})}^{\phi(\vec{x},\beta) = \phi_\alpha(\vec{x})} \prod_{\vec{x},\tau} d\phi(\vec{x}, \tau) \exp\left[-\frac{1}{2} \int_0^\beta d\tau' \int d^3y \phi \mathcal{D} \phi \right].$$
$$(3.10)$$

The differential operator $\mathcal{D}$ is diagonal in Fourier space, and thus the formula

$$\int_{\phi(\vec{x},0) = \phi_\alpha(\vec{x})}^{\phi(\vec{x},\beta) = \phi_\alpha(\vec{x})} \prod_{\vec{x},\tau} d\phi(\vec{x}, \tau) \exp\left[-\frac{1}{2} \int_0^\beta d\tau' \int d^3y \phi \mathcal{D} \phi \right]$$

$$= \text{const}\sqrt{\det \mathcal{D}^{-1}} \qquad (3.11)$$

is applied most efficiently there. In Eq. (3.11) the constancy of the factor refers to independence on β and on volume V. To compute the pressure as $P = \frac{T}{V} \ln Z$ the logarithm needs to be applied to Eq. (3.4). Since $\mathcal{D}^{-1} = D$ and taking into account Eqs. (3.7) and (3.11), up to an inessential constant we have

$$\ln Z = -\frac{1}{2} \ln \prod_{\vec{p},n} (\omega_n^2 + \vec{p}^{\,2} + m^2) = -\frac{1}{2} \sum_{\vec{p},n} \ln (\omega_n^2 + \vec{p}^{\,2} + m^2).$$

$$(3.12)$$

Now, as it stands the logarithm in Eq. (3.12) lacks a scale. Notice that this would continue to be the case in the limit $m \to 0$ where temperature is the only scale available. Thus we need to let $\omega_n^2 + \vec{p}^{\,2} + m^2 \to \beta^2(\omega_n^2 + \vec{p}^{\,2} + m^2)$ in Eq. (3.12). Also, the symbolic sum over $\vec{p}$ is explicitly represented by the integral $\frac{V}{(2\pi)^3} \int d^3p$. Therefore, we have

$$\ln Z = -\frac{1}{2} \sum_{n=-\infty}^{\infty} \frac{V}{(2\pi)^3} \int d^3p \ln[\beta^2(\omega_n^2 + \vec{p}^{\,2} + m^2)]$$

$$= -\frac{VT^3}{(2\pi)^2} \sum_{n=-\infty}^{\infty} \int_0^{\infty} dx\, x^2 \ln[(2\pi n)^2 + x^2 + a^2], \quad (3.13)$$

where $a \equiv \frac{m}{T}$. The logarithm in the integrand of Eq. (3.13) can be rewritten as

$$\ln[(2\pi n)^2 + x^2 + a^2] = \int_1^{x^2+a^2} \frac{dy}{y + (2\pi n)^2} + \ln[1 + (2n\pi)^2]. \quad (3.14)$$

Since the last term in Eq. (3.14) does not depend on temperature it is of no thermodynamical relevance[1] and thus can be ignored. Taking

---

[1] The omission of temperature-independent terms in thermodynamical quantities is common practice. Ultimately, this unsatisfactory situation (gravitation!) is related to the fact that in many four-dimensional interacting quantum field theories an insufficient a priori estimate for the ground state of the system is made. This estimate, which solves the classical, second-order Lagrangian equation of motion, exhibits an additive shift ambiguity of its energy density. Such an ambiguity, which relates to the symmetry

into account Eq. (3.14) in performing the sum over n in Eq. (3.13) first, we have

$$\sum_{n=-\infty}^{\infty} \frac{1}{y + (2\pi n)^2} = \frac{\coth\left[\frac{\sqrt{y}}{2}\right]}{2\sqrt{y}}, \tag{3.15}$$

and thus

$$
\begin{aligned}
\ln Z &= -\frac{VT^3}{(2\pi)^2} \int_0^\infty dx\, x^2 \int_1^{x^2+a^2} dy\, \frac{\coth\left[\frac{\sqrt{y}}{2}\right]}{2\sqrt{y}} \\
&= -\frac{2VT^3}{(2\pi)^2} \int_0^\infty dx\, x^2 \left(\ln\left[\sinh\left(\frac{1}{2}\sqrt{x^2+a^2}\right) \right] \right. \\
&\quad \left. + x\text{-independent terms} \right) \\
&= -\frac{VT^3}{2\pi^2} \int_0^\infty dx\, x^2 \left(\frac{1}{2}\sqrt{x^2+a^2} + \ln[1 - e^{-\sqrt{x^2+a^2}}] \right. \\
&\quad \left. + x\text{-independent terms} \right).
\end{aligned}
\tag{3.16}
$$

The term $\frac{1}{2}\sqrt{x^2+a^2}$ in the integrand of Eq. (3.16) refers to the contribution of vacuum fluctuations. Since a free theory defined on a flat and infinitely extended four-dimensional spacetime is defined to possess no vacuum energy this term and the x-independent contributions to the integrand are ignored. Thus the thermodynamical pressure P of a free theory for a real scalar field is given as

$$P = -\frac{T^4}{2\pi^2} \int_0^\infty dx\, x^2 \ln[1 - e^{-\sqrt{x^2+a^2}}] > 0. \tag{3.17}$$

It is instructive to study the behavior of P for small a analytically. Following [Dolan and Jackiw (1974)], we try an expansion in powers

---

of the equation of motion under global shifts of the potential, no longer persists in SU(2) or SU(3) Yang–Mills theory where the thermal ground state estimate is composed of Bogomolnyi–Prasad–Sommerfield saturated or (anti)selfdual field configurations of zero energy density. Their coarse-grained incarnation satisfies a first-order equation of motion which no longer admits a shift ambiguity of the potential.

of a^2. Taking the limit $a \to 0$ in Eq. (3.17), we have

$$\lim_{a\to 0} P = -\frac{T^4}{2\pi^2} \int_0^\infty dx\, x^2 \ln[1 - e^{-x}] = \frac{\zeta(4)}{\pi^2} T^4 = \frac{\pi^2}{90} T^4, \qquad (3.18)$$

where $\zeta(z) \equiv \sum_{n=1}^\infty n^{-z}$ represents Riemann's zeta function for $\mathrm{Re}\, z > 1$. Next, we consider $\lim_{a\to 0} \partial_{a^2} P$ by differentiating under the integral:

$$\lim_{a\to 0} \partial_{a^2} P = -\lim_{a\to 0} \frac{T^4}{4\pi^2} \int_0^\infty dx \frac{x^2}{\sqrt{x^2 + a^2}} \frac{1}{e^{\sqrt{x^2+a^2}} - 1} \equiv -\lim_{a\to 0} \frac{T^4}{4\pi^2} D(a)$$

$$= -\frac{T^4}{4\pi^2} \int_0^\infty dx \frac{x}{e^x - 1} = -\frac{T^4}{24}. \qquad (3.19)$$

To expand, in turn, the function $D(a)$ in powers of a^2, we need to evaluate $\partial_{a^2}^n D(a)$ $(n = 1,2,3,\dots)$ at $a = 0$. For $n = 1$ we have

$$-\frac{4\pi^2}{T^4} \partial_{a^2}^2 P(a) \equiv I(a) \equiv \partial_{a^2} D(a)$$

$$= \int_0^\infty dx\, x^2 \partial_{x^2} \frac{1}{\sqrt{x^2 + a^2}} \frac{1}{e^{\sqrt{x^2+a^2}} - 1}$$

$$= -\int_0^\infty dx \frac{1}{\sqrt{x^2 + a^2}} \frac{1}{e^{\sqrt{x^2+a^2}} - 1}. \qquad (3.20)$$

It is easy to see that all expressions $\partial_{a^2}^n I(a)$ $(n = 0, 1, 2, 3, \dots)$ diverge at $a = 0$. Thus $\partial_{a^2}^n P(a)$ does not exist at $a = 0$ for $n \geq 2$, and a formal Taylor expansion of $P(a)$ about $a = 0$ possesses diverging coefficients. The expansion can, however, be resummed to yield a finite answer. Technically, this is achieved by considering a modified version of $I(a)$ [Dolan and Jackiw (1974)]:

$$I(a) = -\int_0^\infty dx \frac{1}{\sqrt{x^2 + a^2}} \frac{1}{e^{\sqrt{x^2+a^2}} - 1}$$

$$\to I_\epsilon(a) \equiv -\int_0^\infty dx \frac{x^{-\epsilon}}{\sqrt{x^2 + a^2}} \frac{1}{e^{\sqrt{x^2+a^2}} - 1}, \qquad (3.21)$$

where $0 < \epsilon < 1$. Now, by use of Eq. (3.15) the integrand for $I_\epsilon(a)$ can be split into an infinite sum and a remainder. After integration of each summand in the first contribution, subsequent resummation, combination of this result with the result of integration in the second contribution, and execution of the limit $\epsilon \to 0$ one arrives at [Dolan and Jackiw (1974)]

$$I(a) = -\frac{\pi}{2a} - \frac{1}{2}\ln\frac{a}{4\pi} - \frac{1}{2}\gamma_E + O(a^2), \qquad (3.22)$$

where $\gamma_E \sim 0.577$ is the Euler–Mascheroni constant. Hence the differential equation

$$-\frac{4\pi^2}{T^4}\partial^2_{a^2}P(a) = I(a) \qquad (3.23)$$

subject to the boundary conditions given by Eqs. (3.18) and (3.19) can be solved for $P(a)$. The result is [Dolan and Jackiw (1974)]

$$P(a) = T^4\left(\frac{\pi^2}{90} - \frac{a^2}{24} + \frac{a^3}{12\pi} + \frac{a^4}{64\pi^2}[\ln a^2 - c] + O(a^6)\right), \qquad (3.24)$$

where $c \equiv \frac{3}{2} + 2\ln 4\pi - 2\gamma_E$. From Eq. (3.24) it is now clear that the appearance of the logarithm and odd powers of a is the cause for the nonanalyticity of P in a^2 about the point $a = 0$.

3.1.2   *"Rotation" to real time*

As we have seen, the formulation of finite-temperature field theory in imaginary time implies that in loop integrals, introduced by local interaction monomials beyond quadratic order in the field ϕ in the Lagrangian density [Kapusta (1989);Le Bellac (1991);Landsman and Weert (1987)], instead of frequency *integrals* sums over Matsubara frequencies need to be performed. This formulation therefore exhibits the disadvantage of the impossibility to discern quantum fluctuations, whose presence in free-field propagation (perturbation theory) is independent of the existence of a heat bath, from thermal fluctuations. Since Yang–Mills theory provides for its own thermal ground state, which severely constrains effective quantum fluctuations, such a separation between quantum and thermal fluctuations is required and accomplished by a reformulation of the imaginary time propagator of Eq. (3.7). This correspondence between imaginary and real time for the respective propagators solely occurs in terms of an identity in momentum space, no consideration of an explicit real-time contour is required. In thermal Yang–Mills theory topological field configurations contributing to the ground-state physics can only be addressed in imaginary time. On the other hand, the formulation of the thermodynamics of effective excitations on a different time-integration contour in the action would be inconsistent since the entire partition function of a given field theory is defined on one and the same spacetime. To still be able to discern quantum from thermal fluctuations in the effective theory, we thus need to appeal to an identity in momentum space which is a consequence of analyticity properties of integrands.

To make the connection between the imaginary- and real-time formulation explicit, we study a typical one-loop diagram following the presentation in [Kapusta (1989)]. In imaginary time such a diagram is represented by a frequency sum

$$T \sum_{n=-\infty}^{\infty} S(i\omega_n), \tag{3.25}$$

where the function $S(p_0)$ is analytic within an entire environment of the imaginary axis, and $\omega_n \equiv 2\pi nT$ ($n \in \mathbf{Z}$) is the n-th Matsubara frequency. The integration over the spatial momentum is implicit in (3.25). By use of the theorem of residues the summation in (3.25) can be expressed as a contour integration

$$T \sum_{n=-\infty}^{\infty} S(i\omega_n) = \frac{T}{2\pi i} \oint_C dp_0 \frac{1}{2}\beta \coth\left(\frac{1}{2}\beta p_0\right) S(p_0). \tag{3.26}$$

Notice that the function $\frac{1}{2}\beta \coth\left(\frac{1}{2}\beta p_0\right)$ is analytic except for simple poles at $p_0 = 2\pi nTi$ of residue unity. In Eq. (3.26) the contour C is the set of closed circular curves of radii smaller than πT centered at the poles $p_0 = 2\pi n Ti$, and the integration along these circles runs counter-clockwise. This integration contour can be deformed into a counter-clockwise integration along two lines parallel to the imaginary axis (Re $p_0 = \pm\epsilon, \epsilon > 0$) that lie within the region of analyticity of $S(p_0)$. After a rearrangement of the exponentials of the hyperbolic cotangent and a change of variable $p_0 \to -p_0$ in the β-independent part of the integral one arrives at

$$T \sum_{n=-\infty}^{\infty} S(i\omega_n) = \frac{1}{2\pi i} \int_{-i\infty}^{i\infty} dp_0 \frac{1}{2}[S(p_0) + S(-p_0)]$$

$$+ \frac{1}{2\pi i} \int_{-i\infty+\epsilon}^{i\infty+\epsilon} dp_0 [S(p_0) + S(-p_0)]\frac{1}{e^{\beta p_0} - 1}$$

$$\equiv T \sum_{n=-\infty}^{\infty} S(i\omega_n)\bigg|_{T=0} + T \sum_{n=-\infty}^{\infty} S(i\omega_n)\bigg|_{T>0}. \qquad (3.27)$$

Now, specify $S(p_0)$ as

$$S(p_0) = \frac{1}{(2\pi)^3} \int d^3p \frac{1}{-p_0^2 + \vec{p}^2 + m^2}. \qquad (3.28)$$

Rotating the contour in $T \sum_{n=-\infty}^{\infty} S(i\omega_n)|_{T=0}$ clockwise back to the real axis (with the causal prescription for circumventing the poles at $p_0 = \pm\sqrt{\vec{p}^2 + m^2}$), one arrives for this contribution at

$$T \sum_{n=-\infty}^{\infty} S(i\omega_n)\bigg|_{T=0} = -\frac{1}{(2\pi)^4} \int_{-\infty}^{\infty} dp_0 d^3p \frac{i}{p_0^2 - \vec{p}^2 - m^2 + i\epsilon}. \qquad (3.29)$$

As for $T \sum_{n=-\infty}^{\infty} S(i\omega_n)|_{T>0}$ the contour is closed by a semicircle at infinity circumscribing the right half plane. As a consequence, the contribution of the only pole at $p_0 = \sqrt{\vec{p}^2 + m^2}$ is picked up in this half plane, and we obtain

$$T \sum_{n=-\infty}^{\infty} S(i\omega_n) \Bigg|_{T>0} = -\frac{1}{(2\pi)^3} \int d^3p \frac{1}{\sqrt{\vec{p}^2 + m^2}} \frac{1}{e^{\beta\sqrt{\vec{p}^2+m^2}} - 1}. \quad (3.30)$$

Notice that Eq. (3.30) was obtained by appeal to the theorem of residues, thus it just re-expresses the sum on the left-hand side, and no genuine continuation process is involved.

Equations (3.29) and (3.30) can be summarized as follows:

$$T \sum_{n=-\infty}^{\infty} S(i\omega_n) = 2\pi T \sum_{n=-\infty}^{\infty} \frac{1}{(2\pi)^4} \int d^3p \frac{1}{\omega_n^2 + \vec{p}^2 + m^2}$$

$$= -\frac{1}{(2\pi)^4} \int d^4p \left[\frac{i}{p^2 - m^2} + \delta(p^2 - m^2) \frac{2\pi}{e^{\beta|p_0|} - 1} \right]. \quad (3.31)$$

Thus the following correspondence takes place according to Eq. (3.31)

$$-\frac{1}{\omega_n^2 + \vec{p}^2 + m^2} \longrightarrow \frac{i}{p^2 - m^2} + \delta(p^2 - m^2) \frac{2\pi}{e^{\beta|p_0|} - 1}. \quad (3.32)$$

The right-hand side of Eq. (3.32) is defined as the real-time propagator of the scalar field ϕ in momentum space. It neatly separates into a temperature-independent part and a part expressing that thermal fluctuations are Bose-distributed and on the mass shell $m^2 = p^2$. The expression in Eq. (3.32) for the real-time propagator, when naively applied in diagrammatic expansions, seems to cause mathematical inconsistencies in certain loop diagrams. Consider for example an *interacting* theory of a real scalar field with vertices connecting four lines in one point (a so-called $\lambda\phi^4$-theory). Then in diagrams with n vertices of the form depicted in Fig. 3.1 there is no momentum transfer from the large to the small loops. Thus, due to Eq. (3.32), n-th powers of δ-functions occur in the integration over the loop momentum of the large loop. This is not acceptable mathematically [Dolan and Jackiw (1974)] and has lead to a matrix formulation for

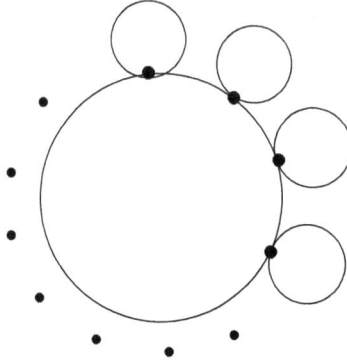

Figure 3.1. A so-called daisy diagram in a $\lambda\phi^4$-theory.

the diagrammatic expansion in real time [Umezawa, Matsumoto, and Tachiki (1982); Niemi and Semenoff (1984)] which essentially amounts to doubling the independent fields with the effect that singularities arising from powers of a δ-function cancel. Such a formalism can be introduced by considering a more elaborate contour for the τ-integration in the action. On the other hand, a generalization to Eq. (3.32), which does not rely on the free-theory approximation, was derived in [Dolan and Jackiw (1974)]. Essentially, the δ-function in Eq. (3.32) is replaced by a function describing the physical energy spectrum of intermediate states at a fixed spatial momentum. This spectral function relates to the imaginary part of the self-energy which adds to the denominator of the $T = 0$ part in Eq. (3.32). No mathematical inconsistencies arise because the spectral function no longer is a δ-function. In Yang–Mills theory a spatial coarse-graining procedure for the estimate of the thermal ground state (see Sec. 5.1) is performed. As far as ground-state physics is concerned, this coarse-graining amounts to an *incomplete* execution of the summation over all field configurations in the partion functtion. It can only be carried out for the contour $0 \le \tau \le \beta$ since no relevant solutions to the selfduality equations are known in real time. In the treatment of effective radiative corrections such a contour cannot be changed. But, again, the inconsistencies of the order-by-order expansion into loops go away by appealing to spectral functions (obtained by resummations in imaginary time).

In real-time *perturbation theory*, with its trivial ground-state estimate, however, the doubling of degrees of freedom due to a deformed contour for the τ integration is the only option to perform fixed-order loop expansions.

3.2 Perturbative Loop Expansion in Thermal Gauge Theory

The perturbative loop expansion of thermodynamical quantities in a four-dimensional Yang–Mills theory at high temperature[2] has been an active playground of investigation for more than thirty years. Perturbation theory, though technically highly involved, is conceptually a well established method of analytic investigation in interacting quantum field theory. It was argued by A. Linde (1980), however, that perturbation theory generates an infrared cutoff which screens the so-called soft magnetic sector too weakly to enforce a convergence of the expansion of thermodynamical quantities. Linde's argument is plausible.

As we shall show in Chapter 5, there is a substantial impact of the nontrivial thermal ground state on the propagation and interaction of gauge modes even at high temperature. This ground state is composed of gauge-field configurations that cannot be approximated by the perturbative expansion. It is precisely these field configurations that provide for spatial correlations giving rise to mass on the effective-theory level. The associated apparent gauge-symmetry breaking does not depend on a small-coupling assumption, leads

---

[2]Motivated by the SU(3) gauge symmetry of Quantum Chromodynamics (QCD), most of the purely perturbative and lattice-gauge-theory investigations were carried out for this gauge group both with and without fundamentally charged fermions (quarks).

to fast convergence of loop expansions, and thus evades the perturbative infrared problem of the soft magnetic sector pointed out by Linde.

To do justice to technically highly impressive work, and also to enable comparison with the nonperturbative results in Chapter 5 we provide here for an brief overview on the philosophy of expanding thermodynamical quantities in terms of a small gauge coupling g. By replacing the renormalization scale μ in Eq. (2.36) by temperature T asymptotic freedom (see Sec. 2.3.3) seems to suggest that for sufficiently high T the fundamental gauge coupling is small, $g \ll 1$, which motivates the perturbative approach.

Perturbative high-temperature expansions are usually carried out in the imaginary-time formulation. A major thrust in activity concerns the dimensional reduction program [Ginsparg (1980); Appelquist and Pisarski (1981)] which "integrates out" the compactified temporal direction. Here only the vanishing Matsubara frequencies give rise to massless three-dimensional gauge modes (zero modes) while finite Matsubara frequencies are associated with massive modes (mass rising linearly in T). To lowest-order in the effective three-dimensional gauge-coupling the effective action S^{3D} of massless three-dimensional modes is given as

$$S^{3D} = \int d^3x \left[\frac{1}{2} \text{tr} \, (F_{ij}^0)^2 + \text{tr} \, [D_i^0, A_4^0]^2 + m_3 \text{tr} \, (A_4^0)^2 \right.$$

$$\left. + \lambda_3 (\text{tr} \, (A_4^0)^2)^2 + \cdots \right]. \qquad (3.33)$$

By a perturbative matching to the full four-dimensional theory (see for example [Schröder (2006)]) the coefficients $m_3, \lambda_3, \ldots$ can be calculated. The expansion in the mass dimension $M > 0$ of local operators built of the zero fields A_i^0 and A_4^0 would be useful if it converged rapidly. Then a truncation at low values of M would provide for a three-dimensional action which approximates the full theory well and therefore would be apt to base lattice-gauge theory simulations upon. Three-dimensional lattice simulations are considerably less "expensive" than those performed in the full four-dimensional theory and provide for some analytical insight.

Let us come back to Linde's argument on four-dimensional perturbation theory which asserts that a small four-dimensional gauge coupling, $g \ll 1$, generates a hierarchy in mass-scales at high temperature. This hierarchy is relied upon in integrating out fluctuations. Momenta of order T [Braaten and Pisarski (1990a,b)], gT, and g^2T are referred to as hard, soft, and ultrasoft, respectively. The applicability of perturbation theory demands that the expectation of the interaction monomial[3] $I \equiv \langle g^2 A^4 \rangle$ in the Yang–Mills Lagrangian is hierarchically smaller than the expectation $K \equiv \langle (\partial A)^2 \rangle$ of the kinetic term. Perturbatively, that is, by an estimate based on the free theory the wave function A is counted as one power of momentum.[4] This applies also to the derivative operation ∂. Thus for hard modes one finds $K \sim T^4$ and $I = g^2T^4$. As a consequence, $K \gg I$, and the perturbative philosophy is selfconsistent. In successively integrating out momentum scales starting from hard towards ultrasoft the above counting is modified by screening of the softer by the harder modes. For soft modes Linde finds: $K \sim g^2T^4$ and $I = g^3T^4$. The result for I is due to a deviation from the above counting introduced by the screening of the four-vertex through hard modes. Thus $K > I$, and perturbation theory may still work for soft modes. As for the ultrasoft sector Linde obtains a weak screening of magnetic modes resulting in $K \sim g^6T^4$ and $I = g^6T^4$. As a result, $K \sim I$ which invalidates the perturbative approach in integrating out these modes. Hence, starting at power g^6, the loop expansion of the pressure, which is minus the free energy and thus a measure for the effective

---

[3]Linde only considers the quartic interaction in the Yang–Mills Lagrangian.
[4]This follows from the fact that the Yang–Mills action is scale-invariant classically. Therefore, any operator of mass dimension unity must be proportional to the external momentum scale.

Lagrangian density, should contain contributions which cannot be computed in perturbation theory.

Historically, the perturbative computation of the Yang–Mills pressure at high temperature up to and including the order $g^6 \ln 1/g$ was performed in the following publications: Up to order g^2 in [Shuryak (1978)], up to order g^3 in [Kapusta (1979)], up to order $g^4 \ln 1/g$ in [Toimula (1983)], up to order g^4 in [Arnold and Zhai (1994)], up to order g^5 in [Zhai and Kastening (1994)], up to order g^6 in [Braaten and Nieto (1996)], and up to order $g^6 \ln 1/g$ in [Kajantie *et al.* (2003)]. In Fig. 3.2 these results are compared for the gauge group SU(3). By the alternating behavior from one level of approximation to the next one it is suggested that perturbative contributions to the pressure of high-temperature Yang–Mills theory do

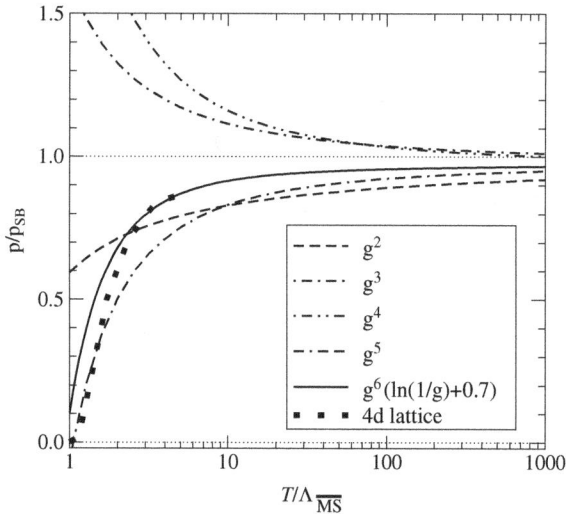

Figure 3.2. The perturbative contributions to the pressure of an SU(3) Yang–Mills gas up to and including orders g^2, g^3, g^4, g^5, and $g^6 \ln 1/g$. Notice the alternating behavior (poor convergence if any) when increasing the order in the perturbative coupling g. Reprinted from [Kajantie *et al.* (2003)] with kind permission by M. Laine and the American Physical Society. Copyright (2003) by the American Physical Society.

not capture its physics in an adequate way. Recent developments therefore address the physics of order g^6 and beyond by subjecting the dimensionally reduced form of the Yang–Mills action to lattice simulations. Notice that the matching of the three-dimensional coefficients to the full theory is done perturbatively, and it is unclear whether the presumed separation of perturbatively accessed hard modes (coefficients) and the soft three-dimensional dynamics supposedly captured by lattice simulations really takes place. To the author's mind this separation is hard to control because the convergence properties of the expansion of Eq. (3.33) in M are unclear.

3.3 Electric Center Symmetry

In this section we would like to discuss a *global* symmetry inherent to a three-dimensional reformulation of the four-dimensional Yang–Mills theory after the compact time-direction of the Euclidean finite-temperature formulation has been integrated out. A central field variable in this three-dimensional effective theory is the Polyakov loop $\mathrm{Pol}(\vec{x}) \in \mathrm{SU(N)}$ which is a Wilson line of the gauge field A_μ along the (compact) time interval: $0 \leq x_4 \equiv \tau \leq \beta$. The nonlocal object $\mathrm{Pol}(\vec{x})$ transforms under four-dimensional gauge transformations, induced by the group element $\Omega(x)$, as $\mathrm{Pol}(\vec{x}) \to \Omega(\tau = 0, \vec{x})\mathrm{Pol}(\vec{x})\Omega^\dagger(\tau = \beta, \vec{x})$. Now, assume that $\Omega(\tau = 0, \vec{x})$ and $\Omega^\dagger(\tau = \beta, \vec{x})$ are distinct elements of the center group Z_N of $\mathrm{SU(N)}$ (see Sec. 1.2). Because the center elements $\Omega(\tau = 0, \vec{x})$ and $\Omega^\dagger(\tau = \beta, \vec{x})$ commute with any $\mathrm{SU(N)}$ group element and thus also with $\mathrm{Pol}(\vec{x})$, the Polyakov loop transforms by multiplication with the center element $\Omega(\tau = 0, \vec{x})\Omega^\dagger(\tau = \beta, \vec{x})$ which may be different from unity. In the three-dimensional theory this multiplication of $\mathrm{Pol}(\vec{x})$ with a center element is induced by an admissible four-dimensional gauge transformation, and therefore we have identified a symmetry of the effective three-dimensional theory.[5]

3.3.1 Polyakov loop: Definition

Let us be more specific by considering the gauge group SU(2). We use the perturbative definition of the gauge field A_μ without absorption of the fundamental coupling g. The Polyakov loop $\text{Pol}(\vec{x})$ is defined as the Wilson line (see Sec. 1.4.2) along the compact time-direction of Euclidean thermal field theory:

$$\text{Pol}(\vec{x}) \equiv \mathcal{P} \exp\left[ig \int_0^\beta d\tau A_4(\tau, \vec{x}) \right]. \tag{3.34}$$

Here $\mathcal{P}$ demands path-ordering. Under gauge transformations with $\Omega(\tau = 0, \vec{x}) = \Omega(\tau = \beta, \vec{x})$ the quantity $\text{tr}\,\text{Pol}(\vec{x})$ is invariant.

Now consider gauge transformations associated with a center jump along the time direction. For SU(2) a simple example of a group element $\Omega^s(\tau)$ inducing such a singular gauge transformation is

$$\Omega^s(\tau) \equiv (2\Theta(\tau - \alpha) - 1)\mathbb{1}_2, \tag{3.35}$$

where the function Θ is defined as

$$\Theta(x) = \begin{cases} 0, & (x < 0), \\ \dfrac{1}{2}, & (x = 0), \\ 1, & (x > 0). \end{cases} \tag{3.36}$$

and $0 < \alpha < \beta$. Obviously, $\Omega^s(\tau = 0) = -\mathbb{1}_2 = -\Omega^s(\tau = \beta)$. According to Eq. (1.3) the periodicity of the gauge field A_μ is, however,

---

[5]The Yang–Mills action $S = \frac{1}{2g^2} \text{tr} \int_0^\beta d\tau d^3x\, F_{\mu\nu}F_{\mu\nu}$ is invariant under all gauge transformations, even those induced by nonperiodic group elements. Admissible gauge transformations are those which do not change the periodicity of the gauge field A_μ they are applied to. Center jumps of the above-assumed form are examples of admissible gauge transformations induced by nonperiodic group elements.

not altered under a gauge transformation induced by Ω^s. Because of Eq. (1.41) and because of the definition of the group center we have

$$\text{Pol}(\vec{x}) \xrightarrow{\Omega} -\text{Pol}(\vec{x}). \qquad (3.37)$$

Since the Yang–Mills action density $S_\beta[A] = \frac{1}{2g^2}\text{tr}\int_0^\beta d\tau d^3x F_{\mu\nu}F_{\mu\nu}$ is invariant also under the singular but periodicity-respecting (and thus admissible) gauge transformation induced by $\Omega^s(\tau)$, the entire partition function

$$Z_\beta = \int_{A_\mu(\tau=0,\vec{x})=A_\mu(\tau=\beta,\vec{x})} \prod_{\tau,\vec{x},b} dA_\mu^b(\tau,\vec{x})\, e^{-S_\beta[A]} \qquad (3.38)$$

is invariant under this gauge transformation. This continues to hold also for SU(N) with a corresponding generalization of the center jump of Eq. (3.35): In an effective three-dimensional theory, obtained by an integration of the time dependence of four-dimensional gauge field configurations, the multiplication by an element of the center group $\mathbf{Z}_N$ is a *global* symmetry. Because in four dimensions this symmetry is associated with a gauge transformation induced by an admissible group element with center jump along compactified Euclidean time the corresponding singularity in the gauge transformation of A_4 [compare with Eq. (1.3)] changes the *electric* flux through the temporal loop. This is the reason why we refer to the associated three-dimensional symmetry as *electric* center symmetry.

3.3.2   Polyakov loop as an order parameter

The Polyakov loop can be considered as an order parameter for deconfinement associated with a dynamical breaking of the electric $\mathbf{Z}_N$ symmetry in the effective three-dimensional theory. It is associated with the free energy of an infinitely heavy test "quark" which transforms under the fundamental representation of the gauge group SU(N). If such a test quark acquires in addition to its infinite mass an infinite free energy by interacting with the gauge fields of a pure SU(N) Yang–Mills theory then this implies that even as a test particle its presence in the thermalized system is rejected. This, however, is equivalent to the statement that an isolated heavy "quark" cannot exist. In a compound state with an antiquark it can, however, exist if the gauge charge of this compound is completely screened at long distances.

We have not yet introduced the concept of fundamental fermions in this book. For the purpose of showing that the Polyakov-loop expectation is an order parameter for deconfinement it is, however, necessary to consider Dirac fermions coupled to the gauge field A_μ. For details of how this coupling is introduced by a fundamental covariant derivative D_μ replacing the ordinary derivative ∂_μ in the free action density $i\bar{\psi}\partial_\mu\gamma_\mu\psi$ of the four-component Dirac[6] spinor $\psi(\tau,\vec{x})$ one may for example consult [Peskin and Schroeder (1995)]. If the field ψ is introduced as a static test-charge then the piece of the action S_{int}, describing the coupling between the gauge field $A_\mu = A_\mu^a t^a$ and the covariantly conserved four-current $j_\mu^a \equiv \bar{\psi}\gamma_\mu t^a\psi$, is given as

$$
S_{\text{int}} = -ig\int_0^\beta d\tau d^3x j_\mu^a A_\mu^a
$$

$$
\overset{\text{static test charge}}{=} -ig\int_0^\beta d\tau d^3x \delta^{(3)}(\vec{x})A_4^1(\tau,\vec{x}) = -ig\int_0^\beta d\tau A_4^1(\tau,\vec{0}). \quad (3.39)
$$

In Eq. (3.39) it is assumed that the gauge-field component $A_4 = A_4^a t^a$ was gauge-rotated into the one-direction of the Lie algebra, and that the heavy test-charge is placed at the spatial origin. The density $j_\mu^a A_\mu^a$ in Eq. (3.39) is not gauge-invariant.[7] In the case of a heavy test-charge the piece $\exp[ig\int_0^\beta d\tau d^3x j_\mu^a A_\mu^a]$ in the weight of the *full* partition function Z'_β of the system *thermal Yang–Mills + static test-charge* must be related to an expression which is manifestly invariant

---

[6]The four 4×4 matrices γ_μ obey the Euclidean Clifford algebra $\{\gamma_\mu, \gamma_\nu\} = 2\delta_{\mu\nu}$.

[7]It is, however, invariant under *global*, that is, spacetime-independent gauge transformations.

under smooth gauge transformations. Comparing Eqs. (3.34) and (3.39), the only gauge-invariant generalization of the exponential taken of minus the right-hand side of Eq. (3.39) leads to the following expression for Z'_β

$$Z'_\beta = \int_{A_\mu(\tau=0,\vec{x})=A_\mu(\tau=\beta,\vec{x})} \prod_{\tau,\vec{x},b} dA^b_\mu(\tau,\vec{x}) \, \mathrm{tr}\, \mathrm{Pol}(\vec{0}) e^{-S_\beta[A]}$$

$$= Z_\beta \times \langle \mathrm{tr}\, \mathrm{Pol}(\vec{0}) \rangle$$

$$\propto \text{(element of center group)} \times \exp(-F/T). \qquad (3.40)$$

Here the average $\langle \cdot \rangle$ is with respect to the weight $e^{-S_\beta[A]}$ introduced by the original Yang–Mills action $S_\beta[A]$, F denotes the free energy of the test-"quark" and Z_β denotes the partition function defined in Eq. (3.38). Since Z_β should be finite[8] the vanishing of $\langle \mathrm{tr}\, \mathrm{Pol}(\vec{0}) \rangle$, indicating that the electric and global center symmetry $\mathbf{Z}_N$ of the effective three-dimensional theory is unbroken,[9] implies by virtue of Eq. (3.40) that $F \to \infty$. This, however, means that test-"quarks" are confined. The other case $\langle \mathrm{tr}\, \mathrm{Pol}(\vec{0}) \rangle \neq 0$, which is N-fold center degenerate (dynamical breaking of the electric center symmetry $\mathbf{Z}_N$), entails that $F < \infty$. In this case an isolated static test-"quark" may exist thus implying deconfinement.

Problems

3.1. Find an explicit expression for the normalization constant N in Eq. (3.2) in a formulation where the time integration is

---

[8]Up to a finite additive constant related to a logarithm of the subvolume of the compact gauge group SU(N) (only gauge transformations respecting periodicity are allowed!) Z_β defines in the infinite-volume limit $V \to \infty$ the thermodynamical pressure P of the theory as $P = \frac{T}{V} \ln Z_\beta$ which is finite.
[9]The transformation in Eq. (3.37) and its generalization to SU(N) then maps zero to zero, and the thermal ground state is left invariant under the action of the electric center symmetry $\mathbf{Z}_N$.

approximated by a Riemann sum over n support points separated equidistantly by Δt.

3.2. Which meromorphic function of z is majorizing $\sum_{n=1}^{\infty} \frac{1}{n^z + a^2}$ (a real) best? Where is the only pole situated? What can one say about the zeros for $a = 0$?

3.3. Find an integral representation of Riemann's zeta function for $\mathrm{Re}\, z > 1$ that makes the statement in Eq. (3.18) explicit.

3.4. Show that $\partial_{a^2}^n I(a)$ ($n = 0, 1, 2, 3, \ldots$) diverge at $a = 0$ where $I(a)$ is defined in Eq. (3.20).

3.5. Perform the steps leading from Eqs. (3.21)–(3.22).

3.6. Solve the differential equation (3.23) to arrive at Eq. (3.24).

3.7. Perform the steps leading from Eqs. (3.26)–(3.27).

3.8. Why is there a modulus sign in the exponential of the integrand in Eq. (3.31)? Argue both physically and mathematically.

3.9. Show that for an SU(2) gauge field configuration A_μ with $A_\mu(\tau = 0, \vec{x}) = A_\mu(\tau = \beta, \vec{x})$ the gauge-transformed field $A_\mu^{\Omega^s}$ is also periodic: $A_\mu^{\Omega^s}(\tau = 0, \vec{x}) = A_\mu^{\Omega^s}(\tau = \beta, \vec{x})$. Modulo the surface $\tau = \alpha$ the group element Ω^s is defined in Eq. (3.35).

3.10. Show that the gauge condition $A_4 = A_4^a t^a \overset{!}{=} A_4^1 t^1$ can always be reached.

References

Arnold, P. and Zhai, C. (1994). The three loop free energy for pure QCD, *Phys. Rev. D*, **50**, 7603.

Appelquist, T. and Pisarski, R. D. (1981). High-temperature Yang–Mills thermodynamics and three-dimensional QCD, *Phys. Rev. D*, **23**, 2305.

Braaten, E. and Nieto, A. (1996). Free energy of QCD at high temperature, *Phys. Rev. D*, **53**, 3421.

Braaten, E. and Pisarski, R. D. (1990a). Soft amplitudes in hot gauge theories: A general analysis, *Nucl. Phys. B*, **337**, 569.

Braaten, E. and Pisarski, R. D. (1990b). Deducing hard thermal loops from Ward identities, *Nucl. Phys. B*, **339**, 310.

Chin, S. A. (1978). Transition to hot quark matter in relativistic heavy-ion collisions, *Phys. Lett. B*, **78**, 552.

Dolan, L. and Jackiw, R. (1974). Symmetry behavior at finite temperature, *Phys. Rev. D*, **9**, 3320.

Feynman, R. P. (1953). Atomic theory of the λ transition in helium, *Phys. Rev.*, **91**, 1291.

Feynman, R. P. and Hibbs, A. R. (1965). *Quantum Mechanics and Path Integrals* (McGraw-Hill).

Ginsparg, P. H. (1980). First order and second order phase transition in gauge theory at finite temperature, *Nucl. Phys. B*, **170**, 388.

Kapusta, J. I. (1989). *Finite-temperature Field Theory* (Cambridge University Press).

Kapusta, J. I. (1979). Quantum chromdynamics at high temperature, *Nucl. Phys. B*, **148**, 461.

Kajantie, K. *et al.* (2003). The pressure of hot QCD up to $g^6 \ln 1/g$, *Phys. Rev. D*, **67**, 105008.

Le Bellac, M. (1991). *Quantum and Statistical Field Theory* (Oxford University Press, Clarendon).

Landsman, N. P. and Weert, C. G. (1987). Real- and imaginary-time field theory at finite temperature and density, *Phys. Rep.*, **145**, 141.

Linde, A. D. (1980). Infrared problem in the thermodynamics of the Yang–Mills gas, *Phys. Lett. B*, **96**, 289.

Migdal, S. (2006). Paradise lost, in *Continuous Advances in QCD 2006*, eds. M. Peloso and M. Shifman (World Scientific).

Niemi, A. J. and Semenoff, G. W. (1984). Finite temperature quantum field theory in Minkowski space, *Ann. Phys.*, **152**, 105.

Peskin, M. E. and Schroeder, D. V (1995). *An Introduction to Quantum Field Theory*, (Addison-Wesley).

Schröder, Y. (2006). Weak-coupling expansion of the hot QCD pressure, *PoS*, JHW2005, 029.

Shuryak, E. V. (1978). Theory of hadronic plasma, *Sov. Phys. JETP*, **47**, 212.

Toimula, T. (1983). The next term in the thermodynamical potential of QCD, *Phys. Lett. B*, **124**, 407.

Umezawa, H., Matsumoto, H. and Tachiki, M. (1982). *Thermo Field Dynamics and Condensed States* (North Holland, Amsterdam).

Zhai, C. and Kastening, B. (1994). The free energy for hot gauge theories with fermions through g^5, *Phys. Rev. D*, **52**, 7232.

Chapter 4

Selfdual Field Configurations

The purpose of this chapter is to introduce the reader to the story on the construction of exact solutions to the Euclidean field equations in Yang–Mills theory. The finite action of these solutions saturates the bound determined by a nontrivial member — the topological charge — of the homotopy groups $\Pi_3(SU(2) = S_3) = \mathbf{Z}$ (pure gauge theory: instantons) or $\Pi_2(SU(2)/U(1) = S_2) = \mathbf{Z}$ (adjoint Higgs model: magnetic monopoles or dyons). The latter feature is implied by their self- or antiselfduality.[1] The instantons of pure Yang–Mills theory have provided for ample playground in mathematics [Atiyah and Singer (1984); Atiyah *et al.* (1978); Donaldson and Kronheimer (1990)] and physics [Adler (1969); Adler and Bardeen (1969); Bell and Jackiw (1969); 't Hooft (1976a,b); Fujikawa (1979, 1980)]. In deconfining Yang–Mills thermodynamics their periodic incarnations, called calorons, are responsible for the emergence of a thermal ground state and the existence of stable (yet unresolvable) magnetic monopoles.

The pioneering work on SU(2) instantons of topological charge $k = \pm 1$, supported by Euclidean spacetime $\mathbf{R}^4$, was performed in [Belavin *et al.* (1975)] (BPST instanton). These solutions to the selfduality equations can be promoted to SU(N) by considering SU(2) embeddings. One year later G. 't Hooft generalized the BPST instanton to arbitrary values of $k \in \mathbf{Z}$ by appealing to a clever change of gauge and an ansatz granting a linear superposition principle

---

[1]If not explicitly stated otherwise the term selfdual will stand for both selfdual and antiselfdual in what follows.

to a scalar function (the instanton prepotential) ['t Hooft (1976a,b)]. 't Hooft's construction provided for sufficient technology to obtain a $k = 1$ instanton *periodic* along one spacetime coordinate. This turned out to be of relevance to the situation at high temperature [Harrington and Shepard (1978)] (Harrington–Shepard caloron). The Harrington–Shepard caloron is particular in that it possesses a trivial Polyakov loop at spatial infinity (a trivial holonomy).

The construction of ['t Hooft (1976a,b)] does not lead to the most general instanton on $\mathbf{R}^4$. One year after the important generalization of ['t Hooft (1976a,b)] to the charge-k case the most general instanton for SU(N) (and other compact gauge groups) was constructed [Atiyah *et al.* (1978); Drinfeld and Manin (1978)] (ADHM construction). The ADHM construction is based on the observation made in [Ward (1977); Atiyah and Ward (1977)] that the analytic vector bundle associated with the selfdual gauge field can be constructed by appealing to a sequence of vector spaces related to sheaf cohomology groups. Remarkably, the ADHM construction of these charge-k instantons on $\mathbf{R}^4$ simply boils down to a problem in linear algebra subject to a quadratic constraint. An insightful generalization of the ADHM construction to the case of a four-torus by Nahm (1980, 1981, 1982, 1983) has elucidated why the task of finding all instantons on $\mathbf{R}^4$ actually is an algebraic problem. Nahm's construction maps one-to-one a selfdual field on the original four-torus onto a selfdual field on the dual four-torus. The selfduality equation for the dual problem reduces to a quadratic constraint for the special case of all periods of the original torus going to infinity, and one is back at the ADHM construction. For the case of three periods of the original torus being infinite, the Nahm duality transformation can be applied to construct the most general caloron out of data imposed on the dual gauge field satisfying an *ordinary* differential equation, the latter being the selfduality equation on the dual torus [Nahm (1983)]. For SU(2) this procedure was explicitly and independently carried out in [Lee and Lu (1998)] and in [Kraan and Van Baal (1998a,b)] for $|k| = 1$ and for a vanishing overall magnetic charge.

In general, the $|k| = 1$ caloron SU(2) possesses a nontrivial Polyakov loop at spatial infinity (nontrivial holonomy) which then does not coincide with an element of the center group $\mathbf{Z}_2$. As a consequence, calorons may exhibit magnetic substructure. This substructure is associated with a BPS magnetic monopole separated by a *finite* spatial distance from its antimonopole. This configuration is very particular because the attraction mediated by the monopoles' magnetic fields is canceled by the repulsion due to the four-component of the gauge field which acts as an adjoint scalar field.

Technically, all knowledge about the properties of calorons of nontrivial holonomy is based on the above-sketched technically and conceptually highly involved chain of successes. Remarkably, this knowledge is not required in the construction and evaluation of the effective theory for deconfining Yang–Mills thermodynamics performed in Chapter 5. Knowing the properties of calorons of nontrivial holonomy is, however, indispensible for the interpretation of the predictions of this effective theory. The consequences of deforming trivial to nontrivial holonomy by the exchange of propagating gauge-field fluctuations between calorons are two-fold: (i) generation of long-lived magnetic charge by the dissociation of calorons of large holonomy into their magnetic monopole constituents and (ii) dimensionless or dimensionful screening of these monopoles by short-lived monopole-antimonopole pairs in calorons of small holonomy or by other long-lived magnetic charges, respectively. Consequence (ii), in turn, implies the condensation of magnetic monopoles at a critical temperature T_c, a process, which invokes a further step in effectively breaking the original gauge symmetry dynamically.

4.1   The BPST Instanton and Multiinstanton Generalization

In this section we discuss for the gauge group SU(2) the construction of the simplest instanton ($|k| = 1$) on $\mathbf{R}^4$ and its generalization to $|k| > 1$. To do this, the notion of a topological current [Chern and Simons (1974)] and the important decomposition of the Yang–Mills action due to E. B. Bogomoln'yi (1976) are introduced. Subsequently, an ansatz yielding an integrable four-dimensional spherically symmetric action density and a gauge potential with one-fold winding at infinity is shown to respect the selfduality equations [Belavin *et al.* (1975)]. This closed-form solution is considered in a gauge-transformed version where the nontrivial topology resides at the center of the lump in action density. In this singular gauge an ansatz is made that allows for a generalization of the case $|k| = 1$ to $|k| > 1$ ['t Hooft (1976a,b)].

4.1.1   *Chern–Simons current and Bogomoln'yi decomposition*

Since for SU(2) $\Pi_3(S_3) = \mathbf{Z}$ is nontrivial it is legitimate to ask whether the classical Euclidean theory on $\mathbf{R}^4$ possesses topologically stabilized finite-action solutions. We work with the convention that the coupling constant g is absorbed into the gauge field A_μ. With Chern and Simons (1974) we define the following (gauge-noninvariant) current

$$K_\mu \equiv \frac{1}{16\pi^2} \epsilon_{\mu\alpha\beta\gamma} \left(A_\alpha^a \partial_\beta A_\gamma^a + \frac{1}{3} \epsilon^{abc} A_\alpha^a A_\beta^b A_\gamma^c \right), \tag{4.1}$$

where $\epsilon_{\mu\nu\alpha\beta}$ denotes the totally antisymmetric tensor ($\epsilon_{4123} = 1$) in four dimensions, and a labels coordinates in the Lie algebra of SU(2): $a, b, c = 1, 2, 3$. It is straightforward to show that

$$\partial_\mu K_\mu = \frac{1}{32\pi^2} F_{\mu\nu}^a \tilde{F}_{\mu\nu}^a, \tag{4.2}$$

where the dual field strength $\tilde{F}_{\mu\nu}$ is defined in Eq. (1.33). The right-hand side of Eq. (4.2) is manifestly gauge invariant. Now, let us assume that a smooth gauge-field configuration A_μ^a on $\mathbf{R}^4$ decays as $|A_\mu^a| \leq 1/|\vec{x}|$ for $|\vec{x}| \to \infty$. Then, according to Gauss' theorem, the spacetime integral over Eq. (4.2) can be expressed as follows

$$
\begin{aligned}
\int d^4x\, \partial_\mu K_\mu &\equiv \int dx_4 \partial_4 \int d^3x\, K_4 + \int dx_4 \int d^3x\, \partial_i K_i \\
&= \int dx_4\, \partial_4 \int d^3x\, K_4 + \int dx_4 \int d\Sigma_i\, K_i \\
&= \int dx_4\, \partial_4 \int d^3x\, K_4 \\
&= \int d^3x\, K_4 \bigg|_{x_4=\infty} - \int d^3x\, K_4 \bigg|_{x_4=-\infty} \\
&\equiv n_{cs}(x_4=\infty) - n_{cs}(x_4=-\infty) \\
&\equiv k = \frac{1}{32\pi^2} \int d^4x\, F_{\mu\nu}^a \tilde{F}_{\mu\nu}^a.
\end{aligned}
\tag{4.3}
$$

In going from the second to the third line in Eq. (4.3) it was used that the element $d\Sigma_i$ and the spatial current components K_i for $|\vec{x}| \to \infty$ behave like $|\vec{x}|^2$ and $1/|\vec{x}|^3$, respectively. Equation (4.3) states that the quantity $\int d^4x\, F_{\mu\nu}^a \tilde{F}_{\mu\nu}^a$ is related to the difference of Chern–Simons numbers n_{cs} [charge of the Chern–Simons current defined in Eq. (4.1)] on the time slices $x_4 = \pm\infty$. That is, whatever smooth and spatially localized deformation of the gauge field away from this boundary is performed, it does not have any influence on the value of k (topological charge or Pontryagin index). The fact that on $\mathbf{R}^4$, indeed, $k \in \mathbf{Z}$ for any *finite-action* configuration, which is smooth everywhere up to isolated points, is verified by classifying its pure-gauge behavior at infinity and its behavior near these isolated singularities (see Secs. 4.1.2 and 4.1.3).

It is straightforward to check that the Euclidean Yang–Mills action on $\mathbf{R}^4$ can be decomposed as

$$
\begin{aligned}
S &= \frac{1}{4g^2} \int d^4x\, F_{\mu\nu}^a F_{\mu\nu}^a \\
&= \frac{1}{4g^2} \int d^4x \left[\pm F_{\mu\nu}^a \tilde{F}_{\mu\nu}^a + \frac{1}{2} \left(F_{\mu\nu}^a \mp \tilde{F}_{\mu\nu}^a \right)^2 \right].
\end{aligned}
\tag{4.4}
$$

Since the first term in the second line of Eq. (4.4) is a topological invariant, solely determined by the boundary behavior of the configuration A_μ, the action reaches its absolute minimum in a given topological sector, if and only if the second, positive semi-definite term vanishes. That is, we obtain a minimal-action configuration within a given topological sector if the (anti)selfduality equations (1.33) are satisfied. For these configurations the action is given as $S = \frac{8\pi^2|k|}{g^2}$.

4.1.2   Instanton in regular gauge

For a regular gauge-field configuration $A_\mu(x)$ to possess finite action on $\mathbf{R}^4$ it is necessary that $A_\mu(x)$ approaches a pure gauge as $|x| \equiv \sqrt{x_\mu x_\mu} \to \infty$: $A_\mu\big|_{|x|=\infty} \to i\Omega\partial_\mu\Omega^\dagger$. For the group element Ω we allow for large deviations from unity along the domain $\partial\mathbf{R}^4 \sim S_3$ at $|x| = \infty$ and consider the two cases

$$\Omega^\pm = \frac{\sigma_\mu^\pm x_\mu}{|x|}, \qquad (4.5)$$

where $\sigma_\mu^\pm \equiv (\mathbf{1}_2, \pm i\vec{\tau})$, and τ_1, τ_2, τ_3 are the Pauli matrices. (The entries of $\sigma_\mu^\pm$ are unit quaternions.) We define the Lie-algebra valued 't Hooft symbols $\eta_{\mu\nu} = \eta^a_{\mu\nu} t_a$ and $\bar{\eta}_{\mu\nu} = \bar{\eta}^a_{\mu\nu} t_a$ (recall that tr $t_a t_b = \frac{1}{2}\delta_{ab}$) as

$$i\eta_{\mu\nu} = \sigma_\mu^+ \sigma_\nu^- - \delta_{\mu\nu}, \qquad i\bar{\eta}_{\mu\nu} = \sigma_\mu^- \sigma_\nu^+ - \delta_{\mu\nu}. \qquad (4.6)$$

It is straightforward to see that the 't Hooft symbols are selfdual and antiselfdual, respectively: $\eta_{\mu\nu} = \tilde{\eta}_{\mu\nu}$ and $\bar{\eta}_{\mu\nu} = -\tilde{\bar{\eta}}_{\mu\nu}$. Moreover, they are antisymmetric: $\eta_{\mu\nu} = -\eta_{\nu\mu}$ and $\bar{\eta}_{\mu\nu} = -\bar{\eta}_{\nu\mu}$.

The pure-gauge configurations $A_\mu^\pm\big|_{|x|=\infty}$, associated with the two possibilities in Eq. (4.5), read explicitly

$$A_\mu^+\big|_{|x|=\infty} = 2\eta_{\mu\nu}\frac{x_\nu}{x^2} \quad \text{or} \quad A_\mu^-\big|_{|x|=\infty} = 2\bar{\eta}_{\mu\nu}\frac{x_\nu}{x^2}. \qquad (4.7)$$

In search of solutions to the selfduality equations (1.33), we may presume a four-dimensional spherically symmetric generalization of the boundary behavior (4.7). Thus we write

$$A^+_\mu(x) = 2\eta_{\mu\nu}f(x^2)\frac{x_\nu}{x^2} \quad \text{or} \quad A^-_\mu(x) = 2\bar\eta_{\mu\nu}f(x^2)\frac{x_\nu}{x^2}, \tag{4.8}$$

where f is a smooth function satisfying $f(x^2 \to \infty) = 1$. Inserting these ansätze into Eq. (1.33), respectively, one obtains the following first-order equation

$$f(1-f) - x^2\frac{df}{dx^2} = 0. \tag{4.9}$$

Equation (4.9) is solved by

$$f(x^2) = \frac{x^2}{x^2 + \rho^2}. \tag{4.10}$$

Here the parameter ρ plays the role of a constant of integration. Obviously, the solution in Eq. (4.10) is smooth if $\rho \in \mathbf{R}_+$ and satisfies the above boundary condition. For the field strength $F^{\pm,a}_{\mu\nu}$, evaluated on

$$A^+_\mu(x) = 2\eta_{\mu\nu}\frac{x_\nu}{x^2 + \rho^2} \quad \text{or} \quad A^-_\mu(x) = 2\bar\eta_{\mu\nu}\frac{x_\nu}{x^2 + \rho^2}, \tag{4.11}$$

one obtains [Shuryak and Schaefer (1996)]

$$F^{+,a}_{\mu\nu} = -\frac{4\eta^a_{\mu\nu}\rho^2}{(x^2 + \rho^2)^2} \quad \text{or} \quad F^{-,a}_{\mu\nu} = -\frac{4\bar\eta^a_{\mu\nu}\rho^2}{(x^2 + \rho^2)^2}. \tag{4.12}$$

Their squares are given as

$$2\,\mathrm{tr}\,F^\pm_{\mu\nu}F^\pm_{\mu\nu} = F^{\pm,a}_{\mu\nu}F^{\pm,a}_{\mu\nu} = \frac{192\rho^4}{(x^2 + \rho^2)^4}. \tag{4.13}$$

Therefore the topological charge k reads

$$k = \frac{1}{32\pi^2}\int d^4x\, F^{\pm,a}_{\mu\nu}\tilde F^{\pm,a}_{\mu\nu} = \pm 1. \tag{4.14}$$

Since the Yang–Mills action is invariant under spacetime translations $x \to x - x_0$ and, of course, under global gauge rotations, the solutions in Eq. (4.11) can be generalized to

$$A_\mu^{+,a}(x; \rho, x_0, \vec{\alpha}) = 2R_b^a(\vec{\alpha})\eta_{\mu\nu}^b \frac{(x - x_0)_\nu}{(x - x_0)^2 + \rho^2} \quad \text{or}$$

$$A_\mu^{-,a}(x; \rho, x_0, \vec{\alpha}) = 2R_b^a(\vec{\alpha})\bar{\eta}_{\mu\nu}^b \frac{(x - x_0)_\nu}{(x - x_0)^2 + \rho^2}, \tag{4.15}$$

where $R_b^a(\vec{\alpha})$ is the matrix of an SO(3) rotation in the fundamental representation (the adjoint of SU(2) with $\vec{\alpha}$ a three-vector associated with the Euler angles). The parameter ρ refers to the freedom of performing spacetime dilatations without changing the classical Yang–Mills action. For the case $|k| = 1$ discussed here this gives a total of eight parameters for the solutions in Eq. (4.15). They span an eight-dimensional Riemannian manifold called *moduli space*. Notice that the solutions in Eq. (4.15) decay as $1/|x|$ for $|x| \to \infty$ which is the reason why their topological charge is picked up by a surface integral of the Chern–Simons current of Eq. (4.1) over an S_3 of infinite radius.

4.1.3   *Singular gauge and multiinstantons*

An important observation due to R. Jackiw and C. Rebbi [Jackiw and Rebbi (1976)] concerns the discrete symmetry of the Yang–Mills action under spacetime inversions

$$x_\mu \to \frac{x_\mu}{x^2}. \tag{4.16}$$

Because of its discreteness there is no continuous parameter (modulus) associated with this symmetry, and therefore it does not generate a variety of new solutions out of a given one. In fact, one can show that the result of applying the inversion of Eq. (4.16) to the solutions of Eq. (4.15) amounts to gauge transformations introducing singularities at the origin (setting $x_0 = 0$ in Eq. (4.15)):

$$A_\mu^+ \to \bar{A}_\mu^+ = \Omega^- A_\mu^+ (\Omega^-)^\dagger + i\Omega^- \partial_\mu (\Omega^-)^\dagger,$$
$$A_\mu^- \to \bar{A}_\mu^- = \Omega^+ A_\mu^- (\Omega^+)^\dagger + i\Omega^+ \partial_\mu (\Omega^+)^\dagger, \tag{4.17}$$

where the group elements $\Omega^{\pm}$ are defined in Eq. (4.5). Because the transformations in Eq. (4.17) are *gauge* transformations (albeit singular) we are assured an invariance of the Yang–Mills action. Thus the same saturation of this action's $|k| = 1$ bound takes place for $\bar{A}_{\mu}^{\pm,a}(x; \rho, x_0, \vec{\alpha})$ as it does for the original configuration $A_{\mu}^{\pm,a}(x; \rho, x_0, \vec{\alpha})$. Since the singular gauge transformations in Eq. (4.17) dewind the gauge field at infinity at the expense of introducing a singularity at the origin the topological charge k is localized there. This prompts the name singular gauge.

The expressions on the right-hand side of Eq. (4.17) can be generalized (appeal to translational symmetry) as

$$\bar{A}_{\mu}^{+,a}(x; \rho, x_0, \vec{\alpha}) = 2R_b^a(\vec{\alpha})\bar{\eta}_{\mu\nu}^b \frac{(x - x_0)_\nu \rho^2}{(x - x_0)^2((x - x_0)^2 + \rho^2)},$$

$$\bar{A}_{\mu}^{-,a}(x; \rho, x_0, \vec{\alpha}) = 2R_b^a(\vec{\alpha})\eta_{\mu\nu}^b \frac{(x - x_0)_\nu \rho^2}{(x - x_0)^2((x - x_0)^2 + \rho^2)}. \quad (4.18)$$

In contrast to the solutions in Eq. (4.15) the (anti)selfdual configurations in Eq. (4.18) decay like $1/|x|^3$ as $|x| \to \infty$ but exhibit a $1/|x - x_0|$ singularity as $x \to x_0$. The surface integral over the Chern–Simons current now picks up contributions at a small S_3 centered about x_0.

The fact that the singular-gauge instantons decay so rapidly towards infinity suggests that some superposition principle be applicable to compose selfdual field configurations of higher topological charge modulus $|k|$ out of those with $k = \pm 1$. G. 't Hooft in ['t Hooft (1976a,b)] noticed that the solutions in Eq. (4.18) can be rewritten as

$$\bar{A}_{\mu}^{+,a}(x; \rho, x_0) = -\bar{\eta}_{\mu\nu}^a \partial_\nu \log\left(1 + \frac{\rho^2}{(x - x_0)^2}\right),$$

$$\bar{A}_{\mu}^{-,a}(x; \rho, x_0) = -\eta_{\mu\nu}^a \partial_\nu \log\left(1 + \frac{\rho^2}{(x - x_0)^2}\right) \quad (4.19)$$

(setting $R_b^a = \delta_b^a$). This led him to make the following generalized ansätze, see also [Jackiw and Rebbi (1976)]:

$$\bar{A}_\mu^{+,a}(x) = -\bar{\eta}_{\mu\nu}^a \partial_\nu \log W,$$

$$\bar{A}_\mu^{-,a}(x) = -\eta_{\mu\nu}^a \partial_\nu \log W. \qquad (4.20)$$

The scalar function $W(x^2)$ in Eq. (4.20) yet needs to be determined by the selfduality Eq. (1.33). Inserting Eq. (4.20) into Eq. (1.33), one obtains

$$F_{\mu\nu} - \tilde{F}_{\mu\nu} = \bar{\eta}_{\mu\nu} \frac{\partial_\alpha \partial_\alpha W}{W},$$

$$F_{\mu\nu} + \tilde{F}_{\mu\nu} = \eta_{\mu\nu} \frac{\partial_\alpha \partial_\alpha W}{W}. \qquad (4.21)$$

Thus for the ansätze in Eq. (4.20) to be selfdual or antiselfdual, respectively, we need to demand that

$$\frac{\partial_\alpha \partial_\alpha W}{W} = 0. \qquad (4.22)$$

A solution to Eq. (4.22) associated with finite action is given by

$$W(x) = \sum_{l=0}^{n} \frac{\rho_l^2}{(x - x_l)^2} \qquad (4.23)$$

with $\rho_l \in \mathbf{R}_+$ and $x_l \in \mathbf{R}^4$. Taking the limits $\rho_0, |x_0| \to \infty$ such that $\frac{\rho_0^2}{x_0^2} \to 1$, the function $W(x)$ reduces to

$$W(x) = 1 + \sum_{l=1}^{n} \frac{\rho_l^2}{(x - x_l)^2}, \qquad (4.24)$$

which is the form of the solution to Eq. (4.22) discovered by 't Hooft (1976a,b). In terms of the ansätze (4.20) the density $s(x) \equiv \frac{1}{2} \mathrm{tr} F_{\mu\nu} \tilde{F}_{\mu\nu}$ reads

$$s(x) = \pm \frac{1}{2} (\partial_\mu \partial_\mu)(\partial_\nu \partial_\nu) \log W(x). \qquad (4.25)$$

In Eq. (4.25) singular expressions of the form $W(x) \sim \rho_l^2/(x-x_l)^2$ are admissible since $s(x)$ remains integrable for $x \to x_l$ (see [Jackiw and Rebbi (1976)]).

Thus, as a functional of $W(x)$, the topological charge k is given as

$$k = \frac{1}{16\pi^2} \int d^4 x \, s(x) = \pm \frac{1}{16\pi^2} \int d^4x (\partial_\mu \partial_\mu)(\partial_\nu \partial_\nu) \log W(x). \quad (4.26)$$

The integral in Eq. (4.26) may be evaluated in a straightforward way [Jackiw and Rebbi (1976)]. One has $k = \pm n$ [Jackiw and Rebbi (1976)]. Again, the selfdual and antiselfdual configurations of (4.20) of topological charge $k = \pm n$ [$W(x)$ given in Eq. (4.23)] can be generalized by multiplying them with a constant SO(3) matrix R_a^b. Notice that this implies that the gauge-field configuration is globally oriented in the Lie algebra: Each individual $k = \pm 1$ instanton or anti-instanton, which for large separation of the centers,

$$\min\{|x_l - x_m|; l \neq m\} \gg \max\{\rho_l\}, \quad (4.27)$$

contributes one unit of topological charge to $k = \pm n$, is oriented alike in su(2). As it turned out, this situation does not represent the most general selfdual or antiselfdual charge-k configuration. However, the solutions (4.20) play an essential role in constructing selfdual or antiselfdual, periodic configurations of charge $k = \pm 1$ in the Euclidean formulation of the finite-temperature Yang–Mills theory: calorons of trivial holonomy [Harrington and Shepard (1978)]. We shall see in Chapter 5, that these calorons are very relevant in constructing the deconfining thermal ground-state estimate.

4.2   Sketch of the ADHM-Nahm Construction

The material presented in this section concerns deep mathematical results whose proper discussion would require a book on its own. Rather than going into the details of the construction, which would imply a substantial deviation from the book's intended line of reasoning, some basic facts are conveyed which prove indispensible for the construction of calorons of nontrivial holonomy and a characterization of their magnetic monopole constituents.

4.2.1   All charge-k instantons on $\mathbf{R}^4$

We have learned in Sec. 4.1.3 about the existence of closed-form expressions for selfdual configurations of topological charge $k \in \mathbf{Z}$ in SU(2) Yang–Mills theory. One may now wonder whether the configurations investigated in Sec. 4.1.3 are generalizable. Thanks to the work of [Ward (1977); Atiyah and Ward (1977)] and of [Atiyah *et al.* (1978); Drinfeld and Manin (1978)] there is a positive answer to this question.

In [Atiyah *et al.* (1978)] a construction of *all* selfdual configurations of charge k in SU(N) (and other compact gauge groups) on the Euclidean spacetime manifold $\mathbf{R}^4$ is given. The reason why this construction completely exhausts the set of all selfdual configurations is linked to the nature of analytic vector bundles associated with the selfdual gauge fields: These field configurations emerge by appealing to a sequence of vector spaces related to sheaf cohomology groups.

Because the ADHM construction was invented by mathematicians, who are very efficient in suppressing redundant notation, it is costumary in the respective literature to absorb a factor of $-i$ into the gauge field A_μ. This renders A_μ traceless *antihermitian* instead of hermitian. In the following we understand that the antihermitian property is delegated to the generators t^a becoming antihermitian: $t^a \rightarrow -it^a$. Therefore, one now has tr $t^a t^b = -\frac{1}{2}\delta^{ab}$, and the coefficients A_μ^a stay real. Modulo a factor of i, the field-strength tensor then reads $F_{\mu\nu} = \partial_\mu A_\nu - \partial_\nu A_\mu + [A_\mu, A_\nu]$ (no factors $\pm i$ in front of the commutator anymore!) at the expense of putting a factor of minus one in front of the action density or topological charge density and omitting a factor i in front of a pure-gauge configuration expressed through a bilinear of a group element. This convention will be used throughout the remainder of this chapter unless stated otherwise.

Let us discuss the case $k > 0$ in SU(N). Bluntly speaking, the selfdual gauge field can be obtained from the kernel of a $2k \times (2k+\text{N})$ matrix $\Delta^\dagger$ where the matrix Δ is given as

$$\Delta \equiv A + BX. \tag{4.28}$$

Here $X \equiv \sigma_\mu^+ x_\mu$, and A, B are $(2k + \text{N}) \times 2k$ matrices. The multiplication of X onto B in Eq. (4.28) is understood as follows: Each of the $2k + \text{N}$ rows of matrix B decays into k subrows of two elements which, when multiplied from the left onto X, again generate subrows of two elements. These compose back into the rows of $2k$ elements of the $(2k+\text{N}) \times 2k$ matrix BX. (For the ADHM construction of configurations with $k < 0$ one uses $X \equiv \sigma_\mu^- x_\mu$ in Eq. (4.28).)

For the construction to go through it is necessary that the $2k \times 2k$ matrix $\Delta^\dagger \Delta$ be invertible and that it commutes with the quaternions, meaning that it is organized into 2×2 blocks, all proportional to $\mathbf{1}_2$, with the real $k \times k$ matrix F of prefactors being invertible. The kernel of $\Delta^\dagger$ then is of dimension N and can be written as a $(2k + \text{N}) \times \text{N}$ dimensional matrix $V(x)$. The columns of $V(x)$ are taken to be orthonormalized:

$$V^\dagger(x)V(x) = \mathbf{1}_\text{N}. \tag{4.29}$$

From V one constructs the charge-k SU(N) gauge field A_μ as

$$A_\mu = V(x)^\dagger \partial_\mu V(x). \tag{4.30}$$

Notice that the configuration A_μ in Eq. (4.30) is antihermitian by virtue of a differentiation of the normalization condition (4.29). The selfduality of the expression in Eq. (4.30) can be checked by direct calculation using the two above-stated conditions on matrix $\Delta^\dagger \Delta$ [Corrigan *et al.* (1978);Christ, Stanton, and Weinberg (1978)]. Namely, one obtains

$$F_{\mu\nu} = iV^\dagger BF^{-1}\eta_{\mu\nu}B^\dagger V, \tag{4.31}$$

where $F^{-1}\eta_{\mu\nu}$ represents the $2k \times 2k$ matrix obtained by multiplying each element of $k \times k$ matrix F^{-1} with the 't Hooft symbol $\eta_{\mu\nu}$ defined in Eq. (4.6). Because of the selfduality of $\eta_{\mu\nu}$ it follows that $F_{\mu\nu}$ in Eq. (4.31) is selfdual.

That the data prescribed by the matrices $\mathcal{A}$ and $\mathcal{B}$ in Eq. (4.28) suffices to generate *all* charge-k selfdual SU(N) configurations was shown in an elementary way in [Corrigan and Goddard (1984)]. In a first step, one notices the fact that the definition of the gauge field A_μ in Eq. (4.30) is unique up to gauge transformations. This may be exploited to transform the matrices $\mathcal{A}$ and $\mathcal{B}$ in Eq. (4.28) as follows:

$$\mathcal{A} \to \mathcal{A}' = \mathcal{Q}\mathcal{A}\mathcal{R}, \quad \mathcal{B} \to \mathcal{B}' = \mathcal{Q}\mathcal{A}\mathcal{R}, \qquad (4.32)$$

where $\mathcal{Q}$ is unitary and of rank $N + 2k$, and $\mathcal{R}$ denotes a nonsingular complex matrix of rank $2k$. Given $\mathcal{A}$ and $\mathcal{B}$, the matrices $\mathcal{Q}$ and $\mathcal{R}$ can be chosen such that $\mathcal{A}'$ and $\mathcal{B}'$ are cast into normal forms as [Corrigan *et al.* (1978)]

$$\mathcal{A}' = \begin{pmatrix} \lambda \\ M \end{pmatrix}, \quad \mathcal{B}' = \begin{pmatrix} 0 \\ 1_{2k} \end{pmatrix}, \qquad (4.33)$$

where λ is an $N \times 2k$, M a $2k \times 2k$, and 0 denotes the $N \times 2k$ matrix with all entries set equal to zero. The condition that the invertible $2k \times 2k$ matrix $\Delta^\dagger \Delta$ needs to commute with the quaternions implies the following conditions for the matrices λ and M:

$$\left(\lambda^\dagger \lambda\right)_{rIsJ} = \kappa_{rs}(1_2)_{IJ}, \quad M_{rIsJ} = (M_\mu)_{rs}(\sigma_\mu^-)_{IJ}. \qquad (4.34)$$

Here the $k \times k$ matrices κ and M_μ are hermitian. The task now is to establish that from *any* selfdual gauge field A_μ of finite action, matrices $\mathcal{A}'$ and $\mathcal{B}'$ of the form (4.33) can be constructed which satisfy the conditions of Eq. (4.34) and that the ADHM construction based on these matrices gives back the original gauge field A_μ.

In order to show this one notices that according to the Atiyah–Singer index theorem [Atiyah and Singer (1968a,b,c, 1971a,b, 1984) the Weyl equation

$$(\sigma_\mu^+)_{IJ}(D_\mu)_{uv}\psi_{vJ} = 0 \qquad (4.35)$$

has k normalizable solutions. Here $u, v = 1, \dots, N$, $D_\mu = \partial_\mu + A_\mu$, and A_μ is selfdual and of finite action. These k negative-chirality zero modes of the full Dirac operator $i\gamma_\mu D_\mu$ can be arranged into a $2N \times k$ matrix Ψ with entries Ψ_{uIr}, $r = 1, \dots, k$. The matrix Ψ be normalized as

$$\int d^4x \Psi^\dagger \Psi = \pi^2 1_k. \qquad (4.36)$$

The $k \times k$ matrices M_μ of Eq. (4.34) now are taken to be

$$M_\mu = -\frac{1}{\pi^2} \int d^4x \, x_\mu \Psi^\dagger \Psi. \qquad (4.37)$$

Furthermore, defining the $N \times 2k$ matrix $\tilde{\Psi}$ by $\tilde{\Psi}_{urJ} \equiv \Psi_{urI}^t \epsilon_{IJ}$, Ψ^t to be the transpose of Ψ on the label I and $\epsilon_{IJ} = \delta_{1I}\delta_{2J} - \delta_{1J}\delta_{2I}$, the asymptotic behavior of $\tilde{\Psi}$ for $|x| \to \infty$ is given as [Corrigan *et al.* (1978)]

$$\tilde{\Psi}_{urJ} \overset{|x| \to \infty}{\longrightarrow} -g_{uv}\lambda_{vrI}\frac{X_{IJ}}{|x|^4}. \qquad (4.38)$$

Here the matrix g is identified with the SU(N) group element that relates to the asymptotic pure-gauge configuration (finite action!) as

$$A_\mu \overset{|x| \to \infty}{\longrightarrow} -(\partial_\mu g)g^\dagger. \qquad (4.39)$$

The matrix λ in Eq. (4.38) is taken to be the $N \times 2k$ submatrix in $\mathcal{A}'$ [see Eq. (4.33)].

One can show by brute force calculation [Corrigan and Goddard (1984)] that the matrices M_μ in Eq. (4.37) and λ in Eq. (4.38) satisfy the constraints (4.34). Finally, one shows that, plugging the associated matrix $\mathcal{A}'$ and matrix $\mathcal{B}'$ of Eq. (4.33) into the ADHM construction returns the original, selfdual configuration A_μ [Corrigan and

Goddard (1984)]. Therefore, for SU(N) all selfdual gauge-field configurations of topological charge k and finite-action are obtained by the ADHM construction.

4.2.2 *The 't Hooft–Polyakov monopole*

To set the stage for a sketch of the Nahm transformation, generalizing the ADHM construction to instantons on the four-torus T^4, we discuss a particular selfdual configuration of finite energy or of finite action on $S_1 \times \mathbf{R}^3$.

For the gauge group SU(2) let us consider an adjoint Higgs model given by the action density (1.44) which, after absorbing the gauge coupling g and a factor of $-i$ into the fields A_μ and ϕ, reads

$$S_{\text{YM,ext}} = -\frac{1}{g^2}\text{tr}\int d^4x \left(\frac{1}{2}F_{\mu\nu}F_{\mu\nu} + D_\mu\phi D_\mu\phi\right) = \int d^4x \mathcal{L}_{\text{YM,ext}}, \quad (4.40)$$

where $D_\mu\phi \equiv \partial_\mu\phi + [A_\mu, \phi]$ and $F_{\mu\nu} = \partial_\mu A_\nu - \partial_\nu A_\mu + [A_\mu, A_\nu]$. Because the gauge-symmetry breaking introduced by $\phi \neq 0$ is SU(2) $\to$ U(1) the consideration of $\Pi_2(\text{SU}(2)/\text{U}(1)) = \Pi_2(S_2) = \mathbf{Z}$ suggests the existence of topologically stabilized, static configurations whose energies are minimal within each topological sector. Three-dimensional solitons of *magnetic* charge with respect to the unbroken gauge group U(1) were discovered independently by ['t Hooft (1974)] and [Polyakov (1974)] by investigating a version of the action density in (4.40) with a quartic Mexican-hat potential $V(\phi)$ added. In the limit $V(\phi) \to 0$ a closed-form expression for this type of magnetic-monopole solution was presented in [Prasad and Sommerfield (1975)] (BPS monopole). It is interesting that this solution can be understood as a particular selfdual and static configuration in a *pure* SU(2) Yang–Mills theory. Originally, W. Nahm discovered his transformation between selfdual configurations of topological charge k and topological charge N, by performing a remarkable limit of the ADHM construction which yields the BPS monopole.

The field equations for the model defined by Eq. (4.40) are

$$D_\mu D_\mu \phi = 0, \quad D_\mu F_{\mu\nu} = [\phi, D_\nu \phi]. \tag{4.41}$$

Assuming the solution to be static and A_4 to vanish, Eqs. (4.41) simplify as

$$D_i D_i \phi = 0, \quad D_i F_{ij} = [\phi, D_j \phi]. \tag{4.42}$$

Now, consider the equation

$$D_i \phi = \pm \frac{1}{2} \epsilon_{ijk} F_{jk} \quad \Leftrightarrow \quad F_{ij} = \pm \epsilon_{ijk} D_k \phi. \tag{4.43}$$

The first of Eqs. (4.42) follows from Eqs. (4.43) by applying the covariant divergence and by appealing to the Bianchi identity (1.32); the second of Eqs. (4.42) is implied by Eqs. (4.43) as follows:

$$D_i F_{ij} = \pm \epsilon_{ijk} D_i D_k \phi = \pm \epsilon_{ijk} \left(\frac{1}{2} \{D_i, D_k\} + \frac{1}{2} [D_i, D_k] \right) \phi$$

$$= \pm \frac{1}{2} \epsilon_{ijk} [F_{ik}, \phi] = \pm \frac{1}{2} \epsilon_{jik} [\phi, F_{ik}] = [\phi, D_j \phi]. \tag{4.44}$$

Notice that the first of Eq. (4.43) represents the selfduality equations of a *pure* and Euclidean Yang–Mills theory if one makes the identification $A_4 = \phi$ and assumes staticity (independence of A_μ on x_4). This implies that the action density of the pure Yang–Mills theory is independent of x_4. For the action to be finite the integration of the action density over x_4 should be constrained to a compact interval, say $0 \le x_4 \le \tilde{\beta}$. The mass scale $\tilde{\beta}^{-1}$ may enter via a boundary condition at spatial infinity:

$$\tilde{\beta} = 2\pi |A_{4,\infty}|^{-1}, \tag{4.45}$$

where $|A_{4,\infty}| \equiv \lim_{|\vec{x}| \to \infty} \sqrt{A_4^a A_4^a}$. Below we give reasons why the proportionality factor of 2π is introduced in Eq. (4.45).

With the above identification, the following ansätze

$$A_4^a = \hat{x}^a \frac{H(r)}{r}, \quad A_i^a = \epsilon_{ij}^a \hat{x}^j \frac{1 - K(r)}{r} \tag{4.46}$$

can be made ['t Hooft (1974); Prasad and Sommerfield (1975)] to solve the selfduality equations

$$\partial_i A_4 + [A_i, A_4] = \pm \frac{1}{2} \epsilon_{ijk} F_{jk} \tag{4.47}$$

of a pure SU(2) Yang–Mills theory on $S_1 \times \mathbf{R}^3$. In Eqs. (4.46) it is understood that $r \equiv |\vec{x}| = \sqrt{x^i x^i}$ and $\hat{x}^a \equiv \frac{x^a}{r}$, and that the real scalar functions H and K depend on r only.

Substituting Eqs. (4.46) into Eqs. (4.47) yields *first-order* ordinary differential equations for $H(r)$ and $K(r)$. In [Prasad and Sommerfield (1975)] second-order equations were considered which follow from Eqs. (4.44) upon the identification $\phi = A_4$ and the subsequent insertion of ansätze (4.46). These second-order equations read

$$r^2 K'' = K(K^2 - 1) + KH^2, \quad r^2 H'' = 2HK^2, \tag{4.48}$$

where a prime denotes differentiation with respect to r. The solutions to Eqs. (4.48) need to be subjected to the following boundary conditions at spatial infinity

$$H_\infty = \pm 2\pi \tilde{\beta}^{-1} r, \quad K_\infty = 0 \tag{4.49}$$

for A_μ to be of finite energy, and one has [Prasad and Sommerfield (1975)]

$$K(r) = \frac{2\pi \tilde{\beta}^{-1} r}{\sinh{(2\pi \tilde{\beta}^{-1} r)}}, \quad H(r) = \pm 2\pi \tilde{\beta}^{-1} r \coth{(2\pi \tilde{\beta}^{-1} r)} \mp 1. \tag{4.50}$$

As we have learned in Sec. 1.5, the original gauge symmetry SU(2) is broken to U(1) by the field A_4 acquiring the expectation $|A_{4,\infty}| = 2\pi \tilde{\beta}^{-1}$ at spatial infinity. (At the point $r = 0$ we have $A_4 = 0$ according to Eqs. (4.46) and (4.50) implying that in the core of this static solution the full gauge symmetry SU(2) is restored.) As a consequence, it should be possible to define an *Abelian* (with respect to the unbroken gauge group U(1)) spatial field strength $\mathcal{F}_{ij}$ associated with the magnetic charge of the static solitonic configuration of Eqs. (4.46) and (4.50). The following definition for the spatial components of the U(1) field-strength tensor $\mathcal{F}_{\mu\nu}$ ('t Hooft tensor) was proposed in ['t Hooft (1974)]:

$$\mathcal{F}_{ij} \equiv \partial_i(\hat{A}_4^a A_j^a) - \partial_j(\hat{A}_4^a A_i^a) - \epsilon^{abc}\hat{A}_4^a\partial_i\hat{A}_4^b\partial_j\hat{A}_4^c, \qquad (4.51)$$

where $\hat{A}_4^a \equiv A_4^a(A_4^b A_4^b)^{-1/2}$. In the gauge, where $A_4 \propto t^3$ (unitary gauge), the tensor $\mathcal{F}_{ij}$ reduces to the usual U(1) expression in terms of the gauge-field components A_i^3. The magnetic field strength $\mathcal{B}_i$ associated with $\mathcal{F}_{ij}$ is given as $\mathcal{B}_i = \frac{1}{2}\epsilon_{ijk}\mathcal{F}_{jk}$. By integrating this flux of the configuration of Eqs. (4.46) and (4.50) over a spatial S_2 of infinite radius, centered at $\vec{x} = 0$, and after letting $A_\mu \to gA_\mu$ one derives the magnetic charge g_m as

$$g_m = \mp\frac{4\pi}{g}. \qquad (4.52)$$

This is an important result because it tells us that gauge coupling g and magnetic charge g_m are *dual* to one another: Increasing the modulus of the former reduces that of the latter and vice versa. The mass m of the magnetic monopole solution in Eqs. (4.46) and (4.50) is equal to the spatial integral over the Euclidean action density of Eq. (4.4) evaluated on the solution Eqs. (4.46) and (4.50). It is given as [Prasad and Sommerfield (1975)]

$$m = \frac{8\pi^2}{g^2}\tilde{\beta}^{-1}. \qquad (4.53)$$

Notice that the monopole mass m *vanishes* if the gauge coupling g diverges. This makes sense since then also its magnetic charge approaches zero. Notice that this is in contrast to the Dirac monopole which is of infinite self-energy in a model where U(1) is considered to be a fundamental gauge symmetry. From Eq. (4.53) it is now clear why a factor of 2π was introduced in relating the mass scales $\tilde{\beta}^{-1}$ and $|A_{4,\infty}|$ in Eq. (4.45): This definition of $\tilde{\beta}$ associates one unit of topological charge, $k = \pm 1$, to the solution given by Eqs. (4.46) and (4.50) [compare with Eq. (4.3)]. Finally, let us compute the holonomy of this configuration, that is, its spatial Polyakov loop at infinity. One has

$$\text{Pol}(r \to \infty) = \exp[\pm \pi i \hat{x} \cdot \vec{\tau}] = \mathbf{1}_2 \cos \pi \pm i \hat{x} \cdot \vec{\tau} \sin \pi$$
$$= -\mathbf{1}_2 \in \mathbf{Z}_2. \tag{4.54}$$

Thus, the BPS monopole of topological charge $k = \pm 1$ represents a very particular case of a caloron configuration with trivial holonomy.

New solutions can be generated from the ones given by Eqs. (4.46) and (4.50) by exploiting the symmetries of the Euclidean Yang–Mills action under spatial translations (shift of the center from $\vec{0}$ to $\vec{x}_0$), and under global U(1) phase changes. New solutions can also be obtained by observing the following [Müller (1986)]: The Minkowskian version of the second-order equations of motion (4.41) turns out to be the second-order, pure Yang–Mills equations in a (1+4)-dimensional Minkowskian spacetime assuming independence on x_4 ($\partial_4 = 0$) and, again, identifying A_4 with ϕ. (As we have seen, under this identification the BPS equations (4.43) turn out to be the selfduality equations of a four-dimensional Euclidean and pure Yang–Mills theory, assuming that $\partial_4 = 0$.) Therefore, a Lorentz boost in the x_0-x_4 plane (x_4 now is the additional space coordinate), which is a symmetry operation in the (1+4)-dimensional theory, generates a new solution when applied to the configuration of Eqs. (4.46) and (4.50) enhanced by $A_0 = 0$. Here the boost is only applied to the five-vector $(0, A_4, \vec{A})$; spacetime arguments are not transformed. Subsequently, the boosted expression for A_0 is taken as the temporal component of the gauge field in 1+3 dimensions, and A_4 is interpreted back as ϕ. After rescaling (a symmetry of the (1+3)-dimensional theory given by Eq. (4.40)) the static dyon solution of [Julia and Zee (1975)] is obtained. This configuration is a monopole of both electric and magnetic charge, and the latter's ratio is determined by the boost parameter.

But let us now return to the magnetic-monopole solution of Eqs. (4.46) and (4.50). The BPS monopole is selfdual on $S_1 \times \mathbf{R}^3$. The ADHM construction, however, was designed for instantons on $\mathbf{R}^4$. The question one may ask is how far the original ADHM construction can be adapted to the case where one of the spacetime dimensions becomes compact. This question was pursued [Nahm (1980)] yielding a beautiful answer: The BPS monopole, which can be understood as the limiting case of a multiinstanton on $\mathbf{R}^4$ [Manton (1978); Adler (1978); Rossi (1979)], actually can be constructed via ADHM if one understands the topological charge k of the multiinstanton as a continuous variable t. Playing the game in dimensionless variables, Nahm set $\tilde{\beta} = 2\pi$ and noticed that the conjugate of the operator Δ in Eq. (4.28), when adapted to ADHM-inspired monopole construction, can be taken as

$$\Delta^\dagger \equiv i\partial_t + X^\dagger \tag{4.55}$$

for the BPS monopole centered at[2] $\tilde{x}_0 = 0$ where, again, $X \equiv \sigma_\mu^+ x_\mu$. This form of $\Delta^\dagger$ satisfies the ADHM constraints on $\Delta^\dagger \Delta$. Namely, $\Delta^\dagger \Delta$ commutes with quaternions and it is invertible on the interval $-\frac{1}{2} \leq t \leq \frac{1}{2}$. Nahm proved the latter statement by explicitly giving the solution to

$$\Delta^\dagger \Delta F(x, t, t') = \delta(t - t') \tag{4.56}$$

subject to the periodic boundary conditions $F(x, \pm\frac{1}{2}, t') = 0$.

The equation $\Delta^\dagger V = 0$ of the ADHM construction is solved by the U(2) group element $V(x, t) = N(\tilde{x}) \exp{(iX^\dagger t)}$. Function $N(\tilde{x})$ follows from the normalization condition $\int_{-\frac{1}{2}}^{\frac{1}{2}} dt \, V^\dagger(x, t) V(x, t) = 1_2$ and is given as

$$N(\tilde{x}) = \left(\frac{\sinh |\tilde{x}|}{|\tilde{x}|} \right)^{-1/2}. \tag{4.57}$$

---

[2]Because the solution is static $(x_0)_4$ may be set equal to zero.

By straightforward calculation one checks that $A_\mu(x) = i \int_{-1/2}^{1/2} dt V^\dagger \partial_\mu V$ represents the solution given by Eqs. (4.46) and (4.50) when setting $\tilde{\beta} = 2\pi$. Notice that in calculating $A_\mu(x)$ the x_4 dependence drops out because it resides in $V(x, t)$ as a U(1) phase.

4.2.3   Nahm's duality transformation

In the adaption of the ADHM construction to the case of the BPS monopole [Nahm (1980)], the operators $\mathcal{A}$ and $\mathcal{B}$, defining the operator Δ in Eq. (4.28) on the dual interval $-\frac{1}{2} \le t \le \frac{1}{2}$, were picked in a very special way. For the gauge group U(N) the much more general case considers a transformation of a charge-k instanton defined on the four-torus T^4 into a charge-N instanton associated with gauge symmetry U(k) on the dual four-torus $\tilde{T}^4$ [Nahm (1981, 1982)]. We define the four-torus T^4 by identifying points in $\mathbf{R}^4$ whose differences coincide with points of the lattice $\{\beta_1 \mathbf{Z} e_1, \beta_2 \mathbf{Z} e_2, \beta_3 \mathbf{Z} e_3, \beta_4 \mathbf{Z} e_4\}$ where the four vectors e_μ form an orthonormal basis of $\mathbf{R}^4$ and $\beta_\mu \in \mathbf{R}_+$. The dual four-torus $\tilde{T}^4$ identifies points in $\mathbf{R}^4$ whose differences coincide with points of the dual lattice $\{\frac{2\pi}{\beta_1} \mathbf{Z} \tilde{e}_1, \frac{2\pi}{\beta_2} \mathbf{Z} \tilde{e}_2, \frac{2\pi}{\beta_3} \mathbf{Z} \tilde{e}_3, \frac{2\pi}{\beta_4} \mathbf{Z} \tilde{e}_4\}$. Here $\tilde{e}_\mu \cdot e_\nu = \delta_{\mu\nu}$, $\forall \mu, \nu$, and $\cdot$ stands for the canonical scalar product on $\mathbf{R}^4$.

    To generalize Nahm's construction of the magnetic monopole via ADHM, the selfdual charge-k gauge field $A_\mu(x)$ is shifted by the (antihermitian) connection $2\pi i z_\mu$ which has no field-strength (pure-gauge). That is, one considers the Weyl operator $D^{z;\pm}(A) \equiv \sigma_\mu^\pm D_\mu^z(A)$ where $D_\mu^z(A) \equiv \partial_\mu + A_\mu + 2\pi i z_\mu$. According to the Atiyah–Singer index theorem [Atiyah and Singer (1968a,b,c, 1971a,b, 1984)] the equation

$$D^{z;-}(A)\psi = 0 \qquad (4.58)$$

has k independent and normalizable solutions $\psi^{z;(i)}$ (zero modes of $D^{z;-}(A)$). These are used to construct a U(k) gauge field $\tilde{A}_\mu(z)$ on the dual torus $\tilde{T}^4$ as follows:

$$\tilde{A}_\mu^{ij}(z) = \int_{T^4} d^4 x \psi^{z;(i)}(x)^\dagger \frac{\partial}{\partial z_\mu} \psi^{z;(j)}(x). \qquad (4.59)$$

The selfduality of $\tilde{A}_\mu(z)$ on $\tilde{T}^4$ is a consequence of the fact that

$$D^{z;-}(A)\,D^{z;+}(A) = D^z_\mu(A)\,D^z_\mu(A)\mathbf{1}_2 + \frac{i}{2}\bar{\eta}_{\mu\nu}[D^z_\mu(A), D^z_\nu(A)] \quad (4.60)$$

$$= D^z_\mu(A)\,D^z_\mu(A)\mathbf{1}_2 + \frac{i}{2}\bar{\eta}_{\mu\nu}\,[F_{\mu\nu}, \cdot]$$

$$= D^z_\mu(A)D^z_\mu(A)\mathbf{1}_2$$

($F_{\mu\nu}$ is selfdual and $\bar{\eta}_{\mu\nu}$ is antiselfdual). Equation (4.60) states that the operator $D^{z;-}(A)D^{z;+}(A)$ commutes with the quaternions. Moreover, it has a trivial kernel in the space of normalizable configurations ψ because the operator $D^{z;+}(A)$ has a trivial kernel [Donaldson and Kronheimer (1990)]. Therefore its inverse $G^z(x,y) \equiv \left(D^{z;-}(A)D^{z;+}(A)\right)^{-1}$ is well-defined. Appealing to Eq. (4.60), one computes the field strength $\tilde{F}_{\mu\nu}(z)$ of the gauge field $\tilde{A}_\mu(z)$ on the dual torus $\tilde{T}^4$ as [Van Baal (1996, 1998)]

$$\tilde{F}_{\mu\nu}(z) = 8\pi^2 \int_{T^4 \times T^4} d^4x d^4y \Psi^z(x)^\dagger G^z(x,y)\eta_{\mu\nu}\Psi^z(y)$$

$$+ 4\pi i \int_{\partial T^4 \times T^4} d\Sigma_\lambda d^4 y \frac{\partial}{\partial z_{[\mu}}\Psi^z(x)^\dagger \sigma^+_\lambda \sigma^-_{\nu]}G^z(x,y)\Psi^z(y), \quad (4.61)$$

where $d\Sigma_\lambda$ denotes the infinitesimal normal to the three-dimensional boundary of T^4, $[\lambda\nu]$ signals antisymmetrization with respect to the indices λ and ν, and Ψ^z is the matrix composed of the zero modes $\psi^{z;(i)}$ as columns. Due to the selfduality of $\eta_{\mu\nu}$ the field strength $\tilde{F}_{\mu\nu}(z)$ would be selfdual if a cancelation of the boundary term in Eq. (4.61) could be assured. In case that all periods of T^4 are finite no boundary terms are generated and $\tilde{A}_\mu(z)$ is selfdual everywhere. Boundary contributions arise at points z where the zero modes $\psi^{z;(i)}(x)$ do not decay sufficiently fast towards infinity if one or more of the periods of T^4 are infinite. Fortunately, such points are isolated so that the U(k) gauge field $\tilde{A}_\mu(z)$ is selfdual almost everywhere. It can be shown that $\tilde{A}_\mu(z)$ indeed is of charge N [Braam and Van Baal (1989)]. Nahm's map of selfdual configurations on T^4 to selfdual configurations of $\tilde{T}^4$ is invertible and squares to unity. The ADHM construction is a special case of the Nahm transformation where T^4 degenerates to $\mathbf{R}^4$ and $\tilde{T}^4$ is a point. Selfduality on the dual manifold (the point) then corresponds to the problem of solving a quadratic constraint as discussed in Sec. 4.2.1.

An interesting limit of the original Nahm transformation is the case where T^4 degenerates into $S^1 \times \mathbf{R}^3$ (from now on: coordinate τ on S^1 and spatial coordinates $\vec{x}$ on $\mathbf{R}^3$). If no net magnetic charge is assumed in the finite-action selfdual configuration A_μ in SU(N) then the Nahm transformation goes through in a straightforward way, and the problem of finding all calorons on $S^1 \times \mathbf{R}^3$ reduces to prescribing selfdual data on the dual circle. The behavior of A_μ at spatial infinity needs to be prescribed in terms of the eigenvalues $\mu_1 \leq t \leq \mu_N$ of the Polyakov loop [Kraan and Van Baal (1998b)]:

$$\text{Pol}(|\vec{x}| \to \infty) \equiv \Omega(\hat{x}) \exp[2\pi i \, \text{diag} \, (\mu_1, \dots, \mu_N)] \Omega^\dagger(\hat{x}), \quad \sum_{p=1}^{N} \mu_p = 0.$$

$$(4.62)$$

In case $\text{Pol}(|\vec{x}| \to \infty)$ coincides with (is different from) an element of the center group $\mathbf{Z}_N$ of SU(N) the caloron that is to be constructed possesses trivial (nontrivial) holonomy [see Secs. 4.3.1 and 4.3.2 for the case of SU(2)]. Since the Nahm-transformed gauge field $\tilde{A}_\mu$ solely depends on the coordinate t of the dual S^1 its field strength $\tilde{F}_{\mu\nu}$ is given as

$$\tilde{F}_{4i} = \frac{d}{dt}\tilde{A}_i + [\tilde{A}_4, \tilde{A}_i], \quad \tilde{F}_{ij} = [\tilde{A}_i, \tilde{A}_j]. \qquad (4.63)$$

Selfduality of $\tilde{F}_{\mu\nu}$ implies the Nahm equations

$$\frac{d}{dt}\tilde{A}_i + [\tilde{A}_4, \tilde{A}_i] - \frac{1}{2}\epsilon_{ijk}[\tilde{A}_j, \tilde{A}_k] = \sum_p (\vec{\alpha}^p)_i \delta(t - \mu_p), \qquad (4.64)$$

where the δ-function inhomogeneities cancel boundary contributions of the selfdual gauge field A_μ at spatial infinity which would violate the selfduality of the gauge field $\tilde{A}_\mu$ at isolated points, see

Eq. (4.61). The problem of constructing a selfdual gauge field A_μ on $S^1 \times \mathbf{R}^3$ by virtue of the (inverse) Nahm transformation reduces to (i) finding data $\tilde{A}_\mu$ that solve the ordinary differential equation (4.64), to (ii) subsequently solving the dual version of the Weyl equation (4.58)

$$D^{x;-}(\tilde{A})\psi^x = 0, \tag{4.65}$$

(considering jumps at $t = \mu_p$) which also is an ordinary differential equation in t, to (iii) normalizing the solutions of Eq. (4.65), and finally to (iv) constructing the selfdual gauge field on $S^1 \times \mathbf{R}^3$ as

$$A_\mu^{ij}(x) = \int dt\, \psi^{x;(i)}(t)^\dagger \frac{\partial}{\partial x_\mu} \psi^{x;(j)}(t). \tag{4.66}$$

4.3 SU(2) Calorons with $k = \pm 1$

From now on we specialize to the case SU(2) and $|k| = 1$ and come back to the case of Euclidean thermodynamics where the gauge theory is subject to a compactified "time" dimension, $0 \le \tau \le \frac{1}{T}$.

To construct a periodic instanton or caloron of charge $k = \pm 1$ and trivial holonomy does not require the machinery of the Nahm transformation. It is this trivial-holonomy caloron [Harrington and Shepard (1978)] that enters into the construction of the thermal ground-state estimate. The reason for this is twofold:

(i) A nontrivial-holonomy caloron [Lee and Lu (1998); Kraan and Van Baal (1998a,b)] is unstable under the quantum noise introduced by trivial-topology fluctuations [Diakonov *et al.* (2004)], and

(ii) its one-loop effective action scales with the spatial volume V leading to total suppression in the limit $V \to \infty$ [Gross, Pisarski and Yaffe (1981)].

4.3.1   *Trivial holonomy*

4.3.1.1   *Construction of the Harrington-Shepard (anti)caloron*

The construction of an SU(2) caloron of trivial holonomy and $|k| = 1$
generalizes the BPS monopole of Sec. 4.2.2 in having the periodic
field configuration A_μ depend on τ. Notice that in taking the limit
$\tilde\beta^{-1} \to 0$ in Eqs. (4.50) the ansätze of Eqs. (4.46) yield $A_\mu \equiv 0$, and the
(smeared) topological charge of the monopole configuration then
is invisible. As we shall see, this is not true for the caloron to be
discussed below: The limit $\beta \to \infty$ gives back the $|k| = 1$ singular-
gauge instanton on $\mathbf{R}^4$.

All that needs to be known for the construction of calorons of
trivial holonomy and of topological charge $k = \pm 1$ is the explicit
expression for the special multiinstanton due to 't Hooft as reviewed
in Sec. 4.1.3. Namely, the periodicity, $A_\mu(\tau = 0, \vec{x}) = A_\mu(\tau = \beta, \vec{x})$, of
the selfdual gauge field $A_\mu(\tau, \vec{x})$ is implied by the periodicity of the
prepotential $W(x)$ of Eq. (4.23). Thus one writes[3]

$$\bar{A}^{+,a}_\mu(x) = -\bar{\eta}^a_{\mu\nu}\partial_\nu \log W,$$

$$\bar{A}^{-,a}_\mu(x) = -\eta^a_{\mu\nu}\partial_\nu \log W \qquad (4.67)$$

and demands periodicity of W. The singular-gauge nature of the
't Hooft ansatz (4.67) implies that topological charge is localized at
the instanton centers. As we shall see below, there is precisely one
location of unit topological charge contained in the slice $0 \leq \tau \leq \beta$.
Like for the singular-gauge instanton on $\mathbf{R}^4$, this guarantees $k = \pm 1$
also for the caloron of trivial holonomy.

The construction of the periodic prepotential was performed
in [Harrington and Shepard (1978)]. To distinguish the caloron from
the multiinstanton on $\mathbf{R}^4$ we write from now on $\Pi(\tau, \vec{x})$ for the

---

[3]We still use the convention that a factor of $-i$ is absorbed into the then
antihermitian generator: $t^a \to -it^a$. Thus the real components $\eta^a_{\mu\nu}$ remain
unchanged while the 't Hooft symbol $\eta_{\mu\nu}$ goes over into $-i\eta_{\mu\nu}$.

prepotential. By construction, the following "mirror sum"

$$\Pi(\tau, \vec{x}; \rho, \beta, x_0) = 1 + \sum_{l=-\infty}^{l=\infty} \frac{\rho^2}{(x - x_l)^2} \quad (4.68)$$

is periodic in τ for $x_l \equiv (\tau_l, \vec{x}_0)$, $\tau_l \equiv \tau_0 + l\beta$, and $0 \leq \tau_0 < \beta$. Only the instanton center at $l = 0$ contributes to the topological charge contained in the slice $0 \leq \tau \leq \beta$: $k = -\frac{1}{32\pi^2} \text{tr} \int_0^\beta d\tau d^3x F_{\mu\nu} \tilde{F}_{\mu\nu} = \pm 1$. Notice that, in the limit $\beta \to \infty$ the prepotential $\Pi(\tau, \vec{x}; \rho, \beta, x_0)$ of Eq. (4.68) approaches the one for a $|k| = 1$ singular-gauge instanton on $\mathbf{R}^4$.

The periodicity of $\Pi(\tau, \vec{x}; \rho, \beta, x_0)$ is made explicit by performing the sum in Eq. (4.68). Specializing to $\tau_0 = 0$ and $\vec{x}_0 = \vec{0}$, this yields

$$\Pi(\tau, \vec{x}; \rho, \beta) = 1 + \frac{\pi\rho^2}{\beta r} \frac{\sinh\left(\frac{2\pi r}{\beta}\right)}{\cosh\left(\frac{2\pi r}{\beta}\right) - \cos\left(\frac{2\pi\tau}{\beta}\right)}, \quad (4.69)$$

where $r \equiv |\vec{x}|$. Obviously, there is a singularity of $\Pi(\tau, \vec{x}; \rho, \beta)$ at $r = \tau = 0$.

Let us now check whether the gauge-field configuration characterized by $\Pi(\tau, \vec{x}; \rho, \beta, x_0)$ in Eq. (4.69) indeed possesses trivial holonomy. To do this, we need to investigate the limit of $A_4^\pm$ towards spatial infinity. One has

$$\Pi(\tau, \vec{x}; \rho, \beta, x_0) \overset{r \to \infty}{=} 1 + \frac{\pi\rho^2}{\beta r}. \quad (4.70)$$

Therefore $\log \Pi(\tau, \vec{x}; \rho, \beta, x_0) \overset{r \to \infty}{=} \frac{\pi\rho^2}{\beta r}$. Since $A_4^\pm$ involves only spatial derivatives of $\log \Pi(\tau, \vec{x}; \rho, \beta, x_0)$, we obtain

$$\lim_{r \to \infty} A_4^\pm(\tau, \vec{x}; \rho, \beta, x_0) \propto \lim_{r \to \infty} \frac{1}{r^2} = 0. \quad (4.71)$$

Therefore the Polyakov loop at spatial infinity coincides with the center element $\mathbf{1}_2$, and the holonomy associated with the Harrington–Shepard caloron is trivial.

4.3.1.2 *Anatomy of the Harrington-Shepard (anti)caloron*

Along the lines of [Gross, Pisarski and Yaffe (1981)], see also [Grandou and Hofmann (2015)], we now discuss how the Harrington-Shepard (anti)caloron (for the remainder of this section simply referred to as (anti)caloron) appears to an observer depending on his distance from the (anti)caloron center at $\tau = r = 0$. This point is the locus of the configuration's topological charge $k = \pm 1$ in the sense that the integral of the Chern-Simons current K_μ in Eq. (4.1) over a three-sphere S_δ^3 of radius δ, which is centered at $\tau = r = 0$, always equals unity for $\delta > 0$. Knowledge about the behavior of the solution away from this singularity is important for the investigations performed in Sec. 9.6.1.

For $|x| \ll \beta$ ($|x| \equiv \sqrt{x^2} \equiv \sqrt{x_\mu x_\mu}$, $x_4 \equiv \tau$) one has

$$\Pi(x) = (1 + \frac{\pi}{3}\frac{s}{\beta}) + \frac{\rho^2}{x^2} + O(x^2/\beta^2), \tag{4.72}$$

where s is defined as

$$s = \pi \frac{\rho^2}{\beta}. \tag{4.73}$$

From Eqs. (4.72) and (4.67) one obtains the following expression for the caloron field strength $F_{\mu\nu} = \frac{1}{2}\epsilon_{\mu\nu\kappa\lambda}F_{\kappa\lambda} \equiv \tilde{F}_{\mu\nu}$

$$F_{\mu\nu}^a = -4\rho'^2 \frac{\bar{\eta}_{\alpha\beta}^a}{(x^2 + \rho'^2)^2} I_{\alpha\mu}I_{\beta\nu} + O(x^2/\beta^4), \tag{4.74}$$

where $I_{\alpha\mu} \equiv \delta_{\alpha\mu} - 2\frac{x_\alpha x_\mu}{x^2}$. At small four-dimensional distances $|x|$ from the center this field strength thus behaves like the one of a singular-gauge instanton with a renormalized scale parameter, $\rho'^2 = \frac{\rho^2}{1+\frac{\pi}{3}\frac{s}{\beta}}$.

For $r \gg \beta$ the selfdual electric and magnetic fields E_i^a and B_i^a of a caloron are static and can be written as

$$E_i^a = B_i^a \sim -\frac{\frac{\hat{x}^a \hat{x}_i}{r^2} - \frac{1}{rs}(\delta_i^a - 3\hat{x}^a \hat{x}_i)}{(1 + \frac{r}{s})^2}, \tag{4.75}$$

where $\hat{x}_i \equiv \frac{x_i}{r}$ and $\hat{x}^a \equiv \frac{x^a}{r}$. For $\beta \ll r \ll s$ Eq. (4.75) simplifies as

$$E_i^a = B_i^a \sim -\frac{\hat{x}^a \hat{x}_i}{r^2},\qquad(4.76)$$

and thus describes a static non-Abelian monopole of unit electric and magnetic charges (dyon). For $r \gg s \gg \beta$ Eq. (4.75) reduces to

$$E_i^a = B_i^a \sim s\,\frac{\delta_i^a - 3\,\hat{x}^a \hat{x}_i}{r^3}.\qquad(4.77)$$

This is the field strength of a static, selfdual non-Abelian dipole field, its dipole moment p_i^a being given as

$$p_i^a = s\,\delta_i^a.\qquad(4.78)$$

4.3.2   Nontrivial holonomy

Let us now turn to the much more involved case of a nontrivial-caloron holonomy. As already mentioned, knowledge about this deformed version of trivial-holonomy selfduality at finite temperature and $|k| = 1$ yields an important constraint on the construction of a useful *a priori* estimate for the deconfining thermal ground state: As a consequence of knowing their classical properties and their behavior under quantum noise, the contribution of nontrivial-holonomy calorons to the *a priori* estimate of the thermal ground state is excluded. Closed-form expressions for the classical field configurations and the one-loop determinant associated with calorons of nontrivial holonomy are also required to interpret radiative corrections in the effective theory for the deconfining phase as discussed in Chapter 5.

The construction of an SU(2) caloron of nontrivial holonomy and $|k| = 1$ represents a strong generalization of the BPS monopole in Sec. 4.2.2. First, its holonomy is nontrivial, and, second, this configuration depends on τ. A constraint on the construction is that the Harrington–Shepard solution of Sec. 4.3.1 is to re-emerge in the limit of trivial holonomy. This is due to the demand that the caloron shall not possess any overall magnetic charge.

Although the path for construction was laid out by W. Nahm's important work [Nahm (1981, 1982, 1983)] it took another fifteen years for the first closed-form expressions of these particular selfdual configurations to surface [Lee and Lu (1998); Kraan and Van Baal (1998a,b)]. Six years later their one-loop quantum weights became available [Diakonov *et al.* (2004)]. This considerable delay may be in part due to the influential paper [Gross, Pisarski and Yaffe (1981)] which — by only invoking the asymptotic spatial behavior of the gauge field — correctly states that there is no contribution of a static, nontrivial holonomy to the partition function in the infinite-volume limit.

In spite of the no-go statement of [Gross, Pisarski and Yaffe (1981)] it turns out that knowledge about the effects of quantum fluctuations on a static caloron holonomy is valuable [Diakonov *et al.* (2004)]. Namely, by considering the temporary creation of a nontrivial holonomy due to absorption of a propagating gauge mode by a caloron of trivial holonomy as an adiabatically slow process, one concludes, based on the work of Diakonov *et al.* (2004), the following: Depending on the size of the holonomy, this process induces either the subsequent dissociation of the system into a separated monopole-antimonopole pair (large holonomy) or a fall-back onto trivial holonomy (small holonomy) (see Sec. 4.4.2). Notice that in both cases the parameter describing the nontrivial caloron holonomy necessarily varies as a function of time.

In sketching the construction of $|k| = 1$ calorons of nontrivial holonomy for the gauge group SU(2) we follow the presentation of [Lee and Lu (1998)]. Let us only discuss the case $k = 1$. A gauge may be picked where the parameter u of mass dimension one enters as follows

$$A_4(\tau, r \to \infty) = -iut^3, \qquad (4.79)$$

and we demand $0 \leq u \leq \frac{2\pi}{\beta}$. On the configuration A_μ with asymptotics (4.79) one has

$$\text{Pol}(r \to \infty) = \exp[iu\beta t^3] = 1_2 \cos \frac{u\beta}{2} + \tau^3 \sin \frac{u\beta}{2} \neq \pm 1_2,$$

$$\left(u \neq 0, \frac{2\pi}{\beta} \right). \qquad (4.80)$$

Thus A_μ is of nontrivial holonomy if $u \neq 0, \frac{2\pi}{\beta}$. The gauge field $\tilde{A}_\mu$ on the dual "torus" is Abelian for $k = 1$. Thus the entries of $\tilde{A}_\mu$ can be considered real numbers. Moreover, the component $\tilde{A}_4$ is not determined by the Nahm equations, and we make the convention $\tilde{A}_4 = 0$.

Let us define three subintervals of the dual torus $-\frac{\pi}{\beta} \leq t \leq \frac{\pi}{\beta}$ by the points $t_{I-II} < t_{II-III}$. These subintervals are labeled as I, II, and III. Were it not for δ-function inhomogeneities at $t_{I-II,II-III}$ the Nahm equations simply would read

$$\frac{d}{dt} \tilde{A}_i = 0. \qquad (4.81)$$

Thus $\tilde{A}_i$ is piecewise constant: $\tilde{A}_i \equiv \tilde{A}_i^{I,II,III}$ for $t \in$ I, II, III, respectively. To accomodate a magnetic monopole and its antimonopole, the dual gauge field $\tilde{A}_i$ needs to jump at $t_{I-II,II-III}$ where $t_{I-II,II-III} = \mp\frac{u}{2}$, respectively. With these conventions the full Nahm equations read

$$\frac{d}{dt} \tilde{A}_i = (\vec{\alpha}^{I-II})_i \delta(t + u/2) + (\vec{\alpha}^{II-III})_i \delta(t - u/2), \qquad (4.82)$$

where the constant vectors $\vec{\alpha}^{I-II}$ and $\vec{\alpha}^{II-III}$ are yet to be determined. The "forward" jump at t_{I-II} is compensated by a "backward" jump at t_{II-III} to guarantee the periodicity[4] of $\tilde{A}_i$. Thus $\tilde{A}_i^{I} = \tilde{A}_i^{III}$.

---

[4]Demanding periodicity up to a U(1) gauge transformation on the dual "torus" would re-introduce a finite value of $\tilde{A}_4$.

The Nahm equations (4.82) are solved by prescribing

$$\tilde{A}_i^{\mathrm{I}} = \tilde{A}_i^{\mathrm{III}} = -(\vec{x}_2)_i, \quad \tilde{A}_i^{\mathrm{II}} = -(\vec{x}_1)_i, \tag{4.83}$$

where $\vec{x}_{1,2}$ are the spatial positions of the monopole and its anti-monopole. Without restriction of generality we may place both monopole and antimonopole on the three-axis. Then the core-separation distance D between them simply is given as $D = (\vec{x}_2 - \vec{x}_1)_3$, and from Eq. (4.82) one reads off the following expressions for $\vec{\alpha}^{\mathrm{I}-\mathrm{II}}$ and $\vec{\alpha}^{\mathrm{II}-\mathrm{III}}$:

$$(\vec{\alpha}^{\mathrm{I}-\mathrm{II}})_i = D\delta_{i3}, \quad (\vec{\alpha}^{\mathrm{II}-\mathrm{III}})_i = -D\delta_{i3}. \tag{4.84}$$

The nontriviality of the kernel of the Weyl operator of Eq. (4.58) at the jump positions $t_{\mathrm{I}-\mathrm{II},\mathrm{II}-\mathrm{III}}$ demands that constant two-component row vectors $a^{\mathrm{I}-\mathrm{II}}$ and $a^{\mathrm{II}-\mathrm{III}}$ as well as constant scalars $\gamma^{\mathrm{I}-\mathrm{II}}$ and $\gamma^{\mathrm{II}-\mathrm{III}}$ exist such that

$$(a^p)^\dagger a^p = \vec{\alpha}^p \cdot \vec{\tau} - i\gamma^p \mathbf{1}_2. \tag{4.85}$$

Here $p = \mathrm{I}\text{-}\mathrm{II}, \mathrm{II}\text{-}\mathrm{III}$, and $\vec{\tau}$ is the three-vector of Pauli-matrices. Equation (4.85) is solved by

$$a^{\mathrm{I}-\mathrm{II}} = (\sqrt{2D}, 0), \quad \gamma^{\mathrm{I}-\mathrm{II}} = iD,$$
$$a^{\mathrm{II}-\mathrm{III}} = (0, \sqrt{2D}), \quad \gamma^{\mathrm{II}-\mathrm{III}} = -iD. \tag{4.86}$$

The Weyl equation, to be solved for the 2×2 matrix function $\psi_x(t)$ and for the two-component row vectors S^p, reads[5]

$$0 = \left[-\frac{d}{dt} + (\tilde{A}_i + (\vec{x})_i)(\vec{\tau})_i + i\tau \right] \psi_x(t) + \sum_p (a^p)^\dagger S^p \delta(t - t_p),$$

$$\tag{4.87}$$

---

[5]Moreover, for SU(2) and $k = 1$ the gauge field $\tilde{A}_\mu$ on the dual interval has topological charge 2. Thus there are two zero modes of the Weyl operator which form the columns of $\psi_x(t)$. Recall also that $\psi_x(t)$ needs to be periodic: $\psi_x(t = -\pi/\beta) = \psi_x(t = \pi/\beta)$.

where the sum is over $p =$ I-II, II-III. To avoid confusion, we stress that τ is the four-component of x while τ_i, $(i = 1, 2, 3)$, are the Pauli-matrices. In addition to solving Eq. (4.87) $\psi_x(t)$ and the S^p need to satisfy the following normalization condition

$$1_2 = \int_{-\pi/\beta}^{\pi/\beta} dt (\psi_x)^\dagger \psi_x + \sum_p (S^p)^\dagger S^p. \tag{4.88}$$

The selfdual gauge field A_μ on $S^1 \times \mathbf{R}^3$ of topological charge $k = 1$ and with a magnetic monopole-antimonopole constituent pair of core separation D is obtained from $\psi_x(t)$ and the S^p as

$$
\begin{aligned}
A_\mu &= \int_{-\pi/\beta}^{\pi/\beta} dt (\psi_x)^\dagger \partial_\mu \psi_x + \sum_p (S^p)^\dagger \partial_\mu S^p \\
&= \frac{1}{2} \int_{-\pi/\beta}^{\pi/\beta} dt \left\{ (\psi_x)^\dagger \partial_\mu \psi_x - \left(\partial_\mu (\psi_x)^\dagger \right) \psi_x \right\} \\
&\quad + \frac{1}{2} \sum_p \left\{ (S^p)^\dagger \partial_\mu S^p - \left(\partial_\mu (S^p)^\dagger \right) S^p \right\}.
\end{aligned}
\tag{4.89}
$$

The expressions in the second and third lines of Eq. (4.89) exploit the normalization condition (4.88). To present the solutions ψ_x^I, ψ_x^{II}, and ψ_x^{III} of Eq. (4.87) subject to the constraint (4.88) in an efficient way, the following variables are introduced:

$$
\begin{aligned}
\vec{y}_1 &\equiv \vec{x} - \vec{x}_1, \quad \vec{y}_2 \equiv \vec{x} - \vec{x}_2, \\
\vec{s}_1 &\equiv u \vec{y}_1, \quad \vec{s}_2 \equiv \left(\frac{2\pi}{\beta} - u \right) \vec{x}_2.
\end{aligned}
\tag{4.90}
$$

One then has

$$
\begin{aligned}
\psi_x^I &= \frac{1}{\sqrt{N_2}} \exp \left[(i\tau + \vec{\tau} \cdot \vec{y}_2)(t + \pi/\beta) \right] C_2, \\
\psi_x^{II} &= \frac{1}{\sqrt{N_1}} \exp \left[(i\tau + \vec{\tau} \cdot \vec{y}_1)t \right] C_1, \\
\psi_x^{III} &= \frac{1}{\sqrt{N_2}} \exp \left[(i\tau + \vec{\tau} \cdot \vec{y}_2)(t - \pi/\beta) \right] C_2,
\end{aligned}
\tag{4.91}
$$

where $C_{2,1}$ are 2×2 matrices, which are determined below, and

$$N_{2,1} \equiv \frac{1}{y_{2,1}} \sinh s_{2,1}. \tag{4.92}$$

Here $y_{2,1} \equiv |\vec{y}_{2,1}|$ and $s_{2,1} \equiv |\vec{s}_{2,1}|$. Notice that $\psi_x^I(t) = \psi_x^{III}(t + 2\pi/\beta)$, which implies that $\psi_x^I(-\pi/\beta) = \psi_x^{III}(\pi/\beta)$.

The two row vectors S^p ($p =$ I–II, II–III), understood as the first and second row of a matrix S, as well as the two matrices $C_{2,1}$ introduced in Eq. (4.91), are all determined by setting the sums of coefficients in front of the two δ-functions in Eq. (4.87) equal to zero and at the same time obeying normalization condition (4.88). By virtue of Eqs. (4.91) the latter assumes the form

$$1_2 = C_2^\dagger C_2 + C_1^\dagger C_1 + S^{I-II} S^{I-II} + S^{II-III} S^{II-III}. \tag{4.93}$$

One obtains the following expression for S

$$S = \frac{1}{\sqrt{\mathcal{N}}} \exp\left[-i\frac{u}{2}\tau\tau_3\right], \tag{4.94}$$

where

$$\mathcal{N} = 1 + \frac{2D}{\mathcal{M}} \left(N_1(\cosh s_2 - (\hat{y}_2)_3 \sinh s_2) + N_2(\cosh s_1 + (\hat{y}_1)_3 \sinh s_1)\right) \tag{4.95}$$

and

$$\mathcal{M} = 2\left(\cosh s_1 \cosh s_2 + \hat{y}_1 \cdot \hat{y}_2 \sinh s_1 \sinh s_2 - \cos\left[\frac{2\pi}{\beta}\tau\right]\right). \tag{4.96}$$

Here $\hat{z}$ indicates a unit vector: $\hat{z} \equiv \frac{\vec{z}}{|z|}$. For C_1, C_2 one has

$$C_1 = \sqrt{\frac{2DN_1}{\mathcal{N}}} \frac{B_1^\dagger}{\mathcal{M}} \left[\exp\left(-\frac{\vec{\tau}}{2} \cdot \vec{s}_2\right) Q_+\right.$$
$$\left. + \exp\left(\frac{\vec{\tau}}{2} \cdot \vec{s}_2\right) Q_-\right] \exp\left(-i\frac{\pi}{\beta}\tau\tau_3\right),$$

$$C_2 = \sqrt{\frac{2DN_2}{\mathcal{N}}} \frac{B_2^\dagger}{\mathcal{M}} \left[\exp\left(\frac{\vec{\tau}}{2} \cdot \vec{s}_1\right) Q_+ + \exp\left(-\frac{\vec{\tau}}{2} \cdot \vec{s}_2\right) Q_-\right], \tag{4.97}$$

where $Q_\pm = \frac{1}{2}(1 \pm \tau_3)$ are projection operators, and the matrices B_1, B_2 are

$$B_1 = \exp\left[i\frac{\pi}{\beta}\tau\right]\exp\left[-\frac{\vec{\tau}}{2}\cdot\vec{s}_1\right]\exp\left[-\frac{\vec{\tau}}{2}\cdot\vec{s}_2\right]$$

$$- \exp\left[-i\frac{\pi}{\beta}\tau\right]\exp\left[\frac{\vec{\tau}}{2}\cdot\vec{s}_1\right]\exp\left[\frac{\vec{\tau}}{2}\cdot\vec{s}_2\right],$$

$$B_2 = \exp\left[i\frac{\pi}{\beta}\tau\right]\exp\left[-\frac{\vec{\tau}}{2}\cdot\vec{s}_2\right]\exp\left[-\frac{\vec{\tau}}{2}\cdot\vec{s}_1\right]$$

$$- \exp\left[-i\frac{\pi}{\beta}\tau\right]\exp\left[\frac{\vec{\tau}}{2}\cdot\vec{s}_2\right]\exp\left[\frac{\vec{\tau}}{2}\cdot\vec{s}_1\right]. \tag{4.98}$$

Finally, substituting into Eq. (4.89) expressions (4.91) and (4.94), subject to Eqs. (4.95)–(4.98), yields a representation of the selfdual gauge field on $S^1 \times \mathbf{R}^3$ as

$$A_\mu(\vec{x}, \tau) = C_1^\dagger V_\mu(\vec{y}_1; u)C_1 + C_2^\dagger V_\mu(\vec{y}_2; \frac{2\pi}{\beta} - u)C_2$$

$$+ C_1^\dagger \partial_\mu C_1 + C_2^\dagger \partial_\mu C_2 + S^\dagger \partial_\mu S, \tag{4.99}$$

where

$$V_4(\vec{x}; u) = \frac{\tau_a}{2i}\hat{x}_a\left(\frac{1}{|\vec{x}|} - u\coth(u|\vec{x}|)\right),$$

$$V_i(\vec{x}; u) = \frac{\tau_a}{2i}\epsilon_{aij}\hat{x}_j\left(\frac{1}{|\vec{x}|} - \frac{u}{\sinh(u|\vec{x}|)}\right). \tag{4.100}$$

Setting $2\pi\tilde{\beta}^{-1} = u$ in Eq. (4.50) and taking into account ansatz (4.46), we identify V_μ of Eq. (4.100) with a BPS monopole [the $-+$ combination in Eq. (4.50)].

It was shown in [Lee and Lu (1998)] that close to the core of the first monopole, that is for $y_1 \ll D$

$$C_2, S \sim \frac{1}{\sqrt{D}} \tag{4.101}$$

and

$$C_1 = \frac{\tau_3 \cosh \frac{s_1}{2} - \vec{\tau} \cdot \hat{y}_1 \sinh \frac{s_1}{2}}{\sqrt{\cosh s_1 - (\hat{y}_1)_3 \sinh s_1}} + \mathcal{O}(1/D). \tag{4.102}$$

Thus C_1 becomes a single-valued unitary matrix. As a consequence, the configuration in Eq. (4.99) is an approximate gauge transform of a BPS monopole for $y_1 \ll D$. Similarly, for $y_2 \ll D$ the matrix C_2 becomes unitary,

$$C_2 = \frac{-\tau_3 \cosh \frac{s_2}{2} - \vec{\tau} \cdot \hat{y}_2 \sinh \frac{s_2}{2}}{\sqrt{\cosh s_2 + (\hat{y}_2)_3 \sinh s_2}} \exp\left[-i(\pi\tau/\beta)\tau_3\right] + \mathcal{O}(1/D), \tag{4.103}$$

and we have

$$C_1, S \sim \frac{1}{\sqrt{D}}. \tag{4.104}$$

Thus one observes again that an approximate gauge transform of a BPS monopole takes place close to $\vec{x}_2$. Compared to the monopole at $\vec{x}_1$ the magnetic charge g_2 of the monopole at $\vec{x}_2$ is reversed, $g_1 = -g_2 = -4\pi$, due to the extra factor $\exp\left[-i(\pi\tau/\beta)\tau_3\right]$ in Eq. (4.103) inducing a *large* gauge transformation. There is a singularity of the solution at the point $(\tau = 0, \vec{x}_{cm})$ where

$$\vec{x}_{cm} \equiv \frac{\beta u}{2\pi}\vec{x}_1 + \left(1 - \frac{\beta u}{2\pi}\right)\vec{x}_2. \tag{4.105}$$

At this point one unit of topological charge is localized. One can show this by expanding the solution about $(\tau = 0, \vec{x}_{cm})$ and by performing the integral of the Chern–Simons current over a small S_3 centered at this point. One may also view the solution (4.99) as the result of a deformation process of a caloron of trivial holonomy of scale ρ to nontrivial holonomy. For the case that $\rho > \beta$ it was shown in [Kraan and Van Baal (1998b)] that the two monopoles become well separated and that their distance D can be expressed in terms of ρ as follows

$$D \equiv \frac{\pi}{\beta}\rho^2 = s. \tag{4.106}$$

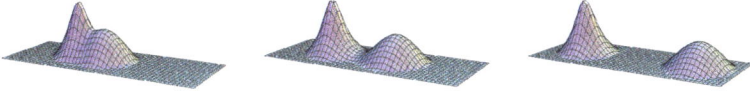

Figure 4.1. Action density of an SU(2) caloron with nontrivial holonomy plotted on a two-dimensional spatial slice. The caloron radius ρ and therefore the monopole-core separation $D = \frac{\pi}{\beta}\rho^2$ increases from left to right while temperature and holonomy are fixed. The peaks of the action density coincide with the core positions of the constituent BPS monopoles. The figures are adopted from [Kraan and Van Baal (1998b)] (with kind permission by Elsevier under licence number 2497060494979).

Relation (4.106) was also obtained in [Lee and Lu (1998)] by considering the limit of trivial holonomy at fixed monopole distance D. Interestingly, the distance D of Eq. (4.106) between the centers of the Abelian magnetic monopole and its antimonopole *coincides* with the spatial scale of charge separation s, defined in Eq. (4.73), to characterize the non-Abelian dipole moment seen at spatial distances $r \gg s$ from the center of a Harrington–Shepard (anti)caloron.

A plot of the action density of a caloron of nontrivial holonomy with varying radius ρ at a fixed temperature and a fixed holonomy is presented in Fig. 4.1. Setting the fundamental gauge coupling g equal to unity, $g = 1$, the masses of the monopoles at $\vec{x}_1$ and $\vec{x}_2$ are given as

$$m_1 = 4\pi u, \quad m_2 = 4\pi \left(\frac{2\pi}{\beta} - u \right), \tag{4.107}$$

respectively. For a large holonomy, that is $u \sim \frac{\pi}{\beta}$, we have $m_1 \sim m_2 \sim 4\pi^2 T \sim 40\, T$. According to Eq. (4.107) one monopole becomes massless in the limit of vanishing holonomy, and, after applying a large gauge transformation, one recovers the Harrington–Shepard solution provided that relation (4.106) is used [Lee and Lu (1998)].

4.4   One-loop Quantum Weights of Calorons

In this section we report on results for the one-loop effective action of a caloron of trivial holonomy, as obtained by Gross, Pisarski, and Yaffe (1981). Subsequently, we sketch the results of

Diakonov *et al.* (2004) for the one-loop quantum weight of a caloron with nontrivial holonomy. Both results are important for a grasp of the microscopics of the ground-state physics in the deconfining high-temperature phase of SU(2) and SU(3) Yang–Mills thermodynamics.

4.4.1   *Harrington–Shepard solution*

The functional determinant about a Harrington–Shepard caloron was calculated in [Gross, Pisarski and Yaffe (1981)]. This corresponds to the approximation that fluctuations about the classical (Euclidean) solution are taken into account up to quadratic order in an expansion of the action which leads to a Gaussian functional integration. In such a semiclassical calculation, zero-modes, which are generated by the symmetries of the classical theory, are integrated out separately from those fluctuations which change the action of the classical caloron.

The result in Gaussian approximation for the quantum weight $\exp[-S_{\text{eff}}]$ of the Harrington–Shepard caloron is given in terms of the effective action as [Gross, Pisarski and Yaffe (1981)]

$$S_{\text{eff}} = \frac{8\pi^2}{\bar{g}^2} + \frac{4}{3}\sigma^2 + 16A(\sigma), \qquad (4.108)$$

where the dimensionless quantity σ is defined as $\sigma \equiv \pi\frac{\rho}{\beta}$ and

$$A(\sigma) \equiv \frac{1}{12}\left[\int_0^\beta d\tau \frac{d^3x}{16\pi^2}\left(\frac{(\partial_\mu \Pi)^2}{\Pi^2}\right)^2 - \int \frac{d^4x}{16\pi^2}\left(\frac{(\partial_\mu \Pi_0)^2}{\Pi_0^2}\right)^2\right].$$

$$(4.109)$$

In Eq. (4.109) the scalar quantities Π and Π_0 are defined in Eqs. (4.69) and (4.23) (setting $W = \Pi_0$, $n = 0$, $\rho_0 = \rho$, and $x_0 = 0$), respectively. The first integral in Eq. (4.109) is over $S_1 \times \mathbf{R}^3$ while the second integral is over $\mathbf{R}^4$. It is worth stating how $A(\sigma)$ behaves in the high- and low-temperature limits $\sigma \to \infty$ and $\sigma \to 0$:

$$A(\sigma) \to -\frac{1}{6}\log\sigma, \quad (\sigma \to \infty), \quad A(\sigma) \to -\frac{\sigma^2}{36}, \quad (\sigma \to 0). \quad (4.110)$$

At a given caloron radius ρ the correction to the classical action $\frac{8\pi^2}{\bar{g}^2}$ thus is large in the high-temperature regime, indicating that the contribution of calorons of trivial holonomy to the partition function is suppressed, while it is small at low temperatures, implying the increasing importance of calorons as the temperature of the system drops. The distinction between high and low temperature can be made by a dynamically generated scale Λ which also determines the ρ dependence of the coupling constant $\bar{g}$ in Eq. (4.108).

4.4.2   Lee–Lu–Kraan–van Baal solution

The calculation of the one-loop quantum weight for a caloron of nontrivial holonomy is much harder than for the trivial-holonomy case. The expressions are so involved that the contribution $\mathcal{Z}_{\text{n.h.}}$ of an isolated, quantum-blurred caloron to the total partition function $\mathcal{Z}$ of the theory has so far only been stated in closed analytical form in the limit

$$\frac{D}{\beta} = \pi \left(\frac{\rho}{\beta}\right)^2 \gg 1. \qquad (4.111)$$

As we shall see in Chapter 5, this is a relevant limit. Apart from the restriction in Eq. (4.111) the result obtained in [Diakonov et al. (2004)] for $\mathcal{Z}_{\text{n.h.}}$ is valid for any value of the holonomy, $0 \le u \le \frac{2\pi}{\beta}$. After the (trivial) integrations over the overall color orientation and time translations are performed one obtains [Diakonov et al. (2004)]

$$\mathcal{Z}_{\text{n.h.}} = C\beta^{-6} \int d^3x_1 \int d^3x_2 \left(\frac{8\pi^2}{\bar{g}^2}\right)^4 \left(\frac{\Lambda e^{\gamma_E}\beta}{4\pi}\right)^{22/3} \left(\frac{\beta}{D}\right)^{5/3}$$

$$\times (2\pi + \beta u\bar{u}D)(uD+1)^{\frac{4}{3\pi}u\beta-1}(\bar{u}D+1)^{\frac{4}{3\pi}\bar{u}\beta-1}$$

$$\times \exp[-VP(u) - 2\pi D P''(u)], \quad (D \gg \beta). \qquad (4.112)$$

[The actual number of integration variables in Eq. (4.112) is four because $\int d^3x_1 \int d^3x_2 = 4\pi \int d^3x \int dD D^2$ where $\vec{x} = \frac{1}{2}(\vec{x}_1 + \vec{x}_2)$.] In Eq. (4.112) $C \sim 1.0314$, $\vec{x}_1$ and $\vec{x}_2$ are the core positions of the monopoles in the classical solution, $\bar{u} \equiv \frac{2\pi}{\beta} - u$, γ_E is the Euler–Mascheroni constant, V denotes the typical spatial volume belonging to the one-caloron system, and Λ is a scale which is invariant under the one-loop renormalization group (dimensional transmutation). The functions $P(u)$ and $P''(u)$ are given as

$$P(u) = \frac{\beta}{12\pi^2}u^2\bar{u}^2 \,,$$

$$P''(u) = \frac{\beta}{\pi^2}\left[\frac{\pi}{\beta}\left(1 - \frac{1}{\sqrt{3}}\right) - u\right]\left[\bar{u} - \frac{\pi}{\beta}\left(1 - \frac{1}{\sqrt{3}}\right)\right]. \quad (4.113)$$

The function $P(u)$ is always positive for $u \neq 0, \frac{2\pi}{\beta}$. The occurrence of the spatial volume V in the exponent in Eq. (4.112) thus implies total suppression of static nontrivial holonomy in the thermodynamical limit $V \to \infty$. As we shall see below, the conclusion that nontrivial holonomy does not play a role in the Yang–Mills partition function is not valid since calorons of nontrivial holonomy are unstable under quantum noise. This, in turn, means that their holonomy is not an externally prescribed, static parameter. Rather, nontrivial holonomy is transient and induced either by the spatial overlap of Harrington–Shepard (anti)calorons (small holonomy, a situation which is collectively described by an effective, pure-gauge configuration, see Sec. 5.1.5) or by the distortions mediated by effective radiative corrections (large holonomy) whose impact on any fundamental (anti)caloron configuration remains, as a matter of principle, unresolved, see Sec. 5.3. Therefore, an (anti) caloron either collapses back onto trivial holonomy (attraction between the magnetic monopole and its antimonopole), or it dissociates (repulsion between the magnetic monopole and its antimonopole) into its magnetic constituents. This can be checked by investigating the second contribution to the exponent in Eq. (4.112). For $0 \leq u \leq \frac{\pi}{\beta}(1 - \frac{1}{\sqrt{3}})$ and for $\frac{\pi}{\beta}(1 + \frac{1}{\sqrt{3}}) \leq u \leq 2\frac{\pi}{\beta}$ the function $P''(u)$ is positive (small holonomy) while it is negative in the complementary range (large holonomy). According to Eq. (4.112) this means that in the former (latter) case the constituent BPS monopoles experience a linear attractive (repulsive) potential.

We now consider the case of a large-holonomy caloron-anticaloron pair in isolation. Let us make an estimate of the typical size of an equilateral tetrahedron whose corners are the positions of magnetic monopoles generated by the dissociation of the caloron and its anticaloron (see Fig. 4.2). The edge length R of the tetrahedron is the typical maximal distance between two BPS monopoles coming out of the caloron or the anticaloron. We assume the holonomy to be maximally nontrivial, $u_{max} = \frac{\pi}{\beta}$. After the creation of such a large holonomy dissociation generates the distance R with probability one.[6]

Thus we may equate the probability for reaching the distance R, governed by Eq. (4.112), with the thermal probability for exciting large holonomy to start with. Since monopoles are at rest shortly

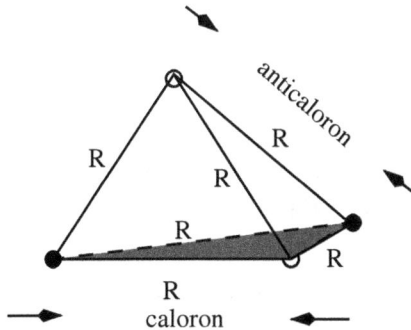

Figure 4.2. The volume spanned by two pairs of BPS monopoles created by the dissociation of a large holonomy caloron-anticaloron pair in isolation.

---

[6]The screening of magnetic charge by the full ensemble is not described by Eq. (4.112) because we assumed the caloron-anticaloron pair to be isolated. Realistically, such a screening takes place in a two-fold manner: renormalization of the magnetic coupling by instable magnetic dipoles (adjacent calorons with small holonomies) and Yukawa screening by other stable monopoles liberated by the dissociation of adjacent large-holonomy calorons (see Sec. 5.4).

after being created the latter probability is roughly given as

$$\exp[-\beta(m_1 + m_2)], \quad m_1 \sim m_2 \sim \frac{4\pi^2}{\beta} \qquad (4.114)$$

[see Eq. (4.107)]. Taking only the exponentially sensitive part of the caloron weight in Eq. (4.112) into account and substituting for V the volume of the tetrahedron, $V = \frac{1}{6\sqrt{2}} R^3$, this translates into the following condition:

$$-\frac{\pi^2}{72\sqrt{2}} \left(\frac{R}{\beta}\right)^3 + \frac{2}{3}\pi \frac{R}{\beta} + 8\pi^2 = 0. \qquad (4.115)$$

There exists only one real and positive solution to this equations. Numerically, we obtain $R \sim 10.1\,\beta$. So on the scale of the inverse temperature the gas of magnetic monopoles is dilute, and our assumption about the isolation of the caloron-anticaloron system is selfconsistent. As we shall see in Sec. 5.5.2, the value of $\sim 21.1\,\beta$ of the typical monopole-antimonopole distance in the thermal ensemble is even larger than $R \sim 10.1\,\beta$ due to screening effects.

While the (extremely small) likelihood for the generation of large-holonomy calorons depends on the value of the holonomy only (and not on the distance D) this is no longer true for a caloron of holonomy close to trivial. Since the latter configuration always collapses back onto trivial holonomy the likelihood for its generation is determined by the caloron weight $\exp[-S_{\text{eff}}]$ with S_{eff} given in Eq. (4.108). A strong dependence of S_{eff} on D (or ρ) at a given temperature exists. In contrast to $\exp[-\beta(m_1 + m_2)] \sim \exp[-8\pi^2]$ the weight $\exp[-S_{\text{eff}}]$ is sizable at S_{eff}'s minimum $\sigma_{\min}$. We conclude that *attraction* between a BPS monopole and its antimonopole (small holonomy) dominates by far the ground-state physics as compared to *repulsion* (large holonomy). Macroscopically, this situation generates a *negative* pressure of the thermal ground-state. We shall compute the temperature dependence of this pressure in Chapter 5. In particular in Sec. 5.1.5 we argue that such a negative pressure is microscopically induced by spatial overlaps of (anti)caloron peripheries described by an effective pure-gauge configuration (small, transient

holonomy shifts which can be interpreted as triggered by soft fundamental radiative corrections) while a density of screened monopoles
emerges due to scattering of effective gauge modes in pointlike effective vertices (large holonomy shifts by deformations of (anti)caloron
cores and subsequent (anti)caloron dissociation, now interpretable
as hard fundamental radiative corrections).

Throughout Chapters 1 through 4 we have gathered all essential ingredients required for the construction of the effective theory for the high-temperature thermodynamics of SU(2) and SU(3)
Yang–Mills theory and for the interpretation of its results.

Problems

4.1. Verify Eq. (4.2). Why is the Chern–Simons current, defined in
Eq. (4.1), not invariant under gauge transformations?

4.2. Using the definitions in Eq. (4.6) and the convention $\epsilon_{1234} = 1$,
show that $\eta_{\mu\nu} = \frac{1}{2}\epsilon_{\mu\nu\alpha\beta}\eta_{\alpha\beta}$ and $\bar{\eta}_{\mu\nu} = -\frac{1}{2}\epsilon_{\mu\nu\alpha\beta}\bar{\eta}_{\alpha\beta}$, that is, the
selfduality of $\eta_{\mu\nu}$ and the antiselfduality of $\bar{\eta}_{\mu\nu}$.

4.3. Verify Eq. (4.7) using Eq. (4.5) and the definition of the 't Hooft
symbols $\eta_{\mu\nu}$ and $\bar{\eta}_{\mu\nu}$.

4.4. Check Eq. (4.9).

4.5. Using Eq. (4.13), verify Eq. (4.14).

4.6. Based on Eqs. (4.5) and (4.17) verify Eq. (4.18) by direct calculation [setting $R_b^a = \delta_b^a$ in Eqs. (4.18)].

4.7. Verify Eq. (4.19) by setting $R_b^a = \delta_b^a$ in Eqs. (4.18).

4.8. Derive Eq. (4.21) from Eqs. (4.20) and (1.33).

4.9. Check that the scalar function $W(x)$ given in Eq. (4.23) indeed
solves Eq. (4.22).

4.10. Appealing to Eq. (4.20), check Eq. (4.25).

4.11. Establish the selfduality of the gauge field defined in Eq. (4.30).
Hint: Use the condition that $\Delta^\dagger \Delta$ [$\Delta^\dagger$ is defined in Eq. (4.28)]
commutes with the quaternions and is invertible. Also, recall
Eq. (4.29). You should end up with Eq. (4.31).

4.12. Compute the topological charge as the four-dimensional integral over the divergence of the Chern–Simons-current [see

Eqs. (4.1) and (4.2)] for the selfdual configuration given by Eqs. (4.46) and (4.50) in the case of a compact dimension $0 \le x_4 \le \tilde{\beta}$. How does the three-dimensional topological current look like and how does its charge? *Hint:* After performing a trivial x_4 integration using Eq. (4.45), employ the three-dimensional form of Gauss' theorem to cast the integral over three-dimensional space of the divergence of the three-current into a two-dimensional surface integral.

4.13. Substituting ansätze (4.46) into the selfduality equation (4.47), derive the first-order equations of motion for the profile functions $K(r)$ and $H(r)$.

4.14. Show that the magnetic charge g_m of the configuration given by Eqs. (4.46) and (4.50) indeed is given as

$$g_m = \mp \frac{4\pi}{g}, \qquad (4.116)$$

where g is the gauge coupling of SU(2) Yang–Mills theory.

4.15. Perform the sum in Eq. (4.68) to obtain the expression of Eq. (4.69).

4.16. Show that the 2×2 matrices given in Eqs. (4.91) indeed solve the Weyl equation (4.87) subject to the Nahm data (4.83) away from the jump positions $t_{\text{I--II}}$ and $t_{\text{II--III}}$.

4.17. Check Eq. (4.93).

4.18. Show that in contrast to the first monopole at $\vec{x}_1$, represented by the solution of Eq. (4.99) [$-+$ combination in Eq. (4.50)], the gauge transformation induced by C_2 given in Eq. (4.103) near the position $\vec{x}_2$ of the second monopole generates the $+-$ combination in Eq. (4.50) and thus a reversed magnetic charge

References

Atiyah, M. F. and Singer, I. M. (1968a). The index of elliptic operators. 1., *Annals Math.*, **87**, 484.

Atiyah, M. F. and Singer, I. M. (1968b). The index of elliptic operators. 2., *Annals Math.*, **87**, 531.

Atiyah, M. F. and Singer, I. M. (1968c). The index of elliptic operators. 3., *Annals Math.*, **87**, 546.

Atiyah, M. F. and Singer, I. M. (1971a). The index of elliptic operators. 4., *Annals Math.*, **93**, 119.

Atiyah, M. F. and Singer, I. M. (1971b). The index of elliptic operators. 5., *Annals Math.*, **93**, 139.

Atiyah, M. F. and Ward, R. S. (1977). Instantons and algebraic geometry, *Phys. Lett. A*, **61**, 81.

Atiyah, M. F., *et al.* (1978). Construction of instantons, *Phys. Lett. A*, **65**, 185.

Atiyah, M. F. and Singer, I. M. (1984). Dirac operator coupled to vector potentials, *Proc. Nat. Acad. Sci.*, **81**, 2597.

Adler, S. L. (1969). Axial vector vertex in spinor electrodynamics, *Phys. Rev.*, **177**, 2426.

Adler, S. L. and Bardeen, W. A. (1969). Absence of higher order corrections in the anomalous axial vector divergence equation, *Phys. Rev.*, **182**, 1517.

Adler, S. (1978). Theory of static quark forces, *Phys. Rev. D*, **18**, 411.

Bell, J. S. and Jackiw, R. (1969). A PCAC puzzle: $\pi^0 \to \gamma\gamma$ in the sigma model, *Nuovo Cim. A*, **60**, 47.

Belavin, A. A., *et al.* (1975). Pseudoparticle solution of the Yang–Mills equations, *Phys. Lett. B*, **59**, 85.

Bogomoln'yi, E. B. (1976). The stability of classical solutions, *Sov. J. Nucl. Phys.*, **24**, 449.

Braam, P. J. and Van Baal, P. (1989). Nahm's transformation for instantons, *Commun. Math. Phys.*, **122**, 267.

Chern, S.-S. and Simons, J. (1974). Characteristic forms and geometric invariants, *Annals Math.*, **99**, 48.

Corrigan, E., *et al.* (1978). A Green's function for the selfdual gauge field, *Nucl. Phys. B*, **140**, 31.

Corrigan, E. and Goddard, P. (1984). Construction of instanton and monopole solutions and reciprocity, *Annals Phys.*, **154**, 253.

Christ, N. H., Stanton, N. K., and Weinberg, E. J. (1978). General selfdual Yang–Mills solutions, *Phys. Rev. D*, **18**, 2013.

Donaldson, S. K. and Kronheimer, P. B. (1990). *The Geometry of Four-Manifolds* (Clarendon Press, Oxford).

Drinfeld, V. G. and Manin, Y. I. (1978). A description of instantons, *Comm. Math. Phys.*, **63**, 177.

Diakonov, D., *et al.* (2004). Quantum weights of dyons and of instantons with trivial holonomy, *Phys. Rev. D*, **70**, 036003.

Fujikawa, K. (1979). Path integral measure for gauge invariant fermion theories, *Phys. Rev. Lett.*, **42**, 1195.

Fujikawa, K. (1980). Path integral for gauge theories with fermions, *Phys. Rev. D*, **21**, 2848; Erratum-*ibid. Phys. Rev. D*, **22**, 1499.

Gross, D. J., Pisarski, R. D., and Yaffe, L. G. (1981). QCD and instantons at finite temperature, *Rev. Mod. Phys.*, **53**, 43.

Grandou, T. and Hofmann, R. (2015). Thermal ground state and nonthermal probes, *Adv. Math. Phys.* **2015**, 197197.

Harrington, B. J. and Shepard, H. K. (1978). Periodic Euclidean solutions and the finite-temperature Yang–Mills gas, *Phys. Rev. D*, **17**, 2122.

Jackiw, R. and Rebbi, C. (1976). Conformal properties of a Yang–Mills pseudoparticle, *Phys. Rev. D*, **14**, 517.

Julia, B. and Zee, A. (1975). Poles with both magnetic and electric charges in non-abelian gauge theory, *Phys. Rev. D*, **11**, 2227.

Kraan, T. C. and Van Baal, P. (1998a). Exact T-duality between calorons and Taub-NUT spaces, *Phys. Lett. B*, **428**, 268.

Kraan, T. C. and Van Baal, P. (1998b). Periodic instantons with non-trivial holonomy, *Nucl. Phys. B*, **533**, 627.

Lee, K. and Lu, C. (1998). SU(2) calorons and magnetic monopoles, *Phys. Rev. D*, **58**, 025011-1.

Manton, N. (1978). Complex structure of monopoles, *Nucl. Phys. B*, **135**, 319.

Müller, K. E. (1986). Lorentz boost. Construction of static multidyon solutions in the Prasad–Sommerfield limit, *Phys. Lett. B*, **177**, 389.

Nahm, W. (1980). A simple formalism for the BPS monopole, *Phys. Lett. B*, **90**, 413.

Nahm, W. (1981). All self-dual multimonopoles for arbitrary gauge groups, CERN preprint TH-3172.

Nahm, W. (1982). *The Construction of all Self-dual Multimonopoles by the ADHM Method* in *Monopoles in Quantum Field Teory*, ed. N. Craigie *et al.* (World Scientific, Singapore), p. 87.

Nahm, W. (1983). Self-dual monopoles and calorons in *Trieste Group Theor. Method 1983*, p. 189.

Polyakov, A. M. (1974). Particle spectrum in the quantum field theory, *JETP Lett.*, **20**, 194.

Prasad, M. K. and Sommerfield, C. M. (1975). An exact solution for the 't Hooft monopole and the Julia–Zee dyon, *Phys. Rev. Lett.*, **35**, 760.

Rossi, P. (1979). Propagation functions in the field of a monopole, *Nucl. Phys. B*, **149**, 170.

Shuryak, E. and Schaefer, T. (1996). Instantons in QCD, *Rev. Mod. Phys.*, **70**, 323, (1998).

't Hooft, G. (1974). Magnetic monopoles in unified gauge theories, *Nucl. Phys. B*, **79**, 276.

't Hooft, G. (1976a). Symmetry breaking through Bell–Jackiw anomalies, *Phys. Rev. Lett.*, **37**, 8.

't Hooft, G. (1976b). Computation of the quantum effects due to a four-dimensional pseudoparticle, *Phys. Rev. D*, **14**, 3432; Erratum-*ibid. Phys. Rev. D*, **18**, 2199.

Van Baal, P. (1996). Instanton moduli for $T^3 \times \mathbf{R}$, *Nucl. Phys. B*, **49**, 238.

Van Baal, P. (1998). Nahm gauge fields for the torus, *Phys. Lett. B*, **448**, 26.

Ward, R. S. (1977). On selfdual gauge fields, *Phys. Lett. A*, **61**, 81.

Chapter 5

The Deconfining Phase

In this chapter the effective theories for the deconfining phase of SU(2) and SU(3) Yang–Mills thermodynamics are derived and applied to the computation of radiative corrections in several quantities. Moreover, we address spatially inhomogeneous, deconfining thermodynamics in an adiabative approximation. We also give interpretations of microscopic concepts, such as the number densities of screened magnetic monopoles and their antimonopoles or the action of an (anti)caloron causing the emergence of the thermal ground state, in terms of the effective quantum dynamics. The driving force behind this development is the fact that perturbation theory and the semiclassical approximation, both operating *a priori* with infinitely resolved fundamental gauge-field configurations, are incapable of saturating the partion function in an essential way: The dynamics of highly resolved degrees of freedom in Yang–Mills theory is too complex to be exhausted by these methods.

The principles of infinite-volume thermodynamics, the existence of certain topologically nontrivial, selfdual configurations on $S_1 \times \mathbf{R}^3$, principles of the renormalization group, and the perturbative renormalizability to any order in the power of the coupling constant, when applied in a combined way, yield an effective theory for the deconfining phase which is conceptually and technically simple. Complications related to the "nervosities" in both the infrared and the ultraviolet sector are integrated into simple configurations of effective fields which form the starting point for an effective loop expansion. The latter seem to exhibit healthy convergence properties.

At a given temperature, the separation between ultraviolet and infrared is induced by an emergent adjoint scalar field ϕ. The existence of ϕ expresses the effect of spatial correlations induced by calorons of trivial holonomy and of topological charge $k = \pm 1$ within a volume of linear dimension $\sim|\phi|^{-1}$. As it turns out, interactions between (anti)calorons facilitated by their spatial tails integrate into a pure-gauge configuration of the *effective* $k = 0$ sector. Momentum transfers $\leq |\phi|^2$ are explicitly taken into account in the effective theory in terms of radiative corrections. This and the fact that the field ϕ leads to the dynamical gauge-symmetry breaking SU(2)$\rightarrow$U(1) or SU(3)$\rightarrow$U(1)^2 (masses for effective off-Cartan fluctuations in a gauge where $\phi \equiv$ const.) imply a rapid convergence of effective loop expansions. The existence of a low-temperature boundary of the deconfining phase is signalled by the divergence of the effective gauge coupling e at T_c comparable to the dynamically generated mass scale Λ. The latter enters as a constant of integration in the derivation of the effective, thermal ground-state estimate.

The derivation of the effective theory for the deconfining phase and computations of radiative corrections in quantities like the polarization tensor for the massless mode in SU(2) or the two-loop and three-loop corrections to the pressure or 2-to-2 scattering of massless modes are performed in detail in this section.

5.1   Deconfining Thermal Ground State

5.1.1   *Two principles imposed by infinite volume Yang–Mills thermodynamics*

Since they impose important constraints onto the construction of an *a priori* estimate of the thermal ground state, we state two basic principles of infinite volume thermodynamics applied to a partial average involving certain gauge fields of vanishing energy–momentum:

(I) In the absence of external sources, a thermodynamical system in the infinite volume limit guarantees, up to admissible gauge transformations, the spatial isotropy and homogeneity of an *effective* local field if this field is *not* associated with the propagation of energy–momentum by *fundamental* gauge fields: The partial ensemble average in combination with a spatial average, leading to the emergence of such a field, yields a nonzero result if and only if this field is a rotational scalar ϕ, and ϕ does not depend on space up to admissible gauge transformations.

(II) Up to admissible gauge transformations the field ϕ must not depend on time.

Principle (I) is selfevident. Principle (II) is a consequence of the fact that ϕ does not possess energy–momentum at any time.

5.1.2   *Coarse-graining and BPS saturation*

In SU(2) and SU(3) Yang–Mills theory a simple description of quantum thermodynamics emerges by subjecting selfdual, nonpropagating (BPS-saturated = selfdual), fundamental field configurations to a spatial coarse-graining procedure. In a step-wise fashion, the derivation of a useful *a priori* estimate for the thermal ground state exploits the nonpropagating nature of selfdual gauge-field configurations[1]: The effective scalar field ϕ emerges upon spatial coarse-graining over the centers of these selfdual configurations and inherits their nonpropagating nature. *Subsequently* infrared effects due to overlapping peripheries (topologically trivial) are cast into an effective pure-gauge configuration a_μ^{gs}. The latter solves the effective Yang–Mills equations subject to a source term provided by the effective, inert field ϕ (see Fig. 5.1). This concept of first integrating out topologically nontrivial configurations in combination with a spatial coarse-graining does not work in field theories with global symmetries because the subsequent coarse-graining process over the topologically trivial sector is much less restricted than in gauge theories.

In SU(2) and SU(3) Yang–Mills theory, the configuration a_μ^{gs} lifts the vanishing *a priori* energy density, associated with the

---

[1]Recall that the Euclidean energy–momentum tensor vanishes identically on selfdual configurations [see Eq. (1.38)].

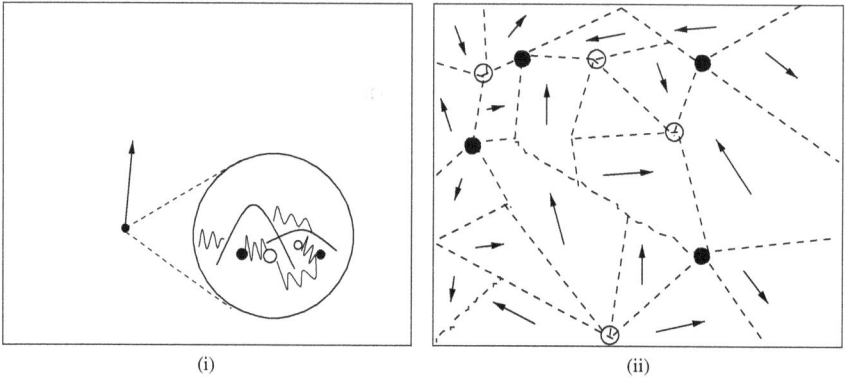

(i) (ii)

Figure 5.1. Spatial coarse-graining over interacting calorons. The *a priori* estimate for the thermal ground state is obtained by averaging over trivial-holonomy calorons of $k = \pm1$ (indicated within the spherical coarse-graining volume) to generate the inert field ϕ and by subsequently solving the effective Yang–Mills equations in ϕ's background to yield the pure-gauge configuration a_μ^{gs}. The arrow in (i) indicates the existence of a gauge where ϕ globally points into one direction of su(2). A typical, spatially domainized configuration (ii) of ϕ emerges upon the inclusion of collective infrared interactions between calorons. If in a given spatial point four or more domains meet then this point is identified as the location of a magnetic monopole [Kibble (1976)] liberated by the dissociation of a caloron deformed to large holonomy. A properly weighted average over all domainizations is obtained in the effective theory by loop-expanding the quantity under investigation (for example the pressure, the energy density, the entropy density, etc.) about the *a priori* estimate of the ground state represented by ϕ and a_μ^{gs}.

nonpropagating effective field ϕ, to a finite value. The reason why ultraviolet fluctuations of $k = 0$ can act as strong holonomy energy-density shifters without deforming (and thus destroying) the nonpropagating effective background ϕ is their nonresolvability at the resolution set by this very background. On the other hand, explicit interactions of the coarse-grained $k = 0$ sector with the field ϕ in the effective theory cannot transfer any energy–momentum to this nonpropagating field.

5.1.3   Inert adjoint scalar field: Phase

Let us first specialize to the gauge group SU(2). The two principles discussed in Sec. 5.1.1 together with the nonpropagating nature of ϕ leave room only for a scalar field of constant modulus describing ground-state thermodynamics: Any other inert Lorentz tensor would break isotropy, and any spacetime dependence of the gauge-invariant modulus $|\phi|$ would break spacetime homogeneity. The scalar field ϕ can consistently participate in the thermodynamics of the entire Yang–Mills system only if it couples to the effective, propagating excitations residing in the effective $k = 0$ sector.[2] Since perturbative renormalizability assures that before interaction with the topologically nontrivial sector the effective $k = 0$ modes are wave-function-renormalized versions of the fundamental $k = 0$ modes and since the effective (as well as the fundamental) $k = 0$ modes thus are in the adjoint representation of the Lie algebra gauge invariance of the effective action, implies ϕ to be in the adjoint representation as well. Finally, since ϕ is a scalar field it must transform homogeneously under gauge transformations.

Spatial coarse-graining over noninteracting calorons of trivial holonomy is performed in two steps. First, one defines and evaluates the kernel of a differential operator $\mathcal{D}$ in terms of a family $\{\hat{\phi}\}$ of adjointly transforming phases. It turns out that the operator $\mathcal{D}$ is linear, of second order, and that it is uniquely determined by $\{\hat{\phi}\}$. Appealing to $|\phi|$'s constancy, the linearity of $\mathcal{D}$ implies that it annihilates the entire field ϕ. The consistency of ϕ's BPS saturation and it solving $\mathcal{D}\phi = 0$ requires a first-order equation for its potential V. The solution to this first-order equation introduces an

---

[2]Assume that this was not the case. Then to any order of the loop expansion of the pressure no mass would emerge via gauge symmetry breaking induced by ϕ, the ground state pressure due to ϕ would exactly vanish, and we would be back at the perturbative loop expansion.

undetermined constant of integration of mass dimension one: the Yang–Mills scale Λ.

What are the options for defining $\{\hat{\phi}\}$ *locally*? Since in a pure Yang–Mills theory the most basic quantity transforming homogeneously under the adjoint representation is the field strength $F_{\mu\nu}$ a local definition would invoke a polynomial in $F_{\mu\nu}$ where the various powers are represented as[3,4]

$$\text{tr}\, t^a F_{\mu\nu} F_{\nu\kappa} F_{\kappa\mu}, \quad \text{tr}\, t^a F_{\mu\nu} F_{\nu\kappa} F_{\kappa\rho} F_{\rho\mu}, \quad \text{etc.} \tag{5.1}$$

or

$$\text{tr}\, \epsilon^{abc} F_{\mu\nu} t^b F_{\nu\kappa} t^c F_{\kappa\mu}, \quad \text{etc.} \tag{5.2}$$

One can show that due to $F_{\mu\nu} \equiv \pm\tilde{F}_{\mu\nu}$ any such locally defined power of $F_{\mu\nu}$ *vanishes identically* [Herbst (2005)].

But if a local definition of $\{\hat{\phi}\}$ is impossible, what are the principles to be honored in a nonlocal construction?

(i) Principles I and II of Sec. 5.1.1.
(ii) The emergence of ϕ is due to an incomplete execution of the partition function by integrating out selfdual configurations only. Since the latter introduce no T dependence into the functional weight (the classical action is $8\pi^2|k|/g^2$) *no explicit T dependence* is allowed to occur in the definition of $\{\hat{\phi}\}$.
(iii) Since $\{\hat{\phi}\}$ is a family of dimensionless, periodic τ dependences and since on the level of selfdual gauge fields no reference is made to any external spatial scale, any integration over moduli of the selfdual configurations in the definition of $\{\hat{\phi}\}$ must come with a *flat* measure. This is also true of the integration over spatial point separations. Moreover, for the same reason Wilson

---

[3]From now on we use hermitian generators normalized as $\text{tr}\, t^a t^b = \frac{1}{2}\delta^{ab}$.
[4]Since a scalar quantitiy is to be defined, covariant derivatives can either be reduced to powers of $F_{\mu\nu}$ or to zero by virtue of the Bianchi identities, the Yang–Mills equations, or the antisymmetry of $F_{\mu\nu}$.

lines between spatial points are along straight lines (no spatial scale to introduce a curvature).

(iv) Since $\{\hat{\phi}\}$ consists of gauge noninvariant functions that are non-trivially periodic in τ, integrations over their gauge orbits and over τ are forbidden.

(v) The isolated selfdual gauge-field configurations that spatial coarse-graining is based upon must be stable under quantum fluctuations. They also must have nonvanishing quantum weight (semiclassical approximation) in the infinite-volume limit in order to contribute to the partition function (trivial holonomy only).

We claim that, in obeying (i)–(v), the following is a unique definition for $\{\hat{\phi}\}$ in su(2) coordinates [Herbst and Hofmann (2004); Hofmann (2005, 2007)]:

$$\{\hat{\phi}^a\} \equiv \sum_{C,A} \mathrm{tr} \int d^3x \int d\rho\, t^a\, F_{\mu\nu}(\tau,\vec{0})\, \{(\tau,\vec{0}),(\tau,\vec{x})\}$$

$$\times F_{\mu\nu}(\tau,\vec{x})\{(\tau,\vec{x}),(\tau,\vec{0})\}, \tag{5.3}$$

where

$$\{(\tau,\vec{0}),(\tau,\vec{x})\} \equiv \mathcal{P} \exp\left[i \int_{(\tau,\vec{0})}^{(\tau,\vec{x})} dz_\mu\, A_\mu(z) \right], \tag{5.4}$$

$$\{(\tau,\vec{x}),(\tau,\vec{0})\} \equiv \{(\tau,\vec{0}),(\tau,\vec{x})\}^\dagger.$$

The Wilson lines in Eq. (5.4) are calculated along the straight spatial line connecting the points $(\tau,\vec{0})$ and $(\tau,\vec{x})$, and $\mathcal{P}$ demands path-ordering. In (5.3) the sum is over the $|k| = 1$ caloron (C) and anticaloron (A) of trivial holonomy (see Sec. 4.3.1), and ρ denotes their instanton scale parameter. Both caloron and anticaloron are spatially centered at $\vec{z}_{C,A} = 0$. On a caloron of trivial holonomy, with the prepotential Π defined in Eq. (4.69), the exponent in Eq. (5.4) reads

$$\int_{(\tau,\vec{0})}^{(\tau,\vec{x})} dz_\mu A_\mu(z)\big|_{C,A} = \pm \int_0^1 ds\, x_i A_i(\tau, s\vec{x})$$

$$= \pm t_b x_b\, \partial_\tau \int_0^1 ds\, \log \Pi(\tau, sr, \rho). \qquad (5.5)$$

Thus the integrand in the exponent of $\{(\tau,\vec{0}),(\tau,\vec{x})\}_{C,A}$ varies along a fixed direction in su(2), and path-ordering can be ignored.

Conditions (i) through (v) are honored by (5.3), and the right-hand side of (5.3) indeed transforms locally and homogeneously. Namely, we have

$$\{(\tau,\vec{0}),(\tau,\vec{x})\} \to \Omega(\tau,\vec{0})\, \{(\tau,\vec{0}),(\tau,\vec{x})\}\, \Omega^\dagger(\tau,\vec{x}),$$

$$\{(\tau,\vec{x}),(\tau,\vec{0})\} \to \Omega(\tau,\vec{x})\, \{(\tau,\vec{x}),(\tau,\vec{0})\}\, \Omega^\dagger(\tau,\vec{0}), \qquad (5.6)$$

$$F_{\mu\nu}(\tau,\vec{x}) \to \Omega(\tau,\vec{x})\, F_{\mu\nu}(\tau,\vec{x})\, \Omega^\dagger(\tau,\vec{x}),$$

and thus

$$\hat{\phi}^a(\tau) \to R^{ab}(\tau)\hat{\phi}^b(\tau), \qquad (5.7)$$

where R^{ab} is the SO(3) matrix

$$R^{ab}(\tau)t^b = \Omega^\dagger(\tau,\vec{0})t^a\Omega(\tau,\vec{0}). \qquad (5.8)$$

Thus all members of $\{\hat{\phi}^a\}$, as defined by (5.3), transform under a time-dependent gauge transformation. That is, the spatial dependence of the microscopic gauge transformation is lost due to spatial coarse-graining.

We are now in a position to explain why $\vec{z}_{C,A} = 0$ in (5.3). Notice, that an extra integration over $\vec{z}_{C,A}$ would need an *explicit* power of β^{-3} to render $\{\hat{\phi}^a\}$ dimensionless which is forbidden by (ii). Thus $\vec{z}_{C,A} \neq 0$ would have to be fixed. But to not violate spatial isotropy the only option is to set $\vec{z}_{C,A} = 0$. Therefore definition (5.3) can be used without restriction of generality. The only modulus of the caloron that may be integrated over is ρ. According to (iii) this integration is subject to a flat measure.

What about calorons with $|k| > 1$ and n-point functions with $n > 2$ as integrands in possible extensions of (5.3)? Besides the translational moduli there are $m > 1$ parameters of dimension length in such configurations. For example, a caloron of trivial holonomy with $k = 2$ has three dimensionful moduli: two scale parameters and the distance between the two locations of its topological charge. Considering an n-point function (n nonlocal factors of the field strength $F_{\mu\nu}$) with $n - 1$ integrations over space and flat-measure integrations over the above m parameters of the caloron, we arrive at a mass dimension $2n - 3(n - 1) - m = 3 - n - m$ of the entire object. To avoid the introduction of explicit powers of β (BPS saturation) in the definition of $\{\hat{\phi}^a\}$ this mass dimension would need to vanish. But for $n \geq 2$ and/or $m > 1$ we have $3 - n - m \neq 0$. Thus higher topological charge as well as higher n-point functions are excluded in the definition of $\{\hat{\phi}\}$, and (5.3) is unique modulo global gauge rotations.

After we have assured that definition (5.3) applies to define the kernel of a differential operator $\mathcal{D}$, which is associated with the phase of the sought-after adjoint scalar field ϕ, we have to perform a lengthy calculation to learn what $\mathcal{D}$ is. Should it turn out that $\mathcal{D}$ is *linear* then this operator would annihilate ϕ together with its phase, and we would be in possession of an equation of motion for ϕ. As we shall see, this turns out to be the case: "Subtle is the Lord but malicious He is not." Being consistent with the fact that ϕ is of vanishing energy density (BPS saturation), this equation of motion finally yields a potential $V(\phi)$, parameterized by a (multiplicative) constant of integration. Assigning a definite value to the latter, in turn, fixes the effective action for the field ϕ.

In explaining the steps of the calculation we closely follow [Hofmann (2007)]. We set temporal shifts of the caloron/anticaloron center equal to zero: $\tau_{C,A} = 0$. Eventually, they are re-instated by letting $\tau \rightarrow \tau + \tau_{C,A}$. Furthermore, we scale all quantities of dimension length to be dimensionless by pulling out a power of β. Those quantities are indicated by a hat symbol. After these conventions are made the Wilson lines $\{(\tau,\vec{0}),(\tau,\vec{x})\}_{C,A}$ evaluate as

$$\{(\tau,\vec{0}),(\tau,\vec{x})\}_{C,A} = \cos g \pm 2it_b \frac{x^b}{r}\sin g, \tag{5.9}$$

where $g = g(\tau,r,\rho) = g(\beta\hat{\tau},\beta\hat{r},\beta\hat{\rho}) \equiv \hat{g}(\hat{\tau},\hat{r},\hat{\rho}) \equiv \int_0^1 ds\,\frac{r}{2}\partial_\tau \log \Pi(\tau, sr,\rho)$. The $+$ and $-$ sign in Eqs. (5.9) relate to the caloron and the anticaloron, respectively. Explicitly, one has

$$\hat{g} = -\pi^2\hat{\rho}^2 \sin(2\pi\hat{\tau}) \int_0^1 ds$$

$$\frac{1}{s}\frac{\sinh(2\pi\hat{r}s)}{[\cosh(2\pi\hat{r}s) - \cos(2\pi\hat{\tau})][\cosh(2\pi\hat{r}s) - \cos(2\pi\hat{\tau}) + \frac{\pi\hat{\rho}^2}{\hat{r}s}\sinh(2\pi\hat{r}s)]}. \tag{5.10}$$

Questions arise as to whether the integral in Eq. (5.10) exists and, if yes, how fast this integral saturates as $\hat{r} > 0$ increases. To address the first question, the potentially problematic point in the domain of integration is $s = 0$ for $\hat{\tau} = k \in \mathbf{Z}$. By Taylor-expanding the sine function in front of the integral in Eq. (5.10) about $\hat{\tau} = k$ and by Taylor-expanding the cosine and the hyperbolic cosine functions in the denominator of the integrand about $\hat{\tau} = k$ and $s = 0$, respectively, one easily checks that the limit $\hat{\tau} \to k$ exists for $\hat{\rho} \geq 0$ and $\hat{r} \geq 0$. As for the saturation property of $\hat{g}$ with increasing $\hat{r}$ one may, for $\hat{r} > \frac{1}{2\pi}$, split the integral in Eq. (5.10) as $\mathcal{I} \equiv \int_0^1 ds \equiv \mathcal{I}_1 + \mathcal{I}_2 \equiv \int_0^{\frac{1}{2\pi\hat{r}}} ds + \int_{\frac{1}{2\pi\hat{r}}}^1 ds = \int_0^1 dz + \int_1^{2\pi\hat{r}} dz$. For the z integration the integrand I is given as

$$I(z,\hat{\rho},\hat{\tau}) \equiv \frac{\sinh z}{z[\cosh z - \cos(2\pi\hat{\tau})][\cosh z - \cos(2\pi\hat{\tau}) + \frac{2(\pi\hat{\rho})^2}{z}\sinh z]}. \tag{5.11}$$

Thus $\mathcal{I}_1$ does not depend on $\hat{r}$. For the z integration in $\mathcal{I}_2$ the integrand I is bounded from above as

$$I(z,\hat{\rho},\hat{\tau}) \leq \frac{2e^z}{(e^z - 2\cos(2\pi\hat{\tau}))^2}, \qquad \forall\hat{\tau},\hat{\rho}\,;z \geq 1. \tag{5.12}$$

Since $\int_1^{2\pi\hat{r}} dz\, I(z, \hat{\rho}, \hat{\tau}) = \int_1^\infty dz\, I(z, \hat{\rho}, \hat{\tau}) - \int_{2\pi\hat{r}}^\infty dz\, I(z, \hat{\rho}, \hat{\tau})$ and since, by virtue of Eq. (5.12), the modulus of the second summand is bounded by $\frac{2}{e^{2\pi\hat{r}} - 2\cos(2\pi\hat{\tau})}$, we are assured a more than exponentially fast saturation in $\hat{r}$. Numerically, $\frac{\int_{2\pi\hat{r}}^\infty dz\, I(z,\hat{\rho},\hat{\tau})}{\mathcal{I}} < 10^{-5}$ for $\hat{r} > 2$ independently of $\hat{\rho}$. Saturation in the limit $\hat{\rho} \to \infty$ is easily checked.

A lengthy calculation [Herbst and Hofmann (2004); Herbst (2005)] shows that the caloron part of the integrand on the right-hand side of (5.3) yields the following expression

$$-i\,\beta^{-2}\frac{32\pi^4}{3}\frac{x^a}{r}\frac{\pi^2\hat{\rho}^4 + \hat{\rho}^2(2 + \cos(2\pi\hat{\tau}))}{(2\pi^2\hat{\rho}^2 + 1 - \cos(2\pi\hat{\tau}))^2} \times F[\hat{g}, \Pi], \qquad (5.13)$$

where the functional F is given as

$$F[\hat{g}, \Pi] = 2\cos(2\hat{g})\left(2\frac{[\partial_\tau \Pi][\partial_r \Pi]}{\Pi^2} - \frac{\partial_\tau \partial_r \Pi}{\Pi}\right)$$

$$+ \sin(2\hat{g})\left(2\frac{[\partial_r \Pi]^2}{\Pi^2} - 2\frac{[\partial_\tau \Pi]^2}{\Pi^2} + \frac{\partial_\tau^2 \Pi}{\Pi} - \frac{\partial_r^2 \Pi}{\Pi}\right). \qquad (5.14)$$

It is easily checked that $F_{\mu\nu}(\tau, \vec{x})_C = F_{\mu\nu}(\tau, -\vec{x})_A$ and that

$$\{(\tau, \vec{0}), (\tau, \vec{x})\}_C = (\{(\tau, \vec{x}), (\tau, \vec{0})\}_C)^\dagger = \{(\tau, \vec{0}), (\tau, -\vec{x})\}_A$$

$$= (\{(\tau, -\vec{x}), (\tau, \vec{0})\}_A)^\dagger. \qquad (5.15)$$

Thus it follows that the anticaloron contribution to the integrand in (5.3) can be obtained by letting $\vec{x} \to -\vec{x}$ in the caloron contribution of Eq. (5.13). Recall, that eventually, the shift $\tau \to \tau + \tau_{C,A}$ needs to be performed in the respective contribution to the total integrand to re-instate the dependence of the integrals in (5.3) on $\tau_{C,A}$.

Due to the appearance of the factor $\frac{x^a}{r}$ in the expression (5.13) an unconstrained angular integration[5] in Eq. (5.3) would yield zero. Thus for the final integration over $\hat{\rho}$ to possess a potentially nonvanishing integrand the radial integral must diverge. But the only term

---

[5]In the spatial integration of (5.3) a transition to spherical coordinates is advantageous.

in the functional $F[\hat{g}, \Pi]$ of Eqs. (5.13) and (5.14) that gives rise to a divergence of the radial integral, is

$$-\sin(2\hat{g})\frac{\partial_r^2 \Pi}{\Pi}, \qquad (5.16)$$

and this divergence is logarithmic. Since only spatial derivatives are involved the term (5.16) arises from magnetic-magnetic correlations. Recall from Sec. 3.2 that it is the magnetic sector whose insufficient screening gives rise to the poor convergence properties of thermal perturbation theory [Linde (1980)].

Let us now show that, indeed, only term (5.16) induces a divergence to the radial integral in (5.3). First of all, no divergence arises for $\hat{r} \to 0$. For $\hat{r} \gg 1$ we have

$$\partial_\tau \Pi(\tau, r) = \beta^{-1} \partial_{\hat{\tau}} \hat{\Pi}(\hat{\tau}, \hat{r}) \xrightarrow{\hat{r} \gg 1} -\frac{(2\pi\hat{\rho})^2}{\beta\hat{r}} \sin(2\pi\hat{\tau}) \exp(-2\pi\hat{r}),$$

$$(5.17)$$

$$\begin{aligned}
\partial_\tau^2 \Pi(\tau, r) &= \beta^{-2} \partial_{\hat{\tau}}^2 \hat{\Pi}(\hat{\tau}, \hat{r}) \\
&\xrightarrow{\hat{r} \gg 1} \frac{2\pi}{\beta} \frac{(2\pi\hat{\rho})^2}{\beta\hat{r}} (4 \sin(2\pi\hat{\tau}) \exp(-4\pi\hat{r}) \\
&\quad - \cos(2\pi\hat{\tau}) \exp(-2\pi\hat{r})).
\end{aligned} \qquad (5.18)$$

Because of the exponential suppression in Eqs. (5.17) and (5.18) all terms in $F[\hat{g}, \Pi]$ (see Eq. (5.14)) containing $\partial_\tau \Pi$ or $\partial_\tau^2 \Pi$ give rise to finite contributions to the radial integral. This also holds for the term with $(\partial_r \Pi)^2$ since

$$\Pi(\tau, r) \equiv \hat{\Pi}(\hat{\tau}, \hat{r}) \xrightarrow{\hat{r} \gg 1} 1 + \frac{\pi\hat{\rho}^2}{\hat{r}} \Rightarrow (\partial_r \Pi(\tau, r))^2 \xrightarrow{\hat{r} \gg 1} \beta^{-2} \frac{\pi^2 \hat{\rho}^4}{\hat{r}^4},$$

$$(5.19)$$

and the integration measure is only quadratic in $\hat{r}$: $d\hat{r}\, \hat{r}^2$. But

$$\partial_r^2 \Pi(\tau, r) \xrightarrow{\hat{r} \gg 1} \beta^{-2} \frac{2\pi\hat{\rho}^2}{\hat{r}^3}, \qquad (5.20)$$

and thus

$$-\int_0^{\hat{R}'} dr\, r^2 \sin(2g)\frac{\partial_r^2 \Pi}{\Pi} \sim -\beta\left(\text{finite} + 2\pi\hat{\rho}^2(\lim_{\hat{r}\to\infty}\sin(2\hat{g}))\int_{\hat{R}}^{\hat{R}'}\frac{d\hat{r}}{\hat{r}}\right),$$

(5.21)

where the $\sim$ sign indicates that the right-hand side approaches the left-hand side more than exponentially fast when increasing $\hat{R} > 1$ and $\hat{R}' \gg \hat{R}$. Obviously, the integral in Eq. (5.21) diverges logarithmically when sending $\hat{R}' \to \infty$.

To summarize, the contribution to $\{\hat{\phi}^a\}$ arising from the caloron is, for $\hat{R}$ sufficiently large, given as

$$\lim_{\hat{R}'\to\infty,\eta'\to 0} i\frac{64\pi^5}{3}\int d\hat{\rho}\,\hat{\rho}^2$$

$$\times\frac{\pi^2\hat{\rho}^4 + \hat{\rho}^2(2 + \cos(2\pi\hat{\tau}))}{(2\pi^2\hat{\rho}^2 + 1 - \cos(2\pi\hat{\tau}))^2}\int d\Omega\frac{x^a}{r}\int_{\hat{R}}^{\hat{R}'}\frac{d\hat{r}}{\hat{r}}\sin(2\hat{g}),$$

(5.22)

where the angular integration in the expression (5.22),

$$\int d\Omega\frac{x^a}{r} = \int_{-1}^{+1}d(\cos\theta)\int_{\alpha_C}^{\alpha_C+2\pi}d\varphi\frac{x^a}{r},\quad (0 \le \alpha_C \le 2\pi),$$

(5.23)

is regularized by introducing a defect/surplus angle $\eta' \ll 1$ for the azimuthal integration in φ: $\alpha_C \to \alpha_C \pm \eta'$ (lower integration limit) and $\alpha_C \to \alpha_C \mp \eta'$ (upper integration limit). This singles out a unit vector $\hat{n}_C \equiv (\cos\alpha_C, \sin\alpha_C, 0)$. Without restriction of generality the angular integral for the anticaloron is also regularized in the x_1x_2-plane[6] at angle α_A. Re-instating $\tau_{C,A}$, where $0 \le \hat{\tau}_C, \hat{\tau}_A \le 1$, we thus

---

[6]A regularization in any other plane would have singled out a unit vector $\hat{n}_A$ which, together with $\hat{n}_C$ would again have spanned a plane.

arrive at

$$\{\hat{\phi}^a\} = \{\Xi_C(\delta^{a1}\cos\alpha_C + \delta^{a2}\sin\alpha_C)\,\mathcal{A}\left(2\pi(\hat{\tau} + \hat{\tau}_C)\right)$$
$$+ \Xi_A(\delta^{a1}\cos\alpha_A + \delta^{a2}\sin\alpha_A)\,\mathcal{A}\left(2\pi(\hat{\tau} + \hat{\tau}_A)\right)\}, \quad (5.24)$$

where the coefficients $\Xi_C, \Xi_A \in \mathbf{R}$ are undetermined and independent of each other. They arise because logarithmically divergent radial integrals[7] multiply vanishing angular integrals, and because the according regularizations in (5.22) are independent of each other. Once $\hat{n}_C$ and $\hat{n}_A$ are fixed, the axis normal to the plane they span — in our case the x_3-axis — can be rotated by a rotation of the cartesian coordinates in which the transition to polar coordinates is performed. But Eq. (5.24) tells us that for $\{\hat{\phi}^a\}$ this amounts to nothing but a global gauge rotation. Thus no breaking of rotational symmetry is introduced by the angular regularizations.

The function $\mathcal{A}(2\pi\hat{\tau})$ in Eq. (5.24) is well approached by

$$\mathcal{A}(2\pi\hat{\tau}) \sim \frac{32\pi^7}{3} \int_0^\xi d\hat{\rho}\,\hat{\rho}^4 \left[\lim_{\hat{r}\to\infty}\sin\left(2\hat{g}(\hat{\tau}, \hat{r}, \hat{\rho})\right)\right]$$
$$\times \frac{\pi^2\hat{\rho}^2 + \cos(2\pi\hat{\tau}) + 2}{\left(2\pi^2\hat{\rho}^2 - \cos(2\pi\hat{\tau}) + 1\right)^2} \quad (\xi \gg 1). \quad (5.25)$$

Since $\hat{g}$ approaches a finite limit as $\hat{\rho} \to \infty$, the integral over $\hat{\rho}$ in Eq. (5.25) diverges cubically for $\xi \to \infty$, and the function $\mathcal{A}(2\pi\hat{\tau})$ rapidly approaches $const_\infty \times \xi^3 \sin(2\pi\hat{\tau})$. Numerically, one obtains $const_\infty = 272.018$.

Already for $\xi = 3$ one has $\frac{const_3 - const_\infty}{const_\infty} = 0.025$, and the functional dependence on $\hat{\tau}$ practically is a sine (see Fig. 5.2). Since there is such a fast saturation towards a sine function the prefactor $272.018 \times \xi^3$ can be absorbed into the undetermined, real numbers $\Xi_{C,A}$ in Eq. (5.24). In this sense the result for $\{\hat{\phi}^a\}$ in Eq. (5.24) becomes independent of the cutoff ξ if ξ is sufficiently large (see

---

[7]Ξ_C, Ξ_A can be made real by a dimensional smearing (see [Herbst and Hofmann (2004)]).

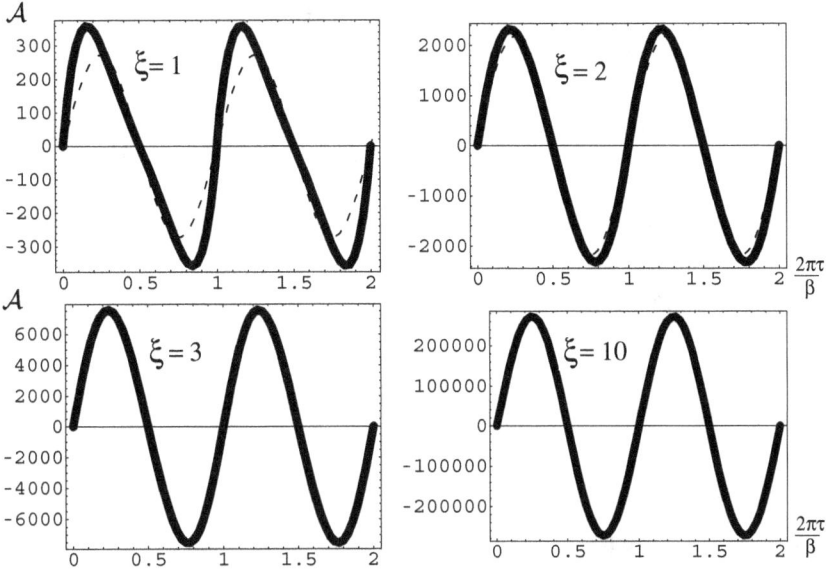

Figure 5.2. The function $\mathcal{A}(\frac{2\pi\tau}{\beta})$ plotted over two periods with different values of ξ. For comparison the function $272\xi^3 \sin(\frac{2\pi\tau}{\beta})$ is plotted as a dashed line.

again Fig. 5.2). On the other hand, because of the ξ^3 dependence of the normalization of $\mathcal{A}(2\pi\hat{\tau})$ in Eq. (5.25) it is clear that the integration is dominated by a small interval including the cutoff ξ. Thus, whatever this cutoff ξ turns out to be, the thermal ground state, partially described by the inert field ϕ, receives contributions from a small band of Harrington–Shepard (anti)caloron scale parameters only. This interval is centered at ξ.

Now, for sufficiently large ξ the set $\{\hat{\phi}^a\}$ represents a two-fold copy of the kernel of the linear differential operator $\mathcal{D} \equiv \partial_\tau^2 + \left(\frac{2\pi}{\beta}\right)^2$ on the space of real, smooth, and period-β functions of τ: Each of the two independent caloron and anticaloron "polarizations", determined by the angles α_C and α_A or, more generally, by the unit vectors $\hat{n}_C$ and $\hat{n}_A$, is annihilated by $\mathcal{D}$, and there are undetermined phase shifts $2\pi\hat{\tau}_{C,A}$ and undetermined amplitudes $|\Xi_{C,A}|$. Thus there are two real parameters for each "polarization" of $\hat{\phi}$. For each "polarization" these two real parameters span completely the kernel of $\mathcal{D}$ in the space of smooth and real functions of τ with period β. Therefore the operator $\mathcal{D}$ is determined uniquely.

5.1.4 *Inert adjoint scalar field: Modulus and potential*

The fact that differential operator $\mathcal{D}$ is linear and that $|\phi| \equiv \sqrt{\frac{1}{2}\,\mathrm{tr}\,\phi^2}$ does not depend on space and time (see Sec. 5.1.1), implies that $\mathcal{D}$ annihilates the entire field $\phi = |\phi|\hat{\phi}$. By virtue of the Euler–Lagrange equation

$$\partial_\tau \frac{\partial \mathcal{L}_\phi}{\partial(\partial_\tau \phi)} - \frac{\partial \mathcal{L}_\phi}{\partial \phi} = 0,$$

this, in turn, implies that the field ϕ possesses a canonic kinetic term $\mathrm{tr}\,(\partial_\tau \phi)^2$ in its effective, Euclidean Langrangian density $\mathcal{L}_\phi$. Moreover, the Euler–Lagrange equation for ϕ and the linearity of $\mathcal{D}$ together imply that the explicit β dependence in $\mathcal{D}$ is replaced by the ϕ-derivative of a potential $V(\phi^2)$. Recall that the selfduality of calorons implies the independence of their action density $\mathcal{L}_\phi$ on temperature. Upon coarse-graining this prohibits any explicit temperature dependence to appear in ϕ's effective action density $\mathcal{L}_\phi$ and, as a consequence, in its equation of motion. Thus the adjoint scalar field ϕ is subject to the Euclidean Lagrangian density

$$\mathcal{L}_\phi = \mathrm{tr}((\partial_\tau \phi)^2 + V(\phi^2)), \tag{5.26}$$

with no explicit β dependence on $V(\phi^2)$. We now determine $V(\phi^2)$. The Euler–Lagrange equations derivable from Eq. (5.26)[8] read [Giacosa and Hofmann (2006)]

$$\partial_\tau^2 \phi^a = \frac{\partial V(|\phi|^2)}{\partial |\phi|^2} \phi^a \ \text{(in components)}$$

---

[8]Notice our notational convention: V is either a scalar-valued function of its scalar-valued argument or a matrix-valued function of its matrix-valued argument. In both cases the functional dependence is identical.

$$\Leftrightarrow \partial_\tau^2 \phi = \frac{\partial V(\phi^2)}{\partial \phi^2} \phi \text{ (in matrix form).} \tag{5.27}$$

Since ϕ's motion is within a plane in the three-dimensional vector space su(2), since $|\phi|$ is independent of space and time, and since ϕ's phase $\hat{\phi}$ is of period β, in τ, one may, without restriction of generality,[9] write the solution to Eq. (5.27) as

$$\phi = 2 |\phi| t_1 \exp\left(\pm \frac{4\pi i}{\beta} t_3 \tau\right). \tag{5.28}$$

The fact that in Eq. (5.28) the motion takes place in the 1,2-plane of SU(2) amounts to a global choice. BPS saturation, or equivalently, the vanishing of the Euclidean energy density and Eq. (5.28) imply that

$$|\phi|^2 \left(\frac{2\pi}{\beta}\right)^2 - V(|\phi|^2) = 0. \tag{5.29}$$

Comparing Eq. (5.27) with $\partial_\tau^2 \phi + \left(\frac{2\pi}{\beta}\right)^2 \phi = 0$, we have

$$\left(\frac{2\pi}{\beta}\right)^2 = -\frac{\partial V(|\phi|^2)}{\partial |\phi|^2}. \tag{5.30}$$

Together, Eqs. (5.29) and (5.30) yield

$$\frac{\partial V(|\phi|^2)}{\partial |\phi|^2} = -\frac{V(|\phi|^2)}{|\phi|^2}. \tag{5.31}$$

Equation (5.31) is a first-order differential equation whose solution reads

$$V(|\phi|^2) = \frac{\Lambda^6}{|\phi|^2}, \tag{5.32}$$

---

[9]Modulo global changes of gauge any combination other than $\Xi_C = \Xi_A$ and $\hat{\tau}_C = \hat{\tau}_A \pm \frac{\pi}{2}$ in Eq. (5.24) (circular polarizations) would lead to a time-dependent modulus of ϕ which is in contradiction with principle II of Sec. 5.1.1.

where Λ denotes an arbitrary mass scale (the Yang–Mills scale). Equations (5.29) and (5.32) imply that

$$|\phi| = \sqrt{\frac{\Lambda^3 \beta}{2\pi}}. \tag{5.33}$$

Hence, modulo a global change of gauge, we have

$$\phi = 2\sqrt{\frac{\Lambda^3 \beta}{2\pi}}\, t_1 \, \exp\left(\pm\frac{4\pi i}{\beta}t_3\tau\right). \tag{5.34}$$

The field ϕ represents a spatially homogeneous background for the dynamics of coarse-grained, propagating gauge fields, and it breaks the gauge symmetry SU(2) down to its Abelian subgroup U(1). As it should, the field ϕ satisfies the following BPS equation

$$\partial_\tau \phi = \pm 2i\, \Lambda^3\, t_3\, \phi^{-1}, \tag{5.35}$$

where $\phi^{-1} \equiv \frac{\phi}{|\phi|^2}$. By differentiating the first-order BPS equation (5.35) with respect to τ and by subsequently appealing to (5.35) once more one easily derives the second-order equation in (5.27).

Because ϕ satisfies both Eqs. (5.35) and (5.27) the usual shift ambiguity in the classical estimate of the ground-state energy density, as allowed by the Euler–Lagrange equation, is absent. We will see later that a critical temperature of $\lambda_c \equiv \frac{2\pi T_c}{\Lambda} = 13.87$ exists which bounds the deconfining phase of SU(2) Yang–Mills thermodynamics from below. Thus one has $\frac{|\phi|^{-1}}{\beta} \geq 8.221 \times (\frac{\lambda}{\lambda_c})^{3/2}, (\lambda \geq \lambda_c)$. But for $\hat{r} = 8.221 \times (\frac{\lambda}{\lambda_c})^{3/2}$ the exponentially suppressed remainder $\int_{2\pi\hat{r}}^{\infty} dz\, I(\hat{p}, \hat{\tau})$ below Eq. (5.12) is a correction of less than one in 10^{22}! Also, setting $\xi = 8.221 \times (\frac{\lambda}{\lambda_c})^{3/2}$ in Eq. (5.25), one sees from Fig. 5.2 that the set $\{\hat{\phi}^a\}$ is deeply saturated.

Thus, with a maximal resolution $|\phi|$ in the effective theory (corresponding to a length scale $|\phi|^{-1}$ up to which short-distance, fundamental field configurations — Harrington–Shepard (anti)calorons — are coarse-grained over to derive the the field ϕ) the infinite-volume limit used to obtain $\{\hat{\phi}^a\}$, and in turn, the differential operator $\mathcal{D}$ is extremely well approximated.

The configuration in Eq. (5.34) cannot be altered by the interaction with effective gauge fields of topological charge $k = 0$. Except for the case of zero energy–momentum transfer in a local vertex involving the field ϕ, which is subjected to infinite resummation (the adjoint Higgs mechanism, see Sec. 5.1.5), we conclude that no other interaction between ϕ and this sector exists in the effective theory. Thus the nonperturbative emergence of the scale Λ is not influenced by the $k = 0$ sector. Compare this with the situation in perturbation theory at $T = 0$ where the Yang–Mills scale Λ is the pole position for the evolution of the fundamental gauge coupling g in the sector with $k = 0$. There, the value of Λ is determined by the value of g at a given resolution $\mu \gg \Lambda$ and two assumptions enter. First, one assumes the perturbative expansion to be sufficiently close to an asymptotic series to justify the low-order truncation of the beta function. Second, one assumes for the regime close to the pole that the perturbative prediction for g's evolution can smoothly be extrapolated from $g \ll 1$ to that regime. Both assumptions are questionable.

Let us now remark on the case SU(3). For this gauge group we write three sets of generators for its SU(2) subgroups as[10]

$$\lambda_1 = \begin{pmatrix} 0 & 1 & 0 \\ 1 & 0 & 0 \\ 0 & 0 & 0 \end{pmatrix}, \quad \lambda_2 = \begin{pmatrix} 0 & -i & 0 \\ i & 0 & 0 \\ 0 & 0 & 0 \end{pmatrix}, \quad \lambda_3 = \begin{pmatrix} 1 & 0 & 0 \\ 0 & -1 & 0 \\ 0 & 0 & 0 \end{pmatrix};$$

(5.36)

$$\bar{\lambda}_1 = \begin{pmatrix} 0 & 0 & 1 \\ 0 & 0 & 0 \\ 1 & 0 & 0 \end{pmatrix}, \quad \bar{\lambda}_2 = \begin{pmatrix} 0 & 0 & -i \\ 0 & 0 & 0 \\ i & 0 & 0 \end{pmatrix}, \quad \bar{\lambda}_3 = \begin{pmatrix} 1 & 0 & 0 \\ 0 & 0 & 0 \\ 0 & 0 & -1 \end{pmatrix};$$

(5.37)

$$\tilde{\lambda}_1 = \begin{pmatrix} 0 & 0 & 0 \\ 0 & 0 & 1 \\ 0 & 1 & 0 \end{pmatrix}, \quad \tilde{\lambda}_2 = \begin{pmatrix} 0 & 0 & 0 \\ 0 & 0 & -i \\ 0 & i & 0 \end{pmatrix}, \quad \tilde{\lambda}_3 = \begin{pmatrix} 0 & 0 & 0 \\ 0 & 1 & 0 \\ 0 & 0 & -1 \end{pmatrix}.$$

(5.38)

---

[10]For historical reasons these generators have an extra factor of two as compared to the generators t_a we have used for SU(2), e.g. $\operatorname{tr} \lambda_1^2 = 2$.

One of these nine generators is a linear combination of the other eight generators. This just reflects the fact that the group manifold of SU(3) locally is not $S_3 \times S_3 \times S_3$ but $S_3 \times S_5$. A set of independent generators is obtained by replacing the two matrices $\bar{\lambda}_3$ and $\tilde{\lambda}_3$ by the single matrix

$$\lambda_8 = \frac{1}{\sqrt{3}}(\bar{\lambda}_3 + \tilde{\lambda}_3) = \frac{1}{\sqrt{3}}\begin{pmatrix} 1 & 0 & 0 \\ 0 & 1 & 0 \\ 0 & 0 & -2 \end{pmatrix} \qquad (5.39)$$

and by keeping the other matrices. The result is the familiar set of Gell-Mann matrices generating the group SU(3).

For the case of SU(3) the field ϕ may wind in each of the above SU(2) algebras for one third of the period β. Except for the points $\tau = 0, \frac{\beta}{3}, \frac{2\beta}{3}$, where it starts to wind within another SU(2) algebra, a solution to the BPS equation

$$\partial_\tau \phi = \pm i \Lambda^3 \begin{cases} \lambda_3 \dfrac{\phi}{|\phi|^2}, & \left(0 \le \tau < \dfrac{\beta}{3}\right) \\ \bar{\lambda}_3 \dfrac{\phi}{|\phi|^2}, & \left(\dfrac{\beta}{3} \le \tau < \dfrac{2\beta}{3}\right) \\ \tilde{\lambda}_3 \dfrac{\phi}{|\phi|^2}, & \left(\dfrac{2\beta}{3} \le \tau < \beta\right) \end{cases} \qquad (5.40)$$

is given as

$$\phi(\tau) = \sqrt{\frac{\Lambda^3}{2\pi T}} \begin{cases} \lambda_1 \exp\left(\mp\dfrac{2\pi i}{\beta}\lambda_3\tau\right), & \left(0 \le \tau < \dfrac{\beta}{3}\right) \\ \bar{\lambda}_1 \exp\left(\mp\dfrac{2\pi i}{\beta}\bar{\lambda}_3\left(\tau - \dfrac{\beta}{3}\right)\right), & \left(\dfrac{\beta}{3} \le \tau < \dfrac{2\beta}{3}\right) \\ \tilde{\lambda}_1 \exp\left(\mp\dfrac{2\pi i}{\beta}\tilde{\lambda}_3\left(\tau - \dfrac{2\beta}{3}\right)\right), & \left(\dfrac{2\beta}{3} \le \tau < \beta\right). \end{cases} \qquad (5.41)$$

Notice that the potential $V = \frac{\Lambda^6}{|\phi|^2}$ is the same on this configuration $\phi(\tau)$ as it is on the configuration (5.34) for the SU(2) case. The gauge-symmetry breaking induced by ϕ in Eq. (5.41) is SU(3) $\rightarrow$ U(1)^2.

For SU(N) with $N \geq 4$ it is not clear in what way calorons of trivial holonomy and of topological charge $|k| = 1$ may introduce an effective gauge-symmetry breaking because fields ϕ_i associated with *independent* SU(2) subgroups may convey a nonmaximal symmetry breaking. Thus the uniqueness of the phase diagram is jeopardized for $N \geq 4$, and therefore we will not discuss this option any further.

5.1.5 *Effective action and a priori estimate of thermal ground state*

So far we were concerned with a spatial coarse-graining procedure over *noninteracting* Harrington–Shepard calorons. Fluctuations in the topologically trivial sector ($k = 0$) have not yet been considered. At first sight, one may object that these fluctuations introduce uncontrollable deviations from the pure BPS situation and that, therefore, our step-wise procedure in integrating out fundamental gauge-field configurations may be useless. Owing to the scale of maximal resolution $|\phi|$ collectively introduced by the $|k| = 1$ sector this, however, turns out not to be the case. We first write an effective action density and argue *a posteriori* that this density is unique at resolution $|\phi|$.

The effective action density, subject to a maximal resolution $|\phi|$ for propagating gauge fields, is given as

$$\mathcal{L}_{\text{eff}}[a_\mu] = \text{tr} \left(\frac{1}{2} G_{\mu\nu} G_{\mu\nu} + (D_\mu \phi)^2 + \frac{\Lambda^6}{\phi^2} \right), \qquad (5.42)$$

where $G_{\mu\nu} = \partial_\mu a_\nu - \partial_\nu a_\mu - ie[a_\mu, a_\nu] \equiv G^a_{\mu\nu} t_a$ denotes the field strength[11] of the *effective*, that is, coarse-grained, propagating,

---

[11]Notice that we have not absorbed the coupling e into the effective gauge field a_μ. A gauge transformation acting on ϕ and a_μ thus reads: $\phi \rightarrow \Omega \phi \Omega^\dagger$ and $a_\mu \rightarrow \Omega a_\mu \Omega^\dagger + \frac{i}{e} \Omega \partial_\mu \Omega^\dagger$.

trivial-topology gauge field $a_\mu = a_\mu^a t_a$, $D_\mu \phi = \partial_\mu \phi - ie[a_\mu, \phi]$, and e is the effective gauge coupling. The Lagrangian density of Eq. (5.42) is valid for both the SU(2) and the SU(3) case.

Why is it true that the effective Lagrangian upon coarse-graining down to resolution $|\phi|$ is uniquely given by Eq. (5.42)? Obviously, $\mathcal{L}_{\text{eff}}$ is gauge-invariant due to the replacement $\partial_\tau \to D_\mu$ in the kinetic term of Eq. (5.26). Also, the presence of the term $\frac{1}{2} G^2$ is demanded by perturbative renormalizability ['t Hooft and Veltman (1972a,b); 't Hooft (1971); Lee and Zinn-Justin (1972)] (see Sec. 2.2). The latter states that integrating out fundamental $k = 0$ fluctuations down to a certain resolution, say $|\phi|$, does not change the form of the Yang–Mills action[12] for $k = 0$ fluctuations below resolution $|\phi|$. But why are there neither local gauge-invariant terms of higher mass dimension nor nonlocal gauge-invariant contributions expressing interactions of the field ϕ with the field a_μ? Nonlocal terms are expandable into powers of D_μ, and thus it suffices to exclude the contribution of such local terms. For example, one could think of a local term $\text{tr}\, G_{\mu\nu} G_{\mu\nu} \left(\frac{\phi}{M}\right)^N$ where M is a mass scale, say

---

[12]Here the $T = 0$ argument for perturbative renormalizability can be used since all $k = 0$ modes that are integrated out are off their mass shell by at least $|\phi|^2 : |p^2| \geq |\phi|^2$. As a consequence, they are not thermalized. Perturbative renormalizability implies the absence of operators with mass-dimension greater than four in an effective action S_μ which is obtained by integrating out $k = 0$ modes from the cut-off scale M_P down to resolution μ. By gauge invariance the form of S_μ and that of the fundamental Yang–Mills action S_{M_P} must coincide. This can be seen as follows: At μ and to any order in perturbation theory all Green's functions are, by perturbative renormalization, made independent of M_P by a proper choice of wave function factor and gauge coupling in S_{M_P}. As a result, S_μ is also independent of M_P. Assume there exists an operator O of mass-dimension $d_O > 4$ in S_μ. Then, by power counting, the coefficient c_O of this operator can be written as $C_0 = \bar{c}_O \mu^{4-d_O}$ where $\bar{c}_O$ is a real, nonvanishing number. Letting $\mu \nearrow M_P$, this, however, contradicts the fact that in S_{M_P} no operators with $d_O > 4$ exist.

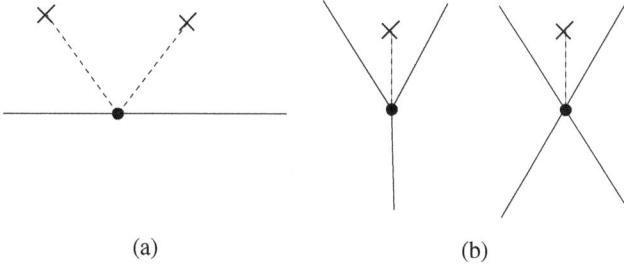

(a)   (b)

Figure 5.3.   Allowed vertex (a) and examples of lowest-mass-dimension operators inducing vertices that are excluded in the effective theory by the inertness of ϕ and perturbative renormalizability (b). Solid lines represent the topologically trivial, effective gauge field a_μ while a dashed line terminating in a cross corresponds to a local insertion of the operator ϕ.

$M = \Lambda$, and N is an odd integer.[13] Another example is

$$\operatorname{tr} \frac{1}{M^{2(3n-2)}} \left(G_{\mu\nu}[D_\mu\phi, D_\nu\phi] \right)^n, \quad (n \geq 1). \tag{5.43}$$

The reason why such a term cannot emerge in the effective theory is the inertness of the field ϕ: Local terms in $\mathcal{L}_{\mathrm{eff}}$ with vertices allowing for momentum exchange between the inert field ϕ and a_μ are excluded (see Fig. 5.3). In other words, a vertex involving ϕ, which is allowed by gauge invariance and which, upon demanding no momentum transfer to ϕ does not reduce to a vertex contained in $\frac{1}{2} G^2$, is excluded by perturbative renormalizability. On the other hand, a vertex that does reduce to one of those represented by $\frac{1}{2} G^2$ upon demanding zero momentum transfer to ϕ is superfluous.[14]

But what about the coupling between a_μ and ϕ contained in $(D_\mu\phi)^2$? As long as no momentum is transferred from a_μ to the field ϕ such a coupling, which is required by gauge invariance, is admissible. For the effective thermal ground state, which may only involve a pure-gauge configuration of the field a_μ, see principle I of

---

[13]Even N would just renormalize the coupling e which is not yet determined.

[14]This would require the mass scale M to be tuned to $|\phi|$.

Sec. 5.1.1, this is obvious. As far as propagating configurations of the field a_μ are concerned, the condition that in the effective theory no momentum is transferred to ϕ represents a nontrivial constraint (see Sec. 5.2.1). We conclude that no terms other than the ones of Eq. (5.42) may contribute to the effective action $\mathcal{L}_{\text{eff}}$.

The fluctuating field a_μ is integrated out loop-expanding the logarithm of the partition function about the free quasiparticle situation. This loop expansion is nontrivial due to the term $ie[a_\mu, a_\nu]$ in $G_{\mu\nu}$ which leads to the occurence of three-vertices and four-vertices. The momentum transfer in these vertices is subject to further constraints imposed by the existence of the maximal resolution $|\phi|$. Finally, the evolution of the *effective* coupling e is determined by the invariance of Legendre transformations between thermodynamic quantities under the applied coarse-graining up to a given loop order (see Sec. 5.2.3).

Let us now discuss the ground-state estimate for the case of SU(2). Apart from (small) radiative corrections (see Sec. 5.4), and modulo global gauge transformations, the *a priori* estimate of the thermal ground state of the effective theory in the deconfining phase is given by the configuration in Eq. (5.34) and the pure-gauge configuration

$$a_\mu^{\text{gs}} = \mp\delta_{\mu 4}\frac{2\pi}{e\beta}\,t_3. \tag{5.44}$$

This is a consequence of the fact that ϕ and a_μ^{gs} are ground-state solutions of the following Euler–Lagrange equation for a_μ as implied by $\mathcal{L}_{\text{eff}}$ in Eq. (5.42):

$$D_\mu G_{\mu\nu} = ie[\phi, D_\nu\phi]. \tag{5.45}$$

Equation (5.45) is solved by ϕ and a_μ^{gs} because $G_{\mu\nu}[a_\kappa^{\text{gs}}] = D^\nu[a_\kappa^{\text{gs}}]\phi \equiv 0$. One can understand the pure-gauge configuration a_μ^{gs} as a consequence of the source term on the right-hand side of the effective Yang-Mills equation (5.45). Indeed, we have shown in Sec. 4.3.1.2 that the HS (anti)caloron in its core region $|x| \ll \beta$ is a four-dimensional (anti)selfdual dipole which transmutes into a three-dimensional (static) (anti)selfdual dipole in the peripheral region $r \gg s \gg \beta$. By the results derived in Sec. 5.3 and the fact that field ϕ is obtained by a spatial coarse-graining over the (anti)caloron core region we conclude that, paradoxically, the inert field ϕ represents the spatially averaged effect of regions in fundamental fields conveying quantum mechanical indeterminism to the interaction of effective fields. On the other hand, field a_μ^{gs}, describing (anti)caloron overlap due to (static) (anti)selfdual dipole fields in the peripheral regions is the effective response of all HS (anti)calorons adjacent to the one HS (anti)caloron explicitly considered in deriving the inert field ϕ and vice versa. Thus the pair ϕ, a_μ^{gs} no longer is subject to a BPS constraint of vanishing energy density. Indeed, the associated action density is given as $\mathcal{L}_{\text{eff}}[a_\mu^{gs}] = \text{tr}\, \frac{\Lambda^6}{\phi^2} = 4\pi\Lambda^3 T,\ (T \equiv \beta^{-1})$, which is interpreted as a temperature-dependent cosmological constant. The overlap between calorons and (anti)calorons as facilitated by their spatial tails, see Sec. 4.3.1.2, lift the energy density ρ^{gs} of the BPS estimate for the ground state from zero to $\rho^{gs} = 4\pi\Lambda^3 T$. Together, ϕ and a_μ^{gs} represent an *a priori* estimate, the bare thermal ground state, which turns out to be quite accurate.

The fact that the ground-state pressure P^{gs} is negative, $P^{gs} = -\rho^{gs}$, is microscopically explained by calorons and anticalorons of small holonomy having their BPS magnetic monopoles-antimonopole constituents [Nahm (1983); Lee and Lu (1998); Kraan and Van Baal (1998a,b)] attract one another under the influence of radiative corrections [Diakonov *et al.* (2004)] which are implied by the dynamics of (anti)caloron overlap (see Sec. 4.4.2).

The "excitation" of large caloron holonomy, which, according to [Diakonov *et al.* (2004)], leads to the dissociation of the associated caloron/anticaloron and hence to the liberation of a screened magnetic monopole and its antimonopole, is extremely rare [Ludescher *et al.* (2008)] (see Sec. 5.5.2). These processes and their influence are collectively described by effective radiative corrections.

The winding gauge, where ϕ is given by Eq. (5.34) and a_μ^{gs} by Eq. (5.44), and the unitary gauge, where $\phi = 2\,|\phi|\, t_3,\ a_\mu^{gs} = 0$, are

connected by a singular but admissible periodic gauge transforma-
tion. Under this gauge transformation the Polyakov loop $\text{Pol}[a_\mu^{\text{gs}}]$ is
transformed from $\text{Pol} = -1_2$ to $\text{Pol} = 1_2$. This very fact points out
the electric Z_2 degeneracy of the thermal ground-state estimate and
thus deconfinement (see Sec. 3.3.2). Let us demonstrate this.

Since $\phi \rightarrow \tilde{\Omega}(\tau)\phi\tilde{\Omega}^\dagger(\tau)$ under a fundamental gauge transformation it is easily checked that, to reach unitary gauge, $\tilde{\Omega}(\tau)$ is
given as

$$\tilde{\Omega}(\tau) = \Omega_{\text{gl}}\, Z(\tau)\, \Omega(\tau), \tag{5.46}$$

where $\Omega(\tau) \equiv \exp[\pm 2\pi i \frac{\tau}{\beta} t_3]$, $Z(\tau) = (2\Theta(\tau - \frac{\beta}{2}) - 1)1_2$, and $\Omega_{\text{gl}} = \exp[i\frac{\pi}{2} t_2]$. The function Θ is defined as

$$\Theta(x) = \begin{cases} 0, & (x < 0), \\ \dfrac{1}{2}, & (x = 0), \\ 1, & (x > 0). \end{cases} \tag{5.47}$$

Thus $\tilde{\Omega}(\tau)$ is periodic albeit not smooth at the point $\tau = \frac{\beta}{2}$. The
periodicity of fluctuations δa_μ, however, is not affected by such a
gauge transformation. Namely, writing $a_\mu = a_\mu^{\text{gs}} + \delta a_\mu$, we have

$$a_\mu \rightarrow \tilde{\Omega}(a_\mu^{\text{gs}} + \delta a_\mu)\tilde{\Omega}^\dagger + \frac{i}{e}\tilde{\Omega}\partial_\mu\tilde{\Omega}^\dagger$$

$$= \Omega_{\text{gl}}\left(\Omega(a_\mu^{\text{gs}} + \delta a_\mu)\Omega^\dagger + \frac{i}{e}\left(\Omega\partial_\mu\Omega^\dagger + Z\partial_\mu Z\right)\right)\Omega_{\text{gl}}^\dagger$$

$$= \Omega_{\text{gl}}\left(\Omega\delta a_\mu\Omega^\dagger + \frac{2i}{e}\delta\left(\tau - \frac{\beta}{2}\right)Z\right)\Omega_{\text{gl}}^\dagger = \Omega_{\text{gl}}\Omega\,\delta a_\mu\,(\Omega_{\text{gl}}\Omega)^\dagger.$$

$$\tag{5.48}$$

Now $\Omega_{\text{gl}}\Omega(\tau = 0) = -\Omega_{\text{gl}}\Omega(\tau = \beta)$, and thus the periodicity of the
fluctuation δa_μ is unaffected by the gauge transformation induced
by $\tilde{\Omega}(\tau)$ (admissibility of this change of gauge). The claimed transformation of the Polyakov loop on a_μ^{gs} is obvious.

We now discuss the case of SU(3). In the background ϕ of Eq. (5.41) the pure-gauge solution to

$$D_\mu G^{\mu\nu} = ie[\phi, D_\nu\phi]. \tag{5.49}$$

reads

$$a_\mu^{gs} = \pm\delta_{\mu 4}\frac{\pi}{e\beta}\begin{cases} \lambda_3, & \left(0 \leq \tau < \frac{\beta}{3}\right) \\[2mm] \bar{\lambda}_3, & \left(\frac{\beta}{3} \leq \tau < \frac{2\beta}{3}\right) \\[2mm] \tilde{\lambda}_3, & \left(\frac{2\beta}{3} \leq \tau < \beta\right). \end{cases} \tag{5.50}$$

On the configuration a_μ^{gs} of Eq. (5.50) the Polyakov loop reads

$$\begin{aligned} \text{Pol}[a_\mu^{gs}] &= \exp\left[\pm i\frac{\pi}{3}\tilde{\lambda}_3\right]\exp\left[\pm i\frac{\pi}{3}\bar{\lambda}_3\right]\exp\left[\pm i\frac{\pi}{3}\lambda_3\right] \\ &= \exp\left[i\frac{\pi}{3}(\pm\tilde{\lambda}_3 \pm \bar{\lambda}_3 \pm \lambda_3)\right]. \end{aligned} \tag{5.51}$$

The $+$ or $-$ sign can be chosen independently for each SU(2) subalgebra. The following combinations are possible:

$$\begin{aligned} \pm(+\tilde{\lambda}_3 + \bar{\lambda}_3 + \lambda_3) &= \pm 2\bar{\lambda}_3, \\ \pm(-\tilde{\lambda}_3 - \bar{\lambda}_3 + \lambda_3) &= \pm 2\tilde{\lambda}_3, \\ \pm(-\tilde{\lambda}_3 + \bar{\lambda}_3 + \lambda_3) &= \pm 2\lambda_3, \\ \pm(+\tilde{\lambda}_3 - \bar{\lambda}_3 + \lambda_3) &= 0. \end{aligned} \tag{5.52}$$

The corresponding values of the Polyakov loop are

$$\text{Pol}_1^\pm = \begin{pmatrix} \exp[\pm\frac{2\pi i}{3}] & 0 & 0 \\ 0 & 1 & 0 \\ 0 & 0 & \exp[\mp\frac{2\pi i}{3}] \end{pmatrix},$$

$$\text{Pol}_2^\pm = \begin{pmatrix} 1 & 0 & 0 \\ 0 & \exp[\pm\frac{2\pi i}{3}] & 0 \\ 0 & 0 & \exp[\mp\frac{2\pi i}{3}] \end{pmatrix}, \tag{5.53}$$

$$\text{Pol}_3^\pm = \begin{pmatrix} \exp[\pm\frac{2\pi i}{3}] & 0 & 0 \\ 0 & \exp[\mp\frac{2\pi i}{3}] & 0 \\ 0 & 0 & 1 \end{pmatrix}, \quad \text{Pol}_4 = 1_3.$$

Pol_4 is a trivial representation of the center group. The set $\text{Pol}_1^\pm$, $\text{Pol}_2^\pm, \text{Pol}_3^\pm$ closes under multiplication with the center elements $\exp[\pm\frac{2\pi i}{3}]1_3, 1_3$. It is a six-dimensional, reducible representation of the center group. The two three-dimensional irreducible representations, which collapse on one another, are spanned by

$$\begin{aligned}
&\frac{1}{3}1_3\left(\text{Pol}_1^\pm + \text{Pol}_2^\mp + \text{Pol}_3^\pm\right), \\
&\frac{1}{3}\exp[\mp\frac{2\pi i}{3}]1_3\left(\text{Pol}_1^\pm + \text{Pol}_2^\mp + \text{Pol}_3^\pm\right).
\end{aligned} \tag{5.54}$$

We conclude that our estimate for the thermal ground state has a Z_3 degeneracy: The electric Z_3 symmetry is dynamically broken and thus we have discussed a *deconfining* phase.

What about a gauge rotation to unitary gauge $a_\mu^{\text{gs}} = 0$ and $\phi = |\phi|\lambda_3$ or $\phi = |\phi|\bar\lambda_3$ or $\phi = |\phi|\tilde\lambda_3$? Such a gauge transformation is induced by the following group element:

$$\tilde\Omega^\dagger = \begin{cases} \exp\left[\mp i\frac{\pi}{\beta}\tau\lambda_3\right]\exp\left[-i\frac{\pi}{4}\lambda_2\right], & \left(0 \le \tau < \frac{\beta}{3}\right) \\[2mm] \exp\left[\mp i\pi\frac{\pi}{\beta}\left(\tau - \frac{\beta}{3}\right)\bar\lambda_3\right]\exp\left[-i\frac{\pi}{4}\bar\lambda_2\right], & \left(\frac{\beta}{3} \le \tau < \frac{2\beta}{3}\right) \\[2mm] \exp\left[\mp i\frac{\pi}{\beta}\left(\tau - \frac{2\beta}{3}\right)\tilde\lambda_3\right]\exp\left[-i\frac{\pi}{4}\tilde\lambda_2\right], & \left(\frac{2\beta}{3} \le \tau < \beta\right). \end{cases} \tag{5.55}$$

By construction $\tilde\Omega^\dagger$ is periodic, at $\tau = \beta$ it jumps back to its value at $\tau = 0$, and thus a fluctuation δa_μ, which is periodic in winding gauge, is also periodic in unitary gauge.

5.2 Free Thermal Quasiparticles

5.2.1 *Quasiparticle spectrum, propagators, and momentum constraints*

In this section we obtain the tree-level mass spectrum in the effective theory, and we formulate constraints on the propagation and interaction of the effective $k = 0$ modes.

We refer to an excitation, which possesses on tree-level a temperature-dependent mass in the effective theory, as a thermal quasiparticle. In general, with each broken generator t_a of the original gauge symmetry a mass m_a is associated (see Sec. 1.5). One has

$$m_a^2 = -2e^2 \text{tr} \, [\phi, t_a][\phi, t_a]. \tag{5.56}$$

For the SU(2) case in unitary gauge, where $\phi = 2|\phi| t_3$, one has

$$m^2 \equiv m_1^2 = m_2^2 = 4e^2 \frac{\Lambda^3}{2\pi T}, \tag{5.57}$$

$$m_3 = 0.$$

For SU(3) we obtain four mass-degenerate directions in su(3). Their mass-squared is $e^2 \frac{\Lambda^3}{2\pi T}$. There are another two degenerate directions of mass-squared $4\, e^2 \frac{\Lambda^3}{2\pi T}$. Imposing unitary gauge in the SU(2) case (gauge condition $\phi = 2\, |\phi| \, t_3 = |\phi| \, \tau_3$, $a_\mu^{gs} = 0$) and in addition Coulomb gauge for the unbroken U(1) subgroup (gauge condition $\partial_i a_i^3 = 0$), one arrives at a completely fixed, physical gauge if one understands that the real-valued gauge function θ in $\Omega_3 \equiv \exp(i\theta t_3)$ vanishes at spatial infinity. Such a gauge-fixing is physical because it exhibits the quasiparticle mass spectrum, the physical number of polarizations — three for $a = 1, 2$ and two for $a = 3$ — and the transversality of free, massless, propagating gauge modes in the field a_μ^3 [unbroken subgroup U(1)]. For SU(3) a Coulomb condition for each of the two massless modes needs to be imposed.

Notice that the number of degrees of freedom before coarse-graining matches those after coarse graining. Namely, for SU(2) one has three species of fundamental, propagating gauge fields ($k = 0$

sector) times two polarizations each plus two species of charge-one scalar magnetic monopoles ($|k| = 1$ sector) before coarse-graining and two species of effectively massive gauge fields times three polarization each plus one species of massless gauge field times two polarizations. In both cases one obtains eight degrees of freedom. For SU(3) one obtains 22 degrees of freedom before and after coarse-graining. In what follows we only consider the SU(2) case.

In unitary–Coulomb gauge and on the level of free quasiparticles the real-time propagators of the fields $a_\mu^{1,2}$ and a_μ^3 are given as [Kapusta (1989)]

$$D_{\mu\nu,ab}^{\text{H}}(p) = -\delta_{ah}\delta_{bh}\tilde{D}_{\mu\nu}\left[\frac{i}{p^2 - m^2 + i0} + 2\pi\delta(p^2 - m^2)n_B(|p_0|/T)\right],$$
(5.58)

$$D_{\mu\nu,ab}^{\text{M}}(p) = -\delta_{a3}\delta_{b3}\left\{P_{\mu\nu}^T\left[\frac{i}{p^2 + i0} + 2\pi\delta(p^2)n_B(|p_0|/T)\right] - i\frac{u_\mu u_\nu}{\vec{p}^2}\right\},$$
(5.59)

where $\tilde{D}_{\mu\nu} = \left(g_{\mu\nu} - \frac{p_\mu p_\nu}{m^2}\right)$, $P_T^{00} = P_T^{0i} = P_T^{i0} = 0$, $P_T^{ij} = \delta^{ij} - p^i p^j/\vec{p}^2$, $u = (1,0,0,0)$ represents the four-velocity of the heat bath, $h = 1,2$ (no sum), and $n_B(x) = 1/(e^x - 1)$ denotes the Bose–Einstein distribution function. Notice that a Wick rotation to real time or a change of signature from Euclidean to Minkowskian is trivial as far as the physics of the ground state estimate of Sec. 5.1.5 and the asscociated dynamical gauge-symmetry breaking are concerned: Both, ground-state pressure and energy density on one hand and quasiparticle mass on the other hand do not depend on Euclidean time.

For the fields $a_\mu^{1,2}$ only thermal propagation occurs, that is, only the term in Eq. (5.58) proportional to n_B contributes. This is explained in Fig. 5.4. Because of the existence of a maximal resolution scale $|\phi|$ the deviation of the four-momentum p in Eq. (5.59) from its free massless shell $p^2 = 0$ is constrained as

$$|p^2| \leq |\phi|^2.$$
(5.60)

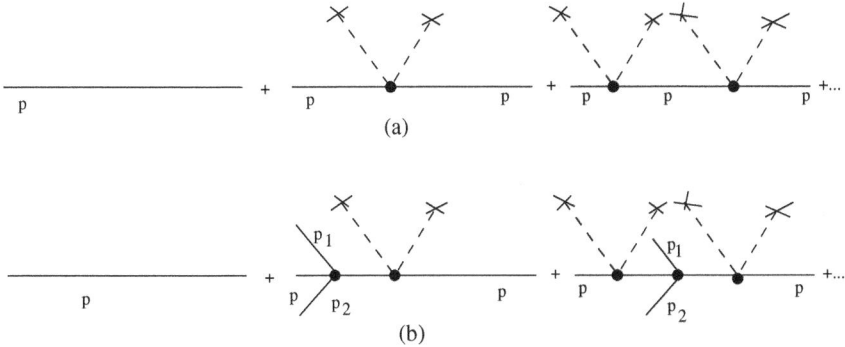

Figure 5.4. (a) The emergence of mass by a resummation of up to infinitely many local insertions of the operator ϕ^2 (up to the two insertions shown only). This process does not impose any momentum transfer to the field ϕ if the external momentum p and all intermediate modes are put on the mass shell $p^2 = m^2$. (b) In case momentum is transferred to the external line in an attempt to move momentum p away from $p^2 = m^2$, momentum is inevitably also transferred to the field ϕ. But this is contradicting the fact that ϕ does not fluctuate. Therefore, in unitary–Coulomb gauge the effective SU(2) gauge fields $a_\mu^{1,2}$ propagate thermally only. The same argument applies to massive modes of SU(3): no nonthermal quantum propagation takes place for tree-level massive thermal quasiparticles.

Notice that condition (5.60) fixes the momentum transfer in a three-vertex by momentum conservation.

The following conditions fix the momentum transfers in a four-vertex:

$$|(p_1 + p_2)^2| \le |\phi|^2, \quad (s \text{ channel});$$
$$|(p_3 - p_1)^2| \le |\phi|^2, \quad (t \text{ channel}); \qquad (5.61)$$
$$|(p_2 - p_3)^2| \le |\phi|^2, \quad (u \text{ channel}).$$

Conditions (5.61) simply state that in each of the 2 by 2 scattering channels the momentum transfer must not resolve the interior of a Harrington-Shepard (anti)caloron where its Euclidean time dependence introduces winding, associated with the occurrence of the fundamental unit of action $\hbar$, see discussion in Sec. 5.3. It is easy to show that the four-vertex is invariant under a permutation of the legs attached to it. This implies that the four-vertex does not distinguish Mandelstam variables s, t, and u in mediating $2 \rightarrow 2$ scattering. The according constraints, however, do. Therefore, 2 by 2 scattering (or such a subprocess within a larger diagram) must be understood as a coherent superposition of the weighted amplitudes of each nontrivial scattering channel s, t, and u where the weights either are 1 (two out of three channels are trivial), $\frac{1}{2}$ (one out of three channels is trivial), or $\frac{1}{3}$ (all three channels are nontrivial) [Krasowski and Hofmann (2014)], see also Sec. 5.4.2.

On the one-loop level in the effective theory (gas of noninteracting thermal quasiparticles and massless excitations) the radiative shift ΔV of the potential $V = \mathrm{tr}\, \frac{\Lambda^6}{\phi^2}$ in Eq. (5.42) due to quantum fluctuations [arising from terms without the factor n_B in Eqs. (5.58) and Eq. (5.59)] is negligibly small.[15] We generously estimate ΔV by the contribution of the massless mode a_μ^3 appropriately weighted by the number of polarizations of all fields a_μ^1, a_μ^2, and a_μ^3:

$$|\Delta V| \le \frac{1}{\pi^2} \int_0^{|\phi|} dp\, p^3 \log\left(\frac{p}{|\phi|}\right) = \frac{|\phi|^4}{16\pi^2} = \frac{\lambda^{-3}}{32\pi^2} V, \qquad (5.62)$$

where

$$\lambda \equiv \frac{2\pi T}{\Lambda}. \qquad (5.63)$$

As we shall see in Sec. 5.2.3, λ is considerably larger than unity when omitting the contribution ΔV. According to Eq. (5.62), such an omission is consistent.

5.2.2 Pressure and energy density of noninteracting excitations

Modulo our present ignorance on the λ dependence of the effective coupling e we are now in a position to write closed analytic

---

[15]Recall that such fluctuations do not occur in the propagator of (5.58) anyway.

expression for the one-loop pressure P and the one-loop energy density ρ of SU(2) Yang–Mills theory in its deconfining phase:

$$P(\lambda) = -\Lambda^4 \left\{ \frac{2\lambda^4}{(2\pi)^6} \left[2\bar{P}(0) + 6\bar{P}(2a) \right] + 2\lambda \right\}, \qquad (5.64)$$

$$\rho(\lambda) = \Lambda^4 \left\{ \frac{2\lambda^4}{(2\pi)^6} \left[2\bar{\rho}(0) + 6\bar{\rho}(2a) \right] + 2\lambda \right\}, \qquad (5.65)$$

where

$$\bar{P}(y) \equiv \int_0^\infty dx\, x^2 \log \left[1 - \exp\left(-\sqrt{x^2 + y^2} \right) \right],$$
$$\bar{\rho}(y) \equiv \int_0^\infty dx\, x^2 \frac{\sqrt{x^2 + y^2}}{\exp\left(\sqrt{x^2 + y^2} \right) - 1}, \qquad (5.66)$$

and

$$a \equiv \frac{m}{2T}. \qquad (5.67)$$

For SU(3) one has

$$P(\lambda) = -\Lambda^4 \left\{ \frac{2\lambda^4}{(2\pi)^6} \left[4\bar{P}(0) + 3\left(4\bar{P}(a) + 2\bar{P}(2a) \right) \right] + 2\lambda \right\} \qquad (5.68)$$

and

$$\rho(\lambda) = \Lambda^4 \left\{ \frac{2\lambda^4}{(2\pi)^6} \left[4\bar{\rho}(0) + 3\left(4\bar{\rho}(a) + 2\bar{\rho}(2a) \right) \right] + 2\lambda \right\}. \qquad (5.69)$$

5.2.3   *Evolution of effective gauge coupling*

On the one-loop level the evolution of the effective coupling e is determined by demanding that ρ is obtained from P by the Legendre transformation[16] $\rho = T\frac{dP}{dT} - P$. The latter is a consequence of the

---

[16]Demanding Legendre transformations to be honored at the effective level the assumption of existence of the Yang–Mills partition function is made. It is conceivable that by the interaction between topological and field configurations with $k = 0$ in an effective action defined at resolution μ larger

existence of the (ultraviolet regularized) partition function formulated in terms of fundamental fields and needs to be obeyed after a reformulation of the same partition function in terms of effective fields. Because the effective theory contains temperature-dependent parameters, m and P^{gs}, a nontrivial condition on the temperature evolution of the effective coupling e is imposed by this Legendre transformation. At the one-loop level, a necessary and sufficient condition for cancelation of implicit temperature dependences to take place in the Legendre transformation is $\partial_m P = 0$. Appealing to Eq. (5.64), one derives the following first-order ordinary differential equation for the SU(2) case:

$$\partial_a \lambda = -\frac{24 \lambda^4 a}{(2\pi)^6} \frac{D(2a)}{1 + \frac{24 \lambda^3 a^2}{(2\pi)^6} D(2a)}, \tag{5.70}$$

where the function D is defined in Eq. (3.19) of Sec. 3.1.1. Equation (5.70) is equivalent to

$$1 = -\frac{24 \lambda^3}{(2\pi)^6} \left(\lambda \frac{da}{d\lambda} + a \right) a\, D(2a). \tag{5.71}$$

For SU(3) one has

$$\partial_a \lambda = -\frac{12\, \lambda^4\, a}{(2\pi)^6} \frac{D(a) + 2\, D(2a)}{1 + \frac{12\, \lambda^3\, a^2}{(2\pi)^6} \left(D(a) + 2\, D(2a) \right)}. \tag{5.72}$$

The evolutions governed by Eqs. (5.70) and (5.72) possess two fixed points each: $a = 0$ and $a = \infty$. The latter fixed point is associated with a critical temperature λ_c of values $\lambda_c = 13.87$ and $\lambda_c = 9.475$

---

than $|\phi|$ cutoffs on the number of higher dimensional operators, irreducible $k = 0$ loops and a limit on the topological charge modulus $|k|$ emerge effectively such that the Yang–Mills partition function can indeed be proven to exist for any $\mu > |\phi|$. It is also conceivable that the existence is not decidable within our present set of quantum field theoretic axioms. In any case, the assumption of existence (and thus the validity of Legendre transformations) appears to be consistent with the existence of the effective theory for the deconfining phase at resolution $|\phi|$ [Hofmann (2007)].

for SU(2) and SU(3), respectively. This fixed point indicates a phase boundary since massive modes decouple by virtue of a diverging effective gauge coupling e meaning that magnetic monopole (test) charges are completely screened and thus massless. On the level of free effective quasiparticles the generation of such isolated screened magnetic charges does not enter into Eqs. (5.70) and (5.72). These charges are described collectively by effective radiative corrections (see Sec. 5.5.2). However, the screening effects imposed on magnetic test charges by instable monopole-antimonopole pairs in calorons of small holonomy are well described by the evolution $e(\lambda)$. Thus at λ_c monopoles and antimonopoles become massless and thus condense into a newly emerging ground state, breaking dynamically the intact Abelian gauge symmetry of effective deconfining Yang–Mills thermodynamics (see Sec. 6.1.3). Notice that because of the negativity of the right-hand sides of Eqs. (5.70) and (5.72) the dependencies $\lambda(a)$ or $a(\lambda)$ are monotonic decreasing. Thus a must fall below unity eventually when increasing λ.

An attractor to the evolution $a(\lambda)$ exists. For SU(2) it is given as $a(\lambda) = 4\sqrt{2}\pi^2\lambda^{-3/2}$ for $\lambda \gg \lambda_c$ and $a(\lambda) \propto -\log(\lambda - \lambda_c)$ for $\lambda \searrow \lambda_c$. Let us show this.

For $a \ll 1$ the expansion of the function $D(2a)$ in powers of a modulo nonanalytic terms can be truncated at zeroth order, recall the discussion in Sec. 3.1.1 leading to Eq. (3.25) [Dolan and Jackiw (1974)]. This simplifies Eq. (5.71) as

$$1 = -\frac{\lambda^3}{(2\pi)^4}\left(\lambda\frac{da}{d\lambda} + a\right)a, \tag{5.73}$$

and the solution, subject to the initial condition $a(\lambda_i) = a_i \ll 1$, is

$$a(\lambda) = 4\sqrt{2}\pi^2\lambda^{-3/2}\left(1 - \frac{\lambda}{\lambda_i}\left[1 - \frac{a_i^2\lambda_i^3}{32\pi^4}\right]\right)^{1/2}. \tag{5.74}$$

Thus for $\lambda \ll \lambda_i$ function $a(\lambda)$ runs into the attractor $a(\lambda) = 4\sqrt{2}\pi^2\lambda^{-3/2}$. Since $a \equiv \frac{m}{2T} = 2\pi e\lambda^{-3/2}$ there is a plateau $e \equiv \sqrt{8}\pi$ in this regime. Because the attractor increases with decreasing λ the condition $a \ll 1$ will be violated at small temperatures. The estimate $14.61 > \lambda_c$ is obtained by setting the attractor equal to unity. Since the true solution in this regime will continue to grow with decreasing λ [negative definiteness of right-hand side of Eq. (5.70)] the right-hand side of Eq. (5.70) will be exponentially suppressed. This verifies the behavior $a(\lambda) \propto -\log(\lambda - \lambda_c)$ for $\lambda \searrow \lambda_c$ and implies, by virtue of Eq. (5.57), a logarithmic singularity at λ_c also for $e(\lambda)$. Numerically, one obtains $\lambda_c = 13.87$, and the behavior of e near λ_c is given as[17]

$$e(\lambda) = -4.59 \log(\lambda - \lambda_c) + 18.42. \qquad (5.75)$$

The fit of Eq. (5.74) was performed for the interval $\lambda_c \leq \lambda \leq 15.0$ (see [Hofmann (2009)]). Notice in Eq. (5.74) the decoupling property of the initial conditions set at a high temperature λ_i from the deconfining thermodynamics at lower temperature: At sufficiently large λ_i the value of a_i in Eq. (5.74) can be varied over a large range without having an effect on the evolution $a(\lambda)$ at lower temperatures. If, when increasing temperature, the initial-condition induced correction to the $\lambda^{-3/2}$ dependence of a in Eq. (5.74) acquires importance then a good reason for this departure from scale invariance (constancy of the coupling e: $e = \sqrt{8}\pi$ for SU(2) and $e = \frac{4}{\sqrt{3}}\pi$ for SU(3)) must exist in a physics model. To the author's mind the only thinkable reason would be a departure from continuous spacetime. The onset of local spacetime structure should be characterized by a typical mass scale M. Common belief is that M coincides with the Planck mass: $M \sim 10^{19}$ GeV.

For SU(3) the equivalent of Eq. (5.72) is

$$1 = -\frac{12\lambda^3}{(2\pi)^6} \left(\lambda \frac{da}{d\lambda} + a \right) \left(aD(a) + 2aD(2a) \right). \qquad (5.76)$$

The high-λ (or small-a) attractor solution to Eq. (5.76) reads $a(\lambda) = \frac{8}{\sqrt{3}}\pi^2 \lambda^{-3/2}$, and the effective coupling reaches a plateau value of

---

[17]The author would like to thank Markus Schwarz for performing this fit to the numerically computed curve.

$e = \frac{4}{\sqrt{3}}\pi$. The value for λ_c is $\lambda_c = 9.475$ [Hofmann (2005); Giacosa and Hofmann (2007a)] where the coupling e and all quasiparticle masses diverge by virtue of a logarithmic pole similar to the one exposed in Eq. (5.75) for the SU(2) case. The constancy of e for $a \ll 1$ signals that the magnetic charge $g = \frac{4\pi}{e}$ of a screened magnetic test monopole is conserved during most of the evolution. For $\lambda \searrow \lambda_c$ both the magnetic charge $g = \frac{4\pi}{e}$ and the mass $M_{mon} \sim \frac{4\pi^2 T}{e}$ of a screened magnetic (test) monopole vanish. The numerical solutions to Eqs. (5.70) and (5.72) can be used to obtain the numerical λ dependences of e which are indictated in Fig. 5.5.

Finally, we may evaluate Eqs. (5.64), (5.65), (5.68), and (5.69) on the numerical solutions to Eqs. (5.71) and (5.76), respectively. The result is depicted in Fig. 5.6. We will show in Secs. 5.4.4 and 5.4.5 that the one-loop expressions (free gas of thermal excitations plus thermal ground state) in Eqs. (5.64), (5.65), (5.68), and (5.69) are accurate on the 1% level. Thus the curves in Fig. 5.6 represent good approximations. Notice the negativity of the pressure shortly above the critical temperatures λ_c which are indicated by the vertical

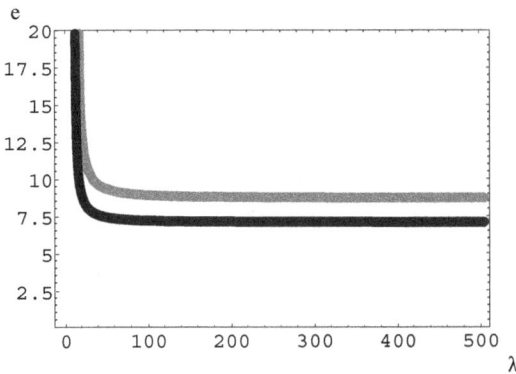

Figure 5.5. The temperature evolution of the effective gauge coupling e in the deconfining phase for SU(2) (gray line) and SU(3) (black line). The gauge coupling diverges logarithmically, $e \propto -\log(\lambda - \lambda_c)$, at $\lambda_c = 13.867$ (SU(2)) and $\lambda_c = 9.475$ SU(3). The respective plateau values are $e = \sqrt{8\pi} \sim 8.89$ and $e = \frac{4}{\sqrt{3}}\pi \sim 7.26$.

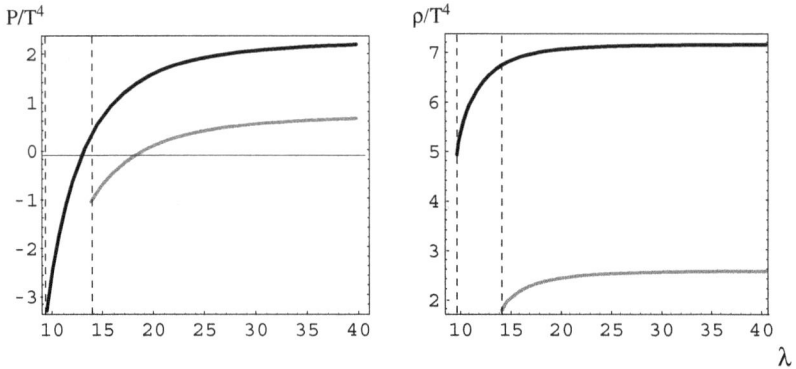

Figure 5.6. Ratio of the pressure P and T^4 (left panel) and of the energy density ρ and T^4 (right panel) for SU(2) and SU(3) (gray and black, respectively) as a function of $\lambda \equiv \frac{2\pi T}{\Lambda}$ on the one-loop level. Vertical dashed lines indicate the phase boundary.

dashed lines. In this region Yang–Mills thermodynamics starts to be ground-state dominated.

5.2.4 *Trace anomaly of energy–momentum tensor*

Let us now discuss how the trace of the energy–momentum tensor $\theta_{\mu\nu}$ acquires a nonvanishing expectation at the level of free quasiparticles in the effective theory. This effect truly is of a nonperturbative nature: Both the thermal ground state and the quasiparticle masses contribute to it. To see this, we define the dimensionless function $h(\lambda)$ [Giacosa and Hofmann (2007a)] as follows

$$h(\lambda) \equiv -\frac{\rho(\lambda) - 3P(\lambda)}{4P^{gs}} = -\frac{\theta_\mu^\mu}{4P^{gs}}. \qquad (5.77)$$

Expanding $h(\lambda)$ [Dolan and Jackiw (1974)] up to quadratic order in a, [Dolan and Jackiw (1974)] one has for SU(2)

$$h(\lambda) \sim 1 + \frac{\lambda^3 a^2(\lambda)}{4(2\pi)^4}. \qquad (5.78)$$

The first term in Eq. (5.78) is due to the thermal ground state estimate while the second term arises from fluctuating, free thermal quasiparticles. Substituting the high-temperature attractor

$a(\lambda) = 4\sqrt{2}\pi^2\lambda^{-3/2}$ $(\lambda \gg \lambda_c)$ into Eq. (5.78), one arrives at $h(\lambda) = \frac{3}{2}$, $(\lambda \gg \lambda_c)$. Recalling that $P^{gs} = -4\pi\Lambda^3 T$ we thus arrive at

$$\frac{\theta^{\mu}_{\mu}}{\Lambda^4} \sim 12\,\lambda \ (\lambda \gg \lambda_c). \tag{5.79}$$

According to Eqs. (5.78) and (5.74) $\frac{\theta^{\mu}_{\mu}}{\Lambda^4}$ exhibits a quadratic correction in λ at very high temperatures, that is, shortly below λ_i. Ignoring this quadratic correction also for SU(3) one obtains $\theta^{\mu}_{\mu} = 24\pi\Lambda^3\,T = 12\,\lambda\Lambda^4$ for $\lambda \gg \lambda_c$.

5.2.5 *Fundamental versus effective gauge coupling*

Knowing the nonperturbatively generated trace anomaly of the energy–momentum tensor (see Sec. 5.2.4), we are now in a position to investigate the temperature dependence of the *fundamental* gauge coupling g. In doing so we follow the presentation given in [Giacosa and Hofmann (2007b)].

To all orders in powers of the coupling g of (Euclidean) SU(N) Yang–Mills perturbation theory at $T = 0$, θ^{μ}_{μ} as an operator is related to the operator $\mathrm{tr}\,F^{\mu\nu}F_{\mu\nu}$ as [Collins, Duncan, and Joglekar (1977); Fujikawa (1980)]

$$\theta^{\mu}_{\mu} = \frac{\beta(g)}{2g}\,F^a_{\mu\nu}F^{\mu\nu,a}, \tag{5.80}$$

where $F^a_{\mu\nu} = \partial_\mu A^a_\nu - \partial_\nu A^a_\mu - gf^{abc}A^b_\mu A^c_\nu$ is the field strength of the *fundamental* Yang–Mills action density $-\frac{1}{4}(F^a_{\mu\nu})^2$ and β is given by the right-hand side of the evolution equation[18] for the *fundamental* gauge coupling g:

$$\mu\,\partial_\mu g = \beta(g) \tag{5.81}$$

---

[18]In this section the same symbol μ is used for two different things: as a Lorentz index and denoting the renormalization scale. We rely on the alertness of the reader to track the proper meaning of the symbol μ from the context in which it is used. Apologies for the inconvenience.

(see Sec. 2.3.3). In Eq. (5.81) the mass scale μ refers to the resolution that is applied to the process at which the value of the coupling g is extracted. In contrast to the chiral anomaly, which is not renormalized owing to its topological nature, the trace anomaly exhibits two resolution-dependent factors: the β-function divided by g and the average of $F^a_{\mu\nu}F^{\mu\nu,a}$.

When using the one-loop expression, $\beta(g) = \mu\partial_\mu g = -bg^3$ ($b = \frac{5N}{48\pi^2}$) of Eq. (2.33), the Landau pole $\mu = \Lambda_L$ occurs in the solution to Eq. (5.81). One has

$$\Lambda_L = \mu_0 \exp\left(-\frac{1}{2bg_0^2}\right) \tag{5.82}$$

[compare with Eq. (2.37)]. Performing a thermal ensemble average over Eq. (5.80), we obtain

$$\rho - 3P = \frac{\beta(g)}{2g} \langle F^a_{\mu\nu}F^{\mu\nu,a}\rangle_T , \tag{5.83}$$

where ρ, P describe the energy density and the pressure of the thermalized Yang–Mills system. For the gauge groups SU(2) and SU(3) they are with good accuracy given by the expressions of Eqs. (5.64), (5.65) and Eqs. (5.68), (5.69), respectively. Notice that in Eq. (5.83) two scales enter: the temperature T, at which the thermal average is calculated, and the scale μ, being the resolution where $\beta(g) = \mu\partial_\mu g$ is evaluated. In the effective theories for deconfining SU(2) and SU(3) Yang–Mills thermodynamics, however, the scales μ and T are not independent but functionally related: $\mu = |\phi| = \sqrt{\frac{\Lambda^3}{2\pi T}}$.

The question remains on how the right-hand side of Eq. (5.83) is evaluated. It is easily shown that the contribution to the average of the action density due to any propagating mode vanishes. Thus for any practical purpose the thermally averaged action density of the *effective* theory is given by the energy density $\rho^{gs} = 4\pi\Lambda^3 T$ of the thermal ground-state estimate. Now, in Eq. (5.83) a thermal average of the *fundamental* and not the effective action density is demanded. We account for this fact by introducing a wave function renormalization factor $f^2(g)$ relating fundamental and effective field strength. Therefore, we have

$$\langle \mathcal{L}_{YM} \rangle_T = \frac{1}{4} \langle F^a_{\mu\nu} F^{\mu\nu,a} \rangle_T = f^2(g) \langle \mathcal{L}_{eff} \rangle_T = f^2(g) \rho^{gs}. \tag{5.84}$$

The function $f(g)$ will be fixed by requiring that for $\lambda \gg \lambda_c$ the fundamental coupling g runs in agreement with perturbation theory. Notice that this assumption only implies that the thermodynamics at high temperature is *dominated* by the coupling between fundamental, propagating modes (gluons). In the effective theory, the thermalized Yang–Mills system can be considered a spatially extended vertex being probed with a selfconsistently adjusting resolution $|\phi|$ owing to temperature T. Propagating fields, which would be resolved at $\mu > |\phi|$, are integrated out. Thus we have:

$$\beta(g) = \mu \, \partial_\mu g = -2T \, \partial_T g = -2\beta_T(g), \tag{5.85}$$

where $\beta_T(g) \equiv T \, \partial_T g$ and $\mu = |\phi|$. Notice that $\beta(g) = \mu \partial_\mu g$ is a positive quantity: Both the trace anomaly $\rho - 3p$ and $\langle F^a_{\mu\nu} F^{a,\mu\nu} \rangle_T$ are positive. The fact that the resolution $\mu = |\phi|$ decreases when increasing T generates a negative function $\beta_T(g)$, in accord with asymptotic freedom for $T \gg T_c$.

Taking into account Eqs. (5.83)–(5.85), we have

$$h(\lambda) = \frac{\rho - 3p}{4\rho^{gs}} = -\frac{\beta_T(g)}{g} f^2(g). \tag{5.86}$$

The graph of the function $h(\lambda)$ is indicated in the right panel of Fig. 5.7. The simple high-T behavior allows to determine function $f(g)$ analytically. We require that the perturbative result $\beta_T(g) = -bg^3$ ($b = \frac{5N}{48\pi^2}$) holds for $g \ll 1$ (or $T \gg T_c$). Then Eq. (5.86) implies that

$$f(g) = \sqrt{\frac{3}{2b}} \frac{1}{g}. \tag{5.87}$$

The evolution equation (5.86) for $g = g(\lambda)$ thus can be recast as:

$$\beta_T(g) = -\frac{2}{3} b h(\lambda) g^3 \Leftrightarrow \partial_\lambda g = -\frac{2}{3} b \frac{h(\lambda)}{\lambda} g^3. \tag{5.88}$$

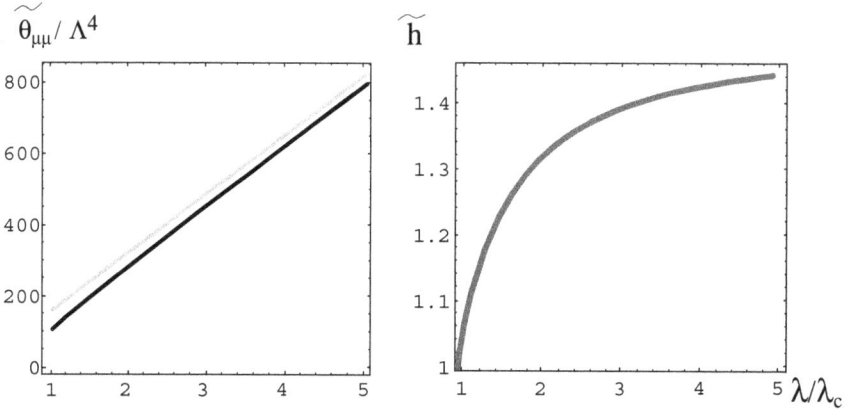

Figure 5.7. The functions $\frac{\tilde{\theta}_\mu^\mu}{\Lambda^4}(\lambda/\lambda_c) \equiv \frac{\theta_\mu^\mu}{\Lambda^4}(\lambda)$ and $\tilde{h}(\lambda/\lambda_c) \equiv h(\lambda)$ [left panel: gray curve represents continuation of large λ/λ_c, black curve is the realistic dependence on λ/λ_c for SU(2)].

From the behavior of $h(\lambda)$ we can immediately infer two interesting properties:

(a) The function $h(\lambda) \simeq \frac{3}{2}$ for $\lambda > 5\lambda_c$. That is, the perturbative equation $\beta_T(g) = -bg^3$ is valid all the way down to $5\lambda_c$. The range of validity of the perturbative treatment for the determination of $g(\lambda)$ thus is rather large.

(b) The function $h(\lambda)$ slowly decreases for decreasing temperatures thus effectively lowering the coefficient b in the perturbative beta function. Therefore the only effect of the highly nonperturbative thermal ground state on the evolution of the fundamental coupling g in comparison with pure one-loop perturbation theory is a mild screening of the perturbative Landau pole: both the residue of the pole and the pole position are slightly decreased.

5.3   Effective Coupling and (Anti)caloron Action

Here we would like to argue as in [Hofmann and Kaviani (2012)] that the plateau value $e = \sqrt{8}\pi$ (in natural units $c = \hbar = 1$) of the effective coupling implies that the action of a typical (anti)caloron, which dominates the emergence of the thermal ground state in the deconfining phase of SU(2) Yang-Mills thermodynamics, actually equals $\hbar$. To do this, we remark that the counting of powers in $\hbar$ in loop expansions carried out in the effective theory takes place as in conventional perturbation theory. Additional subtleties arise because of the existence of a maximal resolution $|\phi|$ which enforces constraints on momentum transfers in four-vertices and on the off-shellness of the massless mode. These constraints have an obvious form in physical unitary-Coulomb gauge, see Sec. 5.2.1.

To make the power counting in $\hbar$ most explicit, we work in units where $k_B = c = \epsilon_0 = \mu_0 = 1$ but $\hbar$ is re-instated as an action. The (dimensionless) exponential

$$- \frac{\int_0^\beta d\tau d^3x \, \mathcal{L}'_{\text{eff}}[a_\mu]}{\hbar}, \tag{5.89}$$

in the weight belonging to fluctuating fields in the partition function[19] thus can, in unitary gauge, be re-cast as

$$- \int_0^\beta d\tau d^3x \, \text{tr} \left(\frac{1}{2}(\partial_\mu \tilde{a}_\nu - \partial_\nu \tilde{a}_\mu - ie\sqrt{\hbar}[\tilde{a}_\mu, \tilde{a}_\nu])^2 - e^2 \hbar [\tilde{a}_\mu, \tilde{\phi}]^2 \right), \tag{5.90}$$

where $\tilde{a}_\mu \equiv a_\mu/\sqrt{\hbar}$ and $\tilde{\phi} \equiv \phi/\sqrt{\hbar}$ are assumed to not depend on $\hbar$ [Brodsky and Hoyer (2011)], see also [Iliopoulos, Itzykson, and Martin (1975)], [Donoghue et al. (2002)], [Bjerrum-Bohr, Donoghue, and Holstein (2003)], [Holstein and Donoghue (2004)]. Notice that because of the terms $\propto \hbar^0$ in (5.90) the units of $\tilde{a}_\mu$ and of $\tilde{\phi}$ are length^{-1}. Thus the coupling e must have the unit of $1/\sqrt{\hbar}$. Together with the results of Sec. 5.2.3 we thus arrive at

$$e = \frac{\sqrt{8}\pi}{\sqrt{\hbar}} \tag{5.91}$$

---

[19]Recall that ϕ is inert. As a consequence, the factor, whose exponent is the potential-part of the effective action, can be pulled out of the partition function and needs not be considered in a discussion of the effective loop expansion.

almost everywhere in the deconfining phase. Because the average over scale parameter ρ depends cubically on the cutoff $\rho \sim |\phi|^{-1} = \xi\beta$ with $\xi \geq 8.221 \times \left(\frac{\lambda}{\lambda_c}\right)^{3/2}$, see text below Eq. (5.25), the action $S_{C,A;\rho\sim|\phi|^{-1}}$ of a just-not-resolved caloron/anticaloron of scale $\rho \sim |\phi|^{-1}$, which thus dominates the emergence of field ϕ, reads

$$S_{C,A;\rho\sim|\phi|^{-1}} = \frac{8\pi^2}{e^2} = \hbar. \tag{5.92}$$

Eq. (5.92) has implications. It suggests that the quantum of action $\hbar$, whose introduction in (5.89) is motivated by the laws of Quantum Mechanics and which should enable a systematic accounting of quantum corrections (number of loops) in the effective theory, coincides with the Euclidean action of a just-not-resolved selfdual (classical) field configuration whose topological charge is fixed to be of modulus unity. That is, the reason for why $\hbar$ really is a constant and that fundamental quantum processes do not invoke integer multiples of this basic unit of action can respectively be traced to constancy of e (with the exception of the logarithmically thin pole at T_c which assures that for increasing T from not far above T_c there is also no dependence on the specific Yang-Mills scale Λ of the SU(2) gauge theory) and the fact that only charge-modulus-unity calorons/anticalorons contribute to the effective thermodynamical consistency of interacting, fundamental topological and plane-wave configurations. Moreover, rather than just multiplying first and second powers of $\hbar^{1/2}$ onto the powers e and e^2 of an *a priori* unknown coupling constant e in three- and four-vertices of a Yang-Mills theory we would infer that these vertices owe their very existence to the presence of just-not-resolved (anti)calorons, that is, (anti)calorons of scale parameter $\rho \sim |\phi|^{-1}$. Recall from Ch. 2 that a perturbative treatment of Yang-Mills theory requires renormalization due to the ultraviolet divergences which associate with loop diagrams. This is because in perturbation theory a local vertex is assumed to persist for arbitrarily high four-momentum transfer. On the contrary, in deconfining, thermal Yang-Mills theory we have just argued that a

Yang-Mills vertex is a local manifestation of topologically nontrivial field configurations – each of whose centers defines the fundamental unit of action $\hbar$ – building up the thermal ground state. But by its very emergence at a prescribed, global temperature T, this thermal ground state cannot be resolved by momentum transfers larger then $|\phi|$ for otherwise it would expose the Euclidean spacetime extendedness of its constituting field configurations: (anti)calorons. Because (anti)calorons certainly do not convey solutions to the Cauchy problem of the Minkowskian Yang-Mills equations we have no choice but to accept them as effective, unresolved introducers of indeterminism into scattering processes via local (effective) Yang-Mills vertices. Thus, our nonperturbative, thermal treatment of Yang-Mills theory, which constructs a ground state out of global (stable) minima of the Euclidean Yang-Mills action in the topological sectors $k = \pm 1$ and subsequently investigates the impact of this ground state on its excitations, naturally accomplishes what now appears to be a mildly artificial albeit highly successful programme in perturbation theory: renormalization.

There is another important implication of Eq. (5.92) concerning the interpretation of U(1) charges in SU(2). Namely, for the fine-structure constant α of Quantum Electrodynamics (QED) to be dimensionless,

$$\alpha = N^{-1}\frac{g^2}{4\pi\hbar},\tag{5.93}$$

the coupling g in Eq. (5.93) must have the unit of $\sqrt{\hbar}$. (N^{-1} denotes a numerical factor related to the mixing of the massless modes belonging to several SU(2) groups, see Sec. 9.6.2.) But this is the case if g is taken to be the electric-magnetically dual to e:

$$g = \frac{4\pi}{e} \propto \sqrt{\hbar}.\tag{5.94}$$

That is, a stable and screened magnetic monopole as liberated by the dissociation of a large-holonomy SU(2) (anti) caloron, see Sec. 5.5.1, or the isolated charge situated in the center of the flux eddy generated by the selfintersection of a center-vortex loop in the confining phase, see Sec. 7.3.2, are interpreted as electric charges in the real world.

5.4 Effective Radiative Corrections

5.4.1 *Loop expansion: General considerations*

In the effective theory for the deconfining phase [compare with Eq. (5.42)] thermodynamical quantities are loop-expanded in terms of effective $k = 0$ fluctuations. This expansion is tied to the real-time formalism because it is subject to constraints formulated in Minkowskian spacetime signature.

If the typical action associated with a loop momentum at increasing loop order L decreases fast then one would expect this expansion to organize itself into a power series in $\hbar^{-1}$ (coefficient of power $L \sim L$-th power of this typical action modulo a combinatorial factor) with good convergence properties. Unlike fixed-order perturbative loop expansions into bubble diagrams order does not coincide with the power of the squared (effective) gauge coupling.

The reason for loop momenta conveying an ever decreasing action with growing loop order L in effective, deconfining Yang–Mills thermodynamics is rooted in the constraints of Eqs. (5.60) and (5.61). As we shall see, the occurrence of a large action associated with a loop momentum is ruled out because each of the L loop momenta needs to obey a constraint on its own off-shellness [see (5.60)] as well as on the momentum transfer to other loops mediated by vertices [see (5.61)]: The more loops there are in a given Feynmann diagram the tighter the constraints on each loop momentum are. We will show in Sec. 5.4.3 that for irreducible loop diagrams the number of independent constraints, starting at a finite loop order L_{max}, exceeds the number of independent loop variables. Thus for $L \geq L_{max}$ the domain for loop integrations is compact and likely empty for L sufficiently large and finite.

5.4.2 Feynman rules in unitary–Coulomb gauge

Let us first state the Feynman rules for the calculation of effective radiative corrections in unitary–Coulomb gauge for the SU(2) case. For the vertices and for SU(N) they are given in Sec. 2.3.1 by letting $g \rightarrow e$, for propagators we need to take into account the ϕ-field induced symmetry breaking in unitary gauge and the finite-temperature effects of the real-time formalism for free fields [Landsman and Weert (1987)]. In the real-time formulation of finite-temperature field theory [Landsman and Weert (1987)] it is possible to discern nonthermal quantum from thermal fluctuations. It is necessary to tell the former from the latter because conditions (5.60) and (5.61) constrain nonthermal quantum fluctuations in the completely fixed and physical unitary–Coulomb gauge.

To conform with the nomenclature of [Schwarz, Hofmann, and Giacosa (2006)] in unitary–Coulomb gauge[20] we refer to the fluctuative su(2) directions 1 and 2 as tree-level heavy (TLH) and to direction 3 as tree-level massless (TLM).

Let us formulate the complete set of Feynman rules for the SU(2) case in momentum space: According to the discussion in Sec. 5.2.1 the free propagator $D^{\mathrm{H}}_{\mu\nu,ab}$ of a TLH mode in unitary gauge is

$$D^{\mathrm{H}}_{\mu\nu,ab}(p) = -\delta_{ah}\delta_{bh}\tilde{D}_{\mu\nu}\left[2\pi\delta(p^2 - m^2)n_B(|p_0|/T)\right], \qquad (5.95)$$

$$\tilde{D}_{\mu\nu} = \left(g_{\mu\nu} - \frac{p_\mu p_\nu}{m^2}\right), \qquad (5.96)$$

where $n_B(x) = 1/(e^x - 1)$ denotes the Bose–Einstein distribution function and $h = 1, 2$ (no sum). For the free TLM mode we have in Coulomb gauge

$$D^{\mathrm{M}}_{\mu\nu,ab}(p) = -\delta_{a3}\delta_{b3}\left\{P^T_{\mu\nu}\left[\frac{i}{p^2} + 2\pi\delta(p^2)n_B(|p_0|/T)\right] - i\frac{u_\mu u_\nu}{\mathbf{p}^2}\right\}, \qquad (5.97)$$

---

[20]Recall the gauge conditions: $\phi = 2\,|\phi|\,t_3$, $a^{\mathrm{gs}}_\mu = 0$, and $\partial_i a^3_i = 0$.

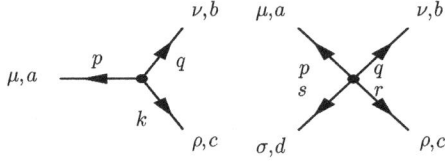

Figure 5.8. Three- and four-vertices.

where

$$P_T^{00} = P_T^{0i} = P_T^{i0} = 0,$$ (5.98)

$$P_T^{ij} = \delta^{ij} - p^i p^j / \mathbf{p}^2.$$ (5.99)

Notice the term $\propto u_\mu u_\nu$ in Eq. (5.97) describing the "propagation" of the temporal U(1) gauge-field component where $u_\mu = (1,0,0,0)$ represents the four-velocity of the heat bath. This term is activated by external sources. Moreover, upon resummation of radiative corrections, it mutates to describe the propagation of electric charge-density waves (see Sec. 5.4.7.6).

The three- and four-gauge-boson vertices [compared with Fig. 5.8], are in principle given as

$$\Gamma^{\mu\nu\rho}_{[3]abc}(p,k,q) = e(2\pi)^4 \delta(p+q+k)\epsilon_{abc}$$

$$\times [g^{\mu\nu}(q-p)^\rho + g^{\nu\rho}(k-q)^\mu + g^{\rho\mu}(p-k)^\nu], \quad (5.100)$$

$$\Gamma^{\mu\nu\rho\sigma}_{[4]abcd} = -ie^2(2\pi)^4 \delta(p+q+s+r)[\epsilon_{abe}\epsilon_{cde}(g^{\mu\rho}g^{\nu\sigma} - g^{\mu\sigma}g^{\nu\rho})$$

$$+ \epsilon_{ace}\epsilon_{bde}(g^{\mu\nu}g^{\rho\sigma} - g^{\mu\sigma}g^{\nu\rho})$$

$$+ \epsilon_{ade}\epsilon_{bce}(g^{\mu\nu}g^{\rho\sigma} - g^{\mu\rho}g^{\nu\sigma})]. \quad (5.101)$$

Recall that the analytical description of the nontrivial ground-state dynamics is facilitated by a spatial coarse-graining down to resolution $|\phi|$. To not resolve a local, effective vertex the maximal off-shellness of effective gauge modes as well as the momentum transfer in four-vertices are constrained as discussed in Sec. 5.2.1: TLM modes may not be further off their "mass" shell than $|\phi|^2$ [compare with (5.60) for the case of a free TLM mode] and TLH modes may only fluctuate thermally, that is, on-shell. Momentum transfers in an effective four-vertex cannot be larger than $|\phi|^2$ in all three Mandelstam variables s, t, and u [compare with (5.61)]. We have discussed in Sec. 5.2.1 that the 4-vertex is blind to permutations of its legs (interchanges of scattering channels). However, the constraints (5.61) distinguish such permutations. Therefore, a coherent superposition of weighted scattering amplitudes in all contributing channels is required. For example,

$$\Gamma^{\alpha\beta\gamma\delta}_{[4]abcd} = \frac{1}{3} \left(\Gamma^{\alpha\beta\gamma\delta}_{[4]abcd}\Big|_s + \Gamma^{\alpha\beta\gamma\delta}_{[4]abcd}\Big|_t + \Gamma^{\alpha\beta\gamma\delta}_{[4]abcd}\Big|_u \right). \qquad (5.102)$$

if all 3 scattering channels contribute. Here the subscripts s, t, or u signal that the vertex is subject to the corresponding s, t, or u channel constraints in (5.61). If, on the other hand, say, the t channel is trivial (figure-eight two-loop contributions to the pressure or polarization tensor of the massless mode on the one-loop level, see Sec. 5.4.4 and Sec. 5.4.7 [21]) then the vertex acts as

$$\Gamma^{\alpha\beta\gamma\delta}_{[4]abcd} = \frac{1}{2} \left(\Gamma^{\alpha\beta\gamma\delta}_{[4]abcd}\Big|_s + \Gamma^{\alpha\beta\gamma\delta}_{[4]abcd}\Big|_u \right). \qquad (5.103)$$

If two channels are trivial then the four-vertex acts in the remaining channel according to Eq. (5.101).

Moreover, according to Landsman and Weert (1987) a loop-diagram is divided by i and by the number of its vertices.

5.4.3   *Loop expansion: Conjecture on its termination*

In this section we give arguments on why the loop expansion for the deconfining Yang–Mills pressure likely terminates at a finite order modulo resummations of one-particle irreducible diagrams for the

---

[21]It is straightfoward to see that the integrand in this case is invariant under $k \to -k$ (k a loop four-momentum) which renders the contributions of s and u channels equal. Thus it suffices to compute the s channel contribution with weighting unity.

polarization tensor. We closely follow [Hofmann (2012)] and [Krasowski and Hofmann (2014)].

5.4.3.1 *Pinch singularities*

Here we would like to point out that the occurrence of powers of delta functions of the same argument, as they appear when real-time expanding the pressure into loops, can be inhibited by appropriate resummations.

At tree-level the propagators of a TLH and a TLM mode are given in Eqs. (5.95) and (5.97), respectively, and the problem of the so-called pinch singularities occurs in ring diagrams such as the one in Fig. 5.9. Let us explain this. Due to the fact that no momentum transfer takes place in TLM propagators across the four-vertices in Fig. 5.9 the thermal part of each TLM line is associated with a delta function of the same argument. If there are n such lines then an n-th power of this delta function occurs. This is unacceptable mathematically.

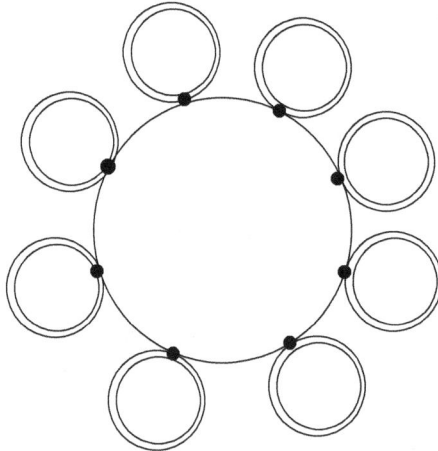

Figure 5.9. A ring diagram as it occurs in the loop expansion of the pressure in SU(2) Yang–Mills thermodynamics. Double (single) lines are associated with TLH- (TLM-) mode propagation.

Resumming one-particle irreducible (1PI) contributions to the polarization tensor, the scalar factor of the tree-level propagators are modified in terms of screening functions $G_H(p)$ and $G_M(p)$ as[22]

$$2\pi\delta(p^2 - m^2)\, n_B(|p_0|/T) \to 2\pi\rho_H(p, T)\, n_B(|p_0|/T),$$

$$\frac{i}{p^2} + 2\pi\delta(p^2)\, n_B(|p_0|/T) \to \frac{i}{p^2 - G_M(p, T)}$$
$$+ 2\pi\rho_M(p, T)\, n_B(|p_0|/T), \quad (5.104)$$

where spectral functions at fixed spatial momentum $\vec{p}$ are defined as

$$\rho_H(p^0, \vec{p}, T) \equiv \frac{1}{\pi} \operatorname{Im} \frac{1}{p^2 - (m^2 + G_H(p, T))}$$
$$= \frac{1}{\pi} \frac{\operatorname{Im} G_H(p, T)}{\left(p^2 - m^2 - \operatorname{Re} G_H(p, T)\right)^2 + \left(\operatorname{Im} G_H(p, T)\right)^2},$$

$$\rho_M(p^0, \vec{p}, T) \equiv \frac{1}{\pi} \operatorname{Im} \frac{1}{p^2 - G_M(p, T)},$$
$$= \frac{1}{\pi} \frac{\operatorname{Im} G_M(p, T)}{\left(p^2 - \operatorname{Re} G_M(p, T)\right)^2 + \left(\operatorname{Im} G_M(p, T)\right)^2}. \quad (5.105)$$

From Eq. (5.105) it follows that if $\operatorname{Im} G \neq 0$ then powers of δ-functions of the same argument relax to powers of finite width (Lorentz-like) peaks of the same argument and thus to mathematically well-defined objects. Compared to the leading contributions, which on the selfconsistent "mass" shell are real (see Sec. 5.4.7.5 and

---

[22]A discussion of the emergence of additional tensor structures $u_\mu u_\nu$ and $p_\mu u_\nu + p_\nu u_\mu$ due to loop effects is not important for our argument. To avoid a contradiction the one-particle irreducible polarization tensor are first computed in real time subject to the "x" case in the constraints (5.60) and (5.61). Subsequently, a "rotation" of the external momentum variable p^0 is performed to imaginary time. Afterwards the resummation is carried out, and finally the result is "rotated" back to real time. Effectively, this amounts to a resummation of insertions of the T-dependent polarization tensor in a way that is identical to the situation at $T = 0$. The part of the propagator containing the Bose–Einstein distribution function n_B is determined subsequently as in Eqs. (5.95) and (5.96).

5.4.7.6), tiny imaginary parts arise in the screening functions for TLM and TLH modes at higher irreducible loop orders. Thus a broadening of delta functions in the sense of Eq. (5.105) indeed takes place. The emergence of screening functions at one-loop order slightly changes the tree-level constraints (5.60) at higher loop order. As we shall see in Secs. 5.4.4 and 5.4.5, the computation of the pressure in a truncation at the two-loop level using tree-level propagators yields results that are accurate on the 1% level.

5.4.3.2 *Loop integration variables versus constraints*

For the exact computation of thermodynamical quantities such as the pressure all diagrams contributing to each mode's full propagator need to be known. Knowing the exact propagator (or the polarization tensor), in turn, fixes the exact dispersion law for each mode. The polarization tensor is a sum over connected bubble diagrams (loop diagrams with no external legs) with one internal line of momentum p cut, such that the diagram remains connected. The thus generated two external lines are amputated subsequently. As a consequence, the vanishing of a connected bubble diagram due to a zero-measure support for its loop-momenta integrations implies that the associated contribution to a polarization tensor is also nil.

We conjecture that *nonvanishing*, connected bubble diagrams enjoy the following property: For the total number V of their vertices we have $V \leq V_{max}$ with $V_{max} < \infty$ provided that all 1PI contributions to the polarization tensors with up to V_{max} many vertices are resummed.

Let us now present our arguments in favor of this claim. The requirement that 1PI contributions to the polarization tensor are resummed assures that (A) at each vertex all constraints in (5.61) are independently operative (subject to slightly modified dispersion laws) and that (B) pinch singularities (powers of delta functions) do not occur because of the broadening of the spectral function of the respective mode's propagator (compare with Sec. 5.4.3.1). We consider the two cases where a connected bubble diagram *solely* contains (i) V_4-many four-vertices and (ii) V_3-many three-vertices.

The relation between the number L of independent loop momenta, the number I of internal lines, and the number V of vertices for planar bubble diagrams is determined by the Euler characteristics for spherical polyhedra. To start with, we only consider planar diagrams which are spherical polyhedra (of genus $g = 0$) with one of the faces removed. In this case one has

$$L = I - V + 1. \tag{5.106}$$

For case (i) and (ii) we have in addition [Weinberg (1995)]

$$I = 2V_4, \quad \text{and} \quad I = \frac{3}{2}V_3, \tag{5.107}$$

respectively. According to (5.60) we thus have in case (i) $2V_4$ constraints (propagators) and, according to (5.61), at least $\frac{1}{2}V_4$ constraints (vertices) on loop momenta. In case (i) this gives a total of $K \geq \frac{5}{2}V_4$ constraints. In case (ii) we have a total of $K = \frac{3}{2}V_3$ constraints (only propagators). Combining Eqs. (5.106) and (5.107), we obtain

$$\text{for (i)}: L = V_4 + 1 \quad \Rightarrow \quad \tilde{K} = 2V_4 + 2,$$
$$\text{for (ii)}: L = \frac{V_3}{2} + 1 \quad \Rightarrow \quad \tilde{K} = V_3 + 2. \tag{5.108}$$

This yields

$$\text{for (i)}: \frac{\tilde{K}}{K} \leq \frac{4}{5}\left(1 + \frac{1}{V_4}\right), \quad \text{for (ii)}: \frac{\tilde{K}}{K} = \frac{2}{3}\left(1 + \frac{2}{V_3}\right). \tag{5.109}$$

The ratio $\tilde{K}/K$ is larger in case (i) ($V_4 \geq 1$) than in case (ii) ($V_3 \geq 2$). Notice that in case (i) the ratio $\frac{\tilde{K}}{K}$ is smaller than unity for $V_4 \geq 5$ while this happens for $V_3 \geq 6$ in case (ii).

The constraints (5.60) and (5.61) are independent inequalities instead of independent equations and thus do not identify independent hypersurfaces (algebraic sets)[23] in a $\tilde{K}$-dimensional Euclidean space $\mathbf{R}^{\tilde{K}}$. Rather, the inequalities (5.60) or (5.61) "fatten" hypersurfaces (algebraic varieties) that would be obtained by setting their right-hand sides equal to zero.[24] As a consequence, the situation $\frac{\tilde{K}}{K} = 1$ fixes a discrete set of compact regions $C_{\tilde{K}}$ in $\mathbf{R}^{\tilde{K}}$ rather than a discrete set of points. If $\frac{\tilde{K}}{K}$ is sufficiently smaller than unity, which is the case for sufficiently large V_4 and/or V_3 according to Eq. (5.109), then the associated diagram does not contribute. Fat hypersurfaces, specified by the number $\kappa \equiv K - \tilde{K}$ of constraints not used up for the determination of $C_{\tilde{K}}$, should have an intersection C_κ such that

$$C_\kappa \cap C_{\tilde{K}} = \emptyset \quad (\kappa \gg 1). \tag{5.110}$$

Notice that according to Eq. (5.109)

$$\frac{1}{2}\tilde{K} \overset{>}{\sim} \kappa \overset{>}{\sim} \frac{1}{4}\tilde{K}, \quad (\tilde{K} \gg 1). \tag{5.111}$$

Although it is not rigorously guaranteed that $C_\kappa \cap C_{\tilde{K}} = \emptyset$ with κ ranging as in (5.111), this is, however, rather plausible.

For completeness let us investigate the generalization of Eq. (5.106) to nonplanar bubble diagrams which can be considered

---

[23]By independent hypersurfaces H_i ($i = 1, \ldots, h \leq \tilde{K}$) in a $\tilde{K}$-dimensional Euclidean space $\mathbf{R}^{\tilde{K}}$ we mean that in a whole environment U of a point in their intersection $\bigcap_{i=1}^{h} H_i$ the normal vectors $\hat{n}_i$ to H_i (computed somewhere on $U \cap H_i$) are linearly independent. If $h = \tilde{K}$ then it follows that $\bigcap_{i=1}^{\tilde{K}} H_i$ is a discrete set of points.
[24]The case of a TLH mode, where only the thermal part of the propagator contributes in terms of a δ-function weight, can be figured as the limit of a fat hypersurface that is subject to a regular weight acquiring zero width but now subject to a singular weight.

polyhedra with one face removed and a nonvanishing number of handles (genus $g > 0$). Equation (5.106) then generalizes as

$$V - I + L + 1 = 2 \longrightarrow V - I + L + 1 = 2 - 2g, \qquad (5.112)$$

where again I is the number of internal lines, L the number of loops, and g represents the genus of the polyhedral surface (the number of handles). The right-hand side of Eq. (5.112) represents the full Euler–L' Huilliers characteristics. Reasoning as above but now based on the general situation of $g \geq 0$, we arrive at

$$\frac{\tilde{K}}{K} \leq \frac{4}{5}\left(1 + \frac{1}{V_4}(1 - 2g)\right), \qquad (V = V_4),$$

$$\frac{\tilde{K}}{K} \leq \frac{2}{3}\left(1 + \frac{2}{V_3}(1 - 2g)\right), \qquad (V = V_3). \qquad (5.113)$$

According to Eq. (5.113) the demand $\frac{\tilde{K}}{K} \leq 1$ for a compact support of the loop integrations is always satisfied for $g \geq 1$ since the number of vertices needs to be positive: $V_4 \geq 0$ and $V_3 \geq 0$. Recall that at $g = 0$ this is true only for $V_4 \geq 5$ and $V_3 \geq 6$, respectively. We thus conclude that bubble diagrams of a topology deviating from planarity are much more severely constrained than their planar counterparts.

5.4.3.3   *One example*

Here we would like to demonstrate how strongly conditions (5.60) and (5.61) constrain the loop momenta by investigating a particular two-loop bubble diagram.

Consider the diagrams in Fig. 5.10. Only thermally propagating TLH modes are involved. Diagram (a) is real while diagram (b) is purely imaginary. Here we only discuss diagram (a).

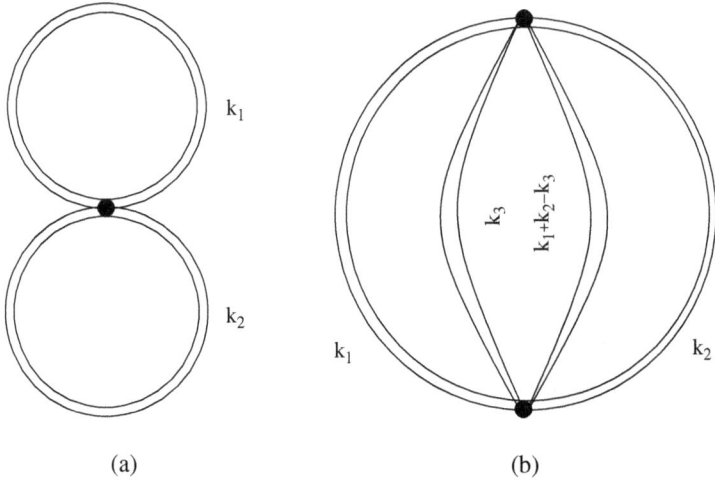

Figure 5.10. (a) Two-loop and (b) three-loop diagram contributing to the pressure in the deconfining phase of SU(2) Yang–Mills thermodynamics. All (double) lines are associated with thermal TLH mode propagation.

We have $\tilde{K} = 4$ and $K = 1$. Thus the region of integration for the radial loop variables cannot be compact. Let us show this explicitly. Before applying the constraints in (5.61) we have[25] [Herbst, Hofmann, and Rohrer (2004)]:

$$|\Delta P_a| = \frac{e^2 \Lambda^4 \lambda^{-2}}{(2\pi)^4} \sum_{\pm} \int dx_1 \int dx_2 \int dz_{12} \frac{x_1^2 x_2^2}{\sqrt{x_1^2 + 4e^2}\sqrt{x_2^2 + 4e^2}}$$

$$\times P_a^{\pm}(x_1, x_2, z_{12}) \, n_B \left(2\pi\lambda^{-3/2}\sqrt{x_1^2 + 4e^2} \right)$$

$$\times n_B \left(2\pi\lambda^{-3/2}\sqrt{x_2^2 + 4e^2} \right). \tag{5.114}$$

---

[25]The reader is referred to Sec. 5.4.4 to see how the expression in Eq. (5.114) emerges through application of the Feynman rules specified in Sec. 5.4.2.

Recall that $\lambda \equiv \frac{2\pi T}{\Lambda}$. We define $\vec{x}_i \equiv \frac{\vec{k}_i}{|\phi|}$, $x_i \equiv |\vec{x}_i|$ $(i = 1, 2)$, $z_{12} \equiv \cos \angle(\vec{x}_1, \vec{x}_2)$, and $P_a^{\pm}(x_1, x_2, z_{12})$ is given as:

$$
P_a^{\pm}(x_1, x_2, z_{12}) \equiv \frac{1}{2} \left(6 - \frac{x_1^2}{4e^2} - \frac{x_2^2}{4e^2} - \frac{x_1^2 x_2^2}{16e^4}(1 + z_{12}^2) \right.
$$

$$
\left. \pm \, 2x_1 x_2 z_{12} \frac{\sqrt{x_1^2 + 4e^2}\sqrt{x_2^2 + 4e^2}}{16e^4} \right). \qquad (5.115)
$$

Applying the constraint $|(k_1 + k_2)^2| \le |\phi|^2$ [see (5.61)], we have

$$
\left| 4e^2 \pm \sqrt{x_1^2 + 4e^2}\sqrt{x_2^2 + 4e^2} - x_1 x_2 z_{12} \right| \le \frac{1}{2}. \qquad (5.116)
$$

Only the minus sign is relevant $(e \ge \frac{1}{2\sqrt{2}})$ in (5.116). Thus the expression within the absolute-value signs is strictly negative, and we have

$$
z_{12} \le \frac{1}{x_1 x_2} \left(4e^2 - \sqrt{x_1^2 + 4e^2}\sqrt{x_2^2 + 4e^2} + \frac{1}{2} \right) \equiv g_{12}(x_1, x_2). \qquad (5.117)
$$

Notice that $\lim_{x_1, x_2 \to \infty} g_{12}(x_1, x_2) = -1$. Apart from a small compact region, where $g_{12}(x_1, x_2) \ge 1$ and which includes the point $x_1 = x_2 = 0$ in the $(x_1 \ge 0, x_2 \ge 0)$-quadrant, the admissible region of x_1, x_2-integration $(-1 \le g_{12}(x_1, x_2) < 1)$ is an infinite strip bounded by the graphs of the two functions

$$
x_2^u(x_1) = \frac{x_1 + 8e^2 + \sqrt{1 + 16e^2}\sqrt{x_1^2 + 4e^2}}{8e^2},
$$

$$
x_2^l(x_1) = \frac{x_1 + 8e^2 - \sqrt{1 + 16e^2}\sqrt{x_1^2 + 4e^2}}{8e^2}. \qquad (5.118)
$$

We conclude that the integration region for radial loop momenta is not compact. Large x_1- and/or x_2- values are, however, Bose-suppressed in Eq. (5.114), and the ratio $\frac{|\Delta P_a|}{P_{1\text{-loop}}}$, as a function of λ, is at most of order 10^{-5} [Schwarz, Hofmann, and Giacosa (2006)] (see also Sec. 5.4.4).

5.4.3.4 *Conclusions*

We have discussed how constraints on loop momenta, which emerge in the effective theory for the deconfining phase of SU(2) Yang–Mills thermodynamics in the physical unitary–Coulomb gauge [Hofmann (2005)], enforce a loop expansion with properties dissenting from those known in perturbation theory. Namely, we have argued that modulo 1PI resummations there is only a finite number of connected bubble diagrams contributing to the expansion of thermodynamical quantities or, by cutting one internal line, to the expansion of the polarization tensor. Our arguments of Sec. 5.4.3.2 apply equally well to the SU(3) case. Because the tree-level quasiparticle mass spectrum is slightly more involved there are mild modifications of the tree-level constraints when going from SU(2) to SU(3).

The reason for the improved convergence properties of loop expansions in the effective theory is clear: The spatial coarse-graining over topological, Euclidean field configurations (core regions: pure quantum pieces, giving rise to the inert field ϕ, peripheral regions: static, (anti)selfdual fields whose overlaps generate an effective pure-gauge configuration a_μ^{gs} which, indeed, is sourced by field ϕ) selfconsistenty generates both a natural resolving power as a function of T and Λ and quasiparticle masses on tree-level. As a consequence, a residual quantum fluctuation in the effective theory typically has a small action as compared to $\hbar$. Moreover, the single-particle action drops rapidly as the collectivity of the quantum fluctuation described by an irreducible loop diagram increases.

5.4.4 *Two-loop corrections to the pressure*

5.4.4.1 *General considerations*

In [Herbst, Hofmann, and Rohrer (2004)] the two-loop corrections to the pressure of an SU(2) Yang–Mills theory in its deconfining phase were calculated omitting the term $\propto u_\mu u_\nu$ in Eq. (5.97). Also, an erroneous evolution equation for the effective coupling $e(\lambda)$ was used in [Herbst, Hofmann, and Rohrer (2004)]. In [Schwarz, Hofmann, and Giacosa (2006)] the term $\propto u_\mu u_\nu$ was taken into account, and the proper evolution equation (5.70) or (5.71) was applied. The intention of this section is to discuss the computation or the estimation of effective two-loop corrections to the pressure in rather great technical detail. We deem necessary since the treatment of radiative effects in Yang–Mills thermodynamics in the effective theory deviates strongly from the established perturbative method.

The two-loop corrections to the pressure are calculated as indicated diagrammatically in Fig. 5.11. One has

$$\Delta P = \Delta P_{\text{nonlocal}} + \Delta P_{\text{local}}. \qquad (5.119)$$

According to the Feynman rules of Sec. 5.4.2 the expressions take the form

$$\Delta P_{\text{local}} = \frac{1}{8i} \int \frac{d^4k}{(2\pi)^4} \frac{d^4p}{(2\pi)^4} \Gamma^{\mu\nu\rho\sigma}_{[4]abcd} D_{\mu\nu,ab}(k) D_{\rho\sigma,cd}(p) \qquad (5.120)$$

and

$$\Delta P_{\text{nonlocal}} = \frac{1}{8i} \int \frac{d^4k}{(2\pi)^4} \frac{d^4p}{(2\pi)^4} \Gamma^{\lambda\mu\nu}_{[3]abc}(p,k,-p-k) \Gamma^{\rho\sigma\tau}_{[3]rst}(-p,-k,p+k)$$
$$\times D_{\lambda\rho,ar}(p) D_{\mu\sigma,bs}(k) D_{\nu\tau,ct}(-p-k). \qquad (5.121)$$

Here $D_{\mu\nu,ab}$ stands for the appropriate TLH and TLM propagator.

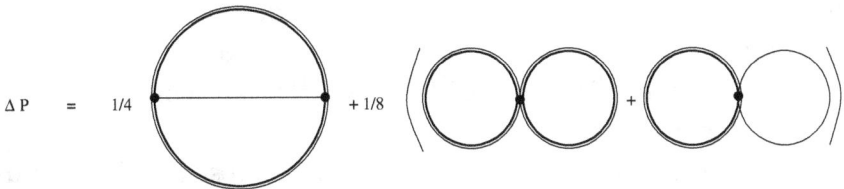

$$\Delta P \quad = \quad 1/4 \quad \bigcirc \quad + 1/8 \left(\bigcirc\bigcirc \quad + \quad \bigcirc\bigcirc \right)$$

Figure 5.11. Two-loop corrections to the pressure. The diagrams to the left (right) are referred to as nonlocal (local) in the text. Double (single) lines are associated with TLH- (TLM-) mode propagation.

After summing over Lie-algebra indices and contracting Lorentz indices the δ-functions associated with the TLH propagators render the integration over the zero components of the loop momenta trivial. The remaining integrations over spatial momenta are performed in spherical coordinates. At this point both constraints (5.60) and (5.61) are used to determine the support for these integrations.

5.4.4.2 *Calculation*

The following calculation of the two-loop corrections to the free quasiparticle pressure in deconfining SU(2) Yang–Mills thermodynamics proceeds along the lines of [Herbst, Hofmann, and Rohrer (2004); Schwarz, Hofmann, and Giacosa (2006)] exploiting the Feynman rules of Sec. 5.4.2.

Let us first introduce a useful convention: Due to the split of the TLM propagator into vacuum (quantum), Coulomb, and thermal contributions in Eq. (5.97), combinations of thermal and vacuum contributions of the TLM propagator arise in Eqs. (5.120) and (5.121). We will consider these contributions separately and denote them by

$$\Delta P^{XYZ}_{\alpha_X \beta_Y \gamma_Z} \quad \text{and} \quad \Delta P^{XY}_{\alpha_X \beta_Y} \tag{5.122}$$

for the nonlocal diagram and the local diagrams in Fig. 5.11, respectively. In Eq. (5.122) capital roman letters take the values H or M, indicating the propagator type (TLH/TLM), and the associated small Greek letters take the values v (vacuum) or t (thermal) or c [Coulomb, the term $\propto u_\mu u_\nu$ in Eq. (5.97)], indicating which part of the propagator is considered. We need to compute ΔP^{HH}_{tt}, ΔP^{HM}_{tt}, ΔP^{HM}_{tv}, ΔP^{HM}_{tc}, ΔP^{HHM}_{ttv}, and ΔP^{HHM}_{ttc}. (ΔP^{HHM}_{ttt} vanishes by momentum conservation.)

The correction ΔP^{HH}_{tt} was computed in [Herbst, Hofmann, and Rohrer (2004)]. [ΔP^{HH}_{vt} and ΔP^{HH}_{vv} do not exist by virtue of Eq. (5.95)]. To perform the contractions in Eqs. (5.120) and (5.121) it is useful to exploit the transversality $P^T_{\mu\nu}$ in the TLM propagator from the start.

The following four relations hold:

$$P_{\mu\nu}^T(q)q^\mu = 0,$$
$$P_{\mu\nu}^T(q)g^{\mu\nu} = -2,$$
$$P_{\mu\nu}^T(q)p^\mu p^\nu = |\vec{p}|^2 - \frac{(\vec{q}\vec{p})^2}{|\vec{q}|^2},$$
$$P_{\mu\nu}^T(q)p^\mu k^\nu = \vec{p}\vec{k} - \frac{(\vec{k}\vec{q})(\vec{p}\vec{q})}{|\vec{q}|^2}. \tag{5.123}$$

Exploiting Eqs. (5.123) and, for historical reasons, *ignoring* the Coulomb part of the TLM propagator, the contractions for local contributions are as follows.

(1) Local, TLH-TLH:

$$\Gamma_{[4]abcd}^{\mu\nu\rho\sigma}\delta_{ah}\delta_{bh}\tilde{D}_{\mu\nu}(p)\delta_{cl}\delta_{dl}\tilde{D}_{\rho\sigma}(k) = -ie^2[\epsilon_{abe}\epsilon_{cde}(g^{\mu\rho}g^{\nu\sigma} - g^{\mu\sigma}g^{\nu\rho}) +$$
$$\epsilon_{ace}\epsilon_{bde}(g^{\mu\nu}g^{\rho\sigma} - g^{\mu\sigma}g^{\nu\rho}) + \epsilon_{ade}\epsilon_{bce}(g^{\mu\nu}g^{\rho\sigma} - g^{\mu\rho}g^{\nu\sigma})] \times$$
$$\delta_{ah}\delta_{bh}\left(g_{\mu\nu} - \frac{p_\mu p_\nu}{m^2}\right)\delta_{cl}\delta_{dl}\left(g_{\rho\sigma} - \frac{k_\rho k_\sigma}{m^2}\right)$$
$$= -\frac{1}{3}ie^2\epsilon_{ace}\epsilon_{ace}[2g^{\mu\nu}g^{\rho\sigma} - g^{\mu\sigma}g^{\nu\rho} - g^{\mu\rho}g^{\nu\sigma}] \times$$
$$\left(g_{\mu\nu} - \frac{p_\mu p_\nu}{m^2}\right)\left(g_{\rho\sigma} - \frac{k_\rho k_\sigma}{m^2}\right)$$
$$= -2ie^2\left(24 - 6\frac{p^2}{m^2} - 6\frac{k^2}{m^2} + 2\frac{p^2k^2}{m^4} - 2\frac{(pk)^2}{m^4}\right), \tag{5.124}$$

where $h = 1,2$ and $l = 1,2$ are summed over.

(2) Local, TLH-TLM:

$$\Gamma_{[4]abcd}^{\mu\nu\rho\sigma}\delta_{ah}\delta_{bh}\tilde{D}_{\mu\nu}(p)\delta_{cl}\delta_{dl}P_{\rho\sigma}^T(k) = -ie^2[\epsilon_{abe}\epsilon_{cde}(g^{\mu\rho}g^{\nu\sigma} - g^{\mu\sigma}g^{\nu\rho}) +$$
$$\epsilon_{ace}\epsilon_{bde}(g^{\mu\nu}g^{\rho\sigma} - g^{\mu\sigma}g^{\nu\rho}) + \epsilon_{ade}\epsilon_{bce}(g^{\mu\nu}g^{\rho\sigma} - g^{\mu\rho}g^{\nu\sigma})] \times$$
$$\delta_{ah}\delta_{bh}\left(g_{\mu\nu} - \frac{p_\mu p_\nu}{m^2}\right)\delta_{cl}\delta_{dl}P_{\rho\sigma}^T(k)$$

$$= -\frac{1}{3} ie^2 \epsilon_{ace} \epsilon_{ace} [2g^{\mu\nu} g^{\rho\sigma} - g^{\mu\sigma} g^{\nu\rho} - g^{\mu\rho} g^{\nu\sigma}] \times$$

$$\left(g_{\mu\nu} - \frac{p_\mu p_\nu}{m^2} \right) p_{\rho\sigma}^T (k)$$

$$= -2ie^2 \left(-12 + 4\frac{p^2}{m^2} + 2\frac{\vec{p}^2 \sin^2\theta}{m^2} \right), \tag{5.125}$$

where $h = 1, 2$ and $l = 1, 2$ are summed over. In Eq. (5.125), θ denotes the angle between $\vec{p}$ and $\vec{k}$.

For the nonlocal diagram we obtain:

$$\Gamma^{\lambda\mu\nu}_{[3]abc}(p, k, q) \Gamma^{\rho\sigma\tau}_{[3]rst}(p, k, q) \delta_{ah} \delta_{rh} \tilde{D}_{\lambda\rho}(p) \delta_{bl} \delta_{sl} \tilde{D}_{\mu\sigma}(k) \delta_{c3} \delta_{t3} P^T_{\nu\tau}(q)$$

$$= e^2 \epsilon_{abc} \epsilon_{rst} [g^{\lambda\mu}(p-k)^\nu + g^{\mu\nu}(k-q)^\lambda + g^{\nu\lambda}(q-p)^\mu] \times$$

$$[g^{\rho\sigma}(p-k)^\tau + g^{\sigma\tau}(k-q)^\rho + g^{\tau\rho}(q-p)^\sigma] \times$$

$$\delta_{ah} \delta_{rh} \delta_{bl} \delta_{sl} \delta_{c3} \delta_{t3} \left(g_{\lambda\rho} - \frac{p_\lambda p_\rho}{m^2} \right) \left(g_{\mu\sigma} - \frac{k_\mu k_\sigma}{m^2} \right) P^T_{\nu\tau}(q), \tag{5.126}$$

where $h = 1, 2$ and $l = 1, 2$ are summed over. To not loose track, we split the calculation into terms $\propto e^2$, $\propto \frac{e^2}{m^2}$ and $\propto \frac{e^2}{m^4}$ and keep $P^T_{\nu\tau}$ uncontracted in a first step. The contraction of structure constants gives an additional factor 2.

The term $\propto e^2$ evaluates as:

$$[g^{\lambda\mu}(p-k)^\nu + g^{\mu\nu}(k-q)^\lambda + g^{\nu\lambda}(q-p)^\mu][g^{\rho\sigma}(p-k)^\tau$$

$$+ g^{\sigma\tau}(k-q)^\rho + g^{\tau\rho}(q-p)^\sigma] g_{\lambda\rho} g_{\mu\sigma} P^T_{\nu\tau}(q)$$

$$= [g_{\rho\sigma}(p-k)^\nu + g^\nu_\sigma(k-q)_\rho + g^\nu_\rho(q-p)_\sigma]$$

$$\times [g^{\rho\sigma}(p-k)^\tau + g^{\sigma\tau}(k-q)^\rho + g^{\tau\rho}(q-p)^\sigma] P^T_{\nu\tau}(q)$$

$$= [4(p-k)^\nu(p-k)^\tau + (p-k)^\nu(k-q)^\tau + (p-k)^\nu(q-p)^\tau$$

$$+ (k-q)^\nu(p-k)^\tau + (k-q)^2 g^{\nu\tau} + (q-p)^\nu(k-q)^\tau$$

$$+ (q-p)^\nu(p-k)^\tau + (k-q)^\nu(q-p)^\tau + (q-p)^2 g^{\nu\tau}] P^T_{\nu\tau}(q)$$

$$= [2p^\nu p^\tau + 2k^\nu k^\tau - 6p^\nu k^\tau + (q-p)^2 g^{\nu\tau} + (k-q)^2 g^{\nu\tau}] P^T_{\nu\tau}(q)$$

$$= 2\left(\vec{p}^2 - \frac{(\vec{p}\vec{q})^2}{|\vec{q}|^2}\right) + 2\left(\vec{k}^2 - \frac{(\vec{k}\vec{q})^2}{|\vec{q}|^2}\right) - 6\left(\vec{p}\vec{k} - \frac{(\vec{p}\vec{q})(\vec{k}\vec{q})}{|\vec{q}|^2}\right)$$
$$- 2(q-p)^2 - 2(k-q)^2. \tag{5.127}$$

Terms proportional to q^ν or q^τ have been omitted after the second-last equality sign in Eq. (5.127) because, when contracted with $P^T_{\nu\tau}(q)$, they vanish. Again, using Eq. (5.123) the expression after the last equality sign in Eq. (5.127) easily follows.

Next we look at the two terms proportional to $\propto \frac{e^2}{m^2}$ [compare with Eq. (5.126)]. The first contribution is:

$$[g^{\lambda\mu}(p-k)^\nu + g^{\mu\nu}(k-q)^\lambda + g^{\nu\lambda}(q-p)^\mu][g^{\rho\sigma}(p-k)^\tau$$
$$+ g^{\sigma\tau}(k-q)^\rho + g^{\tau\rho}(q-p)^\sigma]g_{\lambda\rho}k_\mu k_\sigma P^T_{\nu\tau}(q)$$
$$= [k_\rho k_\sigma(p-k)^\nu + k^\nu k_\sigma(k-q)_\rho + k(q-p)g^\nu_\rho k_\sigma]$$
$$\times [g^{\rho\sigma}(p-k)^\tau + g^{\sigma\tau}(k-q)^\rho + g^{\tau\rho}(q-p)^\sigma]P^T_{\nu\tau}(q)$$
$$= [k^2(p-k)^\nu(p-k)^\tau + k(k-q)(p-k)^\nu k^\tau + k(q-p)(p-k)^\nu k^\tau$$
$$+ k(k-q)k^\nu(p-k)^\tau + (k-q)^2 k^\nu k^\tau + k(q-p)k^\nu(k-q)^\tau$$
$$+ k(q-p)k^\nu(p-k)^\tau + k(q-p)(k-q)^\nu k^\tau$$
$$+ [k(q-p)]^2 g^{\nu\tau}]P^T_{\nu\tau}(q)$$
$$= [k^2 p^\nu p^\tau + q^2 k^\nu k^\tau - 2(kp)p^\nu k^\tau + [k(q-p)]^2 g^{\nu\tau}]P^T_{\nu\tau}(q)$$
$$= k^2\left(\vec{p}^2 - \frac{(\vec{p}\vec{q})^2}{|\vec{q}|^2}\right) + q^2\left(\vec{k}^2 - \frac{(\vec{k}\vec{q})^2}{|\vec{q}|^2}\right) - 2pk\left(\vec{p}\vec{k} - \frac{(\vec{p}\vec{q})(\vec{k}\vec{q})}{|\vec{q}|^2}\right)$$
$$- 2[k(q-p)]^2. \tag{5.128}$$

The second term $\propto \frac{e^2}{m^2}$ is obtained either by direct calculation or by just exchanging $p \leftrightarrow k$ in Eq. (5.128):

$$[g^{\lambda\mu}(p-k)^\nu + g^{\mu\nu}(k-q)^\lambda + g^{\nu\lambda}(q-p)^\mu][g^{\rho\sigma}(p-k)^\tau$$
$$+ g^{\sigma\tau}(k-q)^\rho + g^{\tau\rho}(q-p)^\sigma]g_{\mu\sigma}p_\lambda p_\rho \bar{D}_{\nu\tau}(q)$$

$$= p^2\left(\vec{k}^2 - \frac{(\vec{k}\vec{q})^2}{|\vec{q}|^2}\right) + q^2\left(\vec{p}^2 - \frac{(\vec{p}\vec{q})^2}{|\vec{q}|^2}\right) - 2pk\left(\vec{p}\vec{k} - \frac{(\vec{p}\vec{q})(\vec{k}\vec{q})}{|\vec{q}|^2}\right)$$

$$- 2[p(q-k)]^2. \tag{5.129}$$

Finally, the term $\propto 2\frac{e^2}{m^4}$ is given as

$$[g^{\lambda\mu}(p-k)^\nu + g^{\mu\nu}(k-q)^\lambda + g^{\nu\lambda}(q-p)^\mu][g^{\rho\sigma}(p-k)^\tau$$
$$+ g^{\sigma\tau}(k-q)^\rho + g^{\tau\rho}(q-p)^\sigma]p_\lambda p_\rho k_\mu k_\sigma P_{\nu\tau}^T(q)$$
$$= [(pk)p_\rho k_\sigma(p-k)^\nu + p(k-q)p_\rho k_\sigma k^\nu + k(q-p)p_\rho k_\sigma p^\nu]$$
$$\times [g^{\rho\sigma}(p-k)^\tau + g^{\sigma\tau}(k-q)^\rho + g^{\tau\rho}(q-p)^\sigma]P_{\nu\tau}^T(q)$$
$$= [(pk)(p-k)^\nu + [p(k-q)]k^\nu + [k(q-p)]p^\nu]$$
$$\times [(pk)(p-k)^\tau + [p(k-q)]k^\tau + [k(q-p)]p^\tau]P_{\nu\tau}^T(q)$$
$$= [(kq)^2 p^\nu p^\tau + (pq)^2 k^\nu k^\tau - 2(pq)(kq)p^\nu k^\tau]P_{\nu\tau}^T(q)$$
$$= (kq)^2\left(\vec{p}^2 - \frac{(\vec{p}\vec{q})^2}{|\vec{q}|^2}\right) + (pq)^2\left(\vec{k}^2 - \frac{(\vec{k}\vec{q})^2}{|\vec{q}|^2}\right)$$
$$- 2(pq)(kq)\left(\vec{p}\vec{k} - \frac{(\vec{p}\vec{q})(\vec{k}\vec{q})}{|\vec{q}|^2}\right). \tag{5.130}$$

Now, adding up Eqs. (5.127) through (5.130) (taking care of the correct signs), we have

Eq. $(5.126) = 2e^2[$Eq. $(5.127) -$ Eq. (5.128)

$$- \text{Eq. } (5.129) + \text{Eq. } (5.130)]$$

$$= 2e^2\left\{2\frac{[p(q-k)]^2}{m^2} + \frac{2[k(q-p)]^2}{m^2} - 2(q-p)^2 - 2(k-q)^2\right.$$

$$+ \left[2 - \frac{k^2}{m^2} - \frac{q^2}{m^2} + \frac{(kq)^2}{m^4}\right]\left(\vec{p}^2 - \frac{(\vec{p}\vec{q})^2}{|\vec{q}|^2}\right)$$

$$+\left[2-\frac{q^2}{m^2}-\frac{p^2}{m^2}+\frac{(pq)^2}{m^4}\right]\left(\vec{k}^2-\frac{(\vec{k}\vec{q})^2}{|\vec{q}|^2}\right)$$

$$-\left[6-4\frac{pk}{m^2}+2\frac{(pq)(kq)}{m^4}\right]\left(\vec{p}\vec{k}-\frac{(\vec{p}\vec{q})(\vec{k}\vec{q})}{|\vec{q}|^2}\right)\right\}. \qquad (5.131)$$

Using momentum conservation at the vertices, that is $q = -p - k$, we find:

$$\left(\vec{p}^2-\frac{(\vec{p}\vec{q})^2}{|\vec{q}|^2}\right)=\left(\vec{k}^2-\frac{(\vec{k}\vec{q})^2}{|\vec{q}|^2}\right)=-\left(\vec{p}\vec{k}-\frac{(\vec{p}\vec{q})(\vec{k}\vec{q})}{|\vec{q}|^2}\right)=\frac{\vec{p}^2\vec{k}^2\sin^2\theta}{(\vec{p}+\vec{k})^2}.$$

$$(5.132)$$

Thus we finally arrive at the following expression:
(3) Nonlocal, TLH-TLH-TLM:

$$\Gamma_{[3]abc}^{\lambda\mu\nu}(p,k,q)\Gamma_{[3]rst}^{\rho\sigma\tau}(-p,-k,-q)\delta_{ah}\delta_{rh}\delta_{bl}\delta_{sl}\delta_{c3}\delta_{t3}\tilde{D}_{\lambda\rho}(p)\tilde{D}_{\mu\sigma}(k)P_{\nu\tau}^T(q)$$

$$= 2e^2\left[10p^2+10k^2+16pk-2\frac{k^4}{m^2}-2\frac{p^4}{m^2}-8\frac{p^2(pk)}{m^2}-8\frac{k^2(pk)}{m^2}\right.$$

$$-16\frac{(pk)^2}{m^2}-\frac{\vec{p}^2\vec{k}^2\sin^2\theta}{(p+k)^2}\left(10-3\frac{p^2}{m^2}-3\frac{k^2}{m^2}-8\frac{pk}{m^2}+\frac{p^4}{m^4}\right.$$

$$\left.+\frac{k^4}{m^4}+4\frac{p^2(pk)}{m^4}+4\frac{k^2(pk)}{m^4}+4\frac{(pk)^2}{m^4}+2\frac{p^2k^2}{m^4}\right)\right]. \qquad (5.133)$$

Here we have used the fact that $\Gamma_{[3]}(-p,-k,-q) = -\Gamma_{[3]}(p,k,q)$.

With the contractions of tensor structures at hand, we are now in a position to calculate all two-loop corrections. For ΔP_{tt}^{HH} this is done in detail, for the other contributions we resort to a more compact presentation. We have

$$\Delta P_{tt}^{HH} = \frac{1}{i} \int \frac{d^4p\, d^4k}{(2\pi)^8} \, \Gamma_{[4]hhll}^{\mu\nu\rho\sigma} \tilde{D}_{\mu\nu}(p)\tilde{D}_{\rho\sigma}(k)$$

$$\times (2\pi)\delta(p^2 - m^2)\, n_B(|p_0/T|)(2\pi)\delta(k^2 - m^2)\, n_B(|k_0/T|)$$

$$= -2e^2 \int \frac{d^4p\, d^4k}{(2\pi)^6} \left(24 - 6\frac{p^2}{m^2} - 6\frac{k^2}{m^2} + 2\frac{p^2 k^2}{m^4} - 2\frac{(pk)^2}{m^4} \right)$$

$$\times \delta(p^2 - m^2)n_B(|p_0/T|)\, \delta(k^2 - m^2)n_B(|k_0/T|). \qquad (5.134)$$

The product of δ-functions can be rewritten as

$$\delta(p^2 - m^2)\delta(k^2 - m^2) = \frac{1}{4\sqrt{\vec{p}^2 + m^2}\sqrt{\vec{k}^2 + m^2}} \qquad (5.135)$$

$$\times \left[\delta(p_0 - \sqrt{\vec{p}^2 + m^2})\delta(k_0 - \sqrt{\vec{k}^2 + m^2}) \right.$$

$$+ \delta(p_0 - \sqrt{\vec{p}^2 + m^2})\delta(k_0 + \sqrt{\vec{k}^2 + m^2})$$

$$+ \delta(p_0 + \sqrt{\vec{p}^2 + m^2})\delta(k_0 - \sqrt{\vec{k}^2 + m^2})$$

$$\left. + \delta(p_0 + \sqrt{\vec{p}^2 + m^2})\delta(k_0 + \sqrt{\vec{k}^2 + m^2}) \right]. \qquad (5.136)$$

The expression in the round brackets of the integrand of Eq. (5.130) contains only even products of k and p like p^2, k^2 or pk. Thus, performing the integration over p_0 and k_0 the factors

$$\delta(p_0 - \sqrt{\vec{p}^2 + m^2})\, \delta(k_0 + \sqrt{\vec{k}^2 + m^2}) \quad \text{and}$$

$$\delta(p_0 + \sqrt{\vec{p}^2 + m^2})\, \delta(k_0 - \sqrt{\vec{k}^2 + m^2}),$$

(signs in the argument of δ-functions in (5.136) opposite, crossed-terms) lead to the same result. This is also true for the two uncrossed products of δ-functions with equal signs. After the integration is performed we may therefore set

$$p^2 \to m^2,$$

$$k^2 \to m^2,$$

$$(pk) \to \pm\sqrt{\vec{p}^2 + m^2}\sqrt{\vec{k}^2 + m^2} - \vec{p}\vec{k},$$

$$(pk)^2 \to \vec{p}^2\vec{k}^2 + (\vec{p}^2 + \vec{k}^2)m^2 + m^4 \mp 2\vec{p}\vec{k}\sqrt{\vec{p}^2 + m^2}\sqrt{\vec{k}^2 + m^2} + (\vec{p}\vec{k})^2.$$

$$(5.137)$$

The upper case in (5.137) is obtained when the signs in the arguments of the δ function factors in (5.136) are equal, the lower case when they are opposite.

For local two-loop diagrams only the s-channel constraint of (5.61) represents a restriction on the loop integrations, t- and u-channel conditions are automatically satisfied. Examining the integration constraint

$$|\phi|^2 \geq |(p+k)^2| = |p_0^2 - \vec{p}^2 + k_0^2 - \vec{k}^2 + 2p_0k_0 - 2\vec{p}\vec{k}|$$

$$\to |\phi|^2 \geq |2m^2 \pm 2\sqrt{\vec{p}^2 + m^2}\sqrt{\vec{k}^2 + m^2} - 2\vec{p}\vec{k}|$$

$$\to 1 \geq 2|(2e)^2 \pm \sqrt{x^2 + (2e)^2}\sqrt{y^2 + (2e)^2} - xy\cos\theta| \quad (5.138)$$

shows that only the combinations with opposite signs in (5.136) need to be evaluated. In Eq. (5.138) we have introduced dimensionless variables

$$x = |\vec{p}|/|\phi|, \quad y = |\vec{k}|/|\phi|,$$

$$z = \cos\theta, \quad \lambda^{-3/2} = \frac{|\phi|}{2\pi T}. \quad (5.139)$$

We observe that for the "+" case the difference between the second and third term is always positive in (5.138). Because of the first term, the entire expression is greater than unity for $e \geq \sqrt{8\pi}$. Thus only the "−" case needs to be considered. This is also true for $\Delta P_{tt}^{\mathrm{HM}}$ and $\Delta P_{ttv}^{\mathrm{HHM}}$ though the analytical expressions look different.

Applying Eq. (5.124), ΔP_{tt}^{HH} can be reduced to

$$\Delta P_{tt}^{HH} = -2e^2 \int \frac{d^4p\, d^4k}{(2\pi)^6} \left(24 - 6\frac{p^2}{m^2} - 6\frac{k^2}{m^2} + 2\frac{p^2k^2}{m^4} - 2\frac{(pk)^2}{m^4} \right)$$

$$\times \delta(p_0 - \sqrt{\vec{p}^2 + m^2})\delta(k_0 + \sqrt{\vec{k}^2 + m^2})\frac{n_B(|p_0/T|)n_B(|k_0/T|)}{2\sqrt{\vec{p}^2 + m^2}\sqrt{\vec{k}^2 + m^2}}.$$

Integrating over the zero components by using Eq. (5.137), we arrive at

$$\Delta P_{tt}^{HH} = -2e^2 \int \frac{d^3\vec{p}\, d^3\vec{k}}{(2\pi)^6} \frac{n_B(\sqrt{\vec{p}^2 + m^2}/T)n_B(\sqrt{\vec{k}^2 + m^2}/T)}{2\sqrt{\vec{p}^2 + m^2}\sqrt{\vec{k}^2 + m^2}}$$

$$\times \left[12 - 2\frac{\vec{p}^2}{m^2} - 2\frac{\vec{k}^2}{m^2} - 2\frac{\vec{p}^2\vec{k}^2}{m^4} - 2\frac{(\vec{p}\vec{k})^2}{m^4} \right.$$

$$\left. - 4\frac{\vec{p}\vec{k}}{m^4}\sqrt{\vec{p}^2 + m^2}\sqrt{\vec{k}^2 + m^2} \right]. \tag{5.140}$$

After a change to polar coordinates and an evaluation of the angular integrals the remaining integration measure takes the form $2(2\pi)^2|\vec{p}|^2|\vec{k}|^2d|\vec{p}|d|\vec{k}|d\cos\theta$. As a last step we rescale variables as in Eq. (5.139). The "$-$" case of the constraint of Eq. (5.138) can be re-cast into the following form:

$$-1/2 \leq (2e)^2 - \sqrt{x^2 + (2e)^2}\sqrt{y^2 + (2e)^2} - xyz \leq +1/2. \tag{5.141}$$

Our final result for ΔP_{tt}^{HH} reads

$$\Delta P_{tt}^{HH} = \frac{-2e^2T^4}{\lambda^6} \int dx\, dy\, dz\ \frac{x^2y^2}{\sqrt{x^2 + (2e)^2}\sqrt{y^2 + (2e)^2}}$$

$$\times \left[12 - 2\frac{x^2}{(2e)^2} - 2\frac{y^2}{(2e)^2} - 2\frac{x^2y^2}{(2e)^4} - 2\frac{x^2y^2z^2}{(2e)^4} \right.$$

$$\Delta P^{MH} = \frac{1}{8} \cdot p$$

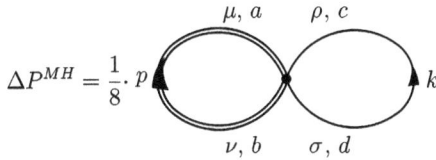

Figure 5.12. The diagram for ΔP^{HM}. A double line denotes the TLH- and a single line TLM-mode propagation.

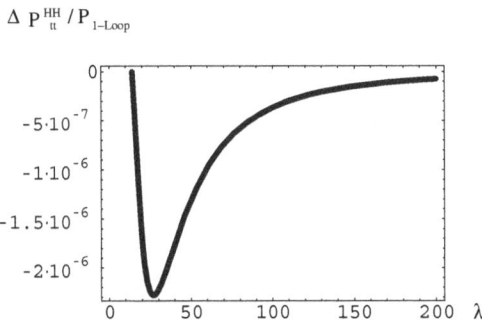

Figure 5.13. $\Delta P_{tt}^{\text{HH}}/P_{\text{1-loop}}$ as a function of λ where $P_{\text{1-loop}}$ is the pressure due to one-loop fluctuations. (The ground-state part in P is subtracted in $P_{\text{1-loop}}$).

$$- 4\frac{xyz}{(2e)^4}\sqrt{x^2 + (2e)^2}\sqrt{y^2 + (2e)^2}\Bigg]$$

$$\times n_B(2\pi\lambda^{-3/2}\sqrt{x^2 + (2e)^2})n_B(2\pi\lambda^{-3/2}\sqrt{y^2 + (2e)^2}),$$

$$(5.142)$$

where the integration is subject to the constraint (5.141). Figure 5.14 depicts the λ dependence of the expression in Eq. (5.142).

Other two-loop corrections $\Delta P_{tt}^{\text{HM}}$, $\Delta P_{tv}^{\text{HM}}$ and $\Delta P_{ttv}^{\text{HHM}}$ are calculated in a similar way. Applying the Feynman rules to the diagram in Fig. 5.12, we have

$$\Delta P_{tt}^{\text{HM}} = \frac{1}{i}\int\frac{d^4p\,d^4k}{(2\pi)^6}\Gamma_{[4]hh33}^{\mu\nu\rho\sigma}\tilde{D}_{\mu\nu}(p)P_{\rho\sigma}^T(k)$$

$$\times n_B(|p_0/T|)n_B(|k_0/T|)\delta(p^2 - m^2)\delta(k^2)$$

where $h = 1, 2$ is summed over. Consider now the following integration constraint:

$$|\phi|^2 \geq |(p+q)^2| = |p_0^2 - \vec{p}^2 + k_0^2 - \vec{k}^2 + 2p_0k_0 - 2\vec{p}\vec{k}|$$

$$\rightarrow |\phi|^2 \geq |m^2 \pm 2|\vec{k}|\sqrt{\vec{p}^2 + m^2} - 2|\vec{p}||\vec{k}| \cos\theta|$$

$$\rightarrow 1 \geq |(2e)^2 \pm 2y\sqrt{x^2 + (2e)^2} - 2xy \cos\theta|. \tag{5.143}$$

Again, the "+" case cannot be satisfied ($e \geq \sqrt{8}\pi$), so only the "−" case needs to be considered. Then $\Delta P_{tt}^{\mathrm{HM}}$ reduces to

$$\Delta P_{tt}^{\mathrm{HM}} = -2e^2 \int \frac{d^4p \, d^4k}{(2\pi)^6} \left(-12 + 4\frac{p^2}{m^2} + 2\frac{\vec{p}^2 \sin^2\theta}{m^2} \right)$$

$$\times \frac{n_B(|p_0/T|)n_B(|k_0/T|)}{2|\vec{k}|\sqrt{\vec{p}^2 + m^2}} \delta(p_0 - \sqrt{\vec{p}^2 + m^2})\delta(k_0 + |\vec{k}|)$$

$$= -\frac{2e^2}{(2\pi)^4} \int d|\vec{p}| \, d|\vec{k}| \, d(\cos\theta) \, \vec{p}^2\vec{k}^2 \left(-8 + 2\frac{\vec{p}^2 \sin^2\theta}{m^2} \right)$$

$$\times \frac{n_B(\sqrt{\vec{p}^2 + m^2}/T)n_B(|\vec{k}|/T)}{|\vec{k}|\sqrt{\vec{p}^2 + m^2}}$$

$$= -\frac{2e^2 T^4}{\lambda^6} \int dx \, dy \, dz \, x^2 y \left(-8 + 2\frac{x^2(1 - z^2)}{(2e)^2} \right)$$

$$\times \frac{n_B(2\pi\lambda^{-3/2}\sqrt{x^2 + (2e)^2})n_B(2\pi\lambda^{-3/2}y)}{\sqrt{x^2 + (2e)^2}}, \tag{5.144}$$

subject to the "-" case in the constraint (5.143). Performing the integration in Eq. (5.144) numerically, we arive at the result of Fig. 5.14. This result is important because, in contrast to the other two-loop corrections to the pressure, the ratio $\Delta P_{tt}^{\mathrm{HM}}/P_{1-\mathrm{loop}}$ does not fall off to zero as $\lambda \rightarrow \infty$. As we shall discuss in Sec. 5.5 this high-temperature behavior is induced by an average over domainizations of the field ϕ, that is, an average density of screened and stable magnetic monopoles arising microscopically due to infrequent dissociations of calorons. The temporary large holonomy of the latter

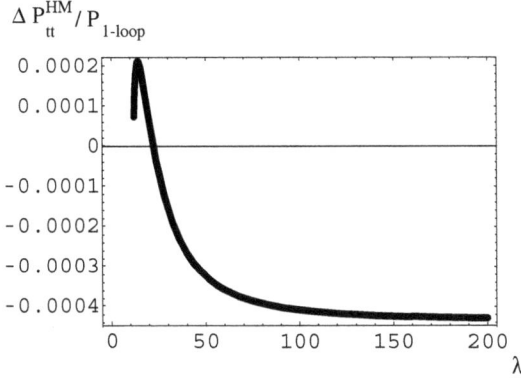

Figure 5.14. $\Delta P_{tt}^{\text{HM}}/P_{1-\text{loop}}$ as a function of λ where $P_{1-\text{loop}}$ is the pressure due to one-loop fluctuations. (The ground-state part in P is subtracted in $P_{1-\text{loop}}$.)

is a consequence of the small momentum transfer mediated by the modes in the effective theory. By definition, this effect cannot be captured by the *a priori* estimate for the thermal ground state but is introduced in an average fashion by an effective radiative correction to the free quasiparticle pressure.

Let us continue by computing $\Delta P_{tv}^{\text{HM}}$. One has

$$\Delta P_{tv}^{\text{HM}} = \frac{1}{i} \int \frac{d^4 p\, d^4 k}{(2\pi)^7} \Gamma_{[4]hh33}^{\mu\nu\rho\sigma} \tilde{D}_{\mu\nu}(p) P_{\rho\sigma}^T(k)\, n_B(|p_0/T|) \frac{i}{k^2} \delta(p^2 - m^2),$$

where $h = 1, 2$ is summed over. After the p_0-integration is performed the integration constraints (5.60) and (5.61) read in this case:

$$|k^2| \le |\phi|^2 \rightarrow |\gamma^2 - y^2| \le 1, \tag{5.145}$$

$$|(p + k)^2| \le |\phi|^2 \rightarrow |(2e)^2 + \gamma^2 - y^2$$

$$\pm 2\gamma\sqrt{x^2 + (2e)^2} - 2xy\cos\theta| \le 1, \tag{5.146}$$

where $y \equiv \frac{|\vec{k}|}{|\phi|}$. The k_0- or γ-integration (γ is the rescaled k_0-component) cannot be performed analytically. Finally, we have

$$\Delta P_{tv}^{\text{HM}} = -2ie^2 \int \frac{d^4 p\, d^4 k}{(2\pi)^7} \left(-12 + 4\frac{p^2}{m^2} + 2\frac{\vec{p}^2 \sin^2\theta}{m^2} \right)$$

$$\times \frac{n_B(\sqrt{\vec{p}^2 + m^2}/T)}{\sqrt{\vec{p}^2 + m^2}} \frac{1}{k^2} \delta(p_0 - \sqrt{\vec{p}^2 + m^2})$$

$$= -4ie^2 \int \frac{d|\vec{p}|\, dk_0\, d|\vec{k}|\, d(\cos\theta)}{(2\pi)^5 \sqrt{\vec{p}^2 + m^2}} \vec{p}^2 \vec{k}^2 \left(-8 + 2\frac{\vec{p}^2 \sin^2\theta}{m^2}\right)$$

$$\times n_B\left(\sqrt{\vec{p}^2 + m^2}/T\right) \frac{1}{k_0^2 - \vec{k}^2}$$

$$= \frac{-4ie^2 T^4}{(2\pi)^5 \lambda^6} \int \frac{dx\, dy\, dy\, dz}{\sqrt{x^2 + (2e)^2}} x^2 y^2 \left(-8 + 2\frac{x^2(1 - z^2)}{(2e)^2}\right)$$

$$\times n_B(2\pi\lambda^{-3/2}\sqrt{x^2 + (2e)^2}) \frac{1}{y^2 - y^2}, \qquad (5.147)$$

subject to the constraints (5.145) and (5.146).

A correction that was omitted in [Herbst, Hofmann, and Rohrer (2004)] but was taken into account in [Schwarz, Hofmann, and Giacosa (2006)] is $\Delta P_{tc}^{\text{HM}}$. Applying Feynman rules to the diagram in Fig. 5.13, yields

$$\Delta P_{tc}^{\text{HM}} = \frac{e^2}{8} \int \frac{d^4k}{(2\pi)^4} \frac{d^4p}{(2\pi)^4}$$

$$\times \left[\epsilon_{fac}\epsilon_{fdb}(g^{\mu\sigma}g^{\nu\rho} - g^{\mu\nu}g^{\rho\sigma}) + \epsilon_{fad}\epsilon_{fbc}(g^{\mu\nu}g^{\rho\sigma} - g^{\mu\rho}g^{\nu\sigma})\right]$$

$$\times \delta_{ah}\delta_{bh}\delta_{cl}\delta_{dl} \left(g_{\mu\nu} - \frac{p_\mu p_\nu}{m^2}\right) 2\pi\, \delta(p^2 - m^2)\, n_B(|p_0|/T) \frac{i u_\rho u_\sigma}{k^2}, \qquad (5.148)$$

where $h = 1, 2$ and $l = 1, 2$ are summed over. After all contractions of indices are performed one has

$$\Delta P_{tc}^{\text{HM}} = \frac{e^2}{2} \int \frac{d^4k}{(2\pi)^4} \frac{d^4p}{(2\pi)^4} \left(2 + \frac{(p_0)^2}{m^2}\right) 2\pi\, \delta(p^2 - m^2)\, n_B(|p_0|/T) \frac{i}{k^2}. \qquad (5.149)$$

Notice that $\Delta P_{tc}^{\text{HM}}$ is manifestly imaginary indicating that it must be canceled by the imaginary part of $\Delta P_{tv}^{\text{HM}}$. (The nonlocal diagram is manifestly real.) Here we resort to an estimate of the modulus

$|\Delta P_{tc}^{HM}|$. This is done as follows. Integrating over p_0 and introducing dimensionless variables as $x \equiv |\mathbf{p}|/|\phi|$, $y \equiv |\mathbf{k}|/|\phi|$, $\gamma \equiv k_0/|\phi|$, and $z \equiv \cos\theta \equiv \cos\angle(\mathbf{p},\mathbf{k})$, we have

$$\Delta P_{tc}^{HM} = \frac{i e^2 \Lambda^4 \lambda^{-2}}{2(2\pi)^5} \sum_{\pm} \int dx\, dy\, dz\, d\gamma \left(3 + \frac{x^2}{4e^2}\right)$$

$$\times \frac{x^2\, n_B(2\pi\lambda^{-3/2}\sqrt{x^2 + 4e^2})}{\sqrt{x^2 + 4e^2}}, \tag{5.150}$$

where $\sum_{\pm}$ refers to the two possible signs of $p_0 \to \pm\sqrt{\mathbf{p}^2 + m^2}$. In dimensionless variables the constraints (5.145) and (5.146) read

$$|\mathbf{k}^2| \le |\phi|^2 \to -1 \le \gamma^2 - y^2 \le 1, \tag{5.151}$$

$$|(p+k)^2| \le |\phi|^2 \to -1 \le 4e^2 \pm 2\sqrt{x^2 + 4e^2}\,\gamma - 2x\,y\,z + \gamma^2 - y^2 \le 1. \tag{5.152}$$

On one hand, to implement both conditions (5.151) and (5.152) exactly is technically involved. On the other hand, neglecting condition (5.152), as was done for ΔP_{tv}^{HM} in [Herbst, Hofmann, and Rohrer (2004)], turns out to be insufficient for a useful estimation of correction ΔP_{tc}^{HM}. Therefore, we fully consider (5.151) and partly implement (5.152) in our calculation. The integrand of Eq. (5.150) is positive definite. Condition (5.152) represents a bound on a positive-curvature parabola in γ. Considering only the minimum $\gamma_{\min}(x) = \mp\sqrt{x^2 + 4e^2}$ of this parabola, relaxes the restrictions on x, y, and z meaning that the integration of the positive definite integrand is over a larger volume than (5.152) actually permits. This is because the minimum of the parabola is greater than -1, see below. Thus we obtain an upper bound for $|\Delta P_{tc}^{HM}|$. Replacing γ by $\gamma_{\min}$, condition (5.152) reads

$$-1 \le x^2 + y^2 + 2x\,y\,z \equiv h(x,y,z) \le 1. \tag{5.153}$$

Because this result is obtained for both signs of $p_0 \to \pm\sqrt{\mathbf{p}^2 + m^2}$ we have $\sum_{\pm} = 2$. Let us now investigate the behavior of the function

$h(x, y, z)$. Notice that $h(x, y, z) > -1$ because $h(x, y, -1) = (x-y)^2 \geq 0$. The upper bound $h(x, y, z) \leq 1$ puts restrictions on the upper limit $\min(1, z_+(x, y))$ for the z-integration where

$$z_+(x, y) \equiv \frac{1 - x^2 - y^2}{2xy}. \tag{5.154}$$

Hence z runs within the range $-1 \leq z \leq \min(1, z_+(x, y))$. The next task is to determine the range for x and y satisfying $-1 \leq z_+(x, y)$. Setting $z_+(x, y) > 1$, we have

$$0 \leq y < \tilde{y}(x) \equiv -x + 1. \tag{5.155}$$

Setting $z_+(x, y) > -1$, one arrives at

$$0 \leq y_-(x) \equiv x - 1 < y < y_+(x) \equiv x + 1. \tag{5.156}$$

The admissible range for x and y is depicted in Fig. 5.15. To obtain limits on the γ-integration we solve condition (5.151) for γ. For $y \geq 1$ we obtain

$$\sqrt{y^2 - 1} \leq \gamma \leq \sqrt{y^2 + 1} \quad \text{or} \quad -\sqrt{y^2 + 1} \leq \gamma \leq -\sqrt{y^2 - 1}, \tag{5.157}$$

and for $0 \leq y < 1$ we obtain

$$-\sqrt{y^2 + 1} \leq \gamma \leq \sqrt{y^2 + 1}. \tag{5.158}$$

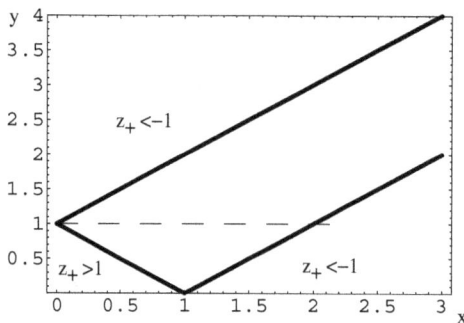

Figure 5.15.   Admissible range for the x- and y-integration. The regions with $z_+ < -1$ are forbidden, the dashed line represents the function $y = 1$.

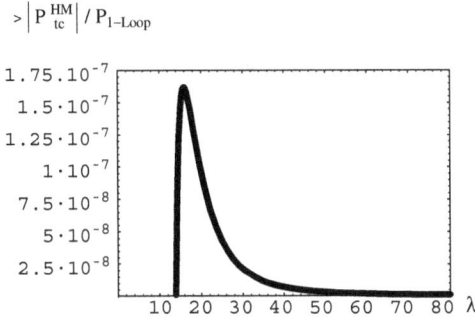

$$> \left| P_{tc}^{HM} \right| / P_{1-\text{Loop}}$$

Figure 5.16. Upper estimate for the modulus $\left| \Delta P_{tc}^{HM} \right| / P_{1-\text{loop}}$ as a function of λ. (The ground-state part in P is subtracted in $P_{1-\text{loop}}$.)

Finally, we arrive at:

$$
\begin{aligned}
\left| \Delta P_{tc}^{HM} \right| < & \left[\int_0^1 dx \int_0^{\tilde{y}} dy \int_{-1}^1 dz \int_{-\sqrt{y^2+1}}^{\sqrt{y^2+1}} dy \right. \\
& + \int_0^1 dx \int_{\tilde{y}}^1 dy \int_{-1}^{z_+} dz \int_{-\sqrt{y^2+1}}^{\sqrt{y^2+1}} dy \\
& + \int_1^2 dx \int_{y_-}^1 dy \int_{-1}^{z_+} dz \int_{-\sqrt{y^2+1}}^{\sqrt{y^2+1}} dy \\
& + 2 \int_0^2 dx \int_1^{y_+} dy \int_{-1}^{z_+} dz \int_{\sqrt{y^2-1}}^{\sqrt{y^2+1}} dy \\
& \left. + 2 \int_2^\infty dx \int_{y_-}^{y_+} dy \int_{-1}^{z_+} dz \int_{\sqrt{y^2-1}}^{\sqrt{y^2+1}} dy \right] \\
& \times \frac{e^2 \Lambda^4 \lambda^{-2}}{(2\pi)^5} \left(3 + \frac{x^2}{4e^2} \right) \frac{x^2 n_B (2\pi\lambda^{-3/2} \sqrt{x^2 + 4e^2})}{\sqrt{x^2 + 4e^2}} .
\end{aligned}
$$

$$(5.159)$$

In Fig. 5.16 a plot of the estimate for $\left| \Delta P_{tc}^{HM} \right|$, which is obtained by numerical integration on the right-hand side of (5.159), is shown as a function of λ.

We now consider the correction ΔP_{tv}^{HM} of Eq. (5.147) which is subject to the integration constraints (5.151) and (5.152). We know that ΔP_{tc}^{HM} is purely imaginary and thus is canceled by the imaginary part of ΔP_{tv}^{HM}. Therefore the interesting quantity is Re ΔP_{tv}^{HM}. We will be content with an upper estimate for $\left|\text{Re } \Delta P_{tv}^{HM}\right|$. We obtain a tight estimate by implementing a relaxed version of (5.152) strictly along the lines developed for $\left|\Delta P_{tc}^{HM}\right|$. Notice that

$$\lim_{\epsilon \to 0} \text{Re } \frac{i}{\gamma^2 - y^2 + i\epsilon} = \lim_{\epsilon \to 0} \frac{\epsilon}{(\gamma^2 - y^2)^2 + \epsilon^2} = \pi \delta(\gamma^2 - y^2) \quad (5.160)$$

in the real part of Eq. (5.147) and that the points $\gamma^2 = y^2$ are not excluded by (5.151). Performing the integrations over azimuthal angles and γ, we thus have (triangle inequality)

$$\left|\text{Re } \Delta P_{tv}^{HM}\right| \leq \frac{e^2}{(2\pi)^4} \frac{\Lambda^4}{2\lambda^2} \int dx\, dy\, dz\, x^2 y \left| -4 + \frac{x^2}{4e^2} - \frac{x^2 z^2}{4e^2} \right|$$

$$\times \frac{n_B \left(2\pi\lambda^{-3/2}\sqrt{x^2 + 4e^2} \right)}{\sqrt{x^2 + 4e^2}}, \quad (5.161)$$

where $h = 1, 2$ is summed over. Treating constraint (5.152) in exactly the same way as in the estimate for $|\Delta P_{tc}^{HM}|$ and carrying out the integrations on the right-hand side of (5.161) numerically, yields the result depicted in Fig. 5.17.

Let us now compute the correction $\Delta P_{ttv}^{HHM} + \Delta P_{ttc}^{HHM}$. In deriving the analytic expression for ΔP_{ttv}^{HHM} we follow [Herbst, Hofmann, and Rohrer (2004)]. One has

$$\Delta P_{ttv}^{HHM} = -\frac{1}{2i} \int \frac{d^4p\, d^4k\, d^4q}{(2\pi)^6} \Gamma_{[3]hl3}^{\lambda\mu\nu}(p, k, q)\Gamma_{[3]hl3}^{\rho\sigma\tau}(-p, -k, -q)$$

$$\times \tilde{D}_{\lambda\rho}(p)\tilde{D}_{\mu\sigma}(k)P_{\nu\tau}^T(q)\, n_B(|p_0/T|)n_B(|k_0/T|)$$

$$\times \frac{i}{(k+p)^2}\delta(p^2 - m^2)\delta(k^2 - m^2)\delta(q + p + k),$$

compare with Fig. 5.19.

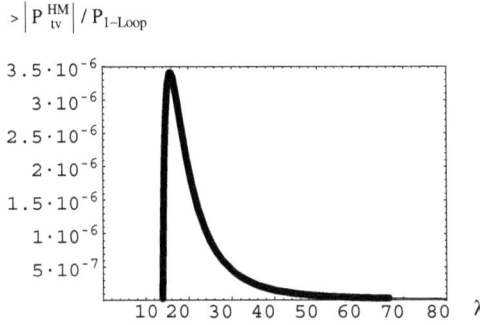

Figure 5.17.   Upper estimate for $\dfrac{|\mathrm{Re}\,\Delta P^{\mathrm{HM}}_{tv}|}{P_{1-\mathrm{loop}}}$ as a function of λ. (The ground-state part in P is subtracted in $P_{1-\mathrm{loop}}$).

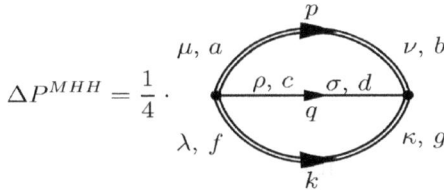

$$\Delta P^{MHH} = \frac{1}{4} \cdot$$

Figure 5.18.   The diagram for ΔP^{HHM}. A double line denotes TLH- and a single line TLM-mode propagation.

The constraint on q^2 reads

$$|q|^2 = |(p+k)^2| = |p^2 + k^2 + 2pk| \le |\phi|^2 \ \rightarrow \tag{5.162}$$

$$|2(2e)^2 \pm 2\sqrt{x^2 + (2e)^2}\sqrt{y^2 + (2e)^2} - 2xy\cos\theta| \le 1. \tag{5.163}$$

This is the same as it is for $\Delta P^{\mathrm{HH}}_{tt}$, so only the "$-$" case needs to be considered:

$$\Delta P^{\mathrm{HHM}}_{ttv} = -\frac{2e^2}{2i} \int \frac{d^4p\,d^4k}{(2\pi)^6}\left[10p^2 + 10k^2 + 16pk - 2\frac{p^4}{m^2} - 2\frac{k^4}{m^2}\right.$$

$$\left. - 8\frac{p^2(pk)}{m^2} - 8\frac{k^2(pk)}{m^2} - 16\frac{(pk)^2}{m^2} - \frac{\vec{p}^2\vec{k}^2\sin^2\theta}{(\vec{p}+\vec{k})^2}\right.$$

$$\times \left(10 - 3\frac{p^2}{m^2} - 3\frac{k^2}{m^2} - 8\frac{pk}{m^2} + \frac{p^4}{m^4} + \frac{k^4}{m^4}\right.$$

$$\left. + 4\frac{p^2(pk)}{m^4} + 4\frac{k^2(pk)}{m^4} + 4\frac{(pk)^2}{m^4} + 2\frac{p^2k^2}{m^4}\right)\Bigg]$$

$$\times \frac{n_B(\sqrt{\vec{p}^2 + m^2}/T)n_B(\sqrt{\vec{k}^2 + m^2}/T)}{2\sqrt{\vec{p}^2 + m^2}\sqrt{\vec{k}^2 + m^2}} \frac{i}{(k+p)^2}$$

$$\times \delta(p_0 - \sqrt{\vec{p}^2 + m^2})\delta(k_0 + \sqrt{\vec{k}^2 + m^2}). \tag{5.164}$$

Using Eq. (5.137), the part $\propto \frac{\vec{p}^2\vec{k}^2 \sin^2\theta}{(\vec{p}+\vec{k})^2}$ in the square brackets reads

$$12 + 4\frac{\vec{p}^2}{m^2} + 4\frac{\vec{k}^2}{m^2} + 4\frac{\vec{p}^2\vec{k}^2(1 - z^2)}{m^4} + 8\frac{|\vec{p}||\vec{k}|z}{m^4}\sqrt{\vec{p}^2 + m^2}\sqrt{\vec{k}^2 + m^2}$$

$$\tag{5.165}$$

after the p_0 and k_0 integrations are carried out. We have introduced $z \equiv \cos\theta$. The remaining part is

$$-16\left[\vec{p}^2 + \vec{k}^2 + \frac{\vec{p}^2\vec{k}^2(1 - z^2)}{m^2} + 2\frac{|\vec{p}||\vec{k}|z}{m^2}\sqrt{\vec{p}^2 + m^2}\sqrt{\vec{k}^2 + m^2}\right].$$

The propagator $1/(p + k)^2$ becomes after rescaling

$$\frac{1}{(p + k)^2} \to \frac{1}{2(2e)^2 - 2\sqrt{x^2 + (2e)^2}\sqrt{y^2 + (2e)^2} - 2xyz}.$$

Thus we have

$$\Delta P_{ttv}^{HHM} = \frac{e^2T^4}{2\lambda^6} \int dx\, dy\, dz \frac{x^2y^2}{\sqrt{x^2 + (2e)^2}\sqrt{y^2 + (2e)^2}}$$

$$\times \frac{n_B(2\pi\lambda^{-3/2}\sqrt{x^2 + (2e)^2})\, n_B(2\pi\lambda^{-3/2}\sqrt{y^2 + (2e)^2})}{(2e)^2 - \sqrt{x^2 + (2e)^2}\sqrt{y^2 + (2e)^2} - xyz}$$

$$\times \left\{16\left[x^2 + y^2 + 2\frac{xyz}{(2e)^2}\sqrt{x^2 + (2e)^2}\sqrt{y^2 + (2e)^2}\right.\right.$$

$$+ \frac{x^2 y^2 (1 + z^2)}{(2e)^2} \Bigg] + \frac{x^2 y^2 (1 - z^2)}{x^2 + y^2 + 2xyz} \left[12 + 4 \frac{x^2}{(2e)^2} + 4 \frac{y^2}{(2e)^2} \right.$$

$$\left. + 4 \frac{x^2 y^2 (1 + z^2)}{(2e)^4} + 8 \frac{xyz}{(2e)^4} \sqrt{x^2 + (2e)^2} \sqrt{y^2 + (2e)^2} \right] \Bigg\},$$

$$(5.166)$$

where the integration is subject to the "$-$" case in the constraint (5.162). Let us now point out a problem in numerically performing this integration. Consider the integrand in Eq. (5.166). The first part in curly brackets contains no singularity and can be integrated numerically without additional thought. The part $\propto \frac{x^2 y^2 (1 - z^2)}{x^2 + y^2 + 2xyz}$ cannot be integrated numerically as it stands since it diverges at $x = y$ and $z = -1$. Complex analysis, that is, the residue theorem, cannot be applied to this problem because we cannot close the contour at infinity due to the integration constraint (5.162). Like we did for $\Delta P_{tc}^{\text{HM}}$, we therefore add $i\epsilon$ ($\epsilon > 0$) to the denominator of the TLM propagator. One needs to prescribe a small value for ϵ and check the numerical convergence of the integral in the limit $\epsilon \to 0$. This convergence indeed takes place.

For $\Delta P_{ttc}^{\text{HHM}}$ the implementation of (5.162) is precisely as it is for $\Delta P_{ttv}^{\text{HHM}}$; only the integrand differs. One has

$$\Delta P_{ttc}^{\text{HHM}} = -\frac{e^2}{4(2\pi)^6} \int d^4k\, d^4p\, d^4q\, \delta(p + k + q)\, \frac{u_\rho u_\sigma}{\vec{q}^2}$$

$$\times [g^{\rho\mu}(p - q)^\lambda + g^{\lambda\rho}(q - k)^\mu + g^{\mu\lambda}(k - p)^\rho]$$

$$\times \left(g_{\mu\nu} - \frac{p_\mu p_\nu}{m^2} \right) \delta(p^2 - m^2)\, n_B \left(\frac{|p_0|}{T} \right)$$

$$\times [g^{\sigma\nu}(p - q)^\kappa + g^{\kappa\sigma}(q - k)^\nu + g^{\nu\kappa}(k - p)^\sigma]$$

$$\times \left(g_{\lambda\kappa} - \frac{k_\lambda k_\kappa}{m^2} \right) \delta(k^2 - m^2)\, n_B \left(\frac{|k_0|}{T} \right). \qquad (5.167)$$

The q-integration is trivial, and the p_0- and k_0-integration over the product $\delta(p^2 - m^2)\,\delta(k^2 - m^2)$ is performed as above [see procedure leading to Eq. (5.137)]. Integrating over azimuthal angles, going over to dimensionless variables, and introducing the abbreviation $S(x, e) \equiv \sqrt{x^2 + 4e^2}$ yields

$$\Delta P_{ttc}^{\mathrm{HHM}} = -\frac{e^2 \Lambda^4 \lambda^{-2}}{4(2\pi)^4} \int dx\,dy\,dz \frac{n_B\left(2\pi\lambda^{-3/2}S(x, e)\right) n_B\left(2\pi\lambda^{-3/2}S(y, e)\right)}{S(x, e)\,S(y, e)}$$

$$\times \frac{x^2 y^2}{x^2 + y^2 + 2x\,y\,z} \left[8\left(4e^2 - \frac{(S(x, e)\,S(y, e) + x\,y\,z)^2}{4e^2}\right) \right.$$

$$-\frac{8e^2 + x^2 + y^2 - 2\,S(x, e)\,S(y, e)}{e^2}\,(S(x, e)\,S(y, e) + x\,y\,z)$$

$$+\frac{8e^2 + x^2 + y^2 + 2\,S(x, e)\,S(y, e)}{16e^4}\,(S(x, e)\,S(y, e) + x\,y\,z)^2$$

$$\left. - 2(8e^2 + x^2 + y^2 - 6S(x, e)\,S(y, e)) \right]. \tag{5.168}$$

As discussed in [Herbst, Hofmann, and Rohrer (2004)] the z-integration in Eq. (5.168) is bounded as

$$-1 \le z \le \max\left(-1, \min\left(1, z_+(x, y)\right)\right), \tag{5.169}$$

where

$$z_+(x, y) \equiv \frac{1}{xy}\left(\frac{1}{2} + 4e^2 - S(x, e)\,S(y, e)\right). \tag{5.170}$$

The next task is to determine the range for x and y such that $-1 \le z_+(x, y) \le 1$. Provided that $x < 10$, $y < 10$ one can solve the condition $z_+(x, y) > 1$ for y. This yields

$$0 \le y \le \tilde{y}(x) \equiv \frac{-(1 + 8e^2)x + \sqrt{1 + 16\,e^2}\,S(x, e)}{8e^2}. \tag{5.171}$$

Notice that the intersection of $\tilde{y}(x)$ with the y- and x-axis is at $y_0 = x_0 = \sqrt{1 + \frac{1}{16e^2}} \sim 1$. Thus our above assumption certainly is satisfied. Setting $z_+(x,y) > -1$ yields

$$y_-(x) \equiv \frac{(1 + 8e^2)x - \sqrt{1 + 16\,e^2}\,S(x,e)}{8e^2} \le y \le y_+$$

$$\equiv \frac{(1 + 8e^2)x + \sqrt{1 + 16\,e^2}\,S(x,e)}{8e^2}. \tag{5.172}$$

Notice the factor $\frac{1}{x^2+y^2+2xyz}$ in the integrand of Eq. (5.168). Upon z-integration this transforms into

$$\sim \log(x^2 + y^2 + 2xyz)\Big|_{-1}^{\max(-1,\min(1,z_+(x,y)))}.$$

For $z = -1$ there is an integrable singularity at $x = y$ presenting a problem for the numerical x- and y-integration. To cope with it we cut out a small band of width 2δ centered at $x = y$ and observe stabilization of the result for $\delta \to 0$. In Fig. 5.19 the region of integration is depicted in the xy-plane. The result of the computation of $\frac{\Delta P_{ttv}^{HHM} + \Delta P_{ttc}^{HHM}}{P_{1-loop}}(\lambda)$ [numerical integration of the right-hand sides on Eqs. (5.166) and (5.168)] is shown in Fig. 5.20.

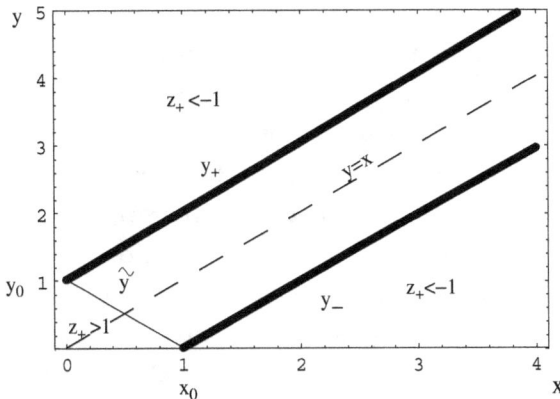

Figure 5.19. The region of integration in the xy-plane corresponding to Eq. (5.168).

$$(\Delta P\,{}^{\mathrm{HHM}}_{\mathrm{ttv}} + \Delta P\,{}^{\mathrm{HHM}}_{\mathrm{ttc}})\,/\,P\,{}_{\mathrm{1-loop}}$$

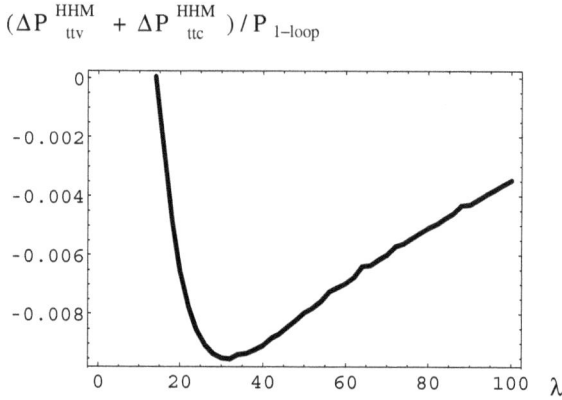

Figure 5.20. $\dfrac{\Delta P^{\mathrm{HHM}}_{ttv}+\Delta P^{\mathrm{HHM}}_{ttc}}{P_{1-\mathrm{loop}}}$ as a function of λ. (The ground-state part in P is subtracted in $P_{1-\mathrm{loop}}$).

5.4.5   Pressure: Three-loop estimates

In this section we provide numerical estimates for the contribution of irreducible three-loop corrections closely following the presentation in [Kaviani and Hofmann (2007)]. By irreducible three-loop we mean that the diagram does not include any line that is dressed by insertions of a finite number of one-loop polarizations. As discussed in Sec. 5.4.3.1, these one-particle reducible contributions to the propagator should be resummed up to infinitely many insertions to avoid the occurrence of pinch singularities [Hofmann (2012)]. Recall that the conjecture of Sec. 5.3.3.2 is that the loop expansion terminates with respect to irreducible diagrams, that is, with respect to all those diagrams which do not yield a one-particle reducible diagram upon performing a cut (in all possible ways) on a single line.

All three-loop irreducible diagrams are depicted in Fig. 5.21. In the following we use the convention as in Fig. 5.21 for labeling the loop momenta. For diagrams A, B, and C the number $\tilde{K}$ of independent, radial loop variables $(p_0, |\vec{p}|)_i$ $(i = 1, 2, 3)$ is $\tilde{K} = 6$, and the number K of independent constraints is $K = 7$. This implies that the support for the loop integrations is either compact or empty (see

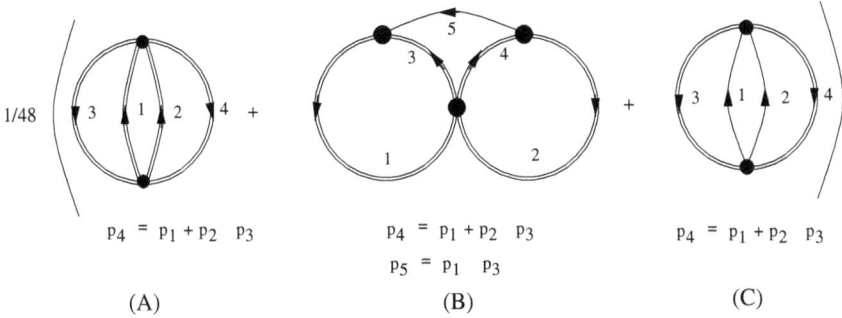

Figure 5.21.   Irreducible three-loop contributions to the pressure. Double (single) lines are associated with the propagators of massive (massless) modes.

the discussion in Sec. 5.4.3.2). As we shall see, the former possibility applies to diagrams A and B while diagram C vanishes because of an empty support for its loop integrations. Let us first discuss diagrams A and B. Upon use of the Feynman rules of Sec. 5.4.2, considering the symmetry factor $\frac{1}{48}$, by appealing to the triangle inequality, using the fact that the modes $a_\mu^{1,2}$ propagate (on-shell) thermally only, integrating over the time-components of the independent loop momenta (momentum conservation), and after a rescaling of the radial components of loop momenta as

$$|\vec{p}_i| \rightarrow x_i \equiv \frac{|\vec{p}_i|}{|\phi|}, \quad (i = 1, 2, 3) \tag{5.173}$$

one arrives at the following estimate for the moduli of the pressure corrections ΔP_A, ΔP_B

$$|\Delta P_{A(B)}| \leq \frac{e^4 \Lambda^4 \lambda^{-2}}{3 \times 2^7 \times (2\pi)^6}$$

$$\times \sum_{l,m,n=1}^{2} \int dx_1 \int dx_2 \int dx_3 \int dz_{12} \int dz_{13} \int_{z_{23,l}}^{z_{23,u}} dz_{23}$$

$$\times \frac{1}{\sqrt{(1 - z_{12}^2)(1 - z_{13}^2) - (z_{23} - z_{12}z_{13})^2}}$$

$$\times \frac{x_1^2 x_2^2 x_3^2}{\sqrt{x_1^2 + 4e^2}\sqrt{x_2^2 + 4e^2}\sqrt{x_3^2 + 4e^2}}$$

$$\times \delta \left(4e^2 + (-1)^{l+m}\sqrt{x_1^2 + 4e^2}\sqrt{x_2^2 + 4e^2} - x_1 x_2 z_{12} \right.$$

$$- \left((-1)^{l+n}\sqrt{x_1^2 + 4e^2}\sqrt{x_3^2 + 4e^2} - x_1 x_3 z_{13} \right)$$

$$\left. - \left((-1)^{m+n}\sqrt{x_2^2 + 4e^2}\sqrt{x_3^2 + 4e^2} - x_2 x_3 z_{23} \right) \right)$$

$$\times |\mathcal{P}_{A(B)}(\vec{x}, \vec{z}, l, m, n)| \, n_B \left(2\pi\lambda^{-3/2}\sqrt{x_1^2 + 4e^2} \right)$$

$$\times n_B \left(2\pi\lambda^{-3/2}\sqrt{x_2^2 + 4e^2} \right) n_B \left(2\pi\lambda^{-3/2}\sqrt{x_3^2 + 4e^2} \right)$$

$$\times n_B \left(2\pi\lambda^{-3/2} \left| (-1)^l \sqrt{x_1^2 + 4e^2} + (-1)^m \sqrt{x_2^2 + 4e^2} \right. \right.$$

$$\left. \left. + (-1)^n \sqrt{x_3^2 + 4e^2} \right| \right), \tag{5.174}$$

where $z_{12} \equiv \cos \angle(\vec{x}_1, \vec{x}_2)$, $z_{13} \equiv \cos \angle(\vec{x}_1, \vec{x}_3)$, and $z_{23} \equiv \cos \angle(\vec{x}_2, \vec{x}_3)$. The functions $\mathcal{P}_A$, $\mathcal{P}_B$ emerge from Lorentz and color contractions and are regular at $x_1 = x_2 = x_3 = 0$ (mass gap for $a_\mu^{1,2}$). We refrain from quoting them here explicitly. In addition, we define:

$$z_{23,u} \equiv \cos |\arccos z_{12} - \arccos z_{13}|,$$
$$z_{23,l} \equiv \cos |\arccos z_{12} + \arccos z_{13}|. \tag{5.175}$$

The integrations in Eq. (5.174) are subject to the following constraints

$$z_{12} \leq \frac{1}{x_1 x_2} \left(4e^2 - \sqrt{x_1^2 + 4e^2}\sqrt{x_2^2 + 4e^2} + \frac{1}{2} \right) \equiv g_{12}(x_1, x_2),$$

$$z_{13} \geq \frac{1}{x_1 x_3} \left(-4e^2 + \sqrt{x_1^2 + 4e^2}\sqrt{x_3^2 + 4e^2} - \frac{1}{2} \right) \equiv g_{13}(x_1, x_3),$$

$$z_{23} \geq \frac{1}{x_2 x_3} \left(-4e^2 + \sqrt{x_2^2 + 4e^2}\sqrt{x_3^2 + 4e^2} - \frac{1}{2} \right) \equiv g_{23}(x_2, x_3).$$

$$\tag{5.176}$$

Notice that for diagram B the constraint (5.60) for the momentum $p_5 = p_1 - p_3$ of the tree-level massless (TLM) mode is the same as the t-channel constraint in (5.61) for the four-vertex. Therefore no extra condition for the (off-shell) momentum p_5 is needed. According to the investigation in Sec. 5.4.3.3 conditions (5.176) together with Eq. (5.175) imply that the support for the integration in x_1, x_2, and x_3 is contained in the compact set $\{x_1, x_2, x_3 \leq 3\}$.

Let us now turn to diagram C. We first consider the case that the momenta p_1 and p_2 of the TLM modes (compare Fig. 5.21), are both off-shell within the constraints dictated by (5.60). In analogy to diagrams A and B one then derives that

$$
|\Delta P_C| \leq \frac{e^4 \Lambda^4 \lambda^{-2}}{3 \times 2^5 \times (2\pi)^8}
$$

$$
\times \sum_{l,m=1}^{2} \int dy_1 \int dx_1 \int dx_2 \int dx_3 \int dz_{12} \int dz_{13} \int_{z_{23,l}}^{z_{23,u}} dz_{23}
$$

$$
\times \frac{x_1^2 x_2^2 x_3^2}{\sqrt{(1 - z_{12}^2)(1 - z_{13}^2) - (z_{23} - z_{12}z_{13})^2}} \left| P_C(\vec{x}, \vec{z}, y_1, l, m) \right|
$$

$$
\times n_B \left(2\pi\lambda^{-3/2}\sqrt{x_3^2 + 4e^2} \right)
$$

$$
\times \frac{n_B \left(2\pi\lambda^{-3/2} \left| (-1)^l \sqrt{x_3^2 + 4e^2} + (-1)^m f_2(\vec{x}, \vec{z}) \right| \right)}{f_2(\vec{x}, \vec{z})\sqrt{x_3^2 + 4e^2}},
$$

$$(5.177)$$

where

$$
f_2(\vec{x}, \vec{z}) \equiv \sqrt{x_1^2 + x_2^2 + x_3^2 + 2x_1 x_2 z_{12} - 2x_1 x_3 z_{13} - 2x_2 x_3 z_{23}},
$$

$$
y_1 \equiv \frac{p_1^0}{|\phi|},
$$

$$(5.178)$$

and $z_{23,l}$, $z_{23,u}$ are defined as in Eq. (5.175). The function $\mathcal{P}_C$ emerges from Lorentz and color contractions and is regular at $x_1 = x_2 = x_3 = 0$ (mass gap for $a_\mu^{1,2}$). The integrations in Eq. (5.177) are subject to the following constraints

$$1 \geq |y_1^2 + y_2^2 - x_1^2 - x_2^2 + 2y_1y_2 - 2x_1x_2z_{12}|,$$
$$1 \geq |y_2^2 - x_2^2 + 4e^2 - (-1)^l 2y_2\sqrt{x_3^2 + 4e^2} + 2x_2x_3z_{23}|,$$
$$1 \geq |y_1^2 - x_1^2 + 4e^2 - (-1)^l 2y_1\sqrt{x_3^2 + 4e^2} + 2x_1x_3z_{13}|,$$
$$1 \geq |y_1^2 - x_1^2|, \quad 1 \geq |y_2^2 - x_2^2|, \tag{5.179}$$

where

$$y_2 = -y_1 + 2(-1)^l \sqrt{x_3^2 + 4e^2} + (-1)^m f_2(\vec{x}, \vec{z}). \tag{5.180}$$

As we shall see below, the constraints in (5.179) imply that the support for the integration in Eq. (5.177) is empty. As a consequence, the cases that one or both of TLM modes in diagram C propagate on-shell also have an empty support. This is because the conditions $|p_1^2|, |p_2^2| \leq |\phi|^2$, which went into (5.179), contain the cases $p_1^2 = 0$ and/or $p_2^2 = 0$. Thus diagram C is the first example of a vanishing irreducible diagram in the loop expansion. As was argued in Sec. 5.4.3.2 one expects that the number of such cases will rapidly increase with increasing loop order, and we anticipate that, starting at a finite loop order, no irreducible diagram will have any support for its loop integrations.

Let us now present estimates obtained by using the Monte Carlo method of integration for the regular integrands of Eqs. (5.174) and (5.177). The x_1-integration is performed analytically in order to eliminate the δ-function in the original integrand. There are eight zeros of the argument of the δ-function in x_1; some of which turn out to be complex and thus can be discarded. As already mentioned, the support in x_2, x_3 for the integral in Eqs. (5.174) is contained in the compact set $\{x_2, x_3 \leq 3\}$ while the support for the integration in z_{12}, z_{13}, z_{23} naturally is contained in the set $\{-1 \leq z_{12}, z_{13} \leq$

$> |\Delta P_A| / P_{1-\text{loop}}$

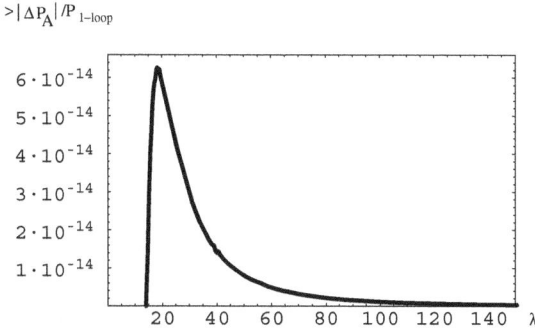

Figure 5.22. An upper estimate for the modulus of the pressure contribution $|\Delta P_A|$ due to diagram A in Fig. 5.21. The plot shows this estimate normalized to the one-loop result $P_{1-\text{loop}}$. (The ground-state part in P is subtracted in $P_{1-\text{loop}}$.)

$+1; z_{23,l} \leq z_{23} \leq z_{23,u}\}$ [see Eq. (5.175)]. Applying the Monte Carlo method, points are chosen randomly in these two compact sets. Any point that satisfies the constraints (5.176) contributes to the integrals. In [Kaviani and Hofmann (2007)] a sample size of 5×10^5 points was used and a typical statistical uncertainty of about 1% was observed in the results. In Fig. 5.22 an estimate for $\frac{|\Delta P_A|}{P_{1-\text{loop}}}$ is shown as a function of (dimensionless) temperature λ. Notice the sudden drop towards zero as a $\lambda \searrow \lambda_c = 13.87$ which is due to the decoupling of the tree-level heavy (TLH) modes $a_\mu^{1,2}$. The functional shape is similar to that of the modulus of the leading two-loop correction depicted in Fig. 5.20. Namely, there is a maximum at $\lambda \sim 20$ and a very rapid decay to the right of this maximum. Notice, however, that the value of the maximum is suppressed by a factor of about 10^{-7} as compared to the smallest two-loop correction (see Fig. 5.16).

In Fig. 5.23 an estimate for $\frac{|\Delta P_B|}{P_{1-\text{loop}}}$ is presented. The maximum of this contribution is comparable to that of the smallest two-loop correction of Fig. 5.16.

For diagram C the compact set $\{x_1, x_2, x_3 \leq R, -R \leq y_1 \leq R, -1 \leq z_{12}, z_{13} \leq +1; z_{23,l} \leq z_{23} \leq z_{23,u}\}$ was chosen in which the Monte Carlo method samples points. R was varied in the range $0.1 \leq R \leq 15$ and samples with up to 6×10^8 points were used. Not

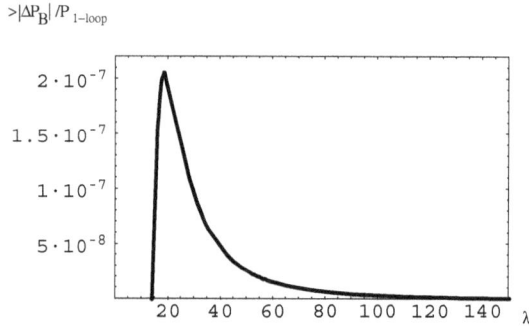

Figure 5.23. An upper estimate for the modulus of the pressure contribution $|\Delta P_B|$ due to diagram A in Fig. 5.21. The plot shows this estimate normalized to the one-loop result $P_{1-\text{loop}}$. (The ground-state part in P is subtracted in $P_{1-\text{loop}}$.)

a single point was found which satisfies all of the conditions (5.179). This is physically suggestive since the diagram describes annihilation or creation of two massive on-shell modes into or out of two massless off-shell modes, respectively. Typically, the off-shellness of a massless mode is comparable to the mass of the massive mode. This very fact, however, is in stark contradiction with condition (5.60). At loop order three we thus have found a first example of an irreducible diagram that vanishes. We expect that this situation occurs more frequently at higher loop orders (see discussion in Sec. 5.4.3.2).

5.4.6 Summary: Loop expansion of pressure

In the previous sections we have described at length how radiative corrections in the effective theory for the deconfining phase of SU(2) Yang–Mills thermodynamics conspire to grant a rapidly converging loop expansion of the pressure. Based on an argument resting on the Euler–L' Huilliers characteristics, the vertex structure of Yang–Mills theory, and a subsequent counting of constraints on *a priori* noncompact integration variables (radial loop momenta) it is conjectured that the expansion into resummed irreducible loops terminates at a finite order. This conjecture was illustrated up to loop number three by numerical estimates.

Again, the cause for nonperturbative convergence, which does not rely on a counting of powers in a small coupling constant,[26] can be traced to the strong constraints imposed by the caloron-based estimate of the thermal ground state on propagating, effective fields: No infrared divergences occur because of a mass gap for two (six) out of three (eight) directions in su(2) (su(3)), and only very little action is associated with each propagating mode due to restrictions on momentum transfers.

It may turn out that a proof of the above conjecture can only be carried out numerically.

5.4.7   One-loop polarization tensor of the massless mode

5.4.7.1   General considerations

After having established estimates indicating evidence for the fast convergence of the expansion of the pressure into irreducible loops subject to resummation of irreducible contributions to the polarization tensors, we now ask the following question: How does a resummation of the polarization tensor for the tree-level massless mode of SU(2) gauge theory influence the propagation properties of this mode within the Yang–Mills plasma? We compute the components of the polarization tensor $\Pi^{\mu\nu}(p_0, \vec{p})$ at the one-loop level, in a first attempt specializing to the case of a free external "photon"[27] on the naive mass shell $p^2 = 0$. Although not selfconsistent this is conceptually insightful and technically interesting because the resolution of integration constraints can be performed analytically. Subsequently, the assumption $p^2 = 0$ is relaxed to selfconsistency by setting $p^2 = G$ where G is the screening function entering the photon's dispersion law. In this case, G is the solution of a genuine gap equation. This

---

[26]Due to the factorial growth of the number of diagrams perturbative loop expansion may at best be asymptotic series.
[27]Motivated by applications we occasionally identify the tree-level massless modes with the photon.

calculation can only be performed numerically. Interestingly, both cases, $p^2 = 0$ and $p^2 = G$, essentially yield identical answers at a temperature $T \sim 2\,T_c$ where the deviations from Abelian behavior peak. In the high-temperature realm, however, the assumption $p^2 = 0$ is too crude.

5.4.7.2   *Prerequistes*

Here we present results first obtained in [Schwarz, Hofmann, and Giacosa (2006)] for the spatially transverse part of the polarization tensor $\Pi^{\mu\nu}$ of the massless mode for the SU(2) case. Owing to the unbroken U(1) gauge invariance, $\Pi^{\mu\nu}$ is four-dimensionally tranverse for any value of p^2 belonging to the TLM mode (or photon),

$$p_\mu \Pi^{\mu\nu} = 0. \tag{5.181}$$

Hence the following decomposition holds

$$\Pi^{\mu\nu} = G(p_0, \vec{p})\, P_T^{\mu\nu} + F(p_0, \vec{p})\, P_L^{\mu\nu}. \tag{5.182}$$

In Euclidean signature one has

$$P_L^{\mu\nu} \equiv \delta^{\mu\nu} - \frac{p^\mu p^\nu}{p^2} - P_T^{\mu\nu}. \tag{5.183}$$

The functions $G(p_0, \vec{p})$ and $F(p_0, \vec{p})$ determine the propagation of the interacting TLM mode. For $\mu = \nu = 0$ Eq. (5.182) yields upon "rotation" to real-time

$$F(p_0, \vec{p}) = \left(1 - \frac{p_0^2}{p^2}\right)^{-1} \Pi^{00}. \tag{5.184}$$

Assuming $\vec{p}$ to be parallel to the z-axis, which is no restriction of generality, we have

$$\Pi^{11} = \Pi^{22} = G(p_0, \vec{p}). \tag{5.185}$$

In imaginary-time the interacting propagator $D^M_{\mu\nu,ab}(p)$ reads

$$D^M_{ab,\mu\nu}(p) = -\delta_{a3}\delta_{b3}\left\{ P_{\mu\nu}^T \frac{1}{p^2 + G} + \frac{p^2}{\vec{p}^2}\frac{1}{p^2 + F} u_\mu u_\nu \right\} \tag{5.186}$$

Notice that for $F = G = 0$ and "rotating" to Minkowskian signature Eq. (5.186) transforms into Eq. (5.97). In [Hofmann (2005)] Π^{00} was calculated for $p_0 = 0$ and in the limit $\vec{p} \to 0$. One has $|\Pi^{00}(0, \vec{p} \to 0)| = |F(0, \vec{p} \to 0)| = \infty$. According to Eq. (5.186) the term $\propto u_\mu u_\nu$ vanishes in this limit.

Letting in Eq. (5.184) $|p_0| \to |\vec{p}|$, we observe that $F \to 0$ provided that Π^{00} remains finite in this limit. One can compute Π^{00} for $p^2 = 0$, and one then sees that this is, indeed, the case. This, in turn, implies that the longitudinal structure (quantum "propagation") in Eq. (5.186) reduces to the free limit, and no longitudinal modes with $p^2 = 0$ propagate. (This would contradict instantaneous propagation mediated by the free limit.) Upon resummation of Π^{00}, which generates the "mass shell" $p^2 = F$, this no longer is the case. We will come back to a discussion of this situation in Sec. 5.4.7.6.

Upon resummation [see Fig. 5.28 on page 207], function G in Eq. (5.186) modifies the dispersion law for the transversely propagating TLM mode as follows [Le Bellac (2000)]:

$$\omega^2(\vec{p}) = \vec{p}^2 + \mathrm{Re}\, G(\omega(\vec{p}), \vec{p}),$$
$$\gamma(\vec{p}) = -\frac{1}{2\omega} \mathrm{Im}\, G(\omega(\vec{p}), \vec{p}), \tag{5.187}$$

where $\omega(\vec{p})$ denotes the real part of the energy of the TLM mode propagating with spatial momentum $\vec{p}$. A finite width $\gamma(\vec{p})$ would introduce a finite lifetime to this photon state. To gain some analytical insight into computing G we first evaluate $\Pi_{11} = G$ for $p_0 = |\vec{p}|$.

5.4.7.3   Calculation of G for $p^2 = 0$

$\Pi^{\mu\nu}$ is the sum of the two diagrams A and B in Fig. 5.24. For $p^2 = 0$ diagram A vanishes. This can be seen as follows. We have

$$\Pi_A^{\mu\nu}(p) = \frac{1}{2i} \int \frac{d^4k}{(2\pi)^4} e^2 \epsilon_{ace}$$

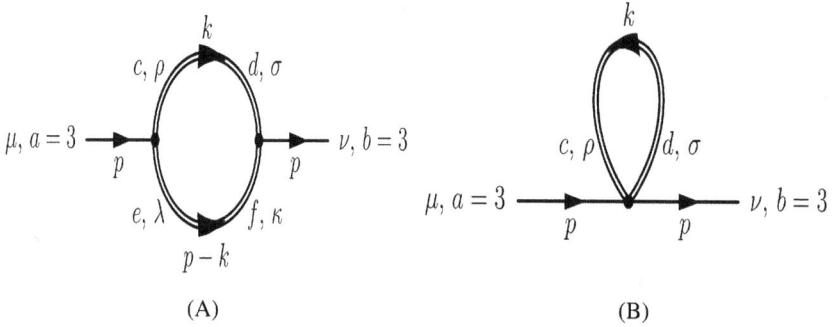

Figure 5.24.   The two diagrams potentially contributing to the TLM-mode polarization tensor.

$$\times [g^{\mu\rho}(-p-k)^{\lambda} + g^{\rho\lambda}(k-p+k)^{\mu} + g^{\lambda\mu}(p-k+p)^{\rho}]$$

$$\times \epsilon_{dbf}[g^{\sigma\nu}(-k-p)^{\kappa} + g^{\nu\kappa}(p+p-k)^{\sigma} + g^{\kappa\sigma}(-p+k+k)^{\nu}]$$

$$\times (-\delta_{ch}\delta_{dh}) \left(g_{\rho\sigma} - \frac{k_{\rho}k_{\sigma}}{m^2} \right)$$

$$\times \left[2\pi\delta(k^2 - m^2) \, n_B(|k_0|/T) \right]$$

$$\times (-\delta_{el}\delta_{fl}) \left(g_{\lambda\kappa} - \frac{(p-k)_{\lambda}(p-k)_{\kappa}}{(p-k)^2} \right)$$

$$\times \left[2\pi\delta((p-k)^2 - m^2) \, n_B(|p_0 - k_0|/T) \right], \tag{5.188}$$

where $h = 1, 2$ and $l = 1, 2$ are summed over. Note that $a = b = 3$. Using $p^2 = 0$ and $k^2 = m^2$, one has $\Pi_A^{\mu\nu}(p)$

$$\Pi_A^{\mu\nu}(p) = ie^2 \int \frac{d^4k}{(2\pi)^2} \left[\left(2kp - 4\frac{(kp)^2}{m^2} \right) g^{\mu\nu} + \left(12 - 2\frac{kp}{m^2} \right) k^{\mu}k^{\nu} \right.$$

$$\left. + \left(-6 + 4\frac{kp}{m^2} \right) (k^{\nu}p^{\mu} + k^{\mu}p^{\nu}) + \left(-5 + \frac{(kp)^2}{m^4} \right) p^{\mu}p^{\nu} \right]$$

$$\times \delta(k^2 - m^2) \, n_B(|k_0|/T) \, \delta((p-k)^2 - m^2) \, n_B(|p_0 - k_0|/T). \tag{5.189}$$

For $p_0 > 0$ the product of δ-functions can be rewritten as

$$\delta(k^2 - m^2) \cdot \delta((p-k)^2 - m^2)$$

$$= \frac{1}{4p_0|\vec{k}|\sqrt{|\vec{k}|^2 + m^2}}$$

$$\times \left[\delta\left(k_0 - \sqrt{|\vec{k}|^2 + m^2}\right) \cdot \delta\left(\cos\theta - \frac{\sqrt{|\vec{k}|^2 + m^2}}{|\vec{k}|}\right) \right.$$

$$\left. + \delta\left(k_0 + \sqrt{|\vec{k}|^2 + m^2}\right) \cdot \delta\left(\cos\theta + \frac{\sqrt{|\vec{k}|^2 + m^2}}{|\vec{k}|}\right) \right], \quad (5.190)$$

where $\theta \equiv \angle(\vec{p}, \vec{k})$. Because $\frac{\sqrt{|\vec{k}|^2 + m^2}}{|\vec{k}|} > 1$ and $-1 \le \cos\theta \le 1$, the argument of the two δ-functions in Eq. (5.190) never vanishes and thus the right-hand of Eq. (5.189) is zero. As a consequence, the photon is stable after one-loop resummation of its polarization tensor ($\gamma = 0$ in Eq. (5.183) for $p^2 = 0$).

Diagram B is associated with the following expression:

$$\Pi_B^{\mu\nu}(p) = \frac{1}{i} \int \frac{d^4k}{(2\pi)^4} (-\delta_{ch}\delta_{dh}) \left(g_{\rho\sigma} - \frac{k_\rho k_\sigma}{m^2}\right) [2\pi\delta(k^2 - m^2)n_B(|k_0|/T)]$$

$$\times (-ie^2)[\epsilon_{abe}\epsilon_{cde}(g^{\mu\rho}g^{\nu\sigma} - g^{\mu\sigma}g^{\nu\rho})$$

$$+ \epsilon_{ace}\epsilon_{bde}(g^{\mu\nu}g^{\rho\sigma} - g^{\mu\sigma}g^{\nu\rho}) + \epsilon_{ade}\epsilon_{bce}(g^{\mu\nu}g^{\rho\sigma} - g^{\mu\rho}g^{\nu\sigma})],$$

$$(5.191)$$

where $h = 1, 2$ is summed over. Note that $a = b = 3$. Applying constraint (5.61) at $p_0 > 0$, $p^2 = 0$, yields:

$$|(p+k)^2| = |2pk + k^2| = \left|2p_0(k_0 - |\vec{k}|\cos\theta) + 4e^2|\phi|^2\right|$$

$$= \left|2p_0\left(\pm\sqrt{k^2 + 4e^2|\phi|^2} - |\vec{k}|\cos\theta\right) + 4e^2|\phi|^2\right| \le |\phi|^2.$$

$$(5.192)$$

For the "+" sign the condition in Eq. (5.192) is never satisfied, for the "−" sign there is a range for p_0 where it is not violated.

We are interested in $\frac{\Pi^{11}}{T^2} = \frac{\Pi^{22}}{T^2} = \frac{G}{T^2}$ as a function of $X \equiv \frac{|\vec{p}|}{T}$ and $\lambda \equiv \frac{2\pi T}{\Lambda}$ when $\vec{p}$ is parallel to the z-axis. Performing the k_0-integration in Eq. (5.191) and introducing dimensionless variables as

$$\vec{y} \equiv \frac{\vec{k}}{|\phi|}, \tag{5.193}$$

we obtain from Eq. (5.191) that

$$\frac{G}{T^2} = \frac{\Pi^{11}}{T^2} = \frac{\Pi^{22}}{T^2} = \frac{e^2}{\pi\lambda^3} \int d^3y \left(-2 + \frac{y_1^2}{4e^2}\right)$$

$$\times \frac{n_B\left(2\pi\lambda^{-3/2}\sqrt{\vec{y}^2 + 4e^2}\right)}{\sqrt{\vec{y}^2 + 4e^2}}, \tag{5.194}$$

where the integration is subject to the following constraint:

$$-1 \le -\lambda^{3/2}\frac{X}{\pi}\left(\sqrt{\vec{y}^2 + 4e^2} + y_3\right) + 4e^2 \le 1. \tag{5.195}$$

In view of constraint (5.195) the integral in Eq. (5.194) is evaluated most conveniently in cylindrical coordinates,

$$y_1 = \rho \cos\varphi, \quad y_2 = \rho \sin\varphi, \quad y_3 = \xi. \tag{5.196}$$

Let us now discuss how the constraint (5.195) is implemented in the ρ- and ξ-integrations. Constraint (5.195) is re-cast as

$$\frac{4e^2 - 1}{\lambda^{3/2}}\frac{\pi}{X} \le \sqrt{\rho^2 + \xi^2 + 4e^2} + \xi \le \frac{4e^2 + 1}{\lambda^{3/2}}\frac{\pi}{X}. \tag{5.197}$$

Notice that Eq. (5.197) gives an upper bound Ξ for ξ: $\xi < \frac{4e^2+1}{\lambda^{3/2}}\frac{\pi}{X} \equiv \Xi$. In contrast, there is no such global lower bound for ξ.

The upper limits for the ρ- and ξ-integration are obtained as follows. Since $\xi < \Xi$ we can square the second part of the inequality

(5.197) and solve for ρ:

$$\rho \leq \sqrt{\left(\frac{\pi}{X}\right)^2 \frac{(4e^2+1)^2}{\lambda^3} - \frac{2\pi}{X} \frac{4e^2+1}{\lambda^{3/2}} \xi - 4e^2} \equiv \rho_M(X, \xi, \lambda).$$

(5.198)

The condition that the expression under the square root in Eq. (5.198) is positive yields the upper limit $\xi_M(X, \lambda)$ for the ξ-integration:

$$\xi \leq \frac{\pi}{2X} \frac{4e^2+1}{\lambda^{3/2}} - 2\frac{X}{\pi} \lambda^{3/2} \frac{e^2}{4e^2+1} \equiv \xi_M(X, \lambda).$$

(5.199)

The lower limit for the ρ-integration is obtained as follows. Upon subtracting ξ from the first part of the inequality (5.197) the result can be squared provided that $\xi < \frac{4e^2-1}{\lambda^{3/2}} \frac{\pi}{X}$. Solving for ρ, we have

$$\rho \geq \sqrt{\left(\frac{\pi}{X}\right)^2 \frac{(4e^2-1)^2}{\lambda^3} - \frac{2\pi}{X} \frac{4e^2-1}{\lambda^{3/2}} \xi - 4e^2} \equiv \rho_m(X, \xi, \lambda).$$

(5.200)

The condition that the expression under the square root in Eq. (5.200) is positive introduces the critical value $\xi_m(X, \lambda)$ for the ξ-integration as:

$$\xi_m(X, \lambda) \equiv \frac{\pi}{2X} \frac{4e^2-1}{\lambda^{3/2}} - 2\frac{X}{\pi} \lambda^{3/2} \frac{e^2}{4e^2-1}.$$

(5.201)

For $-\infty < \xi \leq \xi_m(X, \lambda)$ the lower limit for the ρ-integration is given by $\rho_m(x, \xi, \lambda)$. Notice that according to Eq. (5.201) $\xi_m(X, \lambda)$ is always smaller than $\frac{4e^2-1}{\lambda^{3/2}} \frac{\pi}{X}$ such that our above assumption is consistent. According to Eq. (5.197), the opposite case, $\frac{4e^2-1}{\lambda^{3/2}} \leq \xi \leq \xi_M(X, \lambda)$, leads to $\rho \geq 0$ which does not represent an additional constraint. To summarize, we have

$$\frac{G}{T^2} = \left[\int_{-\infty}^{\xi_m(X,\lambda)} d\xi \int_{\rho_m(X,\xi,\lambda)}^{\rho_M(X,\xi,\lambda)} d\rho + \int_{\xi_m(X,\lambda)}^{\xi_M(X,\lambda)} d\xi \int_0^{\rho_M(X,\xi,\lambda)} d\rho \right]$$

$$\times e^2 \lambda^{-3} \left(-4 + \frac{\rho^2}{4e^2} \right) \rho \, \frac{n_B \left(2\pi \lambda^{-3/2} \sqrt{\rho^2 + \xi^2 + 4e^2} \right)}{\sqrt{\rho^2 + \xi^2 + 4e^2}},$$

(5.202)

where the integral operation indicated in the square brackets is applied to the last line.

5.4.7.4 Results and discussion: G at $p^2 = 0$

In Figs. 5.25 and 5.26 we show plots of $\log_{10} \left| \frac{G}{T^2} \right|$ as obtained by performing the integration in Eq. (5.202) numerically and by appealing to the one-loop evolution of the effective coupling $e = e(\lambda)$, see Sec. 5.2.3. For all λ and those values of X to the right of the dips in Fig. 5.26 $\frac{G}{T^2}$ is real negative (antiscreening). According to the dispersion law in Eq. (5.187) this implies that the energy of a propagating TLM mode is *reduced* as compared to the free case. For X values to

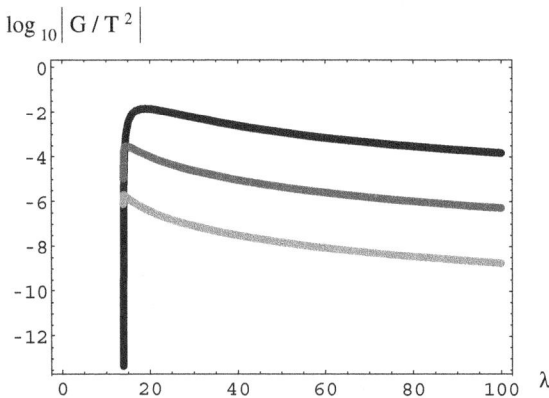

Figure 5.25. $\log_{10} \left| \frac{G}{T^2} \right|$ at $p^2 = 0$ as a function of $\lambda \equiv \frac{2\pi T}{\Lambda}$ for $X = 1$ (black), $X = 5$ (dark gray), and $X = 10$ (light gray).

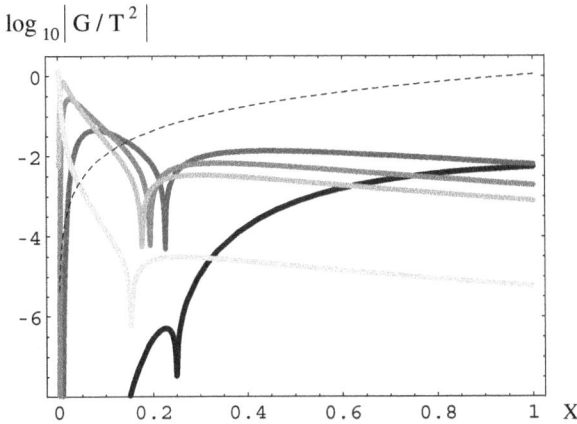

Figure 5.26. $\log_{10}\left|\frac{G}{T^2}\right|$ as a function of $X \equiv \frac{|\vec{p}|}{T}$ for $\lambda = 1.12\,\lambda_c$ (black), $\lambda = 2\,\lambda_c$ (dark gray), $\lambda = 3\,\lambda_c$ (gray), $\lambda = 4\,\lambda_c$ (light gray), $\lambda = 20\,\lambda_c$ (very light gray). The dashed curve is a plot of the function $f(X) = 2\log_{10} X$. TLM modes are strongly screened at X-values for which $\log_{10}\left|\frac{G}{T^2}\right| > f(X)$ $\left(\text{or } \frac{\sqrt{G}}{T} > X\right)$, that is, to the left of the dashed line.

the left of the dips $\frac{G}{T^2}$ is real positive (screening). Figure 5.25 indicates the dependence of $\log_{10}\left|\frac{G}{T^2}\right|$ on λ keeping $X = 1, 5, 10$ fixed. Obviously, the effect on the propagation of TLM modes arising from the interaction with TLH modes is very small (maximum of $\left|\frac{G}{T^2}\right|$ at $X = 1 : \sim 10^{-2}$). As for the high-temperature behavior we observe the following. On one hand, there is clear evidence for a power-like suppression of $\left|\frac{G}{T^2}\right|$ in λ. (The one-loop result for the (quasiparticle) pressure also shows a power-like approach to the Stefan–Boltzmann limit [Hofmann (2005)], see Sec. 8.1. In a similar way, the approach to the limit of vanishing antiscreening is power-like for the TLM mode). Also a sudden drop of $\left|\frac{G}{T^2}\right|$ occurs for $\lambda \searrow \lambda_c = 13.87$. This is indicative of the fact that TLH modes decouple due to their diverging mass. On the other hand, the values of $\log_{10}\left|\frac{G}{T^2}\right|$ at fixed $\lambda > \lambda_c$ are equidistant for (nearly) equidistant values of X. This shows the exponential suppression of $\log_{10}\left|\frac{G}{T^2}\right|$ for $X \geq 1$ and can be understood as follows. For large X Eq. (5.197) demands ξ to be negative and $|\xi|, \rho$ to be large. As a consequence, the square root in Eq. (5.197), which appears as an argument of n_B [see Eq. (5.202)] is large thus implying exponential suppression in X.

Figure 5.26 indicates the dependence of $\log_{10}\left|\frac{G}{T^2}\right|$ on X keeping λ fixed at $\lambda = 1.12\lambda_c$, $\lambda = 2\lambda_c, 3\lambda_c, 4\lambda_c$, and $\lambda = 20\lambda_c$. Only the low-momentum regime is investigated since for $X \sim 1$ the afore-mentioned exponential suppression sets in as can be seen by the linear decrease. For $X < 0.6$ the black curve ($\lambda = 1.12\lambda_c$) is below the curves for $\lambda = 2\lambda_c, 3\lambda_c, 4\lambda_c$ because of the vicinity of λ to λ_c where the mass of TLH modes diverges. The smallness of the curve for $\lambda = 20\lambda_c$ arises due to the above-discussed power suppression. Again, notice the sharp dip occurring for X-values in the range $0.15 \leq X \leq 0.25$. The dip is caused by a change in sign of G: For X to the right of the dip G is negative (antiscreening) while it is positive (screening) to the left. The dashed line is a plot of the function $f(X) = 2\log_{10} X$. The intersection of a curve with $f(X)$ indicates the momentum where the radiatively generated "mass" of the TLM mode is equal to the modulus of its spatial momentum (strong screening) [see Eq. (5.187)]. For $\lambda \sim \lambda_c$ or for $\lambda \gg \lambda_c$ the strong-screening regime shrinks to the point $X = 0$. For $\lambda = 2\lambda_c, 3\lambda_c, 4\lambda_c$ Fig. 5.26 shows that the strong-screening regime has a finite support beginning at $X_s \sim 0.15$.

5.4.7.5   *Calculation of G at $p^2 = G$*

More realistically, the effect of G on shifting photon propagation away from the free mass shell (from $p^2 = 0$ to $p^2 = G$, see Fig. 5.27) should also be considered in the constraint for the momentum transfer in the four-vertex of diagram B in Fig. 5.24. Notice that no change of the integrand in Eq. (5.202) takes place. *A priori* it is not clear whether it is consistent to neglect diagram A also on the modified "mass shell"

$$\omega^2(\vec{p}) \equiv p_0^2(\vec{p}) = \vec{p}^2 + G(p_0(\vec{p}), \vec{p}).\tag{5.203}$$

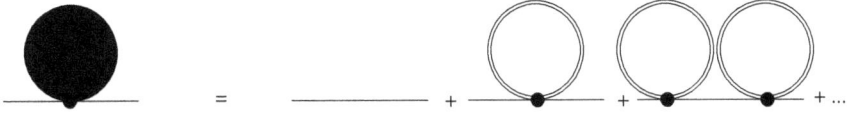

Figure 5.27.   The Dyson series for the resummation of the one-loop irreducible contribution to the polarization tensor of the massless mode. A single line refers to the propagation of a tree-level massless and a double line to the propagation of a tree-level massive mode.

Our strategy, see also [Ludescher and Hofmann (2008)], is to neglect this diagram, thus considering only the integration over the integrand of Eq. (5.202) subject to the modified constraints

$$
\left| \frac{G}{T^2} \frac{\lambda^3}{(2\pi)^2} + \frac{\lambda^{3/2}}{\pi} \left(\pm \sqrt{X^2 + \frac{G}{T^2} \sqrt{\rho^2 + \xi^2 + 4e^2}} - X\xi \right) + 4e^2 \right| \leq 1,
$$

(5.204)

where a sum over $\pm$ is understood when performing the integration. Subsequently, we check for the selfconsistency of the omission of diagram A by evaluating it on the "mass shell" defined by G through diagram B.

In Fig. 5.28 the integrand in Eq. (5.202), subject to the constraints in Eq. (5.204), is plotted for $\lambda = 2\lambda_c$ and $\lambda = 4\lambda_c$. Through the conditions (5.204) the right-hand side of Eq. (5.202) becomes a function of G in contrast to the approximate calculation of Sec. 5.4.7.3 where $p^2 = 0$ was used. The strategy to determine G at a given value of λ in the full calculation thus is to prescribe a value for G in (5.204), then to compute the integral numerically, and to finally list this integral as a function of G. Ultimately, one determines the zero of left-hand minus right-hand side of Eq. (5.202) to find G selfconsistently. Numerically, one may use Newton's method for this task.

In Fig. 5.29 we compare plots of $\log \left| \frac{G}{T^2} \right|$ obtained in the full calculation with those of the approximation $p^2 = 0$, both as a function of X and $Y \equiv \sqrt{X^2 + \frac{G}{T^2}}$ at $\lambda = 2\lambda_c$, $\lambda = 3\lambda_c$, and $\lambda = 4\lambda_c$. Notice the saturation of $\frac{G}{T^2}$ to finite values as $X \to 0$. This is in contrast to the

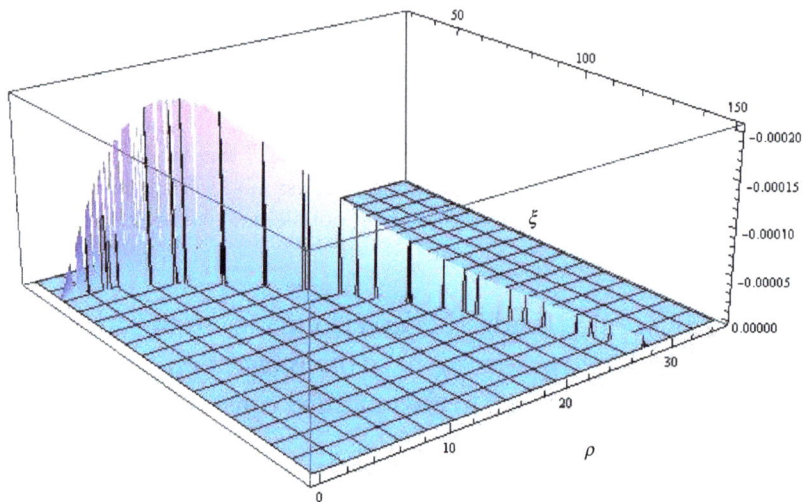

Figure 5.28.   The integrand in Eq. (5.202), subject to the constraints in Eq. (5.204), for $\lambda = 50$, $X = 0.8$, $\frac{G}{T^2} = -0.00146$, and $0 \leq \rho \leq 35$, $30 \leq \xi \leq 150$. This example illustrates the significant constraining power of (5.204), here in the regime of antiscreening.

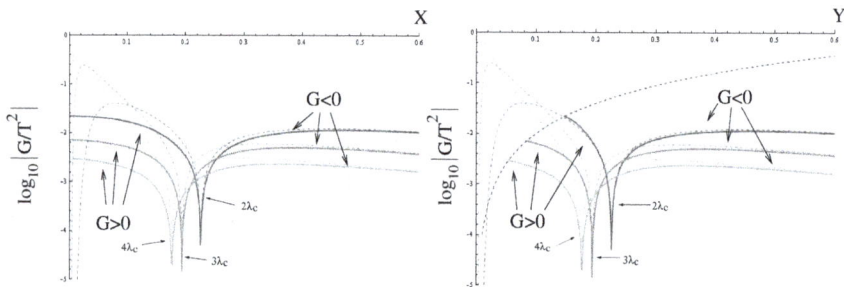

Figure 5.29.   Plots of $\log_{10} |G/T^2|$ in the full calculation (solid gray curves) and for the approximation $p^2 = 0$ (dashed gray curves). The dips in $\log |G/T^2|$ correspond to zeros separating the regime of screening $(G > 0)$ from the regime of antiscrenning $(G < 0)$. The left panel depicts $\log_{10} |G/T^2|$ as a function of X. The right panel shows $\log_{10} |G/T^2|$ as a function of $Y \equiv \sqrt{X^2 + (G/T^2)}$. Here the dashed black curve is the function $2 \log_{10} Y$. In order of increasing lightness of the gray shade the curves correspond to $\lambda = 2\lambda_c$, $\lambda = 3\lambda_c$, and $\lambda = 4\lambda_c$.

result obtained in the approximation $p^2 = 0$. Modes of frequency less than $Y^* = \frac{\sqrt{G}}{T}$ $(x = 0, \lambda)$ do not propagate (imaginary modulus of spatial momentum!) and are evanescent. The regimes of screening (to the left of the zero of $\frac{G}{T^2}$) and antiscreening (to the right of the zero of $\frac{G}{T^2}$) are the same in the results of the full and the approximate calculation. Finally, let us investigate the high-temperature behavior of $Y^*(\lambda)$ which cuts off the spectrum towards low frequencies. We fit the high-temperature behavior of $Y^*(\lambda)$ to a power-law model

$$Y^*(\lambda) = C\lambda^\nu, \quad (\lambda \gg \lambda_c), \tag{5.205}$$

where C and ν are real constants. For the full (approximate) result we obtain $C \sim 20$ and $\nu \sim -3/2$ ($C \sim 2$ and $\nu \sim -2/3$) (see also Fig. 5.30). Thus the approach to Abelian behavior with increasing temperature is much faster for the full calculation than in the approximate result.[28]

So far we have considered diagram B in Fig. 5.24 only. By inspecting the analytic expression in Eq. (5.188), it is clear that if diagram A is finite then it necessarily is purely imaginary. Taking into account that $p^2 = G$, we define in dimensionless Cartesian coordinates the following functions

$$y_3(j) \equiv -\frac{\lambda^{3/2}T^2}{4\pi G}\left(-X\frac{G}{T^2} + (-1)^j\right.$$

$$\times \sqrt{\left(\frac{XG}{T^2}\right)^2 + 4\frac{G}{T^2}\left[\frac{G^2}{4T^4} - 2\pi\lambda^{-3/2}\left(X^2 + \frac{G}{T^2}\right)(y_1^2 + y_2^2 + 4e^2)\right]}\right),$$

---

[28]The decay of $\omega^* = TY^*$ with the square root of T is also seen in the decay of $|\phi|$. Moreover, it occurs in the missing radiation power per unit area ΔR compared to the pure U(1) theory: Since the gap ω^* lies deep inside the Rayleigh–Jeans spectral regime one has $\Delta R = \frac{T}{\pi^2}\int_0^{\omega^*} d\omega\,\omega^2 = \frac{T}{3\pi^2}(\omega^*)^3 \propto T^{-1/2}$. Such a universal decay law strongly points to a high degree of randomness in the emergence of the associated averages.

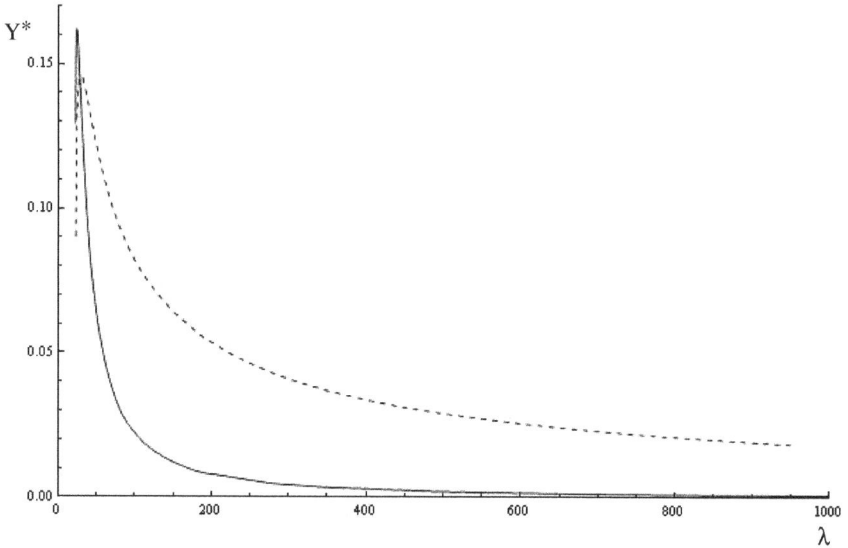

Figure 5.30.   Plots of $Y^*(\lambda) = \frac{\sqrt{G}}{T}(X = 0, \lambda)$ in the full calculation (solid line) and in the approximation $p^2 = 0$ (dashed line).

$$s(j) \equiv \sqrt{y_1^2 + y_2^2 + y_3(j)^2 + 4e^2},$$

$$r(i,j) \equiv 2\pi\lambda^{-3/2}\left((-1)^i s(j)\sqrt{X^2 + \frac{G}{T^2}} - 2\pi\lambda^{-3/2}y_3(j)X\right), \qquad (5.206)$$

where $i, j = 0, 1$. Then we have

$$\frac{\Pi_{11}^A}{T^2} = \frac{\Pi_{22}^A}{T^2}$$

$$= \frac{ie^2}{(2\pi)^2}\sum_{i,j=0}^{1}\int_{-\infty}^{\infty}dy_1\int_{-\infty}^{\infty}dy_2$$

$$\times\left\{\left[-2\,r(i,j) - \left(\frac{\lambda^{3/2}}{2\pi}\right)^2\frac{r(i,j)^2}{4e^2} + 7\frac{G}{T^2} - \left(\frac{\lambda^{3/2}}{2\pi}\right)^2\left(\frac{G}{2eT^2}\right)^2\right]\right.$$

$$+ \left[48 \, \pi^2 \lambda^{-3} - 2 \frac{r(i,j)}{4e^2} - 3 \frac{G}{4e^2 \, T^2} \right.$$

$$\left. + \left(\frac{\lambda^{3/2}}{2\pi} \right)^2 \left(\frac{G}{4e^2 \, T^2} \right)^2 \right] y_1 \, y_2 \Bigg\}$$

$$\times \frac{\pi}{\left| 2 \, \lambda^{3/2} y_3(j) \sqrt{X^2 + \frac{G}{T^2}} \right|} n_B (2\pi \lambda^{-3/2} s(j)) \, n_B$$

$$\times \left(\left| \sqrt{X^2 + \frac{G}{T^2}} + (-1)^i 2\pi \lambda^{-3/2} s(j) \right| \right), \tag{5.207}$$

and no further constraints need to be imposed [Ludescher *et al.* (2008)]. The integral in Eq. (5.207) was computed numerically by using the result for G (determined by Π_{11}^B) for various temperatures λ [Ludescher *et al.* (2008)]. A constantly vanishing result was obtained. Thus even in the result of the full calculation, no finite photon width[29] emerges at one-loop order, and the problem of computing G is selfconsistently solved by computing Π_{11}^B only.

5.4.7.6 *Propagation of longitudinal modes*

Let us now come back to the question about the nature of plasma modes that are associated with the propagation of the field a_0^3 described by the term $\propto u_\mu u_\nu$ in

$$D_{\mu\nu,ab}^M(p) = -\delta_{a3}\delta_{b3} \left\{ P_{\mu\nu}^T \frac{1}{G - p^2} + \frac{p^2}{\vec{p}^2} \frac{1}{F - p^2} u_\mu u_\nu \right\} \tag{5.208}$$

in Euclidean signature.

---

[29]This statement solely relates to the fact that a single tree-level massless mode of the effective theory is incapable of producing a pair of tree-level heavy modes. The collective deposition of energy by *fundamental*, diagonal gauge modes into the generation of monopole-antimonopole pairs, which are not resolved in the effective theory and thus are part of the thermal ground state (see Sec. 5.5) does, however, take place. These processes are effectively descibed by the screening regime $G > 0$ (See Sec. 5.4).

Notice that the field strength $\partial_i a_0^3$ of these modes indeed points in the direction of propagation set by the spatial momentum vector $\vec{p}$ prompting the name "longitudinal". On shell, that is, for $p^2 = F$ the relation

$$F(p_0, \vec{p}) = \left(1 - \frac{p_0^2}{p^2}\right)^{-1} \Pi^{00} \tag{5.209}$$

between the 00-component of the polarization tensor $\Pi^{\mu\nu}$ and screening function F simply becomes

$$T^2 X^2 \equiv \vec{p}^2 = \Pi^{00}. \tag{5.210}$$

To determine function F on-shell we first consider diagram B in Fig. 5.24 and subsequently check that diagram A vanishes for $F = p^2$. The expression for Π^{00} from diagram B reads

$$\frac{\Pi^{00}}{T^2} = 2e^2 \lambda^{-3} \int d\xi \int d\rho \left(3 + \frac{\rho^2 + \xi^2}{4e^2}\right) \rho$$

$$\times \frac{n_B \left(2\pi \lambda^{-3/2} \sqrt{\rho^2 + \xi^2 + 4e^2}\right)}{\sqrt{\rho^2 + \xi^2 + 4e^2}}. \tag{5.211}$$

In Eq. (5.211) the integrations are subject to the following constraint

$$\left| \frac{F}{T^2} \frac{\lambda^3}{(2\pi)^2} + \frac{\lambda^{3/2}}{\pi} \left(\pm\sqrt{X^2 + \frac{F}{T^2}} \sqrt{\rho^2 + \xi^2 + 4e^2} - X\xi\right) + 4e^2 \right| \leq 1, \tag{5.212}$$

where a sum over $\pm$ is understood when the integration is performed. Together Eqs. (5.210)–(5.212) yield at any given value of X an implicit equation for F which can be solved numerically. One then checks, that, indeed, diagram A of Fig. 5.24 vanishes on the shell $F = p^2$ (F real).

In dimensionless quantities the dispersion law reads

$$Y^2 \equiv \frac{\omega^2}{T^2} = X^2 + \frac{F}{T^2}. \tag{5.213}$$

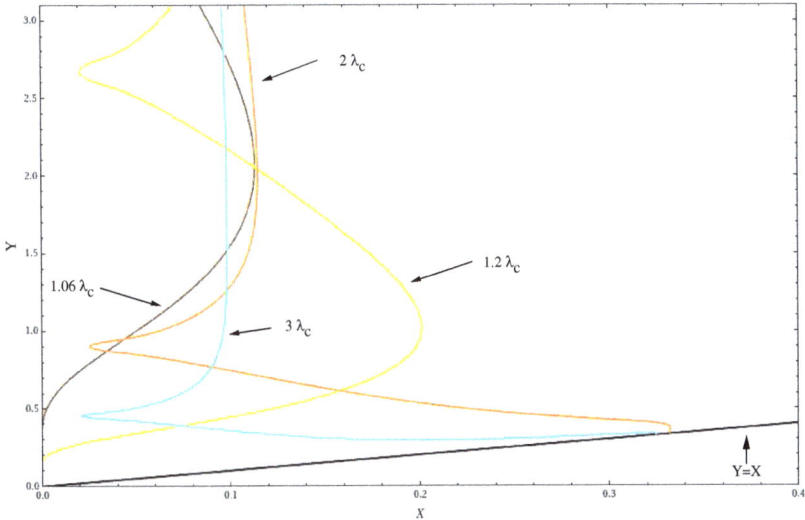

Figure 5.31.   Plots of the dispersion law $Y = \sqrt{X^2 + \frac{F}{T^2}}$ for longitudinal modes as a function of X for temperatures $\lambda = 1.06\,\lambda_c$, $\lambda = 1.2\,\lambda_c$, $\lambda = 2.0\,\lambda_c$, and $\lambda = 3.0\,\lambda_c$. The straight line is a plot of $Y = X$.

In Fig. 5.31 the positive-energy part of this dispersion law[30] is shown for temperatures $\lambda = 1.06\,\lambda_c$, $\lambda = 1.2\,\lambda_c$, $\lambda = 2.0\,\lambda_c$, and $\lambda = 3.0\,\lambda_c$. Notice the existence of various branches, separated by divergences of the group velocity. In Chapter 9 we link deconfining SU(2) Yang–Mills theory to the physics of thermalized photon propagation. As we shall see, this postulate implies a critical temperature $T_c \sim 2.73\,\text{K}$ and demands a magnetic-electric dual interpretation, see also Sec. 5.3. Thus the longitudinally propagating Abelian electric field in SU(2) Yang–Mills theory becomes a magnetic field in the thermal photon plasma. These long-wavelength magnetic modes do not convey energy in interactions with electric charges. Thus no direct thermalization of these modes through radiator walls takes place. They do, however, influence the direction of motion of unbound electric charges. In interaction with electric

---

[30]The author would like to thank Carlos Falquez for performing the numerical calculations.

charges prevailing inside galaxies the longitudinal magnetic modes of the Cosmic Microwave Background may have seeded an amplification (dynamo) mechanism for magnetic fields at, say, redshift $z \equiv \frac{T}{T_c} - 1 \sim 1$ [see Kronberg (1994) and Falquez, Hofmann and Baumbach (2011)].

5.4.8   *Considerations on one-loop TLM-TLM scattering*

Here we would like to discuss a particular non-bubble, one-loop process, namely, the scattering amplitude of two TLM modes. This is a truly radiatively induced process which does not exist on tree level. Our presentation closely follows [Krasowski and Hofmann (2014)].

5.4.8.1   *Notational conventions*

Let us set up our notational conventions.

Vertices. Vertices are discussed on different levels. Some arguments require only formulas that are valid independently of the sort of attached gauge modes. In such cases we just use the symbols p_1, p_2 etc. for four-momenta. On a level, where we would like to distinguish between TLH and TLM modes, we use R, S, P for the four-momenta of TLH modes and p, q for the four-momenta of TLM modes. In applications to the actual scattering process we employ a and b to indicate the four-momenta of incoming TLM modes. Outgoing TLM modes are labeled by c and d. Four-momenta of TLH modes in the loop are denoted by u and v.

Feynman diagrams. Throughout Sec. 5.4.8 the propagation of TLH modes is indicated by double lines in Feynman diagrams while TLM modes are represented by single lines.

Scattering channels. An overall channel in photon-photon scattering is labeled by the Mandelstam variable S,T or U (captial letters). On the other hand, for scattering channels associated with a given four-vertex Mandelstam variables are in lower case letters. In this way, one specific configuration can be written in a short-hand notation. For example, "Stu" describes overall S-channel scattering with the t-channel realized at the first four-vertex and the u-channel at the second four-vertex.

5.4.8.2 Exclusion of three-vertices

At the three-vertex a TLM mode always connects to two TLH modes (oppositely charged w.r.t. U(1)∈ SU(2)). For later use let us now check whether the photon can be an external, on-shell particle of positive energy. Because of four-momentum conservation and the on-shellness of the TLH modes the following conditions apply:

$$R^2 = S^2 = m^2,$$
$$p^2 = 0,$$
$$(R + S)^2 = p^2 = 0. \tag{5.214}$$

The energy of the TLH modes can be positive or negative, depending on the direction of energy flow in the loop of the overall scattering diagram, see Fig. 5.32.

Therefore, one has

$$2m^2 \pm 2\sqrt{|\mathbf{R}| + m^2}\sqrt{|\mathbf{S}|^2 + m^2} - 2\mathbf{RS} = 0$$

upon squaring $\Rightarrow$

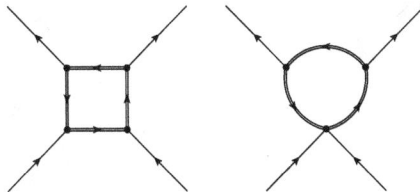

Figure 5.32.   Box diagram (left) and penguin diagram (right). Both possibilities are excluded to contribute to TLM-TLM scattering.

$$\left(|\mathbf{R}|^2 + m^2 \right) \left(|\mathbf{S}|^2 + m^2 \right) - m^4 - (\mathbf{RS})^2 + 2m^2 \mathbf{RS} = 0$$

$$\Rightarrow \quad m^2 \left(|\mathbf{R}|^2 + |\mathbf{S}|^2 + 2\,|\mathbf{R}|\,|\mathbf{S}| \cos \angle(\mathbf{R}, \mathbf{S}) \right)$$

$$+ |\mathbf{R}|^2 |\mathbf{S}|^2 \left(1 - \cos^2 \angle(\mathbf{R}, \mathbf{S}) \right) = 0. \qquad (5.215)$$

The minimal value of each of the two overall summands in the last equation of (5.215) is zero: The first summand vanishes for $\cos \angle(\mathbf{R}, \mathbf{S}) = -1$ and $|\mathbf{R}| = |\mathbf{S}|$, the second summand is zero for $\cos \angle(\mathbf{R}, \mathbf{S}) = \pm 1$. So the only configuration satisfying Eq. (5.215) is $\angle(\mathbf{R}, \mathbf{S}) = \pi$ and equal energy of the TLM modes. Since in a loop integral the thus allowed integration over angular variables is over a hypersurface of measure zero we conclude that three-vertices do not contribute to the overall TLM-TLM mode one-loop scattering.

5.4.8.3 *Constraints on four-vertex*

Here we consider the scattering kinematics in a four-vertex.

Specializing to the situation of Fig. 5.33 and introducing $x \in \{p, P\}$ to cover both diagrams, we have

$$s = \left| (R + S)^2 \right| \leq |\phi|^2 , \qquad (5.216)$$

$$t = \left| (x - R)^2 \right| \leq |\phi|^2 , \qquad (5.217)$$

$$u = \left| (x - S)^2 \right| \leq |\phi|^2 . \qquad (5.218)$$

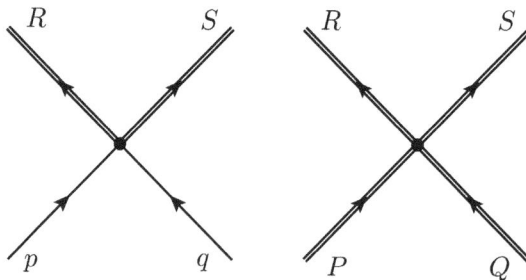

Figure 5.33. TLM-TLM to TLH-TLH and TLH-TLH to TLH-TLH scattering. These are the cases that need to be distinguished in all scattering channels.

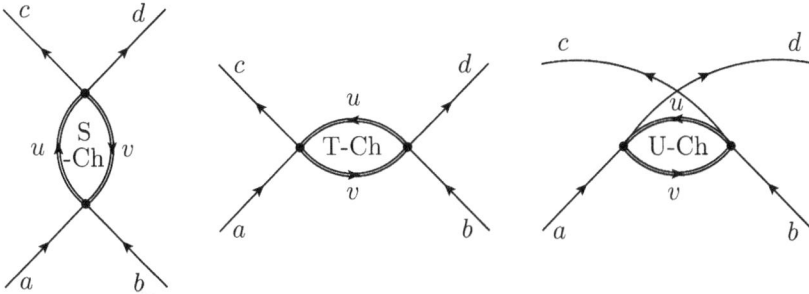

Figure 5.34. The overall scattering channels for the only admissible class of diagrams (two four-vertices) in one-loop photon-photon scattering. Here a,b,c, and d are the four-momenta of the TLM modes, and u and v denote the loop four-momenta of the TLH modes.

There are $3^3 = 27$ possible scattering channel combinations in the three overall channels for one-loop TLM-TLM scattering, compare with Fig. 5.34. (Three scattering channels for each of the two four-vertices times three overall scattering channels for TLM-TLM scattering.)

5.4.8.4 *Energy-flow constraints*

Let us now investigate all possible constraints that can occur in one-loop TLM-TLM scattering in view of energy flow. To do this, it is advantageous to work with dimensionless quantities. We will use two different normalizations. A tilde on top of a four-momentum component x marks normalizations with respect to the modulus $|\phi|$ of the inert field ϕ, a hat denotes normalization with respect to the temperature T, e.g.

$$\tilde{x} \equiv \frac{x}{|\phi|} = \frac{2e}{m}x = \frac{\lambda^{\frac{3}{2}}}{2\pi}\frac{x}{T} \equiv \frac{\lambda^{\frac{3}{2}}}{2\pi}\hat{x}, \qquad (5.219)$$

where, again, the dimensionless temperature λ is defined as

$$\lambda \equiv \frac{2\pi T}{\Lambda}. \qquad (5.220)$$

Furthermore, the dimensionless mass $\tilde{m} \equiv \frac{m}{|\phi|}$ is given as

$$\tilde{m} = 2e . \tag{5.221}$$

Recall that $e \geq \sqrt{8\pi}$.

For a four-vertex connecting two TLM modes (four-momenta $\tilde{p}$, $\tilde{q}$) with two TLH modes (four-momenta $\tilde{R}$, $\tilde{S}$) simple combinatorics allows the following possibilities of momentum transfer through the vertex, redundant by four-momentum conservation and constrained by the maximal resolution $|\phi|$ of the effective theory:

$$\left|\left(\tilde{R} \pm \tilde{S}\right)^2\right| \leq 1 \tag{5.222}$$

$$\left|\left(\tilde{p} \pm \tilde{R}\right)^2\right| \leq 1 \tag{5.223}$$

$$\left|\left(\tilde{p} \pm \tilde{S}\right)^2\right| \leq 1 \tag{5.224}$$

$$\left|\left(\tilde{q} \pm \tilde{R}\right)^2\right| \leq 1 \tag{5.225}$$

$$\left|\left(\tilde{q} \pm \tilde{S}\right)^2\right| \leq 1 \tag{5.226}$$

$$\left|(\tilde{p} \pm \tilde{q})^2\right| \leq 1 . \tag{5.227}$$

Both, TLM and TLH modes are on shell. Therefore, we have

$$\tilde{R}_0 = \pm\sqrt{\left|\tilde{\mathbf{R}}\right|^2 + \tilde{m}^2} , \tag{5.228}$$

$$\tilde{S}_0 = \pm\sqrt{\left|\tilde{\mathbf{S}}\right|^2 + \tilde{m}^2} , \tag{5.229}$$

$$\tilde{p}^2 = 0 , \tag{5.230}$$

$$\tilde{q}^2 = 0 . \tag{5.231}$$

From Eqs. (5.221) and (5.228) through (5.231) it follows that certain sign combinations of R_0 and S_0 are forbidden. Let us now classify these excluded cases. We consider

$$1 \geq \left| \left(\tilde{R} \pm \tilde{S} \right)^2 \right| = \left| 2\tilde{m}^2 \pm 2\tilde{R}_0 \tilde{S}_0 \mp 2\tilde{\mathbf{R}}\tilde{\mathbf{S}} \right|$$

$$= \left| 2\tilde{m}^2 \pm 2 \left(\pm \sqrt{\left|\tilde{\mathbf{R}}\right|^2 + \tilde{m}^2} \right) \left(\pm \sqrt{\left|\tilde{\mathbf{S}}\right|^2 + \tilde{m}^2} \right) \mp 2\tilde{\mathbf{R}}\tilde{\mathbf{S}} \right| \qquad (5.232)$$

and

$$1 \geq \left| \left(\tilde{p} \pm \tilde{R} \right)^2 \right| = \left| \tilde{m}^2 \pm 2\tilde{p}_0 \tilde{R}_0 \mp 2\tilde{\mathbf{p}}\tilde{\mathbf{R}} \right|$$

$$= \left| \tilde{m}^2 \pm 2\tilde{p}_0 \left(\pm \sqrt{\left|\tilde{\mathbf{R}}\right|^2 + \tilde{m}^2} \right) \mp 2\tilde{\mathbf{p}}\tilde{\mathbf{R}} \right| . \qquad (5.233)$$

In the following discussion cases (5.222) are treated in terms of (5.232) while cases (5.223) through (5.226) are covered [31] by (5.233). To proceed, note that

$$\left| \left(\pm \sqrt{\left|\tilde{\mathbf{R}}\right|^2 + \tilde{m}^2} \right) \left(\pm \sqrt{\left|\tilde{\mathbf{S}}\right|^2 + \tilde{m}^2} \right) \mp \tilde{\mathbf{R}}\tilde{\mathbf{S}} \right| \geq \tilde{m}^2 \qquad (5.234)$$

and

$$\left| p_0 \left(\pm \sqrt{\left|\tilde{\mathbf{R}}\right|^2 + \tilde{m}^2} \right) \mp \tilde{\mathbf{p}}\tilde{\mathbf{R}} \right| \geq 0 \qquad (5.235)$$

Inequality (5.234) is true because of the Cauchy-Schwarz inequality applied to the two vectors $\left(\left|\tilde{\mathbf{R}}\right|, \tilde{m} \right)$ and $\left(\left|\tilde{\mathbf{S}}\right|, \tilde{m} \right)$ with the canonical scalar product of $\mathbb{R}^2$ and the fact that $\left| \cos \angle \tilde{\mathbf{R}}\tilde{\mathbf{S}} \right| \leq 1$. Inequality (5.235) is selfevident. From (5.234), $\tilde{m}^2 > 1$ (see (5.221)), and (5.232)

---

[31] We do not distinguish R and S or p and q. For the argument it is only important that the former are associated with TLH modes and the latter with TLM modes.

it follows that

$$\mathrm{sgn}\left(\tilde{R}_0\right) = \pm\mathrm{sgn}\left(\tilde{S}_0\right)$$

are forbidden for cases (5.222), respectively. Because of (5.235), $\tilde{m}^2 > 1$, and (5.233) it is clear that

$$\mathrm{sgn}\left(\tilde{R}_0\right) = \pm 1$$

are forbidden for cases (5.223), respectively (and for cases (5.225), respectively). In addition, the respective cases (5.224) and (5.226) ($\tilde{R}$ replaced by $\tilde{S}$) are also excluded. Obviously, no implication for the signs of $\tilde{R}_0$ or $\tilde{S}_0$ arises from (5.227).

5.4.8.5   *TLM-TLM scattering*

Let us now assess the kinematic options for one-loop TLM-TLM scattering. To do this, we first apply the results of Sec. 5.4.8.4 to each of the overall scattering channels depicted in Fig. 5.34. This excludes a vast majority of energy-flow combinations. In the overall S-channel all combinations for the two four-vertices can be excluded by analytical arguments. For the overall T-channel and the overall U-channel the analytical treatment leaves four possible combinations. These are analyzed numerically based on Monte-Carlo simulations.

At a given vertex, we use the following book keeping device for excluded combinations of the two loop energies u_0 and v_0. Every cell in a 2×2 table accounts for one possible combination of energy flow:

$\tilde{u}_0 > 0;\ \tilde{v}_0 > 0$	$\tilde{u}_0 > 0;\ \tilde{v}_0 < 0$
$\tilde{u}_0 < 0;\ \tilde{v}_0 > 0$	$\tilde{u}_0 < 0;\ \tilde{v}_0 < 0$

.

An "X" in a given cell signals that the corresponding combination is forbidden.

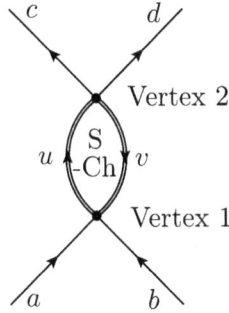

Figure 5.35. The overall S-channel.

Energy flow: Overall S-channel

The lower vertex in Fig. 5.35 is denoted by number 1, the upper one by number 2. At a given vertex the constraint on a scattering channel can be expressed in a twofold way because of total four-momentum conservation across the vertex. For example, the constraint at the first vertex $\left|(\tilde{a}+\tilde{b})^2\right| \le 1$ can be rewritten as $\left|(\tilde{u}-\tilde{v})^2\right| \le 1$. To exclude all sign combinations of the energies in the loop momenta one needs to look at either form and combine both statements [32].
Let us now visualize the constraints at vertex 1 in terms of their tables:

$$\text{s-ch: } 1 \ge \left|(\tilde{a}+\tilde{b})^2\right| = \left|(\tilde{u}-\tilde{v})^2\right| \to$$

$$\text{t-ch: } 1 \ge \left|(\tilde{a}-\tilde{u})^2\right| = \left|(\tilde{b}+\tilde{v})^2\right| \to$$

$$\text{u-ch: } 1 \ge \left|(\tilde{a}+\tilde{v})^2\right| = \left|(\tilde{b}-\tilde{u})^2\right| \to$$

For vertex 2 we obtain:

[32]The two 2 × 2 tables, obtained from each of the forms expressing the vertex constraint, are put on top of one another.

Table 1.   Forbidden combinations of energy flow (marked with a X) in all scattering channel combinations of vertex 1 and vertex 2 in the overall S-channel.

Vertex 2 \ Vertex 1	s-ch.	t-ch.	u-ch.
s- ch.	·X / X·	XX / XX	XX / XX
t-ch.	XX / XX	X· / XX	X· / XX
u-ch.	XX / XX	X· / XX	X· / XX

$$\text{s-ch: } 1 \geq \left|(\tilde{u} - \tilde{v})^2\right| = \left|(\tilde{c} + \tilde{d})^2\right| \rightarrow \boxed{\begin{array}{cc} & X \\ X & \end{array}},$$

$$\text{t-ch: } 1 \geq \left|(\tilde{u} - \tilde{c})^2\right| = \left|(\tilde{v} + \tilde{d})^2\right| \rightarrow \boxed{\begin{array}{cc} X & \\ X & X \end{array}},$$

$$\text{u-ch: } 1 \geq \left|(\tilde{u} - \tilde{d})^2\right| = \left|(\tilde{v} + \tilde{c})^2\right| \rightarrow \boxed{\begin{array}{cc} X & \\ X & X \end{array}}.$$

In the diagram for the overall scattering channel (Fig. 5.35) the constraints on the two vertices have to be satisfied simultaneously: If one combination of energy flow and scattering channel is forbidden by one constraint then it is not allowed at all. Thus it is suggested to use a 3×3 table for visualization of excluded scattering-channel and energy-flow combinations within a given overall channel. Each of the cells of this 3×3 table, corresponding to a certain combination of scattering channels at vertex 1 and vertex 2, is obtained by placing the two respective 2×2 tables on top of one another. For the overall S-channel we thus are left with Tab. 1.

Therefore, only six possible combinations remain. Two of them can be excluded analytically.

Exclusion of configuration Sss

For the configuration Sss two combinations of energy flow are allowed by Tab. 1. To exclude them, we proceed by a more detailed analysis of the general situation expressed in (5.227) and (5.228) to (5.231). For our specific cases one has

$$\left| \left(\tilde{a} + \tilde{b} \right)^2 \right| = \left| 2\tilde{a}\tilde{b} \right| = 2 \left| \tilde{a}_0 \tilde{b}_0 (1 - \cos \left(\angle \tilde{a}\tilde{b} \right)) \right| \le 1$$

$$\Rightarrow -\frac{1}{2} \le \tilde{a}_0 \tilde{b}_0 (1 - \cos \left(\angle \tilde{a}\tilde{b} \right)) \le \frac{1}{2}$$

$$\Rightarrow -\frac{1}{2\tilde{a}_0 \tilde{b}_0} \le (1 - \cos \left(\angle \tilde{a}\tilde{b} \right)) \le \frac{1}{2\tilde{a}_0 \tilde{b}_0} \, . \qquad (5.236)$$

The first inequality $-\frac{1}{2\tilde{a}_0 \tilde{b}_0} \le (1 - \cos \angle \tilde{a}\tilde{b})$ is always satisfied because $\left| \cos \angle \tilde{a}\tilde{b} \right| \le 1$ and because $a_0, b_0 > 0$. We also have

$$\tilde{u}^2 = \tilde{m}^2 \, , \qquad (5.237)$$

$$\tilde{v}^2 = \tilde{m}^2 \, . \qquad (5.238)$$

In addition, four-momentum conservation holds. Thus eight independent entries of the two four-momenta $\tilde{u}$ and $\tilde{v}$ are reduced to four. The on-shellness of each mode leads to an additional reduction from four to two: $|\tilde{u}|$ is determined by $\tilde{a}, \tilde{b}$, and the orientation $\mathbf{e_u} = \frac{\mathbf{u}}{|\mathbf{u}|} = \frac{\tilde{\mathbf{u}}}{|\tilde{\mathbf{u}}|}$ (two angles). One has

$$0 = \tilde{v}^2 - \tilde{m}^2 = \left(\tilde{u} - \tilde{a} - \tilde{b} \right)^2 - \tilde{m}^2 = -2\tilde{u} \left(\tilde{a} + \tilde{b} \right) + 2\tilde{a}\tilde{b}$$

$$= -2 \left(\pm\sqrt{|\tilde{\mathbf{u}}|^2 + \tilde{m}^2} \right) \left(\tilde{a}_0 + \tilde{b}_0 \right) + 2\tilde{u} \left(\tilde{a} + \tilde{b} \right) + 2\tilde{a}\tilde{b}$$

$$\Rightarrow \left(\pm\sqrt{|\tilde{\mathbf{u}}|^2 + \tilde{m}^2} \right) \left(\tilde{a}_0 + \tilde{b}_0 \right) = |\tilde{\mathbf{u}}| \left(\tilde{a} + \tilde{b} \right) \mathbf{e_u} + \tilde{a}\tilde{b}$$

$$\Rightarrow \left(|\tilde{\mathbf{u}}|^2 + \tilde{m}^2 \right) \left(\tilde{a}_0 + \tilde{b}_0 \right)^2$$

$$= |\tilde{u}|^2 \left(\left(\tilde{a} + \tilde{b}\right) e_u\right)^2 + \left(\tilde{a}\tilde{b}\right)^2 + 2|\tilde{u}| \left(\tilde{a}\tilde{b}\right) \left(\tilde{a} + \tilde{b}\right) e_u$$

$$\Rightarrow 0 = |\tilde{u}|^2 \left(\left(\tilde{a}_0 + \tilde{b}_0\right)^2 - \left(\left(\tilde{a} + \tilde{b}\right) e_u\right)^2\right)$$

$$+ |\tilde{u}| \left(-2\left(\tilde{a}\tilde{b}\right)\left(\tilde{a} + \tilde{b}\right) e_u\right)$$

$$+ \left(\tilde{m}^2 \left(\tilde{a}_0 + \tilde{b}_0\right)^2 - \left(\tilde{a}\tilde{b}\right)^2\right)$$

$$\Rightarrow |\tilde{u}|_{1/2} = \frac{\left(\tilde{a}\tilde{b}\right)\left(\tilde{a} + \tilde{b}\right) e_u}{\left(\tilde{a}_0 + \tilde{b}_0\right)^2 - \left(\left(\tilde{a} + \tilde{b}\right) e_u\right)^2}$$

$$\pm \frac{\left(\tilde{a}_0 + \tilde{b}_0\right)\sqrt{-\tilde{m}^2(\tilde{a}_0 + \tilde{b}_0)^2 + (\tilde{a}\tilde{b})^2 + \tilde{m}^2((\tilde{a} + \tilde{b})e_u)^2}}{(\tilde{a}_0 + \tilde{b}_0)^2 - ((\tilde{a} + \tilde{b})e_u)^2}.$$

$$(5.239)$$

In the computation of the amplitude only an integration over the orientation e_u remains when it comes to loop integration. The solutions $|\tilde{u}|_{1/2}$ in Eq. (5.239) must be real and positive. This implies that valid configurations satisfy the following inequality (argument of square root in Eq. (5.239) must be positive):

$$- \tilde{m}^2 \left(\tilde{a}_0 + \tilde{b}_0\right)^2 + \left(\tilde{a}\tilde{b}\right)^2 + \tilde{m}^2 \left(\left(\tilde{a} + \tilde{b}\right) e_u\right)^2 \geq 0$$

$$\Rightarrow \quad - \tilde{m}^2 \left(\tilde{a}_0^2 + \tilde{b}_0^2 + 2\tilde{a}_0\tilde{b}_0\right) + \tilde{a}_0^2\tilde{b}_0^2 \left(1 - \cos\left(\angle ab\right)\right)^2$$

$$+ \tilde{m}^2 \left(\tilde{a}_0^2 + \tilde{b}_0^2 + 2\tilde{a}_0\tilde{b}_0 \cos\left(\angle ab\right)\right) \left(e_u e_{a+b}\right)^2 \geq 0. \quad (5.240)$$

The values of $e_u e_{a+b}$, at which $\angle ab$ is least constrained, are $e_u e_{a+b} = \pm 1$. For these values the inequality reads as

$$\tilde{a}_0^2\tilde{b}_0^2 \left(1 - \cos\left(\angle ab\right)\right)^2 - 2\tilde{m}^2\tilde{a}_0\tilde{b}_0 \left(1 - \cos\left(\angle ab\right)\right) \geq 0. \quad (5.241)$$

To proceed, we need the roots $\left(1 - \cos\left(\angle ab\right)\right)_{1/2}$ of the equation

$$\tilde{a}_0^2\tilde{b}_0^2 \left(1 - \cos\left(\angle ab\right)\right)^2 - 2\tilde{m}^2\tilde{a}_0\tilde{b}_0 \left(1 - \cos\left(\angle ab\right)\right) = 0. \quad (5.242)$$

These roots are given as

$$\left(1 - \cos\left(\angle\mathbf{ab}\right)\right)_1 = 0, \tag{5.243}$$

$$\left(1 - \cos\left(\angle\mathbf{ab}\right)\right)_2 = \frac{2\tilde{m}^2}{\tilde{a}_0\tilde{b}_0}. \tag{5.244}$$

Inequality (5.241) either is solved by values of $\left(1 - \cos\left(\angle\mathbf{ab}\right)\right)$ that are smaller than the first solution or larger than the second one because with respect to the variable $\left(1 - \cos\left(\angle\mathbf{ab}\right)\right)$ the left hand side of (5.241) is a parabola with positive curvature. $\left(1 - \cos\left(\angle\mathbf{ab}\right)\right)_1 < 0$ can not be satisfied, and therefore every valid $\left(1 - \cos\left(\angle\mathbf{ab}\right)\right)$ has to be larger than $\frac{2\tilde{m}^2}{\tilde{a}_0\tilde{b}_0}$:

$$\left(1 - \cos\left(\angle\mathbf{ab}\right)\right) \geq \frac{2\tilde{m}^2}{\tilde{a}_0\tilde{b}_0}. \tag{5.245}$$

We can now compare the two requirements for $\left(1 - \cos\left(\angle\mathbf{ab}\right)\right)$.

* From on-shellness and momentum conservation at vertex 1 (Eq. (5.245)):

$$\left(1 - \cos\left(\angle\mathbf{ab}\right)\right) \geq \frac{2\tilde{m}^2}{\tilde{a}_0\tilde{b}_0}.$$

* From the s-channel momentum transfer constraint at vertex 1 (Eq. (5.236)):

$$\left(1 - \cos\left(\angle\mathbf{ab}\right)\right) \leq \frac{1}{2\tilde{a}_0\tilde{b}_0}.$$

The upper bound is smaller than the lower bound because $\tilde{m} = 2e \geq 2\sqrt{8\pi} > 1$, and so we conclude that there are no configurations satisfying the momentum transfer constraint, on-shellness of the TLH modes, and momentum conservation simultaneously. Thus the combination of s and s from the two four-vertices to the overall S-channel is excluded in TLM-TLM scattering, and we are left with four unexcluded configurations:

Vertex 1 / Vertex 2	s-ch.	t-ch.	u-ch.
s- ch.	_ X / X _	X X / X X	X X / X X
t-ch.	X X / X X	X _ / X X	X _ / X X
u-ch.	X X / X X	X _ / X X	X _ / X X

Energy flow: Overall T- and U-channels

Similar arguments as used for the overall S-channel can be applied to the overall T- and U-channels, compare with Fig. 5.37. The following considerations do not depend on the specific values of the external momenta but only on the fact that they are on shell. The overall U-channel thus is constrained in the same way as the overall T-channel. (One just interchanges the external momenta c and d, see Fig. 5.34). Let us now put forward the arguments for the T-channel.

The exclusion tables for the constraints at vertex 1 can be obtained by referring to Sec. 5.4.8.4:

$$\text{s-ch: } 1 \geq \left|(\tilde{a} - \tilde{v})^2\right| = \left|(\tilde{c} - \tilde{u})^2\right| \rightarrow \boxed{\begin{array}{cc} & X \\ X & X \end{array}} ,$$

$$\text{t-ch: } 1 \geq \left|(\tilde{a} - \tilde{c})^2\right| = \left|(\tilde{v} - \tilde{u})^2\right| \rightarrow \boxed{\begin{array}{cc} & X \\ X & \end{array}} ,$$

$$\text{u-ch: } 1 \geq \left|(\tilde{a} + \tilde{u})^2\right| = \left|(\tilde{v} + \tilde{c})^2\right| \rightarrow \boxed{\begin{array}{cc} X & X \\ X & \end{array}} .$$

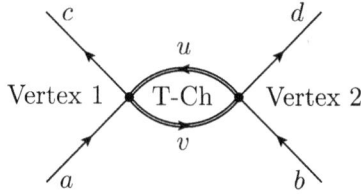

Figure 5.36. Overall T-channel. The vertex on the left side is labeled by number 1, the one on the right side by number 2.

Table 2. Forbidden combinations of energy flow in all scattering-channel combinations of the overall T-channel (and U-channel) by momentum transfer constraints.

Vertex 2 \ Vertex 1	s-ch.		t-ch.		u-ch.	
s- ch.	X	X	X	X	X	X
	X	X	X		X	
t-ch.		X		X	X	X
	X	X			X	
	X		X		X	
u-ch.		X		X	X	X
	X	X	X	X	X	X

The constraints at vertex 2 read:

$$\text{s-ch: } 1 \geq \left|\left(\tilde{v}+\tilde{b}\right)^2\right| = \left|\left(\tilde{u}+\tilde{d}\right)^2\right| \rightarrow \boxed{\begin{array}{cc} X & X \\ X & \end{array}},$$

$$\text{t-ch: } 1 \geq \left|\left(\tilde{v}-\tilde{u}\right)^2\right| = \left|\left(\tilde{b}-\tilde{d}\right)^2\right| \rightarrow \boxed{\begin{array}{cc} & X \\ X & \end{array}},$$

$$\text{u-ch: } 1 \geq \left|\left(\tilde{v}-\tilde{d}\right)^2\right| = \left|\left(\tilde{b}-\tilde{u}\right)^2\right| \rightarrow \boxed{\begin{array}{cc} & X \\ X & X \end{array}}.$$

We now superimpose all combinations of the tables for the two vertices to obtain the result shown in Tab. 2.

Table 3.   Forbidden combinations of energy flow in all scattering-channel combinations of the overall T-channel (and U-channel) by momentum transfer constraints and energy momentum conservation.

Vertex 2 \ Vertex 1	s-ch.		t-ch.		u-ch.	
s- ch.	X	X	X	X	X	X
	X	X	X	X		
t-ch.	X	X	X	X	X	X
	X	X	X	X	X	X
u-ch.			X	X	X	X
	X	X	X	X	X	X

That is, for the overall T-channel (or for the overall U-channel) eight possible configurations cannot yet be eliminated.

<u>Exclusion of t-channels in the overall T-channel</u>

The same argumentation that excluded the combination Sss relying on a momentum transfer constraint, energy momentum conservation, and on the on-shellness of the loop modes, exclude the combinations Tts, Ttt, Ttu, Tst and Tut. Thus, we obtain an updated version of the exclusion Tab. 2 in terms of Tab. 3.

Again, only four configurations are left unexcluded.

5.4.8.6   *Monte-Carlo analysis of remaining cases*

For 4 out of the 36 configurations in each of the overall S-, T-, and U-channels no analytical exclusion could be performed. These remaining cases thus are treated numerically. To obtain an estimate on numerical precision in sampling the according algebraic varieties in these non-excluded combinations we also sample the analytically excluded Sss configuration. To proceed, a suitable set of non-redundant variables must be defined. As we will see, it is possible to parametrize the overall scattering process by the variables (referring to Fig. 5.34) a_0, b_0, the energies of the incoming photons, the angle

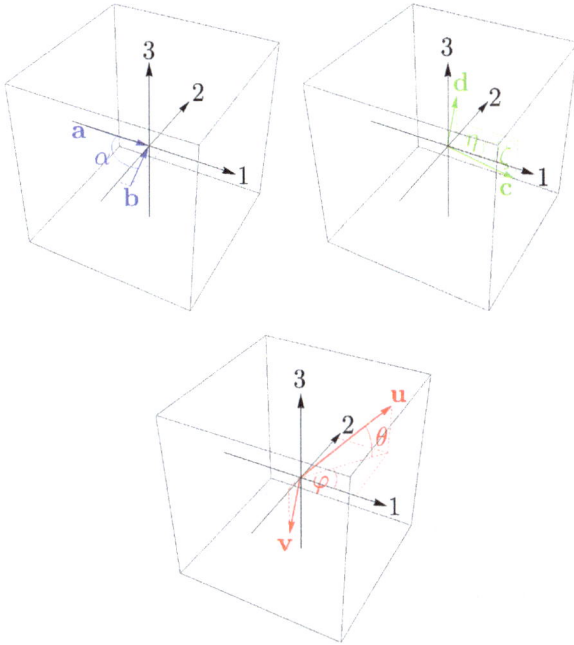

Figure 5.37.   A visualization of the involved three-momenta (redundant variables). The incoming momenta are blue, the internal ones are red, and the outgoing momenta are green.

$\alpha \equiv \angle \mathbf{ab} \in [0, 2\pi]$ between their three-momenta, and the angles $\zeta \in \left[-\frac{\pi}{2}, \frac{\pi}{2}\right]$ and $\eta \in [0, 2\pi]$, that are necessary to describe the spatial orientation of one of the outgoing particles. Moreover two angles, $\theta \in \left[-\frac{\pi}{2}, \frac{\pi}{2}\right]$ and $\varphi \in [0, 2\pi]$, are sufficient to describe the kinematic state of the internal particles, see Fig. 5.37.

In the following we relate the involved four-momenta to these parameters. First, we exploit four-momentum conservation and that the external photons are on shell:

$$a = \begin{pmatrix} a_0 \\ a_0 \\ 0 \\ 0 \end{pmatrix}, \, b = \begin{pmatrix} b_0 \\ b_0 \cos(\alpha) \\ b_0 \sin(\alpha) \\ 0 \end{pmatrix}, \, c = \begin{pmatrix} c_0 \\ c_0 \cos(\zeta)\cos(\eta) \\ c_0 \cos(\zeta)\sin(\eta) \\ c_0 \sin(\zeta) \end{pmatrix},$$

$$d = a + b - c \tag{5.246}$$

Eq. (5.246) implies

$$
0 = d^2 = \left(a + b - c\right)^2 = 2ab - 2ac - 2bc
$$
$$
= 2(a_0 b_0 - \left(a_0 + b_0\right) c_0 - a_0 b_0 \cos\left(\alpha\right)
$$
$$
+ c_0 \left(b_0 \cos\left(\alpha - \eta\right) + a_0 \cos\left(\eta\right)\right) \sin\left(\zeta\right))
$$
$$
\Rightarrow c_0 = \frac{a_0 b_0 (1 - \cos\left(\alpha\right))}{a_0 + b_0 - b_0 \cos\left(\alpha - \eta\right) \sin\left(\zeta\right) - a_0 \cos\left(\eta\right) \sin\left(\zeta\right)} ,
$$

independently of the overall scattering channel S, T, or U. Momenta of the massive modes, however, are dependent on S, T, or U. Both internal particles are on shell. For the overall S-channel and for given incoming momenta $\tilde{a}, \tilde{b}$ the absolute value of $|u|$ loop three-momentum $\mathbf{u}$ is related to its orientation $\mathbf{e_u}$, which depends on θ and φ, as given in Eq. (5.239). We have

$$
|\tilde{u}|_{1/2}^S = \frac{\left(\tilde{a}\tilde{b}\right)\left(\tilde{a} + \tilde{b}\right)\mathbf{e_u}}{\left(\tilde{a}_0 + \tilde{b}_0\right)^2 - 2\left(\left(\tilde{a} + \tilde{b}\right)\mathbf{e_u}\right)^2}
$$
$$
\pm \frac{\left(\tilde{a}_0 + \tilde{b}_0\right)\sqrt{-\tilde{m}^2\left(\tilde{a}_0 + \tilde{b}_0\right)^2 + \left(\tilde{a}\tilde{b}\right)^2 + \tilde{m}^2\left(\left(\tilde{a} + \tilde{b}\right)\mathbf{e_u}\right)^2}}{\left(\tilde{a}_0 + \tilde{b}_0\right)^2 - \left(\left(\tilde{a} + \tilde{b}\right)\mathbf{e_u}\right)^2} .
$$

$$(5.247)$$

The expressions for the overall T- and U-channels can be obtained by interchanging momenta: The T-channel relates to the S-channel by letting $a \to c$ and $b \to -a$; the U-channel relates to the S-channel by $a \to d$ and $b \to -a$. Clearly, the energy u_0 is determined by on-shellness, and the other internal four-momentum v is completely determined by four-momentum conservation. Whether or not a given set of parameter values satisfies all constraints is tested by inserting it into Eqs. (5.246) and (5.247) before, in turn, the four-vertex constraints are probed at a given value of the dimensionless temperature λ. Because λ, a_0, b_0 are not bounded from above their to-be-tested values need to be limited in the numerical procedure.

We require $\lambda_c = 13.867 \leq \lambda \leq 100$ and $\hat{a}_0 = \frac{a_0}{T}, \hat{b}_0 = \frac{b_0}{T} \leq 100$. Here λ_c is the critical temperature for the deconfining-preconfining phase transition where $m \to \infty$ and massive modes thus decouple, see Sec. 5.2.3. Parameter values are sampled randomly in a conditioned way, and parameter sets satisfying the constraints are counted.

In the overall S channel, 87 out of 6.144×10^{10} tested parameter sets satisfied the constraints. This number is suppressed by a factor $\sim$ thirteen compared to the overall T-channel where 1110 out of 6.144×10^{10} sets satisfied the constraints. Because of the afore mentioned symmetry of the constraints under the exchange of outgoing photons T- and U-channel should yield identical results. The number quoted for the T-channel represents the average of all four remaining combinations in Tsu and Tus. We conclude from the S- versus T- plus U- channel comparison that the former is supported by less than 5% of the integration volume of the latter two channels. Practically, this excludes the overall S-channel. We also investigated the distribution of valid parameter values. For example, Fig. 5.38 depicts the abundance of the sum of hits over all non-excluded channel combinations as a function of λ. Two things are worth pointing out. First, we see that processes with $\lambda > 40$ are very rare. The highest temperature associated with a valid configuration is $\lambda = 63.32$, and this is an extreme outlier. Second, the abundance is rapidly decaying for $\lambda \leq 18.8$. In fact, no contribution at a temperature smaller than $\lambda = 18.15$ was ever detected.

Another interesting distribution is the abundance of energies relative to temperature T, $\hat{a}_0$ and $\hat{b}_0$, as shown in Fig. 5.39. As anticipated, there is no obvious difference between the distributions of $\hat{a}_0$ and $\hat{b}_0$. The constraints seem to imply an upper bound of about 40. Configurations with values of $\hat{a}_0$ and $\hat{b}_0$ above this bound are anyway strongly suppressed by the Bose-Einstein distributions associated with external, thermalized photons.

To explore the numerical data further, we zoom into the θ-φ plane about valid parameter-value combinations ($\lambda, \alpha, \eta, \zeta, \tilde{a}_0$ and $\tilde{b}_0$). It was numerically not possible to resolve the associated varieties

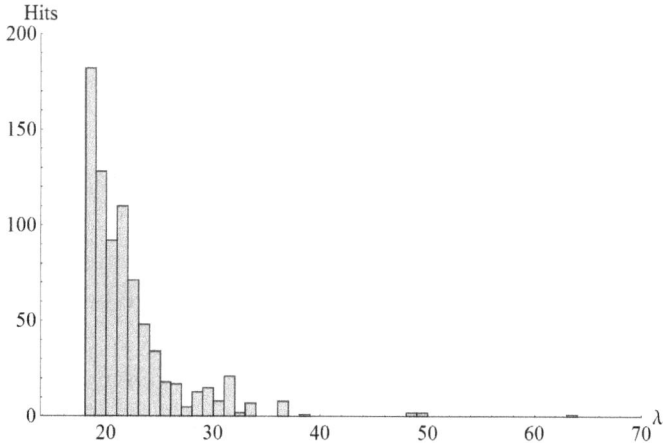

Figure 5.38. This histogram shows the distribution of the sum of hits over all tested channel combinations as a function of dimensionless temperature λ in the range from $\lambda_c = 13.867$ to 100.

of valid configurations about a particular, pivotal one at once. Therefore we imposed a relaxation of the constraints to broaden the region of valid configurations. The relaxation is implemented by virtue of a softening factor Υ implemented as $1 \to \Upsilon$ on the right hand sides in Eqs. (5.222) to (5.227). (The momentum transfer must only be smaller than $\Upsilon |\phi|^2$ in the relaxed as opposed to the physical situation.) For the combination shown in Fig. 5.41, representative of all other non-excluded combinations, there are series of four region plots. The first (top left) depicts a region of interest of size $\Delta\theta = 0.4$ and $\Delta\varphi = 0.8$ which is centered around the pivotal hit detected during the Monte Carlo test. This region is blown up into the second plot (top right) where, in turn, a centered region of interest of size $\Delta\theta = 0.04$ and $\Delta\varphi = 0.08$ is shown. Again, the latter is blown up into the third region plot (bottom left). The region of interest marked here and shown in full size in plot four (bottom right) has an extent of $\Delta\theta = 0.004$ and $\Delta\varphi = 0.008$. This last plot represents the physical situation with $\Upsilon = 1$. Values of the softening factor Υ are chosen to point out the nature of the actual, physical variety. The contour

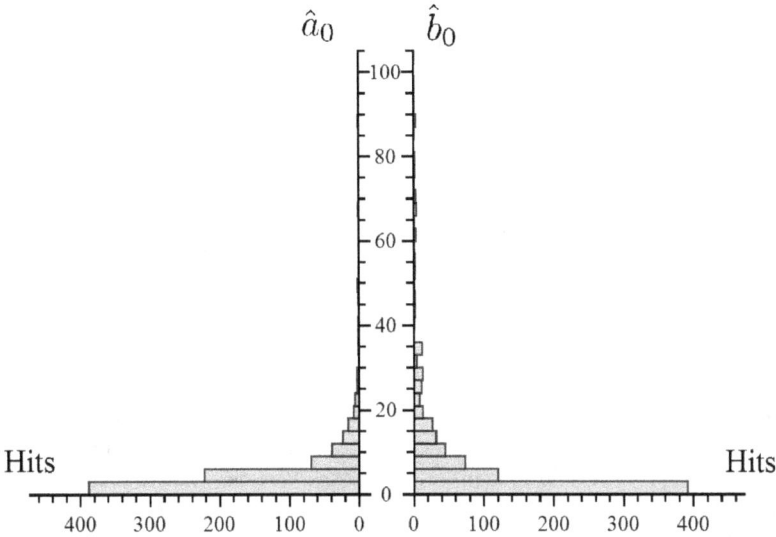

Figure 5.39. Comparison of the distributions of the two incoming energies, normalized with respect to temperature T. Hits in each bin represent the sum over all tested channel combinations.

plots in the θ-φ plane signal the different constraints by a color code as explained in the caption of Fig. 5.40.

5.4.8.7   Summary on one-loop TLM-TLM scattering

We have investigated systematically how the unitary-gauge constraints of the effective theory for deconfining SU(2) Yang-Mills thermodynamics limit the contributions of loop momenta to the amplitude for one-loop TLM-TLM scattering. Only one type of Feynman diagrams with two four-vertices is admissible to mediate this process, and a large part of channel and energy-sign combinations for the scattering through these vertices is analytically excluded relying on a subset of all energy-flow and momentum-transfer constraints. Out of a total of 108 scattering-channel and energy-sign configurations for the loop four momenta 12 configurations cannot be excluded analytically. These remaining cases do not give rise to pair creation or annihilation (practically, there is an exclusion of the

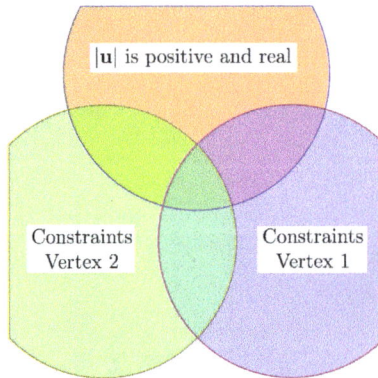

Figure 5.40. This region plot depicts color coded representations of the effects of single and combined constraints. The varieties marked in orange are associated with real and positive solutions $|\mathbf{u}|$ to Eq. (5.247). Varieties, where the momentum transfer constraints in the first vertex are satisfied, are marked in green, and those, where the momentum transfer constraints are fulfilled at the second vertex, are indicated in blue. The valid variety is represented by the intersection of these three varieties.

overall S-channel) and were analyzed by Monte-Carlo sampling subject to all constraints. The associated hit densities decay very rapidly with temperature. We have also investigated the admissible algebraic variety in the vicinity of a selected, pivotal Monte-Carlo hit to demonstrate how filamentous it is.

One may, at first sight, object that an analysis of allowed regions for the loop integration is not sufficient to draw a conclusion about the actual smallness of the integral for the amplitude since singular integrands may arise on the filamentous integration regions. This is not the case, however, because we consider a reduced integration manifold, obtained after singular distributions in the original integrand are integrated out. (On-shell conditions, associated with δ-distributions in the full $4l$-dimensional space of loop integration at loop order l, are implemented in the analysis of vertex conditions from the start. Thus the integration manifold considered here is lower than $4l$-dimensional. Integrands are either regular, or if singular, thanks to their integrability can be made regular by changes of variables.)

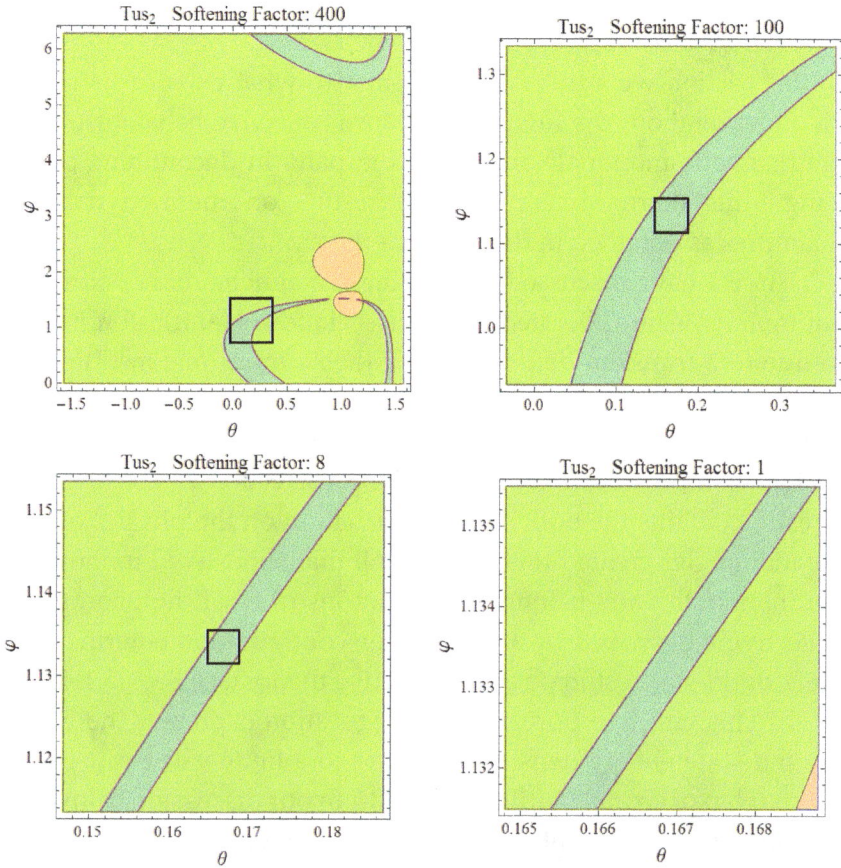

Figure 5.41. A visualization of the constraints in the θ-φ plane in the overall T-channel with scattering channels u and s for $|\mathbf{u}|_2$ and the following parameter values of the pivotal configuration: $\lambda = 25.727$, $\tilde{a}_0 = 36.5239$, $\tilde{b}_0 = 11.0003$, $\alpha = 1.53077$, $\eta = 5.82265$, $\zeta = 2.09548$.

All in all, our results suggest that TLM-TLM scattering within the deconfining SU(2) Yang-Mills plasma is feeble. Finally, a higher dimensional generalization (more than two four-vertices per diagram) of the technology of energy-flow exclusion developed in the present work may turn out to be key to proving the termination of the expansion of thermodynamical quantities into irreducible bubble diagrams at a finite loop order as conjectured in Sec. 5.4.3.

5.5   Stable, Screened Magnetic Monopoles

In this section we address the question to what extent it is possible to learn about the high-temperature collective behavior of stable magnetic monopole-antimonopole pairs in deconfining SU(2) Yang–Mills thermodynamics from results on thermodynamical quantities as obtained in the effective theory.

As we have discussed before, these magnetic monopoles and antimonopoles are liberated by the dissociation of (anti)calorons due to due to a transient deformation of their central regions, thereby introducing sizable shifts of the holonomy. These processes are collectively described by radiative corrections in the effective theory. At high temperature a single two-loop correction to the pressure already contains the bulk of the information on the physics of stable monopoles created this way. Recall that short-lived monopole-antimonopole pairs belonging to calorons of small holonomy are collectively described by a pure-gauge configuration entering into the ground-state estimate of the effective theory (see Secs. 5.1.4 and 5.1.5). This effective puregauge configuration expresses the lift of ground-state energy density from zero to a finite value caused by the overlap of spatial (anti)caloron tails. In our discussion below we heavily rely on the results of [Ludescher *et al.* (2008)].

5.5.1   *Remarks on screening of static magnetic charge*

Here we would like to discuss the principal difference in magnetic charge screening as described by the one-loop approximation of free quasiparticles in the effective loop expansion of the pressure and a high-temperature radiative correction at two loops.

Before we do this, let us point out the following. On distances larger than the minimal spatial length [Hofmann (2005)]

$$|\phi|^{-1} = \frac{(13.87)^{3/2}}{2\pi T_c} \sqrt{\frac{T}{T_c}} \qquad (5.248)$$

and *not* taking into account radiatively liberated stable and isolated monopole-antimonopole pairs (free quasiparticle approximation), the depletion of a single magnetic *test* charge by instable magnetic monopoles is, for $T \gg T_c$, described by the effective, electric coupling $e = \sqrt{8}\pi$ [high-temperature solution of Eq. (5.71)]. Recall that the (constant) value of e signals that for $T \gg T_c$ the thermodynamics of these modes solely is determined by topology and temperature. This fact may be conceived as a nonperturbative manifestation of asymptotic freedom. Both the constancy $e = \sqrt{8}\pi$ and the two-loop correction to the pressure divided by T^4 are approached in power-like speed in T (see Figs. 5.6 and 5.14). Notice that asymptotic constancy of e is not to be confused with the behavior of the *fundamental* gauge coupling [Giacosa and Hofmann (2007b)] (see Sec. 5.2.5) which runs logarithmically.

At high temperature the holonomy-independent sum $m_{>|\phi|^{-1}}$ of the masses of the BPS monopole and BPS antimonopole (BPS mass), being the constituents of a (anti)caloron of nontrivial holonomy, is given as [Hofmann (2005); Lee and Lu (1983); Kraan and Van Baal (1998a,b)]

$$m_{>|\phi|^{-1}} = \frac{8\pi^2 T}{e} = \sqrt{8}\pi\, T. \tag{5.249}$$

Notice that Eq. (5.249) describes the situation of a pair of BPS-saturated monopole and antimonopole placed as *test* charges into a surrounding where (anti)calorons of small holonomy generate short-lived magnetic dipoles. Effectively, this leads to a *finite multiplicative charge renormalization* for each of these test particles as compared to the naive vacuum. Here no departure from the BPS limit is implied, and the U(1) gauge field of the test (anti)monopole still has infinite range. A linear superposition of U(1) monopole potentials then leads to a dipole form which for large distances R decays like $1/R^3$. In the effective theory, this corresponds to the approximation of a free and massless excitation and two massive *free* thermal quasiparticles plus the tree-level estimate of the thermal ground state.

On the level of radiative corrections, however, we are concerned with the physics of screened, stable magnetic charges whose average distance $n^{-1/3}$ at high temperature is given by $n^{-1/3} = c/T$. Here c is a positive, real, and *dimensionless*[33] constant which we will determine below, and n denotes the spatial density of isolated and screened monopole-antimonopole pairs. Thus

$$|\phi|\, n^{-1/3} = 2\pi c \left(\frac{T_c}{13.87\,T}\right)^{3/2}, \qquad (5.250)$$

which is smaller than unity for sufficiently high T. In fact, an estimate implies that for $T \geq 1.91\,T_c$ isolated and screened (anti)monopoles are not resolved in the effective theory. This estimate uses the high-temperature value $e = \sqrt{8}\,\pi$. A more careful investigation shows that isolated and screened (anti)monopoles actually are never fully resolved in the effective theory for the deconfining phase. However, as we have seen in Sec. 5.4.7, their effect on the propagation of the tree-level massless gauge mode is sizable at temperatures a few times T_c. The mass of a stable and screened magnetic monopole-antimonopole pair no longer is given by a simple charge renormalization in the BPS mass formula (5.249). Rather, screening introduces a mass scale associated with an exponentially decaying U(1) potential. Again, by nonperturbative asymptotic freedom this mass scale is proportional to temperature for $T \gg T_c$.

To make contact with degrees of freedom, whose collective long-distance effects are detectable (antiscreening and screening of effective gauge-field propagation) but which never appear explicitly in the effective theory, we may consider the situation of thermally fluctuating, free excitations (one-loop truncation of loop expansion of the pressure) as the asymptotic starting point. Small interactions between these excitations, as they are described by

---

[33]Owing to asymptotic freedom on the nonperturbative level, the presence of the Yang–Mills scale Λ is forgotten by the system at large T.

radiative corrections,[34] introduce interesting physics but do not change the energy density of the effective, asymptotic one-loop situation. The important implication then is that radiative corrections to this one-loop situation must be *canceled* by the physics of unresolved degreesof freedom on the level of thermodynamic quantities. By the result of [Diakonov *et al.* (2004)], which shows that (anti)calorons dissociate upon strong quantum deformation into isolated magnetic monopole-antimonopole pairs, we are led to conclude that these degrees of freedom are, indeed, stable, and screened magnetic monopoles and antimonopoles. In an adjoint Higgs model topologically stabilized monopoles occur at the vertex of four or more domains of the adjoint Higgs field in three-dimensional space [Kibble (1976)]. This suggests that the high-temperature two-loop correction of Fig. 5.14 should be interpreted as the *negative* of a correction to the free quasiparticle pressure obtained by a thermal average over *all* domainized configurations of the field ϕ. Indeed, the result of such a thermal average is homogeneous in space. To perform the average in terms of ϕ-field domainizations is hard. To compute an effective two-loop diagram, however, is rather simple, see Eqs. (5.140), (5.141), and (5.142).

After these preparations we now are in a position to compute the average mass m of screened and stable monopole-antimonopole pairs on distances of the order of $n^{-1/3}$. Since the spatial coarse-graining, which determines the field ϕ, saturates exponentially fast on distances of a few $\beta \equiv 1/T$ [Herbst and Hofmann (2004), Hofmann (2007)] (see Sec. 5.1.3), we expect that

$$\frac{m}{m_{>|\phi|^{-1}}} = O(1) \tag{5.251}$$

which, indeed, is the case. Let us show this.

---

[34]Recall that for $T \gg T_c$ it is only the two-loop diagram in Fig. 5.12, involving on-shell a massless and either of the two massive modes, that survives [see Fig. 5.14].

As already mentioned, at large temperatures the two-loop diagram for the pressure involving a four-vertex connecting one massless with either of the two massive modes is the *only surviving radiative* correction Schwarz, Hofmann, and Giacosa (2006) (see Sec. 5.3.4). For $T \gg T_c$ one obtains (compare with Fig. 5.14)

$$\frac{\Delta P_{2-\text{loop}}}{P_{1-\text{loop}}} = -4.39 \times 10^{-4}, \tag{5.252}$$

where $\Delta P_{2-\text{loop}} \equiv P_{2-\text{loop}} - P_{1-\text{loop}}$. Since for $T \gg T_c$ one has Hofmann (2005) (see Eq. (5.63))

$$P_{1-\text{loop}} = \frac{4}{45}\pi^2 T^4 \tag{5.253}$$

it follows from Eq. (5.252) that $\Delta P_{2-\text{loop}}$ is also proportional to T^4. Now $\Delta P_{2-\text{loop}}$ and the associated energy density $\Delta\rho_{2-\text{loop}}$, which are both negative, must obey the following relation[35]

$$\Delta\rho_{2-\text{loop}} = T\frac{d\Delta P_{2-\text{loop}}}{dT} - \Delta P_{2-\text{loop}}. \tag{5.254}$$

Equation (5.254) and the fact that $\Delta P_{2-\text{loop}} \propto T^4$ imply an equation of state

$$\Delta\rho_{2-\text{loop}} = 3\,\Delta P_{2-\text{loop}}, \tag{5.255}$$

and Eq. (5.252) tells us that the thermal energy density $\Delta\rho_{M+A}$, attributed to the presence of unresolved, screened, and stable pairs of (no longer BPS saturated) magnetic monopoles and antimonopoles,

---

[35]Legendre transformations are linear and thus hold for each hierarchical order in the effective loop expansion separately.

is given as[36]

$$\frac{\Delta\rho_{M+A}}{T^4} \equiv \frac{1}{2\pi^2} \int_\mu^\infty dy \sqrt{y^2 - \mu^2}\, y^2\, n_B(y) \overset{!}{=} -\frac{\Delta\rho_{2-\text{loop}}}{T^4}$$

$$= 4.39 \times 10^{-4} \times \frac{4}{15}\pi^2, \tag{5.256}$$

where $\mu \equiv m/T$ and $n_B(y) \equiv 1/(\exp(y) - 1)$. The only positive solution μ of Eq. (5.256) is numerically given as $\mu = 10.1224$. Thus we have

$$\frac{m}{m_{>|\phi|^{-1}}} = 1.139. \tag{5.257}$$

Equation (5.257) expresses the fact that the effect of all other, screened, unresolved, and stable (anti)monopoles, generated by the dissociation of (anti)calorons of large holonomy as induced by the scattering of tree-level massless modes of the effective theory (screening range $G > 0$, see Sec. 5.4.7.5) off the tree-level massive ones, is a lift of the mass m of a given pair to a value about 14% above the BPS bound. Again, the physical difference between screening by unstable intermediary monopole-antimonopole pairs (BPS level) and screening by stable (anti)monopoles is that the magnetic field of a stable monopole has infinite range (no mass-scale, multiplicative charge renormalization) in the former case while Yukawa screening occurs in the latter case.

5.5.2   Monopole density, monopole-antimonopole distance, and screening length

Relying on $\mu = 10.1224$ [see Eq. (5.256)] we may now compute the hypothetic (since (anti)monopoles are not resolved) monopole

---

[36]Since the process of (anti)caloron dissociation, creating pairs of isolated and screened magnetic monopoles and antimonopoles that are subject to an exact, overall charge neutrality, is irreversible, the density of these objects is sharply fixed at a given temperature. Thus there is no chemical potential associated with monopole-antimonopole pairs.

density n at high temperature as

$$n = \frac{T^3}{2\pi^2} \int_{\mu}^{\infty} dy \, y \sqrt{y^2 - \mu^2} \, n_B(y) = 9.799 \times 10^{-5} T^3. \qquad (5.258)$$

For reasons of symmetry there is a *universal* mean monopole-antimonopole distance $\bar{d}$. That is, the distance between a monopole and its antimonopole, both stemming from the dissociation of the same (anti)caloron, is, on average, the same as the distance between a monopole adjacent to an antimonopole which did not originate from the same (anti)caloron. From Eq. (5.258) we have

$$\bar{d} = n^{-1/3} = 21.691 \, \beta. \qquad (5.259)$$

Thus the constant c in Eq. (5.250) is given as $c = 21.691$, and the right-hand side of (5.250) is smaller than unity for $T > 2 \, T_c$.

By virtue of Eq. (5.258) the hypothetic magnetic screening length l_s (for a derivation see Korthals-Altes (2004))[37] reads

$$l_s = \frac{1}{g} \sqrt{\frac{T}{n}} = \frac{e}{4\pi} (21.691)^{3/2} \beta = 71.43 \, \beta, \qquad (5.260)$$

where the magnetic coupling g is given as $g = \frac{4\pi}{e}$. Thus we have

$$\frac{l_s}{\bar{d}} = 3.293. \qquad (5.261)$$

Equation (5.261) tells us that monopoles and antimonopoles are not far separated on the scale of their screening length. Moreover, the impact of all other stable and screened monopoles and anti-monopoles on a given stable pair is small due to efficient cancelations of mutual attraction or repulsion. This is expressed by the small lift of the BPS mass $m_{>|\phi|^{-1}}$ to m [compare with Eq. (5.257)].

---

[37] The screening length l_s is a measure for the typical spatial decay rate of the Yukawa potential associated with a screened monopole.

5.5.3   Area law for spatial Wilson loop?

Here we review the arguments of Korthals-Altes (2004) on how an area law for the spatial string tension occurs if a good separation of monopoles and antimonopoles on the scale of their magnetic screening length l_s can be assumed. As shown in the previous section, such a good separation does not take place in deconfining SU(2) Yang–Mills thermodynamics. It is however, instructive to go through the derivation to discuss the essential feature implying an area law: individual and independent contributions of unscreened magnetic flux by monopoles and antimonopoles adjacent to the surface spanned by the quadratic spatial contour of side-length L.

By virtue of the length-scale l_s one considers a slab of thickness $2l_s$ containing magnetic quasiparticles. These quasiparticles contribute a mean magnetic flux through the minimal surface spanned by the aforementioned contour. Only quasiparticles that are contained within the slab are responsible for the flux. A more refined treatment introducing no such constraint but taking into account a Yukawa-like potential for the static magnetic field sourced by each of these objects yields similar results.

Technically speaking, one exponentiates the normalized flux $\Phi_{l=1}$ of a single (anti)monopole through the minimal surface spanned by the quadratic contour in the limit $L \to \infty$ as

$$V(L)|_{l=1} \equiv \exp\left(2\pi i \frac{\Phi_{l=1}}{Q}\right), \qquad (5.262)$$

where Q is the magnetic charge. Depending on the sign of Q and on the location with respect to the minimal surface one has $\Phi_{l=1} = \pm\frac{1}{2}$ and thus

$$V(L)|_{l=1} \equiv -1. \qquad (5.263)$$

The next step is to assume that the probability $P(l)$ for the occurrence of l charges within the slab is given by the Poisson distribution

$$P(l) = \frac{\bar{l}^l}{l!} \exp(-\bar{l}), \qquad (5.264)$$

where $\bar{l}$ is the mean value of the number of magnetic charges contained inside the slab. Thus one obtains

$$\bar{V}(L) = \sum_{l=0}^{\infty} P(l)(-1)^l = \exp(-2\bar{l}). \tag{5.265}$$

Since $\bar{l} = 4A(L)l_s n(T)$, where $A(L)$ is the minimal surface and $n(T)$ is the (temperature-dependent) density of monopoles (or anti-monopoles or monopole-antimonopole pairs), we observe that the quantity $-\ln \bar{V}(L)$ exhibits area law with the spatial string tension σ_s given as

$$\sigma_s = 8 l_s n(T). \tag{5.266}$$

However, since in deconfining SU(2) Yang–Mills thermodynamics by virtue of Eq. (5.261) monopoles and antimonopoles have an essential overlap of their magnetic potentials on the scale of the screening length l_s, one cannot assume an independent and individual contribution of magnetic flux by each (anti)monopole through the minimal surface spanned by the contour. Therefore the occurence of an area law for the spatial Wilson loop is questionable.

5.5.4 *Spatial Wilson loop in effective variables*

5.5.4.1 *Generalities on spatial Wilson loop in the effective theory*

Here we present a computation of the spatial Wilson loop in the gauge-field variables of the effective theory for the deconfining phase of SU(2) Yang–Mills thermodynamics. We closely follow [Ludescher *et al.* (2009)].

To compute a Wilson loop in effective variables associated with a given, finite resolution μ is in general different from what the definition in fundamental variables demands. However, at finite temperature in the deconfining phase of the Yang–Mills theory there is only one resolution scale where the effective action minimally differs from the fundamental action, namely at $\mu = |\phi|$. Here topologically trivial and nontrivial field configurations are separated modulo mass generation and constraints on the hardness of quantum fluctuations of the former imposed by the latter. Since an ensemble average over stable and screened magnetic monopoles is generated by the radiative corrections in the effective theory (see Sec. 5.5.2), it is suggestive that the spatial Wilson loop be evaluated on propagating, radiatively dressed gauge modes, which collectively convey the magnetic flux. The effective theory, indeed, describes the average flux sourced by the associated ensemble of magnetic monopoles and antimonopoles This statement is purely intuitive. No proof for the equivalence of the fundamental and the effective Wilson loop is presently available.

In computing the spatial Wilson loop we consider the exponentiated one-effective-gauge-boson exchanges within the spatial quadratic contour C of side-length L, and we use an expansion of the N-point functions in the effective theory [Bassetto, Nardelli, and Soldati (1991)]. Because of the rapid numerical convergence of the effective loop expansion (see Secs. 5.4.3–5.4.5), we are content here with a resummation of the one-loop polarization tensor for the TLM mode mode into its effective propagator. Moreover, we only consider the tree-level propagators of the two TLH modes. The exchange of these three modes is subsequently exponentiated to take into account the trivial part of higher N-point contributions (see Fig. 5.42). Pointing already towards physics applications, we will denote the TLM modes by γ and the TLH modes by $V^{\pm}$ in what follows. With the above restrictions the logarithm of $W[C]$ is written as

$$\ln W[C] = -\frac{1}{2} C_F \oint dx_\mu dy_\nu \, D_{\mu\nu}(x - y), \qquad (5.267)$$

where C_F denotes the Dynkin index in the fundamental representation, defined as the normalization factor of the generators of the algebra (i.e. $C_F \equiv \frac{1}{2}$), and

$$D_{\mu\nu} = \sum_{a=1}^{3} D_{\mu\nu}^{(a)} \qquad (5.268)$$

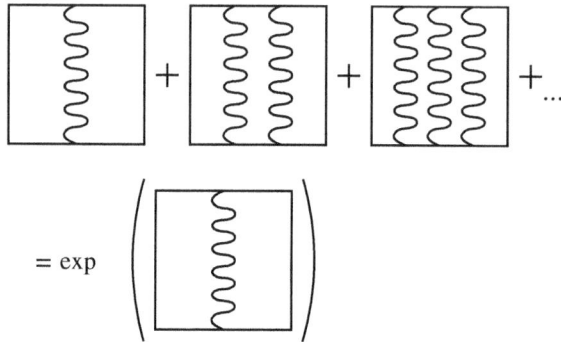

Figure 5.42. Illustration of the diagrammatic approach. Every summand in this formula represents an entire class of N-gauge-boson exchange diagrams, which is encoded in terms of a (suppressed) combinatorial factor.

is the sum of the tree-level propagators ($V^\pm$ corresponds to $a = 1, 2$) and the one-loop resummed propagator (γ corresponds to $a = 3$).

In spacetime the tree-level propagators for $a = 1, 2$ are defined as the Fourier transforms of their momentum space counterparts

$$D_{\mu\nu}^{1,2}(\beta, x - y) = -\int \frac{d^4 p}{(2\pi)^4} e^{-ip(x-y)} \left(g_{\mu\nu} - \frac{p_\mu p_\nu}{m^2}\right)$$
$$\times 2\pi\delta(p^2 - m^2)n_B(\beta|p_0|). \tag{5.269}$$

The one-loop resummation of the transverse, magnetic part[38] of the polarization tensor (screening function G) into the dressed γ propagator (see Eq. 5.182), yields the following result

$$D_{\mu\nu}^3(\beta, x - y) = -\int \frac{d^4 p}{(2\pi)^4} e^{-ip(x-y)} \left[P_{\mu\nu}^T \left(\frac{i}{p^2 - G(p^0, \vec{p}) + i0} \right.\right.$$
$$\left.\left. + 2\pi\delta(p^2 - G(p^0, \vec{p}))n_B(\beta|p_0|) \right) - i\frac{u_\mu u_\nu}{\vec{p}^2} \right], \tag{5.270}$$

---

[38] We may ignore electric screening effects since we study the *spatial* Wilson loop.

where the transverse projection operator is given as

$$P_{00}^T(p) = P_{0i}^T(p) = P_{i0}^T(p) = 0,$$

$$P_{ij}^T(p) = \delta_{ij} - \frac{p_i p_j}{\vec{p}^2}. \tag{5.271}$$

We calculate the contour integral in the 1,2-plane, that is $x_0 = y_0 = x_3 = y_3 = 0$, and consider at first an arbitrary propagator $D_{\mu\nu}(p)$. Explicit expressions for $D_{\mu\nu}^{1,2}$ ($V^\pm$ gauge modes) and for $D_{\mu\nu}^3$ (γ mode) are substituted later. One obtains [Keller (2008)]

$$\ln W[C] = -\frac{1}{4} \int \frac{d^4 p}{(2\pi)^4} \oint \oint dx_\mu dy_\nu D_{\mu\nu}(p)\, e^{-ip(x-y)} \big|_{x_0=y_0=x_3=y_3=0}$$

$$= -4 \int \frac{d^4 p}{(2\pi)^4} \sin^2\left(\frac{p_1 L}{2}\right) \sin^2\left(\frac{p_2 L}{2}\right)$$

$$\times \left(\frac{D_{11}}{p_1^2} - \frac{D_{12}}{p_1 p_2} - \frac{D_{21}}{p_1 p_2} + \frac{D_{22}}{p_2^2} \right). \tag{5.272}$$

5.5.4.2   Part due to tree-level massive modes

Here we present the result for the contribution of the $V^\pm$ propagators to the logarithm of the spatial Wilson loop. Inserting the $V^\pm$ propagators of Eq. (5.269) into Eq. (5.272) and summing over both contributions, we have

$$\ln W[C]_{V^\pm} = \frac{1}{\pi^3} \int d^3 p \, \frac{\sin^2\left(\frac{p_1 L}{2}\right) \sin^2\left(\frac{p_2 L}{2}\right)}{\sqrt{p_1^2 + p_2^2 + p_3^2 + m^2}}$$

$$\times n_B(\beta\sqrt{p_1^2 + p_2^2 + p_3^2 + m^2}) \left(\frac{1}{p_1^2} + \frac{1}{p_2^2} \right). \tag{5.273}$$

Rescaling the momenta p_i and the squared $V^\pm$ mass m^2 in Eq. (5.273) to dimensionless quantities, and using $e = \sqrt{8}\pi$, one obtains

$$\hat{p}_i = p_i \cdot L, \tag{5.274}$$

$$\hat{m}^2 = \frac{m^2}{T^2} = \frac{(2e)^2}{T^2}\frac{\Lambda}{2\pi T} = \frac{128\pi^4}{\lambda^3}, \tag{5.275}$$

where L is the side-length of the spatial quadratic contour C. To eventually perform the limit $L \to \infty$, we introduce the dimensionless parameter τ as

$$\tau = T \cdot L. \tag{5.276}$$

Equation (5.273) is then recast as

$$\ln W[C]_{V^{\pm}} = \frac{1}{\pi^3} \int d\hat{p}_1 d\hat{p}_2 d\hat{p}_3 \frac{\sin^2\left(\frac{\hat{p}_1}{2}\right)\sin^2\left(\frac{\hat{p}_2}{2}\right)}{\sqrt{\hat{p}_1^2 + \hat{p}_2^2 + \hat{p}_3^2 + \frac{128\pi^4}{\lambda^3}\tau^2}}$$

$$\times n_B \left(\frac{\sqrt{\hat{p}_1^2 + \hat{p}_2^2 + \hat{p}_3^2 + \frac{128\pi^4}{\lambda^3}\tau^2}}{\tau}\right)\left(\frac{1}{\hat{p}_1^2} + \frac{1}{\hat{p}_2^2}\right). \tag{5.277}$$

From Eq. (5.277) it is obvious that in the limit $\tau \to \infty$ the contribution to $\frac{\ln W[C]}{\tau^2}$ of the $V^{\pm}$ modes is nil, and we no longer need to discuss their potential impact on the spatial string tension.

5.5.4.3   *Thermal part due to massless mode*

Let us now discuss the contribution of the thermal part of the γ mode to $\ln W[C]$. In Sec. 5.4.7.5 we have computed the screening function G selfconsistently on the radiatively induced "mass shell" $p^2 - G(p^0, \vec{p}) = 0$. Inserting the thermal part of the magnetically dressed γ propagator of Eq. (5.270) into Eq. (5.272), we obtain

$$\ln W[C]_{\gamma}^{\text{th}} = -\frac{1}{2\pi^3} \int d\hat{p}_1 d\hat{p}_2 d\hat{p}_3 \frac{\sin^2\left(\frac{\hat{p}_1}{2}\right)\sin^2\left(\frac{\hat{p}_2}{2}\right)}{\sqrt{\hat{p}_1^2 + \hat{p}_2^2 + \hat{p}_3^2 + \hat{G}\tau^2}}$$

$$\times n_B \left(\frac{\sqrt{\hat{p}_1^2 + \hat{p}_2^2 + \hat{p}_3^2 + \hat{G}\tau^2}}{\tau}\right)\left(\frac{1}{\hat{p}_1^2} + \frac{1}{\hat{p}_2^2}\right), \tag{5.278}$$

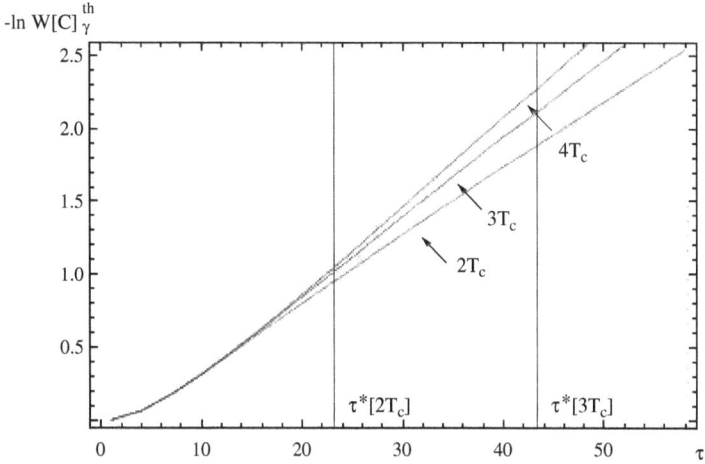

Figure 5.43. Plots of $-\ln W[C]_\gamma^{th}$ as a function of $\tau = T \cdot L$. The vertical lines correspond to the value of L coinciding with the minimal length scale $|\phi|^{-1}$ in the effective theory at a given temperature. The left vertical line is for $T = 2T_c$, the right vertical line for $T = 3T_c$, and the vertical line for $T = 4T_c$ would be at $\tau^*[4T_c] = 65.77$, and thus the latter is not contained in the figure.

where $\hat{G} \equiv \frac{G}{T^2}$. In Fig. 5.43 plots of $-\ln W[C]_\gamma^{th}$ for $T = 2T_c, 3T_c$, and $4T_c$ are shown as functions of τ. Clearly, for each temperature and for large τ we observe a *perimeter* law at the level of approximation used.[39] For τ considerably below τ^*, which is associated with the maximal resolution $|\phi|$ [we set $L = |\phi|^{-1}$ in Eq. (5.276)], we do, however, observe curvature in $-\ln W[C]_\gamma^{th}$.

---

[39] It is likely that this approximation captures all essential physics due to the rapid convergence in the number of external legs in N-point functions. Notice that in a one-loop diagram making up a radiative correction to γ's N-point function there are $N - 1$ independent external four-momenta p_i all of which are subject to the constraint $|p_i^2| \leq |\phi|^2$. With increasing N this should rapidly suppress the contribution of one-loop corrections to γ's N-point function. Moreover, the expansion into the number M of irreducible loops at a given N is conjectured to terminate at a finite order (see Sec. 5.4.3).

5.5.4.4 Quantum part due to massless mode

We now turn to the contribution to $\ln W[C]_\gamma$ of the quantum part of γ's dressed propagator. Inserting the quantum part of the magnetically dressed γ propagator of Eq. (5.270) into Eq. (5.272), we obtain

$$\ln W[C]_\gamma^{\mathrm{vac}} = -\frac{i}{4\pi^4} \int d^4\hat{p} \, \sin^2\left(\frac{\hat{p}_1}{2}\right) \sin^2\left(\frac{\hat{p}_2}{2}\right)$$

$$\times \frac{1}{\hat{p}^2 - \hat{G}\tau^2 + i\epsilon}\left(\frac{1}{\hat{p}_1^2} + \frac{1}{\hat{p}_2^2}\right), \qquad (5.279)$$

where $\hat{G} \equiv \frac{G}{T^2}$ and $\hat{p}_0 \equiv p_0 L$. Notice that in this case the screening function G receives contributions from both diagrams A and B in Fig. 5.24 because the only constraint on γ's momentum is $|p^2 - \mathrm{Re}\,G(p^0, \vec{p})| \leq |\phi|^2$. In Fig. 5.44 the imaginary part of G (due to diagram A) is plotted as a function of $X_0 \equiv \frac{p_0}{T}$ and of $X \equiv \frac{p}{T}$ at $T = 2\,T_c$. Clearly, $\mathrm{Im}\,G$ is nonvanishing only within a small region centered at the origin of the X_0–X-plane. Figure 5.44 thus suggests

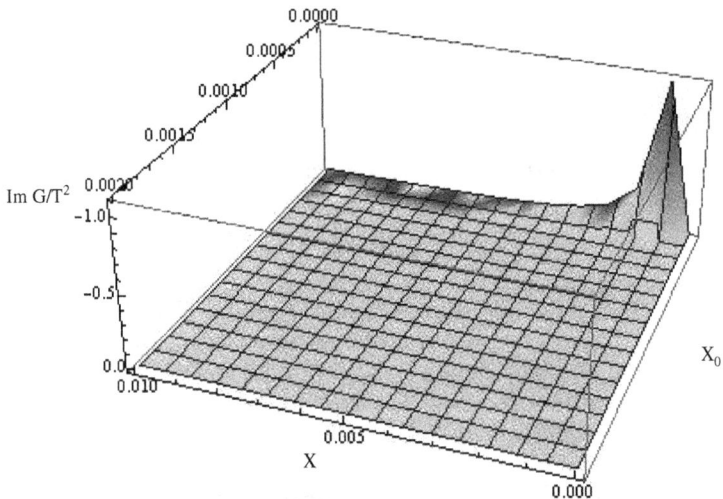

Figure 5.44. $\mathrm{Im}\,G/T^2$, as orginating from diagram A in Fig. 5.24, as a function of $X_0 \equiv \frac{p_0}{T}$ and of $X \equiv \frac{|\vec{p}|}{T}$ at $T = 2\,T_c$.

that the modulus of the factor $1/(\hat{p}^2 - \hat{G}\tau^2 + i0)$ in Eq. (5.279) can be estimated by the situation of free γ propagation. That is, we approximately may evaluate the integral in Eq. (5.279) by setting $G = 0$. To do this, one may imagine the condition $|p^2| \leq |\phi|^2$ to be implemented in such a way that a strongly decaying analytic factor (for $|p^2| > |\phi|^2$) is introduced to enable the application of the theorem of residues. One then obtains

$$\ln W[C]_{\gamma}^{\mathrm{vac}} = \frac{1}{2\pi^3} \int d^3\hat{p} \, \sin^2\left(\frac{\hat{p}_1}{2}\right) \sin^2\left(\frac{\hat{p}_2}{2}\right)$$

$$\times \frac{1}{\sqrt{\hat{p}_1^2 + \hat{p}_2^2 + \hat{p}_3^2}} \left(\frac{1}{\hat{p}_1^2} + \frac{1}{\hat{p}_2^2}\right). \qquad (5.280)$$

The integral in Eq. (5.280) is UV divergent. Introducing a UV cutoff, one sees that the part of the spectrum, where $G > 0$, contributes a real function that is monotonically decreasing in τ. For $G < 0$ and sufficiently large τ the associated contribution is imaginary with a modulus that also represents a function which is monotonic decreasing in τ. Furthermore, the UV divergence itself does not depend on τ. Thus we may refrain from considering this quantum contribution any further.

5.5.4.5 *Summary of results for effective spatial Wilson loop*

To summarize, the only nontrivial contribution to $\ln W[C]_{\gamma}$ arises from the thermal part of the resummed γ-propagator as investigated in Sec. 5.5.4.3. There we observe that for hypothetical values of L smaller than the minimal length $|\phi|^{-1}$ in the effective theory an area law emerges. This is qualitatively in line with lattice investigations using the Wilson action at finite lattice spacing [Borgs (1985)] in the sense that an artificial resolution scale is introduced to probe the system at small spatial distances.

In lattice simulations the typical spatial resolution — the inverse spatial lattice spacing — even at temperatures a few times T_c is considerably larger then the natural[40] resolution $|\phi|$ of continuum and infinite-volume Yang–Mills thermodynamics [Manousakis and Polonyi (1987);Bali *et al.* (1993)]. Lattice simulations are usually performed with the Wilson action at *finite* values of the lattice spacing a. However, due to the contribution of topologically nontrivial field configurations to the partition function a renormalization-group-evolved perfect lattice action certainly is more complicated than the Wilson action obtained from the continuum Yang–Mills action by naive discretization. Using the Wilson action, the scaling regime for the fundamental coupling hardly makes any reference to bulk properties of the highly nonperturbative ground state physics (trace anomaly, see Sec. 5.2.4). So what we have indirectly argued for in this section is that the physics of screened and isolated magnetic (anti)monopoles is very sensitive to a mild resolution dependence of the *partition function* as it is artificially introduced by the Wilson action. In principle, this action should be modified by nonperturbative effects yielding the perfect lattice action. The latter, however, is extremely hard to generate at finite temperatures.

Given this observation and the principle problem of comparison between the fundamental Wilson loop and its effective counterpart it is not surprising that an area law for the spatial Wilson loop is measured in lattice simulations (in accord with a three-dimensional strong-"coupling" argument Borgs (1985)) subject to the Wilson action. Notice that the introduction of a finite spatial lattice spacing still working with the Wilson action actually is likely to act physically (which it should not) in separating monopoles from antimonopoles. This could be the reason for measuring a net magnetic flux through the spatial contour in lattice simulations. We stress that lattice simulations using the Wilson action are interesting and important because they strongly point to certain aspects of the

---

[40]By "natural" we mean that at maximal resolution $|\phi|$ the effective action for deconfining Yang–Mills thermodynamics is of the simple form of Eq. (5.42).

highly nonperturbative thermal ground-state physics. However, we suspect that they are not sufficiently adapted to describe the subtle effects attributed to the dissociation of (anti)calorons of large holonomy taking place in infinite-volume continuum Yang–Mills thermodynamics. To turn this into a rigorous statement for the spatial Wilson loop in effective variables the above-mentioned correspondence between the effective and the fundamental description would have to be proved. Moreover, a precise estimate for the contribution to the integration over the spatial contour of N-point functions with higher internal loop number would have to be obtained.

5.5.5   Improved ground state estimate by thermal resummation

As a final point on the physics of stable, screen, and unresolved monopoles, we would like to exhibit a surprising fact concerning energy density and pressure associated with the resummation of the transverse (magnetic) part of the polarization tensor of the tree-level massless mode.[41]

As we have seen in Sec. 5.1.5, a direct contribution to the linear temperature dependence of the pressure P^{gs} and the energy density ρ^{gs} of the thermal ground-state estimate arises via the potential $V(\phi) = \text{tr}\,\frac{\Lambda^6}{\phi^2} = 4\pi\Lambda^3 T$. Thus

$$\frac{\rho^{gs}}{T^4} = -\frac{P^{gs}}{T^4} = 2(2\pi)^4\,\lambda^{-3} \sim 3117.09\,\lambda^{-3}, \qquad (5.281)$$

where $\lambda \equiv \frac{2\pi T}{\Lambda}$, and the Yang–Mills scale Λ is related to the critical temperature T_c as $\Lambda = \frac{2\pi}{13.87}\,T_c$. There are, in addition, corrections $\Delta P^{gs,1-\text{loop}}$ and $\Delta\rho^{gs,1-\text{loop}}$ to P^{gs} and ρ^{gs}, respectively, arising from free and thermal fluctuations of tree-level massive modes [Giacosa and Hofmann (2007a)]. These corrections are linear in T. Namely, at

---

[41]Since longitudinal modes are not directly thermalized we do not consider them here.

high temperature one has

$$\frac{\Delta\rho^{\text{gs},1-\text{loop}}}{T^4} = \frac{\Delta P^{\text{gs},1-\text{loop}}}{T^4} = -\frac{1}{4}a^2 = -2(2\pi)^4\,\lambda^{-3} \sim -3117.09\,\lambda^{-3}.$$

(5.282)

Here, $a \equiv \frac{2e|\phi|}{T} = \frac{8\sqrt{2}\pi^2}{\lambda^{3/2}}$. Notice that by virtue of the linear dependence on temperature of $P^{\text{gs}} + \Delta P^{\text{gs},1-\text{loop}}$ the Legendre transformation yields $\rho^{\text{gs}} + \Delta\rho^{\text{gs},1-\text{loop}} = 0$ which, indeed, is the case [compare Eqs. (5.281) and (5.282)].

Let us now demonstrate that at high temperature a linear T dependence of the pressure and energy density of the thermal ground-state estimate [see Sec. 5.1.5] is also generated by the selfconsistent one-loop propagation of the massless modes albeit subject to a hierarchically smaller coefficient compared to that of Eq. (5.282) and (5.281). Recalling that there is a leading quartic dependence of $\Delta P^{2-\text{loop}}$ on high temperature[42] [see Eq. (5.252)] this comes as a surprise. While the correction $\Delta P^{2-\text{loop}}$ of Eq. (5.252) is interpreted as the loss in pressure of a thermal gas of massless photons due to the generation of calorons of large holonomy and their subsequent dissociation into screened and stable monopole-antimonopole pairs [Ludescher *et al.* (2009)], the linear dependence that is due to the selfconsistent resummation of the one-loop polarization tensor represents a positive correction to the 1-loop pressure. That is, *radiatively* a ground-state contribution emerges by infinite resummation of those effective fluctuations that couple directly to the bare thermal ground-state via their tree-level quasiparticle mass. In other words, the selfconsistent propagation of tree-level massless modes through a sea of stable and screened but unresolved monopole-antimonopole pairs exhibits a ground-state component.

---

[42]Two-loop diagrams for pressure corrections other than the one involving a four-vertex between the thermal parts of a tree-level massless and a tree-level massive mode exhibit a leading power in T which is smaller than four [Schwarz, Hofmann, and Giacosa (2006); Ludescher *et al.* (2009)].

The selfconsistently modified dispersion law for transversally propagating, tree-level massless modes was obtained in [Ludescher and Hofmann (2008)] (see also Sec. 5.4.7.5), and is expressed in terms of the screening function $G(T, |\vec{p}|)$ as

$$p_0^2 = \vec{p}^2 + G(T, |\vec{p}|) \quad \Leftrightarrow \quad Y^2 = X^2 + \frac{G}{T^2}(\lambda, X), \tag{5.283}$$

where p_0 is the mode's energy, $\vec{p}$ its spatial momentum, and $X \equiv \frac{|\vec{p}|}{T}$, $Y \equiv \frac{p_0}{T}$. Recall that on the "mass shell" of Eq. (5.283) there is no contribution to the polarization tensor arising from diagram A depicted in Fig. 5.24. Recall that, this indicates that pairs of tree-level heavy modes cannot be excited by a single tree-level massless mode. Still, for $G > 0$ propagating energy (tree-level massless mode) is invested in the creation of unresolved monopole-antimonopole pairs, that is, in a lift of the energy density of the thermal ground-state estimate. This will be shown below explicitly.

Resummation of the polarization tensor into a free quasiparticle contribution, $P^{1-\mathrm{loop,res}}$, to the total pressure due to radiatively dressed tree-level massless modes is performed by connecting the external legs on the left-hand side of Fig. 5.27. For pressure and energy density one has:

$$\frac{P^{1-\mathrm{loop,res}}}{T^4} = \frac{1}{\pi^2} \int_0^\infty dX \, \frac{X^2 \sqrt{X^2 + G/T^2}}{e^{\sqrt{X^2 + G/T^2}} - 1},$$

$$\frac{\rho^{1-\mathrm{loop,res}}}{T^4} = -\frac{1}{\pi^2} \int_0^\infty dX \, X^2 \log\left(1 - e^{-\sqrt{X^2 + G/T^2}}\right). \tag{5.284}$$

We define the corrections $\frac{\Delta P^{1-\mathrm{loop,res}}}{T^4}$ and $\frac{\Delta \rho^{1-\mathrm{loop,res}}}{T^4}$ to the contributions of free, *massless* particles as

$$\frac{\Delta P^{1-\mathrm{loop,res}}}{T^4} \equiv \Delta \bar{P} \equiv \frac{P^{1-\mathrm{loop,res}}}{T^4} - \frac{\pi^2}{45},$$

$$\frac{\Delta \rho^{1-\mathrm{loop,res}}}{T^4} \equiv \Delta \bar{\rho} \equiv \frac{\rho^{1-\mathrm{loop,res}}}{T^4} - \frac{\pi^2}{15}. \tag{5.285}$$

In Fig. 5.45 both $\Delta \bar{P}$ and $\Delta \bar{\rho}$ are depicted as functions of λ within the high-temperature interval $100 \leq \lambda \leq 500$ (recall that $\lambda_c = 13.87$).

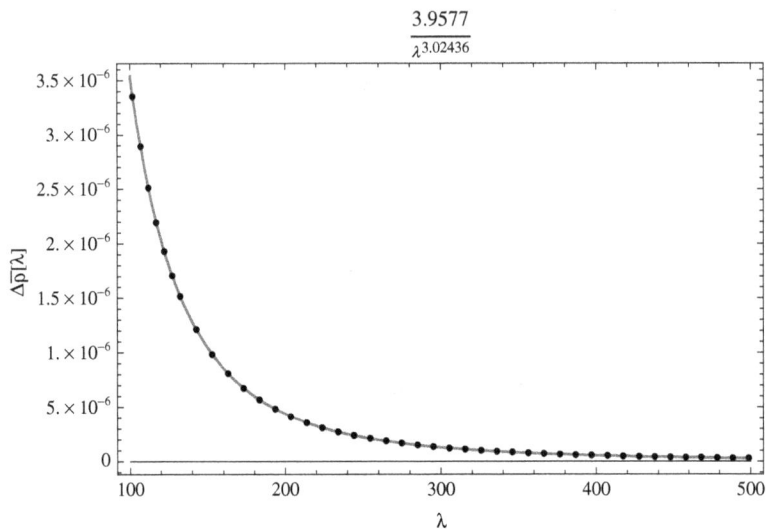

(a)

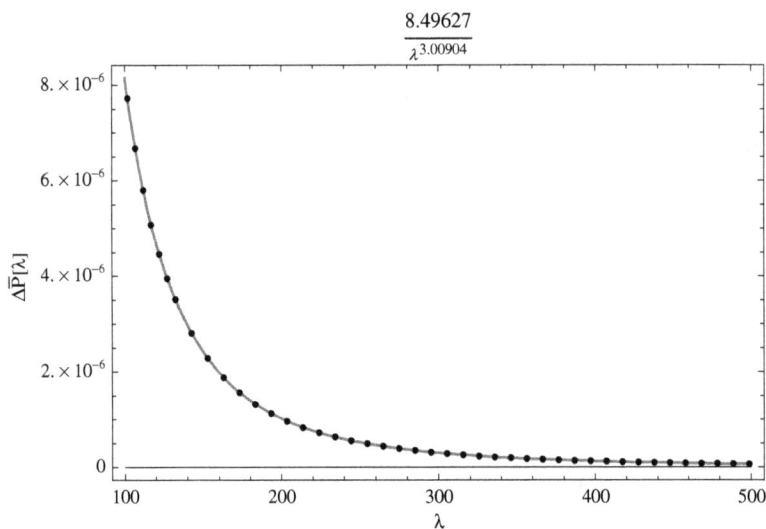

(b)

Figure 5.45. The high-temperature λ dependence of (a) $\Delta\bar{\rho}$ and (b) $\Delta\bar{P}$. Dots correspond to computed values, the lines are fits to this data.

An excellent fit to the following power laws

$$\Delta \bar{P} \equiv c_P \lambda^{\delta}, \quad \Delta \bar{\rho} \equiv c_{\rho} \lambda^{\gamma} \tag{5.286}$$

reveals that

$$c_P = 8.49627, \quad \delta = -3.00904, \quad c_{\rho} = 3.9577, \quad \gamma = -3.02436. \tag{5.287}$$

Equation (5.287) expresses the following: Through infinite and selfconsistent resummation the power and even the sign of the temperature dependence of a fixed-order correction to a given thermodynamical quantity is profoundly changed. By resumming the transverse part of the one-loop polarization of the massless modes, we generate a linear, indirectly ground state induced correction which is about three orders of magnitude smaller than the *a priori* estimate given by ρ^{gs} [compare Eqs. (5.281), (5.286), and (5.287)]. In calculating the contribution of higher irreducible loop orders to the polarization tensor we observe a hierarchial decrease and expect a termination of the loop expansion at a finite order [Hofmann (2012); Kaviani and Hofmann (2007)] (see Sec. 5.3.3) such that a consideration of such small corrections to the resummation process practically does not change our result obtained here by one-loop resummation.

Corrections $\Delta P^{1-\text{loop,res}}$ and $\Delta \rho^{1-\text{loop,res}}$ are not thermodynamically selfconsistent. Namely, defining $\Delta \rho_L^{1-\text{loop,res}}$ to be the Legendre transform of $\Delta P^{1-\text{loop,res}}$,

$$\Delta \rho_L^{1-\text{loop,res}} \equiv T \frac{d \Delta P^{1-\text{loop,res}}}{dT} - \Delta P^{1-\text{loop,res}}, \tag{5.288}$$

we observe from Fig. 5.46 that by no means $\Delta \rho_L^{1-\text{loop,res}} = \Delta \rho^{1-\text{loop,res}}$ (or $\Delta \bar{\rho}_L = \Delta \bar{\rho}$); even their signs are different. This is in contrast to the correction $\Delta P^{2-\text{loop}}$ of Eq. (5.252) which is by far the dominating two-loop correction to the pressure at high

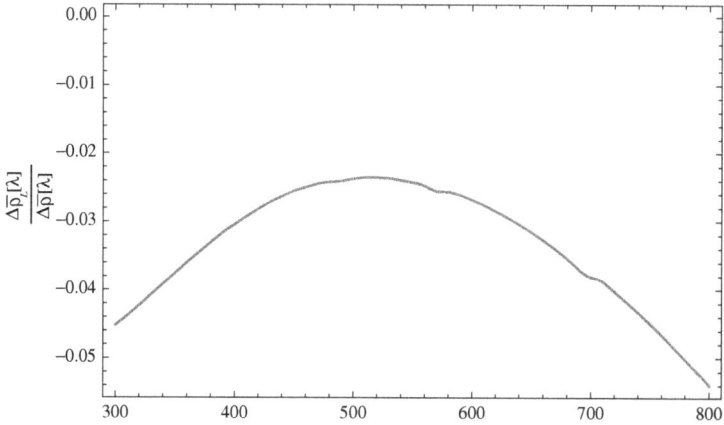

Figure 5.46. The λ dependence of the ratio $\frac{\Delta\bar{\rho}_L}{\Delta\bar{\rho}}$ where $\Delta\rho_L^{1-\text{loop,res}} \equiv T^4\,\Delta\bar{\rho}_L$ is defined in Eq. (5.288) and $300 \leq \lambda \leq 800$. The figure clearly expresses that the corrections to pressure and energy density of a thermal gas of massless particles due to one-loop resummation are thermodynamically not selfconsistent. Radiative corrections attached to tree-level massive modes and the off-shell contribution of the tree-level massless mode are missing.

temperature and thus is thermodynamically selfconsistent.[43] Thermodynamical selfconsistency after one-loop resummation is only expected to occur when corrections due to a shift in dispersion law of the tree-level massive modes are taken into account at resummed one-loop level and after considering the effect of the cut nonlocal two-loop diagram of Fig. 5.18 (see Fig. 5.20), with the tree-level massless line being off-shell.[44] To summarize, we have seen numerically

---

[43]The correction $\Delta\rho^{2-\text{loop}}$, defined by Legendre transformation of $\Delta P^{2-\text{loop}}$, is interpreted as an investment into the emergence of a mass density of stable monopole-antimonopole pairs in [Ludescher *et al.* (2009)].

[44]Small deviations from thermodynamical selfconsistency after taking these contributions into account arise from the use of tree-level quasi-particle consistency at a higher radiative order. Since the contributions to the pressure arising from radiative corrections follow a large hierarchy [Hofmann (2012); Kaviani and Hofmann (2007)] we are assured that the demand for thermodynamical selfconsistency at a higher loop order introduces small changes to the evolution of e which in turn induces very small changes to the radiative corrections themselves. For practical purposes it

that in resumming at high temperature the transverse part of the one-loop polarization tensor of the thermalized tree-level massless mode selfconsistently on this mode's thermal "mass shell" a linear dependence on temperature is generated in the corresponding corrections to the pressure and the energy density of the thermal gas of massless particles. This is remarkable because a two-loop correction to the pressure, which dominates all fixed-order radiative corrections at high-temperature, depends quartically on temperature. A linear dependence on temperature of the pressure correction is suggestive for ground-state physics [Hofmann (2005, 2007); Giacosa and Hofmann (2007a)] influencing the thermal propagation of the tree-level massless mode.

5.6   Thermomagnetic Effect

In this final section of the chapter we would like to address the question of how far the effects of a deformation of the deconfining thermalized SU(2) Yang–Mills system by a static, spatial temperature distribution can be described analytically.[45] In this context it may be worthwhile to point out that in condensed matter physics a temperature gradient applied across a medium comprising of mobile electric charge carriers can induce an electric field (Seebeck effect, thermoelectric effect) [Nolas, Sharp, and Goldsmid (2001)]. Since the *a priori* estimate of the deconfining thermal SU(2) Yang–Mills ground state comprises of an entire mass spectrum (see Sec. 5.1) of short-lived magnetic charges[46] we would expect the occurrence of a similar but dual phenomenon: A static temperature gradient induces a static *magnetic* field by virtue of the ground-state "medium" being populated by *magnetic* charge carriers.

---

thus suffices to work at one-loop selfconsistency of thermodynamical quantities to judge the selfconsistency of radiative corrections.

[45]The work presented in this section was performed by Carlos Falquez and the author.

[46]Recall that these charges are with respect to the unbroken U(1) subgroup of SU(2) (see Sec. 5.5).

5.6.1   *Adiabatic approximation*

To arrive at analytical results in describing the thermomagnetic effect with respect to the U(1) subgroup of SU(2) we resort to an adiabatic approximation for the effective ground-state physics. Namely, we assume that the spatial variation of the thermal ground state, induced by the temperature distribution $T(\vec{x})$, can be described by letting $T \to T(\vec{x})$ in the expression for the modulus of the field ϕ: $|\phi|(\Lambda, T) = \sqrt{\frac{\Lambda^3}{2\pi T}}$ [compare with Eq. (5.33)]. Notice that, as a result of inhomogeneous and anisotropic ground-state physics, the contribution of massless modes to an overall, static U(1) magnetic field $\vec{B}$ occurs via small *radiative* corrections induced by tree-level heavy (TLH) modes.[47] These radiative corrections determine the dispersion law for massless modes which is modified via the position dependence of the mass of the TLH modes induced by $|\phi|(\Lambda, T(\vec{x}))$. Because of the smallness of this effect we neglect the influence of thermal excitations in generating the magnetic field $\vec{B}$. Thus in unitary gauge $\phi^a = 2\,\delta^{a3}\,|\phi|$ we have $\vec{B} = \mathbf{rot}\,\vec{a}^3$ where $\vec{a}^3$ is a solution to the Minkowskian Yang–Mills equations subject to the source term determined by $|\phi|(\Lambda, T(\vec{x}))$.

For the adiabatic approximation to be reliable the energy density $\rho^B(\vec{x}) \equiv \frac{1}{2}\,\vec{B}^2$ of the magnetic field $\vec{B}$ locally needs to be smaller than the energy density $\rho^{gs}(\vec{x}) = 4\pi\Lambda^3\,T(\vec{x})$ of the thermal ground-state estimate (see Sec. 5.1.5). That is, for a reliable adiabatic approximation we need to demand that

$$R_\rho \equiv \frac{\rho^B}{\rho^{gs}}(\vec{x}) = \frac{\vec{B}^2}{8\pi\Lambda^3\,T(\vec{x})} < 1. \qquad (5.289)$$

---

[47]In the adiabatic approximation, applied to free thermal quasiparticles, the contribution to this magnetic field is generated by the same fluctuation integral as for constant temperature by letting $T \to T(\vec{x})$ in the Bose distribution. Thus the contribution of free massless modes yields the same vanishing contribution to $\vec{B}$ as in the case $T = \text{const}$.

5.6.2 Minkowskian Yang–Mills equations for static fields in unitary gauge

The effective Yang–Mills equations read

$$D_\mu G^{\mu\nu} = ie\left[\phi, D^\nu\phi\right] \tag{5.290}$$

[compare with Eq. (5.49)]. Again, $D_\mu\cdot \equiv \partial_\mu\cdot -ie\,[a_\mu,\cdot]$ and $G^{\mu\nu} \equiv \partial^\mu a^\nu - \partial^\nu a^\mu - ie\,[a^\mu, a^\nu]$. Moreover, we make the useful convention that the signature of the metric be $(-,+,+,+)$. Equations (5.290) imply the following equation for the gauge field a_μ

$$\begin{aligned}
&\partial_\mu\left(\partial^\mu a^\nu - \partial^\nu a^\mu\right) - 2ie\,\partial_\mu\left[a^\mu, a^\nu\right] + ie\left[\partial_\mu a^\mu, a^\nu\right] \\
&- ie\left[\partial_\nu a_\mu, a^\mu\right] - ie\left[\phi, \partial^\nu\phi\right] - e^2\left[a_\mu, [a^\mu, a^\nu]\right] \\
&+ e^2\left[\phi, [\phi, a^\nu]\right] = 0.
\end{aligned} \tag{5.291}$$

For $\nu = 0$ and $\nu = k$ ($k = 1,2,3$) Eq. (5.291) yields

$$\begin{aligned}
&\partial^2 a_0 - 2ie\,\partial_j\left[a_j, a_0\right] + ie\left[\partial_j a_j, a_0\right] \\
&- e^2\left[a_\mu, [a^\mu, a_0]\right] + e^2\left[\phi, [\phi, a_0]\right] = 0,
\end{aligned} \tag{5.292}$$

$$\begin{aligned}
&\partial^2 a_k - \partial_k\partial_j a_j - 2ie\,\partial_j\left[a_j, a_k\right] + ie\left[\partial_j a_j, a_k\right] \\
&- ie\left[\partial_k a_\mu, a^\mu\right] - e^2\left[a_\mu, [a^\mu, a_k]\right] + e^2\left[\phi, [\phi, a_k]\right] = 0,
\end{aligned} \tag{5.293}$$

respectively. Here $\partial^2 \equiv \partial_i\partial_i$, and all latin indices run from 1 to 3. Decomposing a_μ as $a_\mu = a_\mu^a t^a$, assuming stationarity (no dependence on x^0), and working in unitary gauge $\phi^a = 2\,\delta^{a3}\,|\phi|$, one derives from Eqs. (5.292) and (5.293) the following four independent equations:

$$\begin{aligned}
&\left(\partial^2 + 2ie\,a_j^3\partial_j + ie\,\partial_j a_j^3 - e^2\,a_\mu^3 a^{\mu,3} + 4e^2\,|\phi|^2\right) a_0^+ \\
&- ie\left(a_0^3\partial_j + 2\partial_j a_0^3\right) a_j^+ = e^2\left(a_0^+ a_\mu^- - a_0^- a_\mu^+ - a_0^3 a_\mu^3\right) a^{\mu,+},
\end{aligned} \tag{5.294}$$

$$\left(\partial^2 - 2e^2\, a_\mu^+ a^{\mu,-}\right) a_0^3 + e^2\left(a_0^+ a_\mu^- + a_0^- a_\mu^+\right) a^{\mu,3}$$

$$= -ie\left(a_0^- \partial_j + 2\partial_j a_0^-\right) a_j^+ + c.c., \tag{5.295}$$

$$\left(\partial^2 + 2ie\, a_j^3 \partial_j + ie\, \partial_j a_j^3 - e^2\, a_\mu^3 a^{\mu,3} + 4e^2\, |\phi|^2\right) a_k^+$$

$$- ie\left(a_k^3 \partial_j + 2\partial_j a_k^3\right) a_j^+ + ie\left(a_\mu^3 \partial_k - \partial_k a_\mu^3\right) a^{\mu,+} - \partial_k \partial_j a_j^+$$

$$= e^2\left(a_k^+ a_\mu^- - a_k^- a_\mu^+ - a_k^3 a_\mu^3\right) a^{\mu,+}, \tag{5.296}$$

$$\left(\partial^2 - 2e^2\, a_\mu^+ a^{\mu,-}\right) a_k^3 + e^2\left(a_k^+ a_\mu^- + a_k^- a_\mu^+\right) a^{\mu,3} - \partial_k \partial_j a_j^3$$

$$= -ie\left(a_k^- \partial_j + 2\partial_j a_k^-\right) a_j^+ - ie\, a_\mu^- \partial_k a^{\mu,+} + c.c. \tag{5.297}$$

The symbol *c.c.* stands for the complex conjugate of the explicit expressions on the right-hand side of Eqs. (5.295) and (5.297), and

$$a_\mu^\pm \equiv \frac{1}{\sqrt{2}}\left(a_\mu^1 \pm i a_\mu^2\right), \quad a^{\mu,\pm} \equiv \frac{1}{\sqrt{2}}((a^\mu)^1 \pm i(a^\mu)^2), \quad a^{\mu,3} \equiv (a^\mu)^3.$$

$$\tag{5.298}$$

5.6.3 *Simple static configurations with a U(1) magnetic field*

In search for a static, U(1) magnetic field let us make the simplifying assumption that

$$a_0^3 = a_k^\pm = 0. \tag{5.299}$$

Then we have $a_\mu^\pm a_\mu^3 = 0$. By virtue of this condition and by direct application of the conditions (5.299), Eqs. (5.295) and (5.296) are trivially satisfied ($0 = 0$). From (5.294) we have

$$\left[\partial^2 + 2ie\, a_j^3 \partial_j + ie\partial_j a_j^3 - e^2\, a_j^3 a_j^3 + 4e^2\, |\phi|^2\right] a_0^+ = 0. \tag{5.300}$$

Using $f\partial^2 f = \frac{1}{2}\partial^2 f^2 - (\mathbf{grad}\ f)^2$ (f a smooth, scalar function), this can be re-cast as

$$e^2\left(4\,|\phi|^2 - a_j^3 a_j^3\right) - \left(\frac{\partial_j a_0^+}{a_0^+}\right)^2 + \frac{1}{2}\partial^2 a_0^{+2}/a_0^{+2}$$

$$+ie\,\partial_j\left(a_0^{+2} a_j^3\right)/a_0^{+2} = 0. \tag{5.301}$$

Writing $a_0^+ \equiv A\,e^{i\varphi}$, where A is real-positive and φ is real, we obtain the following two real equations from Eq. (5.301):

$$4e^2\,|\phi|^2 = e^2\,|\vec{a}^3|^2 - \frac{\partial^2 A}{A} + \partial_j\varphi\left[\partial_j\varphi + 2e\,a_j^3\right], \tag{5.302}$$

$$e\,\partial_j a_j^3 + 2e\,a_j^3\frac{\partial_j A}{A} + \partial^2\varphi + 2\partial_j\varphi\frac{\partial_j A}{A} = 0. \tag{5.303}$$

Substituting Eq. (5.299) into Eq. (5.297) and using $a_0^+ \equiv A\,e^{i\varphi}$ again, we have

$$[\partial^2 + 2e^2\,A^2]a_k^3 - \partial_k\partial_j a_j^3 = -\,2e\,A^2\partial_k\varphi$$

or

$$2e\,A^2\left[e\,a_k^3 + \partial_k\varphi\right] = \partial_k\partial_j a_j^3 - \partial^2 a_k^3. \tag{5.304}$$

From Eq. (5.303) it follows that

$$\partial_j\left[A^2\left(e\,a_j^3 + \partial_j\varphi\right)\right] = 0.$$

Thus there is a vector field $\vec{V}$ such that

$$A^2\left(e\,a_k^3 + \partial_k\varphi\right) = \mathbf{rot}_k\vec{V}. \tag{5.303'}$$

Moreover, from Eq. (5.304) we infer that

$$2e\,\mathbf{rot}_k\vec{V} = \partial_k\partial_j a_j^3 - \partial^2 a_k^3. \tag{5.304'}$$

Therefore, we may write

$$2e\,\vec{V}_k = \mathbf{rot}_k\vec{a}^3 = \vec{B}_k. \tag{5.305}$$

Equation (5.305) can be used to re-write Eq. (5.302) as follows:

$$4e^2 \, |\phi|^2 = e^2 \, |\vec{a}^3|^2 - \frac{\partial^2 A}{A} + \partial_j \varphi \frac{\mathbf{rot}_j \vec{B}}{e \, A^2} - \left(\partial_j \varphi \right)^2 . \tag{5.302'}$$

5.6.4   Profiles

We now specialize to the situation that *all* fields depend on $x_3 \equiv z$ only. From Eqs. (5.304') and (5.305) it then follows that

$$2e \, \mathbf{rot}_k \vec{V} = \mathbf{rot}_k \vec{B} = (\delta_{k3} - 1) \, \partial_z^2 a_k^3. \tag{5.304''}$$

Thus the magnetic field $\vec{B}$ is perpendicular to the direction of the gradient to the temperature profile $T = T(z)$, and $a_3^3 = -\frac{1}{e} \partial_z \varphi$. This condition on a_3^3 implies that $\varphi = \varphi(z)$ can be set equal to zero [U(1) gauge parameter]. Equation (5.303') specializes to

$$e \, A^2 a_{1,2}^3 = \mathbf{rot}_{1,2} \, \vec{V}. \tag{5.306}$$

Combining Eq. (5.306) with Eq. (5.304'') yields

$$A^2 = -\frac{\partial_z^2 a_{1,2}^3}{2e^2 \, a_{1,2}^3}. \tag{5.307}$$

Equation (5.307) can be used to determine $A(z)$ from given profiles $a_1(z)$ and $a_2(z)$ which, in turn, determine the magnetic field $\vec{B}$ [see Eq. (5.305)] as

$$\vec{B}_1 = -\partial_z a_2^3, \quad \vec{B}_2 = \partial_z a_1^3. \tag{5.308}$$

Once $A(z)$ is fixed, we may appeal to the specialization of Eq. (5.302'),

$$4e^2 \, |\phi|^2 = 4e^2 \, \frac{\Lambda^3}{2\pi T} = e^2 \, |\vec{a}^3|^2 - \frac{\partial^2 A}{A}, \tag{5.302''}$$

to determine $|\phi|^2(z)$ and thus $T(z)$. Resorting to a power-law dependence on z,

$$a_1^3 = b_1 \, z^{\alpha_1}, \quad a_2^3 = b_2 \, z^{\alpha_2}, \quad (b_{1,2}, \alpha_{1,2} \text{ real}), \tag{5.309}$$

it follows from Eq. (5.307) that $\alpha_1 = \alpha_2 \equiv \alpha$. Because A is real Eq. (5.307) also implies that

$$\alpha(\alpha - 1) < 0 \quad \Rightarrow \quad 0 < \alpha < 1. \tag{5.310}$$

Let us work with the central value $\alpha = \frac{1}{2}$ in what follows. From Eq. (5.307) we then have

$$A^2 = \frac{1}{8e^2\, z^2}. \tag{5.311}$$

Substituting this into Eq. (5.302''), we obtain

$$T(z) = \frac{2\Lambda^3}{\pi}\, \frac{z^2}{b^2\, z^3 - \frac{2}{e^2}}, \tag{5.312}$$

where $b \equiv \sqrt{b_1^2 + b_2^2}$. By virtue of Eq. (5.312) positivity of temperature implies that

$$z > \left(\frac{2}{e^2\, b^2}\right)^{1/3} = \left(\frac{1}{4\pi^2\, b^2}\right)^{1/3}, \tag{5.313}$$

where the plateau value $e = \sqrt{8}\,\pi$ was used ($T > 2\, T_c$, see Sec. 5.2.3). Finally, the ratio R_ρ in Eq. (5.289) evaluates as

$$R_\rho = \frac{\hat{b}^2}{64}\, \frac{\hat{b}^2\hat{z}^3 - \frac{1}{4\pi^2}}{\hat{z}^3}, \quad (T > 2\, T_c), \tag{5.314}$$

where dimensionless variables $\hat{b} \equiv \frac{b}{\Lambda^{3/2}}$, $\hat{z} \equiv \Lambda z$ have been introduced.

Let us now distinguish the two cases $\hat{b} > \sqrt{8}$ and $\hat{b} \leq \sqrt{8}$: It is easily checked that for $\hat{b} > \sqrt{8}$ the range of $\hat{z}$-values, where $0 < R_\rho < 1$ (positivity of temperature and energy density, and adiabaticity), shrinks rapidly towards the point $\hat{z} = \left(\frac{1}{4\pi^2\hat{b}^2}\right)^{1/3}$ with increasing $\hat{b}$ (see Fig. 5.47). At this point we have $R_\rho = 0$. Since in the effective theory for the deconfining phase the smallest resolvable distance $\hat{z}_c$ is given as $\hat{z}_c \equiv \frac{\Lambda}{|\phi|(T_c)} = \sqrt{13.87} \sim 3.72$ we conclude from Fig. 5.47 that the thermomagnetic effect is unresolved. (The

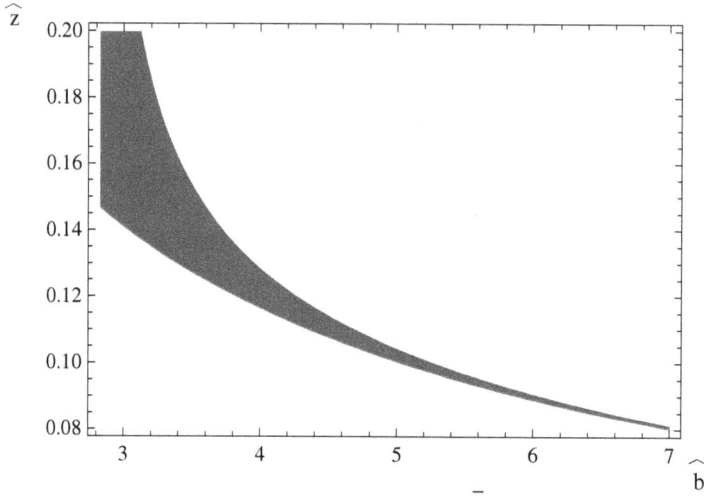

Figure 5.47. Shaded region depicts, as a function of $\hat{b} > \sqrt{8}$, the range of $\hat{z}$ values allowed by positivity of temperature and energy density, and adiabaticity.

small range of $\hat{b}$ values just above $\sqrt{8}$, where $\hat{z}$ may become larger than $\hat{z}_c$, causes $\frac{T}{T_c}$ to fall below unity.) For the case $0 \leq \hat{b} \leq \sqrt{8}$ the conditions $\hat{z} > \left(\frac{1}{4\pi^2 \hat{b}^2}\right)^{1/3}$ and $\frac{T}{T_c} > 1$ represent the only restrictions on $\hat{z}$. In Fig. 5.48 the ratios R_ρ and $\frac{T}{\Lambda}$ are plotted for $0.1 \leq \hat{b} < 1$ and $\left(\frac{1}{4\pi^2 \hat{b}^2}\right)^{1/3} \leq \hat{z} \leq 10$, and it is clear that ranges for $\hat{b}$ and $\hat{z}$ values exist where $\frac{T}{T_c}$ is considerably larger than unity and where the thermomagnetic effect is resolved in the effective theory. Notice, however, the smallness of R_ρ.

Because of an electric-magnetic dual interpretation of the postulate that deconfining SU(2) Yang–Mills theory describes thermal photon gases (see Chapter 9) the thermomagnetic effect of the SU(2) theory actually would represent a *thermoelectric* effect in the real world.

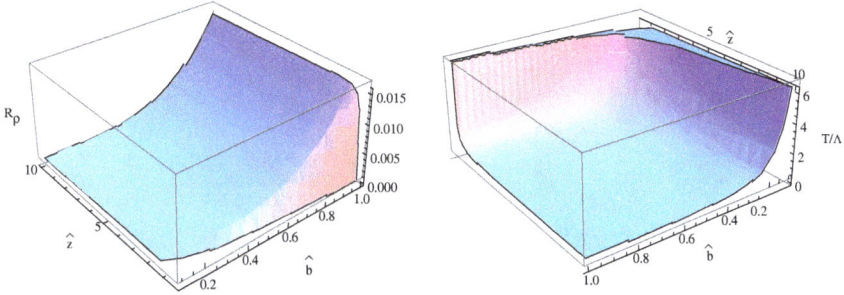

Figure 5.48.   The ratios R_ρ (left panel) and $\frac{T}{\Lambda} \sim 2.21 \frac{T}{T_c}$ (right panel) as functions of $\hat{b}$ and $\hat{z}$ for $0.1 \le \hat{b} \le 1$ and $\left(\frac{1}{4\pi^2 \hat{b}^2}\right)^{1/3} \le \hat{z} \le 10$.

Problems

5.1. Show that the local densities of Eqs. (5.1) and (5.2) vanish identically on selfdual field configurations and thus cannot function as local order parameters for a topology-induced gauge symmetry breaking on the level of fundamental fields.

5.2. Check Eqs. (5.9) and (5.10).

5.3. Why does the limit $\hat{\rho} \to \infty$ exist for $\hat{g}$ as defined in Eq. (5.10) and how fast is it approached?

5.4. Using the result of Eq. (5.9), perform the computation of the contribution of the Harrington–Shepard caloron to the integrand in (5.3) leading to Eqs. (5.13) and (5.14). *Hints:* Pull out the r- and angle-independent factor

$$\frac{[\partial_\tau \Pi(\tau,\vec{0})]^2}{\Pi^2(\tau,\vec{0})} - \frac{2}{3}\frac{\partial_\tau^2 \Pi(\tau,\vec{0})}{\Pi(\tau,\vec{0})}$$

$$= -\frac{16\pi^4}{3}\frac{\rho^2}{\beta^2}\frac{\pi^2\rho^2 + \beta^2\left(2 + \cos\frac{2\pi\tau}{\beta}\right)}{\left(2\pi^2\rho^2 + \beta^2\left(1 - \cos\frac{2\pi\tau}{\beta}\right)\right)^2} \qquad (5.315)$$

which arises from the field strength at $(\tau,\vec{0})$. The algebra trace over the product of Wilson lines and field strength at $(\tau,\vec{x})$ is performed by using the following relations for the

Pauli matrices τ_a and the components $\eta_{\mu\nu}^a$, $\bar{\eta}_{\mu\nu}^a$ of the 't Hooft symbols:

$$[\tau^a, \tau^b] = 2i\epsilon_{abc}\tau^c,$$
$$\tau^a\tau^b = \delta_{ab} + i\epsilon_{abc}\tau^c, \tag{5.316}$$
$$\tau^a\tau^b\tau^c = \delta_{ab}\tau^c - \delta_{ac}\tau^b + \delta_{bc}\tau^a + i\epsilon_{abc}.$$

The traces of products of Pauli matrices are

$$\operatorname{tr} \tau^a = 0,$$
$$\operatorname{tr} \tau^a\tau^b = 2\delta_{ab},$$
$$\operatorname{tr} \tau^a\tau^b\tau^c = 2i\epsilon_{abc},$$
$$\operatorname{tr} \tau^a\tau^b\tau^c\tau^d = 2 \left(\delta_{ab}\delta_{cd} + \delta_{ad}\delta_{bc} - \delta_{ac}\delta_{bd}\right),$$
$$\operatorname{tr} \tau^a\tau^b\tau^c\tau^d\tau^e = 2i \left(\delta_{ab}\epsilon_{cde} + \delta_{cd}\epsilon_{abe} - \delta_{cd}\epsilon_{abd} + \delta_{de}\epsilon_{abc}\right).$$

The 't Hooft symbols η and $\bar{\eta}$ read in components as

$$\eta_{\mu\nu}^a = \epsilon_{a\mu\nu} + \delta_{a\mu}\delta_{\nu4} - \delta_{a\nu}\delta_{\mu4},$$
$$\bar{\eta}_{\mu\nu}^a = \epsilon_{a\mu\nu} - \delta_{a\mu}\delta_{\nu4} + \delta_{a\nu}\delta_{\mu4}. \tag{5.317}$$

The 't Hooft symbol η satisfies a number of relations:

$$\eta_{\mu\nu}^a\eta_{\mu\nu}^b = 4\delta_{ab},$$
$$\eta_{\lambda\mu}^a\eta_{\lambda\nu}^a = 3\delta_{\mu\nu},$$
$$\eta_{\mu\nu}^a\eta_{\mu\nu}^a = 12,$$
$$\eta_{\lambda\mu}^a\eta_{\lambda\nu}^b = \delta_{ab}\delta_{\mu\nu} + \epsilon_{abc}\eta_{\mu\nu}^c, \tag{5.318}$$
$$\epsilon_{abc}\eta_{\mu\nu}^b\eta_{\kappa\lambda}^c = \delta_{\mu\kappa}\eta_{\nu\lambda}^a + \delta_{\nu\lambda}\eta_{\mu\kappa}^a - \delta_{\mu\lambda}\eta_{\nu\kappa}^a - \delta_{\nu\kappa}\eta_{\mu\lambda}^a,$$
$$\epsilon_{abc}\eta_{\mu\nu}^b\eta_{\mu\lambda}^c = 2\eta_{\nu\lambda}^a,$$
$$\eta_{\mu\nu}^a\eta_{\nu\lambda}^b\eta_{\lambda\mu}^c = 4\epsilon_{abc}.$$

The relations (5.318) hold also for $\bar{\eta}$. Moreover, one has

$$\eta^a_{\mu\nu}\eta^a_{\kappa\lambda} = \delta_{\mu\kappa}\delta_{\nu\lambda} - \delta_{\mu\lambda}\delta_{\nu\kappa} + \epsilon_{\mu\nu\kappa\lambda},$$

$$\bar{\eta}^a_{\mu\nu}\bar{\eta}^a_{\kappa\lambda} = \delta_{\mu\kappa}\delta_{\nu\lambda} - \delta_{\mu\lambda}\delta_{\nu\kappa} - \epsilon_{\mu\nu\kappa\lambda},$$

$$\epsilon_{\lambda\mu\nu\sigma}\eta^a_{\rho\sigma} = \delta_{\rho\lambda}\eta^a_{\mu\nu} + \delta_{\rho\nu}\eta^a_{\lambda\mu} + \delta_{\rho\mu}\eta^a_{\nu\kappa},$$

$$-\epsilon_{\lambda\mu\nu\sigma}\bar{\eta}^a_{\rho\sigma} = \delta_{\rho\lambda}\bar{\eta}^a_{\mu\nu} + \delta_{\rho\nu}\bar{\eta}^a_{\lambda\mu} + \delta_{\rho\mu}\bar{\eta}^a_{\nu\kappa},$$

$$\eta^a_{\mu\nu}\bar{\eta}^b_{\mu\nu} = 0.$$

$$(5.319)$$

5.5. Check that no divergence arises in the r-integration (measure $dr\, r^2$) over the integrand given in Eq. (5.13) for $r \to 0$.

5.6. Derive from the BPS equation (5.35) the second-order equations (5.27). What constraint does the BPS condition [ϕ is a solution to Eq. (5.35)] impose onto the kernel $\{\hat{\phi}\}$? Argue in terms of the number of parameters associated with each space.

5.7. Derive Eqs. (5.64) and (5.65).

5.8. Derive Eqs. (5.70) and (5.72).

5.9. Show the equivalence of Eq. (5.70) with Eq. (5.71). Derive Eq. (5.76), which is equivalent to (5.72) for the SU(3) case, directly from the demand that the free quasiparticle energy density ρ be related to the free quasiparticle pressure P by virtue of the Legendre transformation $\rho = T\frac{dP}{dT} - P$.

5.10. Solve Eq. (5.76) for $a \ll 1$.

5.11. Compute in terms of a_i and λ_i the quadratic correction to the expression in Eq. (5.79) that arises when Eq. (5.74) is substituted into Eq. (5.78). Such a correction is commonly detected in lattice simulations of the trace anomaly and is likely related to the fact that initial conditions for the downward evolution in temperature are set at rather moderate temperatures (typically $T_i \sim 5\, T_c$).

5.12. Verify that Eq. (5.79) also holds for SU(3).

5.13. Show that the action density of the effective theory vanishes for each mode propagating on-shell. Compare the average action density of maximally off-shell massless modes [see Eq. (5.60)] with the action density of the thermal ground-state estimate. What is your conclusion?

5.14. For $N = 2, 3$ solve Eq. (5.88) numerically, subject to the boundary condition that the location of the Landau pole coincides with λ_c. Considering this solution at $\lambda = 10 \lambda_c$, extract an initial condition for the solution to the purely perturbative evolution equation $\lambda \partial_\lambda g = -bg^3$ and solve it. For each case, $N = 2, 3$, compare the two curves.

5.15. Determine the functions $\mathcal{P}_A$, $\mathcal{P}_B$ in Eq. (5.174).

5.16. Compute Π^{00} for the TLM mode on the naive mass shell $p^2 = 0$ by appealing to the definition of Eq. (5.183).

5.17. Show by also considering a discrete symmetry of the integrand in Eq. (5.202) that the conditions (5.204) indeed express the s-channel constraint (5.61) at $p^2 = G$.

5.18. Convince yourself that the magnetic screening length l_s, defined in Eq. (5.260), as well as the average distance $\bar{d}$ between a monopole and an antimonopole, defined in Eq. (5.259), are much smaller than $|\phi|^{-1}$ for $T > 3 T_c$.

5.19. In unitary gauge $\phi^a = 2 \delta^{a3} |\phi|$ derive Eqs. (5.294)–(5.297) from Eqs. (5.292) and (5.293) by assuming stationarity and by appealing to the definitions in Eqs. (5.298).

References

Bali, G. S., *et al.* (1993). Spatial string tension in the deconfined phase of the (3+1)-dimensional SU(2) gauge theory, *Phys. Rev. Lett.*, **71**, 3059.

Bassetto, A., Nardelli, G. and Soldati, R. (1991). *Yang–Mills Theories in Algebraic Non-Covariant Gauges: Canonical Quantization and Renormalization* (World Scientific Publishing Company).

Bjerrum-Bohr, N. E. J., Donoghue, J. F., and Holstein, B. R. (2003), Quantum corrections to the Schwarzschild and Kerr metrics, *Phys. Rev. D*, **68**, 084005. [Erratum-ibid. D, **71**, 069904 (2005)]

Borgs, C. (1985). Area law for spatial wilson loops in high-temperature lattice gauge theory, *Nucl. Phys. B*, **261**, 455.

Brodsky, S. J. and Hoyer, P. (2011). The $\hbar$ expansion in quantum field theory. *Phys. Rev. D* **83**, 045026.

Collins, J. C., Duncan, A. and Joglekar, S. D. (1977). Trace and dilatation anomalies in gauge theories, *Phys. Rev. D*, **16**, 438.

Diakonov, D., *et al.* (2004). Quantum weights of dyons and of instantons, *Phys. Rev. D*, **70**, 036003.

Dolan, L. and Jackiw, R. (1974). Symmetry behavior at finite temperature, *Phys. Rev. D*, **9**, 3320.

Donoghue, J. F. et al. (2002), Quantum corrections to the Reissner-Nordström and KerrâŁ"Newman metrics, *Phys. Lett. B*, **529**, 132.

Falquez, C., Hofmann R., and Baumbach, T. (2012), *Quant. Matt.*, **1**, 153.

Fujikawa, K. (1980). Comment on chiral and conformal anomalies, *Phys. Rev. Lett.*, **44**, 1733.

Giacosa, F. and Hofmann, R. (2006). Thermal ground state in deconfining Yang–Mills thermodynamics, *Progr. Theor. Phys.*, **118**, 759.

Giacosa, F. and Hofmann, R. (2007a). Linear growth of the trace anomaly in Yang–Mills thermodynamics, *Phys. Rev. D*, **76**, 085022.

Giacosa, F. and Hofmann, R. (2007b). Nonperturbative screening of the Landau pole, *Phys. Rev. D*, **77**, 065022.

Herbst, U. and Hofmann, R. (2004). Emergent inert adjoint scalar field in SU(2) Yang–Mills thermodynamics due to coarse-grained topological fluctuations, arXiv:hep-th/0411214v4.

Herbst, U., Hofmann, R. and Rohrer, J. (2004). SU(2) Yang–Mills thermodynamics: Two-loop corrections to the pressure, *Acta Phys. Pol. B*, **36**, 881.

Herbst, U. (2005). The deconfining phase of SU(2) Yang–Mills thermodynamics, Diploma thesis, Universität Heidelberg, arXiv:hep-th/0506004.

Hofmann, R. (2005). Nonperturbative approach to Yang–Mills thermodynamics, *Int. J. Mod. Phys. A*, **20**, 4123 (2005); Erratum-*ibid. A*, **21**, 6515.

Hofmann, R. (2007). Yang–Mills thermodynamics, arXiv:0710.0962v3.

Hofmann, R. (2009). Onset of magnetic monopole-antimonopole condensation, Onset of magnetic monopole-antimonopole condensation, *ISRN High Energy Phys.*, **2012**, Article ID 853692.

Hofmann, R. (2012). Loop expansion in Yang–Mills thermodynamics, *Braz. J. Phys.*, **42**, 110.

Hofmann, R. and Kaviani, D. (2012). The quantum of action and finiteness of radiative corrections: Deconfining SU(2) Yang-Mills thermodynamics, *Quant. Matt.* **1**, 41.

Holstein, B. R. and Donoghue, J. F. (2004), Classical Physics and Quantum Loops. *Phys. Rev. Lett.*, **93**, 201602.

't Hooft, G. (1971). Renormalization of massless Yang–Mills fields, *Nucl. Phys. B*, **33**, 173.

't Hooft, G. and Veltman, M. (1972a). Regularization and renormalization of gauge fields, *Nucl. Phys. B*, **44**, 189.

't Hooft, G. and Veltman, M. (1972b). Combinatorics of gauge fields, *Nucl. Phys. B*, **50**, 318.

Iliopoulos, J., Itzykson, C., and Martin, A. (1975). Functional methods and perturbation theory, *Rev. Mod. Phys.* **47**, 165.

Kapusta, J. I. (1989). *Finite-temperature Field Theory* (Cambridge University Press).

Kaviani, D. and Hofmann, R. (2007). Irreducible three-loop contributions to the pressure in Yang–Mills thermodynamics, *Mod. Phys. Lett. A*, **22**, 2343.

Keller, J. (2008). Gauge-invariant two-point correlator of energy density in deconfining SU(2) in Yang–Mills thermodynamics, Diploma thesis, Universität Heidelberg, arXiv:0801.3961.

Kibble, T. W. B. (1976). Topology of cosmic domains and strings, *J. Phys. A*, **9**, 1387.

Korthals-Altes, C. (2004). Quasi-particle model in hot QCD, arXiv: hep-ph/0406138v2.

Kraan, T. C. and Van Baal, P. (1998a). Exact T-duality between calorons and Taub-NUT spaces, *Phys. Lett. B*, **428**, 268.

Kraan, T. C. and Van Baal, P. (1998b). Periodic instantons with non-trivial holonomy, *Nucl. Phys. B*, **533**, 627.

Krasowski, N. and Hofmann, R. (2014). Quantum corrections to the Reissner-Nordstr'om and Kerr-Newman metrics, *Annals Phys.*, **347**, 287.

Kronberg, P. P. (1994). Extragalactic magnetic fields, *Rept. Prog. Phys.* **57**, 325.

Landsman, N. P. and Weert, C. G. (1987). Real- and imaginary-time field theory at finite temperature and density, *Phys. Rep.* **145**, 141.

Lee, B. W. and Zinn-Justin, J. (1972). Spontaneously broken gauge symmetries. I. Preliminaries, *Phys. Rev. D*, **5**, 3121.

Lee, K. and Lu, C. (1983). SU(2) calorons and magnetic monopoles, *Phys. Rev. D*, **58**, 025011-1.

Le Bellac, M. (2000). *Thermal Field Theory*, (Cambridge University Press).

Linde, A. D. (1980). Infrared problem in thermodynamics of the Yang–Mills gas, *Phys. Lett. B*, **96**, 289.

Ludescher, J. and Hofmann, R. (2008). Thermal photon dispersion law and modified black-body spectra, *Annalen d. Physik*, **18**, 271, (2009).

Ludescher, J., *et al.* (2008). Spatial Wilson loop in continuum, deconfining SU(2) Yang–Mills thermodynamics, *Annalen. d. Physik*, **19**, 102, (2010).

Manousakis, E. and Polonyi, J. (1987). Nonperturbative length scale in high temperature QCD, *Phys. Rev. Lett.*, **58**, 847.

Nahm, W. (1983). Self-dual monopoles and calorons in *Trieste Group Theor. Method 1983*, p. 189.

Nolas, G. S., Sharp, J., and Goldsmid, J. (2001). *Thermoelectrics: Basic Principles and New Materials Developments*, Springer Series in Materials Science, Vol. 45, (Springer, 2001).

Schwarz, M., Hofmann, R., and Giacosa, F. (2006). Radiative corrections and the one-loop polarization tensor of the massless mode in SU(2) Yang–Mills thermo-dynamics, *Int. J. Mod. Phys. A*, **22**, 1213.

Weinberg, S. (1995). *The Quantum Theory of Fields I* (Cambridge University Press).

Chapter 6

The Preconfining Phase

Lowering the temperature in the deconfining phase, the effective coupling e "forgets" about the initial condition set at a high temperature and eventually runs into a logarithmic pole at the dimensionless temperatures $\lambda_c = 13.87$ and $\lambda_c = 9.475$ for the gauge groups SU(2) and SU(3), respectively. A stable magnetic monopole, liberated by the dissociation of a caloron or anticaloron of large holonomy, possesses a typical mass $M_{\mathrm{mon}} \sim \frac{4\pi^2 T}{e}$ and a magnetic charge $g = \frac{4\pi}{e}$ modulo small screening effects induced by the presence of other stable monopoles and antimonopoles (see Sec. 5.4.1). Recall that this expression for the monopole mass is a consequence of the screening of magnetic charge g induced by short-lived magnetic dipoles. Thus for $\lambda \searrow \lambda_c$ stable magnetic monopoles and antimonopoles become massless through screening, and their collective coupling to the tree-level massless mode[1] of the effective theory for the deconfining phase vanishes: Magnetic monopoles become free and massless particles at λ_c. In the thermal average of the overall magnetic flux a divergence of the monopole-antimonopole pair's Bose distribution at vanishing mass and zero momentum (monopole-antimonopole condensation) cancels a zero of their magnetic flux through a two-dimensional sphere of infinite radius. As a consequence, in the limit $e \to \infty$ the thermally averaged magnetic flux solely depends on

---

[1]For SU(3) two independent species of monopoles and their antimonopoles couple to their respective tree-level massless modes surviving the dynamical gauge symmetry breaking induced by the thermal ground-state estimate.

the angle between the Dirac strings of the two magnetic charges in the monopole-antimonopole pair. This angle (periodic variable) is proportional to Euclidean time τ and parameterizes the phase of an inert[2] complex scalar field φ.

As in the deconfining phase, the modulus of φ and its potential V in the effective Euclidean action density follow by demanding consistency of BPS saturation and stationarity of the action. As a consequence both quantities $|\varphi|$ and V, a mass scale $\bar{\Lambda}$ enters as an integration constant. The scale $\bar{\Lambda}$ can be expressed in terms of the scale Λ of the deconfining phase.[3] Similar to the deconfining phase microscopic interactions between (anti)monopoles, mediated by propagating U(1) (dual) gauge modes at momentum transfer larger than $|\varphi|$, collectively emerge after spatial coarse-graining as a pure-gauge configuration $a_\mu^{D,gs}$. Since a free U(1) quantum field theory is trivially renormalizable the effective U(1) gauge field obeys the same form of the action density after coarse-graining as the "fundamental" (dual) gauge fields does. Moreover, the coupling between the field φ (coarse-grained, condensed monopoles and antimonopoles) and a_μ^D is minimally gauge-invariant. This coupling induces a mass of the effective U(1) gauge field by the Abelian Higgs mechanism, which is determined by $|\varphi|$ and the magnetic coupling g. Since the magnetic g turns out to be finite, the exception being an isolated point in temperature, this does not imply complete confinement: The dual gauge field, albeit massive, still propagates. For this reason why we refer to the phase, where the thermal ground state is represented by a monopole condensate, as *preconfining* phase.

The validity of Legendre transformations in the effective theory enforces an evolution equation for the magnetic coupling g subject

---

[2]The fact that φ is obtained by a spatial average over zero-energy and zero-pressure magnetic dipoles of noninteracting monopole and antimonopole constituents implies that the field φ does not propagate.

[3]This is a consequence of the fact that the pressure is continuous across a thermal phase boundary.

to the initial condition $g(\lambda_c) = 0$. This evolution quickly runs into a logarithmic pole where the phase of preconfining thermodynamics is terminated.[4] It turns out that $T_{c'}$ is by about 10% smaller than the critical temperature T_c of the deconfining–preconfining transition. Thus the preconfining phase occupies a narrow region in the phase diagram.

The purpose of the present chapter is to analyze in a quantitative way the physics that was sketched above.

6.1 Condensation of Magnetic Monopole-Antimonopole Pairs

6.1.1 *Geometric considerations*

In this section we consider the magnetic charge g of a static, noninteracting monopole as an integral of its U(1) magnetic field, contained in the 't Hooft tensor $\mathcal{F}_{\mu\nu}$ (see Eq. (4.51)), over a closed two-dimensional spatial surface. In other words, the magnetic charge g is obtained as

$$g = \int d\Sigma_{\mu\nu}\, \mathcal{F}_{\mu\nu} = \pm\frac{4\pi}{e}, \qquad (6.1)$$

where $d\Sigma_{\mu\nu}$ denotes the differential area element of a closed two-dimensional spatial surface trapping the monopole. Here the core size of the monopole is assumed to be much smaller than the minimal distance between its center of mass and the two-dimensional surface.

---

[4]Formerly collapsing Abrikosov–Nielsen–Olesen vortex loops or center-vortex loops (CVLs), which are unresolved in the effective theory, then become thin and therefore quasistable. These solitons carry magnetic flux whose direction is two-fold degenerate. Therefore a two-fold degeneracy is associated with the electric dipole moment of each soliton. Since there is a large hierarchy between the mass of the CVL with zero selfintersection and the scale $\tilde{\Lambda}$ characterizing the (ground state dominated) energy density of the Yang–Mills system at $T_{c'}$, the deconfining ground state becomes instable against decay into these particles as $T \searrow T_{c'}$: A nonthermal "re-heating" of the system (Hagedorn downward transition) takes place.

In unitary gauge, where the orientation of the adjoint Higgs field in su(2) (in our case that of the A_4-component of a caloron of nontrivial holonomy) is "combed" into a fixed direction, the flux F of the magnetic field $B_i = \mp \frac{\hat{x}_i}{e(x_1^2+x_2^2+x_3^2)} = \frac{1}{2}\epsilon_{ijk}\mathcal{F}_{jk}$ ($\hat{x}_i \equiv \frac{x_i}{|\vec{x}|}$) is compensated by a Dirac string in position space: For a monopole centrally trapped by a sufficiently large S_2 the magnetic flux through this S_2 vanishes since the flux contribution of the Dirac string cancels that of the hedgehog magnetic field.

If a static monopole or antimonopole lies outside an S_2 of infinite radius a finite distance b away from the latter's surface and if its Dirac string does not pierce this surface then the magnetic flux F (see Fig. 6.1) calculates as

$$
\begin{aligned}
F &= \int_{\text{plane}} d\Sigma_{\mu\nu}\,\mathcal{F}_{\mu\nu} \\
&= \pm\frac{1}{e}\int_0^{2\pi} d\beta \int_{-\infty}^{\infty} dr\, r\cos\alpha(\vec{B},\vec{b})\frac{1}{r^2+b^2} \\
&= \pm\frac{4\pi}{e}b \int_0^{\infty} dr\,\frac{r}{(r^2+b^2)^{3/2}} \\
&= \pm\frac{4\pi}{e}.
\end{aligned}
\tag{6.2}
$$

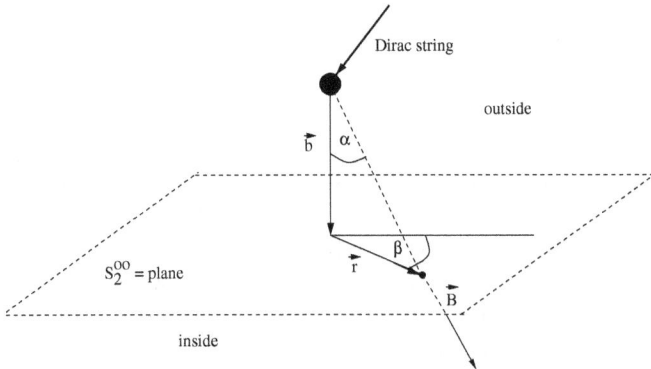

Figure 6.1.   A magnetic monopole in unitary gauge outside an S_2 with infinite radius. The Dirac string does not pierce the surface.

6.1.2 *Derivation of the phases of macroscopic complex scalar fields*

There is one species of magnetic monopoles and antimonopoles in the SU(2) case while there are two independent species for SU(3).

SU(2) case

In unitary gauge we consider an isolated system of a monopole and an antimonopole, which both are at rest and do not interact, outside of an S_2 with *infinite* radius, S_2^∞ (see Fig. 6.2). We characterize their Dirac strings by unit vectors $\hat{x}_m$ and $\hat{x}_a$ which point away from the cores of the monopole and the antimonopole respectively. Let P be the plane perpendicular to S_2^∞ such that the intersection line $L = P \cap S_2^\infty$ coincides with the intersection line of S_2^∞ with the plane D spanned by $\hat{x}_m$ and $\hat{x}_a$. (The case, where $D = S_2^\infty$ is inessential for what follows.) Whether or not this system contributes to the magnetic flux through S_2^∞ depends on the angle $\delta = \angle(\hat{x}_m, \hat{x}_a)$ and on the angle γ which the projection of $\hat{x}_m$ onto P forms with L. A net

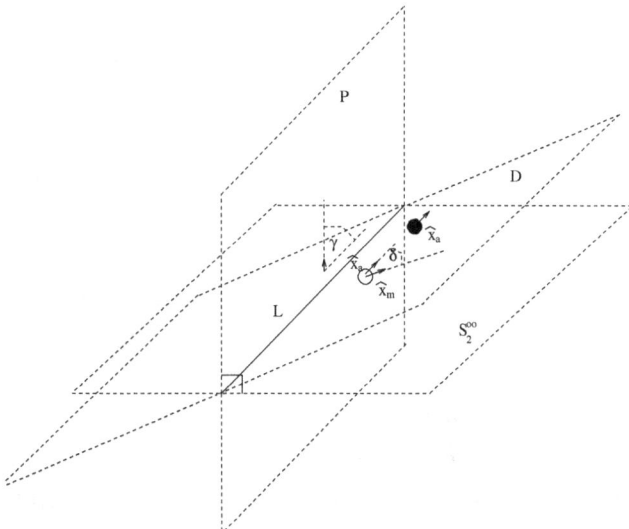

Figure 6.2. A magnetic monopole and its antimonopole placed outside of an S_2^∞. For an explanation see text.

magnetic flux through S_2^∞ takes place if and only if either $\hat{x}_m$ or $\hat{x}_a$ *alone* pierces S_2^∞. For a given angle δ the angle γ is uniformly distributed. In the absence of a heat bath the probability of measuring a flux $\frac{4\pi}{e}$ or a flux $-\frac{4\pi}{e}$ through S_2^∞ thus is given as $\frac{\delta}{2\pi}$. We conclude that for a given angle δ the average positive or negative flux $\bar{F}_\pm$ through S_2^∞ reads

$$\bar{F}_\pm = \pm\frac{\delta}{2\pi}\frac{4\pi}{e} = \pm\frac{2\delta}{e} \quad (0 \le \delta \le \pi). \tag{6.3}$$

Notice that δ and $2\pi - \delta$ generate the same average flux $\bar{F}_\pm$, thus the restriction $0 \le \delta \le \pi$ in Eq. (6.3).

So far we have discussed the flux through S_2^∞ which is generated by an isolated monopole-antimonopole system with no interactions. To derive the phase of the macroscopic complex field φ describing the Bose condensate of such systems we couple the isolated system to the heat bath, project onto zero-spatial-momentum states of the monopole-antimonopole system such that each constituent does not carry any spatial momentum[5] and perform the massless limit $e \to \infty$ which takes place for $T \le T_c$. (The case of a state of zero-spatial-momentum with opposite and finite momenta of the constituents, which is induced by interactions between the monopole and its antimonopole in the pair, may generate a closed and instable magnetic flux line in the condensate. This situation takes place when a large number of (anti)calorons of large holonomy dissociate into monopole-antimonopole pairs almost simultaneously inside a small spatial volume. On the coarse-grained level, the thermal average over these interacting unstable flux loops, which collapse as soon as they are created, will later be described by an effective pure-gauge configuration.)

The monopole and antimonopole are generated by the dissociation of a (anti)caloron of large holonomy, and the sum of monopole and antimonopole mass, M_{m+a}, is, after screening and in

---

[5]This requirement assures that the system does not possess any energy and pressure in the limit $e \to \infty$.

a holonomy-independent way, given as

$$M_{m+a} = \frac{8\pi^2}{e\beta} \qquad (6.4)$$

[compare with Eq. (4.107) for the unscreened case]. The thermally averaged flux $\bar{F}_{\pm,th}(\delta)$ system at zero spatial momentum and at a given, finite value of the coupling e is obtained as

$$\bar{F}_{\pm,th}(\delta) = 4\pi \int d^3p \, \delta^{(3)}(\vec{p}) n_B(\beta |E(\vec{p})|) \bar{F}_{\pm}$$

$$= \pm \frac{8\pi\delta}{e} \int d^3p \frac{\delta^{(3)}(\vec{p})}{\exp\left[\beta\sqrt{M_{m+a}^2 + \vec{p}^2}\right] - 1}. \qquad (6.5)$$

After setting $\vec{p} = 0$ in n_B and by appealing to Eq. (6.4), the small-mass expansion of the denominator of n_B reads

$$\lim_{\vec{p}\to 0} \left(\exp\left[\beta\sqrt{M_{m+a}^2 + \vec{p}^2}\right] - 1\right)$$

$$= \frac{8\pi^2}{e} \left(1 + \frac{1}{2}\frac{8\pi^2}{e} + \frac{1}{6}\left(\frac{8\pi^2}{e}\right)^2 + \cdots\right). \qquad (6.6)$$

Using Eq. (6.6), the limit $e \to \infty$ can now safely be performed in Eq. (6.5). We have

$$\lim_{e\to\infty} \bar{F}_{\pm,th}(\delta) = \pm\frac{\delta}{\pi} \quad (0 \le \delta \le \pi). \qquad (6.7)$$

Because δ is an angle, the right-hand side of Eq. (6.7) defines the argument of complex[6] and periodic functions f such that

$$\left\{\frac{\varphi}{|\varphi|}\left(\frac{\delta}{\pi}\right)\right\} \subset \left\{f\left(\frac{\delta}{\pi}\right)\right\}. \qquad (6.8)$$

---

[6]After spatial coarse-graining only a complex and inert scalar field φ can couple to the dual gauge modes: φ is composed of constituents that are energy- and pressure-free, thus φ does not propagate, and spatial isotropy, homogeneity, and thermal equilibrium demand φ to be a scalar of modulus $|\varphi|$ which is constant in spacetime.

Since f's argument was obtained by a projection onto zero spatial momentum, thus guaranteeing the inertness (no propagation because of zero energy and momentum) of any quantity emerging by a spatial average over the heat-bath coupled monopole-antimonopole systems, the only admissible nontrivial periodic dependences of the functions f are those on the Euclidean time τ. Thus we may set $\frac{\delta}{\pi} = \frac{\tau}{\beta}$. Since the functions f are periodic, $f(\tau = 0) = f(\tau = \beta)$, they can be expanded into Fourier series,

$$f\left(\frac{\tau}{\beta}\right) = \sum_{n=-\infty}^{n=\infty} f_n \exp\left[2\pi in\frac{\tau}{\beta}\right], \tag{6.9}$$

where f_n are (complex) constants. Only $n = \pm 1$ are allowed in Eq. (6.9) since the physical situation generating the continuous parameter δ does not repeat itself for $0 \le \delta \le \frac{\pi}{m}, \frac{\pi}{m} \le \delta \le \frac{2\pi}{m}, \dots, (m > 1)$: Higher-charge monopoles are absent. [Recall that only (anti)calorons of large holonomy and topological charge modulus *unity* dissociate into pairs of winding number modulus *one* monopoles and anti-monopoles in the deconfining phase.] We conclude that the equation of motion satisfied by the functions f is:

$$Df\left(\frac{\tau}{\beta}\right) \equiv \partial_\tau^2 f\left(\frac{\tau}{\beta}\right) + \left(\frac{2\pi}{\beta}\right)^2 f\left(\frac{\tau}{\beta}\right) = 0. \tag{6.10}$$

Since $|\varphi|$ is spatially homogeneous not only the phase $\frac{\varphi}{|\varphi|}$ but the entire field φ is annihilated by the linear differential operator $D = \partial_\tau^2 + (\frac{2\pi}{\beta})^2$ introduced in Eq. (6.10).

Since f_1 and f_{-1} in Eq. (6.9) represent two complex constants, which can be chosen freely, the associated complex, two-dimensional kernel of the differential operator D on the space of complex-valued functions of τ determines the operator D in Eq. (6.10) uniquely.

SU(3) case

Since the two U(1) factors surviving the dynamical symmetry breakdown induced by the field ϕ in the deconfining phase are independent of one another there is no coupling between the two

associated monopole species. As a consequence, there are two independent monopole condensates, described by complex and inert
scalar fields φ_1 and φ_2 whose phases are both annihilated by the
uniquely determined differential operator D of Eq. (6.10).

6.1.3   Onset of monopole condensation

As an interlude we are in this section concerned with the transition
between deconfining and preconfining SU(2) Yang–Mills thermodynamics. In approaching this transition from above in a specified
way, we ask the question: what the critical value for the effective coupling e at which the thermal ground state of the deconfining phase
re-arranges to attribute mass to a formerly massless gauge mode
[condensation of magnetic (anti)monopoles]? Technically speaking,
the answer to this question is obtained in a surprisingly simple way
if a connection between effective and fundamental degrees of freedom is made. We will consider two limiting cases in letting $T \searrow T_c$:
an infinitely fast and a slow approach.

Recall that, starting from the *a priori* estimate of the deconfining
thermal ground state, the computation of thermodynamic quantities
is organized into a rapidly converging loop expansion in terms of
effective gauge-field fluctuations of trivial topology (see Sec. 5.4.3).
At high temperature this expansion indicates the presence of unresolved stable and screened (anti)monopoles of a number density
$\propto T^3$. At temperatures not far above the critical temperature T_c the
collective dynamics of these defects induces, depending on momentum, screening or antiscreening of the tree-level massless, effective
gauge mode (see Sec. 5.4.7).

Recall also that the λ dependence of e describes the change
in screening of a test charge due to instable magnetic dipoles arising from (anti)calorons whose holonomy is only mildly deformed
away from trivial: When increasing λ away from λ_c, e approaches
the plateau $e \equiv \sqrt{8}\pi$ rapidly, and near λ_c one has

$$e(\lambda) = -4.59 \log (\lambda - \lambda_c) + 18.42 \quad (\lambda_c \leq \lambda \leq 15.0) \qquad (6.11)$$

(see Sec. 5.2.3).

Due to the high-temperature radiative corrections in the effective theory for the deconfining phase the mass m_{M+A} of a noninteracting but screened monopole-antimonopole test system[7] is lifted away from the BPS bound:

$$m_{M+A} = m_M + m_A = \frac{8\pi^2 T}{e(\lambda)} + \text{non-BPS}. \tag{6.12}$$

Comparing with Eq. (5.257) of Sec. 5.5.1, the term "non-BPS" in Eq. (6.12) amounts to a 13.9% correction introduced by the presence of all other stable screened but unresolved magnetic monopole-antimonopole pairs which are collectively generated by effective radiative corrections. This correction is comparable to the "photon's" magnetic screening mass m_m, which, due to weak coupling, is calculable in perturbation theory. Indeed, an order-of-magnitude estimate matches rather well with the result extracted from the two-loop correction of the pressure at high temperature: To lowest order in the magnetic coupling $g = \frac{4\pi}{e}$ we have $m_m = \frac{4\pi T}{\sqrt{3}e}$ [Andersen (1997)]. Compared to the BPS term in Eq. (6.12) this is a correction of about 10% at high temperature.

Based on the discussion presented by Huang (1963) of thermalized, noninteracting Bose particles with mass m a relation was formulated in Nieto (1970) between the total number density n and the density n_0 of particles residing in the condensate. For statistical weight unity [only one species of monopole-antimonopole pairs occurs in an SU(2) Yang–Mills theory] one has

$$n_0 = n - n_c$$

$$\equiv n - \frac{T^3}{2\pi^2}\mu^2 \sum_{l=1}^{\infty} e^{l\mu} \frac{K_2(l\mu)}{l}, \tag{6.13}$$

---

[7]For a test system immersed in an effective thermal ensemble including the *a priori* ground-state estimate of Sec. 5.1.5 and noninteracting gauge-field excitations [see Sec. 5.2.1] fundamental, short-lived magnetic dipoles do not introduce a mass scale, so screening only occurs via a renormalization of the coupling constant, and, modulo this coupling-constant renormalization, the BPS mass formula Eq. (4.107) can be applied.

where $\mu \equiv m/T$, and $K_2(x)$ is the modified Bessel function of the second kind. At the onset of Bose condensation, where n_0 is yet zero, the total number density n is given by the number density n_{fr} of freely fluctuating particles

$$n = n_{\mathrm{fr}} \equiv \frac{T^3}{2\pi^2} \int_0^\infty dx \frac{x^2}{e^{\sqrt{x^2+\mu^2}} - 1}. \qquad (6.14)$$

So in the fully thermalized system, which takes place if T is *slowly* lowered towards T_c, we have at the onset of Bose condensation

$$\int_0^\infty dx \frac{x^2}{e^{\sqrt{x^2+\mu_c^2}} - 1} = n_c = \mu_c^2 \sum_{l=1}^\infty e^{l\mu_c} \frac{K_2(l\mu_c)}{k}. \qquad (6.15)$$

Equation (6.15) determines the critical *ratio* μ_c between mass and temperature at which Bose condensation starts to occur. We remark that in conventional statistical systems and in the absence of interactions the mass m of a given species of bosonic particles is a predetermined quantity which does not depend on temperature. The solution μ_c to Eq. (6.15) thus determines the critical temperature T_c for Bose condensation as $T_c = \frac{m}{\mu_c}$. In deconfining Yang–Mills thermodynamics, however, a pair of a stable and screened magnetic monopole and its antimonopole owes its very existence to the presence of a heat bath of a given temperature. The corresponding relation between mass and temperature [see Eq. (6.12)] and μ_c (see Eq.(6.15)] determine the critical temperature for condensation to be the solution to the following equation

$$T_c = \frac{m(T_c)}{\mu_c}. \qquad (6.16)$$

Notice that in the limit $\mu_c \to 0$ Eq. (6.15) yields the identity $2\zeta(3) = 2\zeta(3)$ where $\zeta(z)$ is Riemann's zeta function. Since for increasing μ_c the left-hand side of Eq. (6.15) is monotonic decreasing and the right-hand side is monotonic increasing it follows that $\mu_c = 0$ is the only solution. Thus in deconfining SU(2) Yang–Mills thermodynamics a slow approach of T_c from above implies that only massless monopoles and antimonopoles condense into a new ground state. The point in temperature, which marks the onset of this process, is the position λ_c of the logarithmic pole in e as described by Eq. (6.11).

Alternatively, one may ask what happens in the limit where T_c is rapidly approached from above. As we will see, such an adiabatic (sudden) approximation fully takes into account the static screening effects imposed by the system at high temperature but neglects the influence of propagating dual gauge modes not far above T_c. Since the pole in the coupling $e(\lambda)$ [see Eq. (6.11)] is logarithmic (reflecting the fact that the Yang–Mills scale Λ nonperturbatively interferes with the dynamics of fundamental propagating gauge modes only shortly above λ_c [Giacosa and Hofmann (2007)]) we may consider the high-temperature limit to extract the dependence on temperature of the density n_{as} of *interacting* but statistically equilibrated monopole-antimonopole pairs. Recall that n_{as} is extracted from a particular two-loop correction to the quasiparticle pressure as calculated in the effective theory (see Sec. 5.4.4) and that one has

$$n_{as} = (21.691)^{-3} T^3 \sim 9.8 \times 10^{-5} T^3. \tag{6.17}$$

At high temperature, stable and screened (anti)monopoles are non-relativistic [see Eq. (6.12)]. If temperature is lowered towards T_c in a sufficiently rapid way then the metamorphosis of stable and screened (anti)monopoles into almost massless, stable and strongly screened monopoles and antimonopoles takes place rapidly enough to not affect their nonrelativistic nature.[8] The condensation condition (6.15) thus modifies as

$$\frac{n_{as}}{T^3} = (21.691)^{-3} = \zeta(3/2) \left(\frac{\mu_c}{2\pi}\right)^{3/2}, \tag{6.18}$$

---

[8]To catch up in velocity (anti)monopoles must interact via the exchange of dual gauge modes that are close to their mass shell and therefore propagate at a speed close to the velocity of light. In contrast to the portion of magnetic screening induced by an increased activity of unstable monopole-antimonopole pairs, described by the effective theory's thermal ground-state estimate, this part of the thermalization of (anti)monopoles thus requires a finite amount of time.

where the expression to the far right is obtained by considering the limit $\mu_c \gg 1$ of n_c or, equivalently, of $1/2\pi^2$ times the right-hand side of (6.15). To summarize, Eq. (6.18) determines μ_c in a situation where highly nonrelativistic, stable and screened (anti)monopoles are adiabatically fast deprived of their mass by cooling (instantaneously intensified screening by unstable dipoles) so that no time is available for them to start moving.

The solution to Eq. (6.18) is $\mu_c = 7.04 \times 10^{-3}$. Note the amusing fact that the nonrelativistic nature of monopole-antimonopole pairs is assured by the large-mass limit of n_c while the solution to Eq. (6.18) actually corresponds to a small mass on the scale of temperature. The resolution of this apparent puzzle is grounded in the fact that the sudden approximation employed does not admit a thermodynamical interpretation: The expression for n_{as} at $T \gg T_c$ is analytically continued down to T_c.

Using

$$\mu = \left(8\pi^2 + \frac{4\pi}{\sqrt{3}} \right) e^{-1} \qquad (6.19)$$

[compare with Eq. (6.12) and the paragraph below Eq. (5.74)], we obtain

$$e_c = 1.225 \times 10^4. \qquad (6.20)$$

The result of the sudden approximation in Eq. (6.20) represents a lower bound on values of e_c occurring for finite-velocity approaches $\lambda \searrow \lambda_c$. In the latter cases μ_c must take values in between the extremes obtained for zero and infinite velocity,

$$0 \leq \mu_c \leq 7.04 \times 10^{-3} \quad \text{or} \quad \infty \geq e_c \geq 1.225 \times 10^4. \qquad (6.21)$$

Let us summarize this section. We have considered two scenarios for the onset of magnetic monopole-antimonopole condensation at the deconfining–preconfining transition in SU(2) Yang–Mills thermodynamics: Infinitely slow and infinitely fast downward approach of T_c. In the former situation, we have shown that pairs of stable monopoles and antimonopoles start to condense when they become massless, that is, at the pole position λ_c for the effective coupling e. This is a consequence of the fact that due to the irreversibility of the (anti)monopole creation process [dissociation of large-holonomy (anti)calorons] and due to overall charge neutrality [pairwise creation in an infinite spatial volume] the chemical potential associated with pairs is nil. In the latter situation, we obtain a lower bound on the value e_c of the critical coupling for a finite-velocity approach of T_c.

6.1.4   Coarse-grained and free monopoles: BPS saturation and Euler–Langrange

Let us now come back to the discussion of a coarse-grained condensate of monopoles and antimonopoles that we have started in Sec. 6.1.2. After the unique determination of the differential operator D, which annihilates the phase of the complex scalar field φ, we now ask what its Euclidean action density $\mathcal{L}_\varphi$ and modulus are. According to the results of Sec. 6.1.2 we know that, modulo an inessential constant phase shift, the field φ takes the following form

$$\varphi = |\varphi| \exp\left[\pm 2\pi i \frac{\tau}{\beta}\right]. \tag{6.22}$$

Here the gauge-invariant quantity $|\varphi| = \sqrt{\varphi^*\varphi}$ is spacetime-independent. (Again, the field φ, which is void of energy and pressure, does not fluctuate, and, accordingly, a spacetime dependence of $|\varphi|$ would violate the spacetime homogeneity of one-point functions involving the field φ. But this is forbidden by thermal equilibrium in an infinite spatial volume.)

Let us now turn to the discussion of the Euclidean action density $\mathcal{L}_\varphi$ in the absence of any monopole-antimonopole interaction. No explicit temperature dependence may appear in $\mathcal{L}_\varphi$ on the level of noninteracting monopoles and antimonopoles since the mass of a static monopole is nil in the limit $e \to \infty$. One may thus write

$$\mathcal{L}_\varphi = \frac{1}{2}\partial_\tau \varphi^* \partial_\tau \varphi + \frac{1}{2}V(|\varphi|^2), \tag{6.23}$$

where $V(|\varphi|^2)$ is a gauge-invariant potential to be determined. The Euler–Lagrange equation, which follows from the action density in (6.23), reads

$$\partial_\tau^2 \varphi = \frac{\partial V(|\varphi|^2)}{\partial|\varphi|^2}\varphi \overset{\text{Eq. (6.22), }\varphi\neq 0}{\Longleftrightarrow} \left(\frac{2\pi}{\beta}\right)^2 = -\frac{\partial V(|\varphi|^2)}{\partial|\varphi|^2}. \tag{6.24}$$

On the other hand, the field φ is BPS-saturated (vanishing of the Euclidean energy density because it is obtained by averaging over static and massless monopoles and antimonopoles). Equations (6.23) and (6.22) thus imply that

$$|\varphi|^2 \left(\frac{2\pi}{\beta}\right)^2 - V(|\varphi|^2) = 0. \tag{6.25}$$

Together, Eqs. (6.24) and (6.25) yield

$$\frac{\partial V(|\varphi|^2)}{\partial|\varphi|^2} = -\frac{V(|\varphi|^2)}{|\varphi|^2}. \tag{6.26}$$

The solution to the first-order equation (6.26) reads

$$V(|\varphi|^2) = \frac{\bar{\Lambda}^6}{|\varphi|^2}, \tag{6.27}$$

where $\bar{\Lambda}$ is a mass scale which enters as a constant of integration. Substituting Eq. (6.27) into Eq. (6.25) yields

$$|\varphi| = \sqrt{\frac{\bar{\Lambda}^3}{2\pi T}} = \sqrt{\frac{\bar{\Lambda}^3 \beta}{2\pi}}. \tag{6.28}$$

The quantity $|\varphi|$ sets the scale of maximal resolution in the effective theory for the preconfining phase. An $S_{2,R=|\varphi|^{-1}}$ separating a monopole in the interior from its antimonopole in the exterior (or vice versa) experiences the same magnetic flux as an $S_{2,R=\infty}$ since in the condensate the monopole-antimonopole distance and their core-size is nil. Thus the monopole and the antimonopole cannot probe the finite curvature of $S_{2,R=|\varphi|^{-1}}$ and the infinite-surface limit, used to derive in Sec. 6.1.2 the operator D, is trivially saturated in the spatial coarse-graining process.

6.2   The Dual Gauge Field

So far we have discussed how noninteracting, massless magnetic monopoles conspire to form an inert scalar field[9] φ upon a spatial average. We now ask how the topologically trivial propagating, and dual U(1) [U(1)^2 for SU(3)] gauge field(s) is (are) modified under this coarse-graining and how they describe microscopic monopole-antimonopole interactions.

6.2.1   *Effective action for the preconfining phase and thermal ground state*

Since the field a^3_μ of the deconfining phase does not interact with itself the coarse-grained field a^D_μ is subject to the same form of the free Abelian action. Also, local minimal U(1) gauge invariance demands that $\partial_\tau \to \mathcal{D}_\mu \equiv \partial_\mu + iga^D_\mu$ where g denotes the magnetic coupling whose running with temperature will be determined in Sec. 6.2.2. Moreover, there are no higher-mass-dimension operators in the effective action for the fields φ and a^D_μ. Such operators would couple the field φ to propagating modes of the field a^D_μ opening the possibility for momentum transfers smaller than $|\varphi|^2$. This would violate the inertness (BPS saturation) of φ. Thus, φ and a^D_μ are coupled minimally. The entire effective action for the preconfining phase of an SU(2) Yang–Mills theory thus reads

$$S = \int_0^\beta d\tau \int d^3x \left[\frac{1}{4} G^D_{\mu\nu} G^D_{\mu\nu} + \frac{1}{2}(\mathcal{D}_\mu \varphi)^* \mathcal{D}_\mu \varphi + \frac{1}{2} \frac{\bar{\Lambda}^6}{|\varphi|^2} \right], \quad (6.29)$$

where $G^D_{\mu\nu} \equiv \partial_\mu a^D_\nu - \partial_\nu a^D_\mu$, and, in contrast to the deconfining phase, there are no residual radiative corrections to be integrated out.

Keeping in mind the inertness of the field φ, the Euler–Lagrange equations for the gauge field a^D_μ are given as

$$\partial_\mu G^D_{\mu\nu} = ig \left[(\mathcal{D}_\nu \varphi)^* \varphi - \bar{\varphi} \mathcal{D}_\nu \varphi^* \right]. \quad (6.30)$$

---

[9]For SU(3) there are two such fields.

By virtue of $\mathcal{D}_\nu \varphi = 0$ the pure-gauge configuration $a_\mu^{D,\mathrm{gs}} = \mp \frac{2\pi}{g\beta}\delta_{\mu 4}$ solves Eq. (6.30). Inserting $a_\mu^{D,\mathrm{gs}}$ and φ into (6.29), one reads off the ground-state energy density and pressure as $\rho^{\mathrm{gs}} = -P^{\mathrm{gs}} = \pi \bar{\Lambda}^3 T$. That is, the pure-gauge configuration $a_\mu^{D,\mathrm{gs}}$ lowers (raises) the vanishing pressure (energy density) of massless but free condensed magnetic monopoles to a definite negative (positive) and temperature-dependent value. Hence we associate the name thermal ground-state with the field configurations $a_\mu^{D,\mathrm{gs}}$ and φ. Since the dual gauge field, albeit massive, is free, this *a priori* estimate of the thermal ground state is exact.[10] Recall that in the deconfining phase small radiative corrections do alter the *a priori* estimate for the deconfining ground state (see Sec. 5.5.5).

Let us now address the Polyakov loop evaluated on $a_\mu^{D,\mathrm{gs}}$. We have

$$\mathrm{Pol}[a_\mu^{D,\mathrm{gs}}] = 1 \tag{6.31}$$

independently of whether we evaluate it in the gauge where $a_\mu^{D,\mathrm{gs}} = \mp \frac{2\pi}{g\beta}\delta_{\mu 4}$ or in any other admissible gauge. Let us show this. The field φ remains periodic under $\varphi \to \Omega^\dagger \varphi$ (admissible change of gauge) if and only if $\Omega = \exp[i(2\pi n \frac{\tau}{\beta} + \alpha(\vec{x}))]$ where $n \in \mathbf{Z}$, and α is a real function of space only. Hence $a_\mu^D \to a_\mu^D + \frac{2\pi n}{g\beta}\delta_{\mu 4} + \frac{\partial_j \alpha(\vec{x})}{g}\delta_{\mu j}$ under Ω, and thus the periodicity of a_μ^D is (trivially) assured. (Here $j = 1, 2, 3$.) In particular, one has

$$a_\mu^{D,\mathrm{gs}} \to \left(\mp\frac{2\pi}{g\beta} + \frac{2\pi n}{g\beta}\right)\delta_{\mu 4} + \frac{\partial_j \alpha(\vec{x})}{g}\delta_{\mu j}$$
$$= \frac{2\pi(n \mp 1)}{g\beta}\delta_{\mu 4} + \frac{\partial_j \alpha(\vec{x})}{g}\delta_{\mu j}. \tag{6.32}$$

Thus in any admissible gauge the Polyakov loop evaluated on $a_\mu^{D,\mathrm{gs}}$ is unity, $\mathrm{Pol}[a_\mu^{D,\mathrm{gs}}] = \exp[ig\int_0^\beta d\tau\, a_4^{D,\mathrm{gs}}] = 1$, and the electric $\mathbf{Z}_2$

---

[10]Again, the influence of the quantum fluctuations of a_μ^D on ρ^{gs} and P^{gs} is negligible, see Sec. 6.2.2.

degeneracy of the thermal ground state, which occured in the deconfining phase, no longer persists in the preconfining phase.

In a way completely analogous to SU(2) one derives for SU(3) the following effective action for the preconfining phase:

$$
S = \sum_{l=1}^{2} \int_0^\beta d\tau \int d^3x \left[\frac{1}{4} G^D_{\mu\nu,l} G^D_{\mu\nu,l} + \frac{1}{2} \left(D_{\mu,l}\varphi_l \right)^* D_{\mu,l}\varphi_l + \frac{1}{2} \frac{\bar\Lambda^6}{|\varphi_l|^2} \right].
$$

(6.33)

Since SU(3)→U(1)^2 in the deconfining phase there are now two independent species of magnetic monopoles, dual gauge fields, $a^D_{\mu,1}$, $a^D_{\mu,2}$, and monopole-antimonopole condensates, the latter represented by inert complex scalar fields φ_1, φ_2. The magnetic coupling g and the scale $\bar\Lambda$ are universal, $a^{D,gs}_{\mu,1} = a^{D,gs}_{\mu,2} = \mp \frac{2\pi}{g\beta}\delta_{\mu4}$, and the thermal ground-state energy density and pressure are given as $\rho^{gs} = -P^{gs} = 2\pi\bar\Lambda^3 T$. The Polyakov loop, evaluated on the thermal ground-state configurations $a^D_{\mu,1}$, $a^D_{\mu,2}$, is unity in any admissible gauge also for SU(3). This shows that the electric $\mathbf{Z}_3$ degeneracy of the ground-state estimate in the deconfining phase no longer persists in the preconfining phase.

6.2.2 Free quasiparticle excitations, scale matching, and running magnetic coupling

In the effective theory for the preconfining phase, as described by action (6.29), the dynamical breaking of the residual gauge symmetry U(1) [for SU(2)] and U(1)^2 [for SU(3)] is manifested in terms of a quasiparticle mass m for the dual gauge field. One has $m = g|\varphi| = g|\varphi_1| = g|\varphi_2| = aT$ where $a = 2\pi g\bar\lambda^{-3/2}$. Let us show this: In unitary gauge, $\varphi = |\varphi| = \varphi_{1,2}$ and $a^{D,gs}_\mu = a^{D,gs}_{\mu,1} = a^{D,gs}_{\mu,2} = 0$, the relation $m = g|\varphi| = g|\varphi_{1,2}|$ for the mass of the fluctuations δa^D_μ, $\delta a^D_{\mu,1}$, and $\delta a^D_{\mu,2}$ can be read off from (6.29) and (6.33), respectively (Abelian Higgs mechanism), and $a = 2\pi g\bar\lambda^{-3/2}$ then follows from Eq. (6.28) and the definition $\bar\lambda \equiv 2\pi T/\bar\Lambda$.

Because the field φ is inert and since no "photon" selfinteraction occur in an Abelian gauge theory, the excitations in the

effective theory for the deconfining phase are *free* thermal quasiparticles. Notice that the number of degrees of freedom before coarse-
graining matches those after coarse-graining. Namely, for SU(2) one
has before coarse-graining one species of propagating gauge field
times two polarizations plus one species of center-vortex loop (see
Chapter 7). Effectively, that is, after coarse-graining, one has one
species of massive, dual gauge field times three polarizations. Thus,
one obtains three degrees of freedom *before* and three degrees of
freedom *after* coarse-graining. For SU(3) one obtains six degrees of
freedom before and six degrees of freedom after coarse-graining.

Let us now estimate the contribution $-\Delta V$ of quantum
fluctuations to the thermodynamic pressure: For both SU(2) and
SU(3) one obtains in close analogy to the deconfining phase the
following estimate for the modulus of the ratio $\frac{\Delta V}{V}$:

$$\left| \frac{\Delta V}{V} \right| \leq \frac{\bar{\lambda}^{-3}}{24\pi^2}. \tag{6.34}$$

As we shall see, $\bar{\lambda} \geq 7.075$ [SU(2)] and $\bar{\lambda} \geq 6.467$ [SU(3)]. Thus
$-\Delta V$ is a correction of order 10^{-5} to the thermal ground-state result
$-V$. From the results of Sec. 5.2.3 [see Fig. 5.7] and appealing to the
continuity of the pressure across a thermal phase boundary we know
that the ground-state pressure dominates the total pressure in the
preconfining phase (see also below).

So far we have considered the scale $\bar{\Lambda}$, which enters as a constant of integration in Eq. (6.27), as a free parameter of the preconfining phase. But once the scale Λ of the deconfining phase is set
by a high-temperature experiment the scale $\bar{\Lambda}$ is also fixed. This is a
consequence of the following relations:

$$\bar{\Lambda} = \left(4 + \frac{\lambda_c^3}{720\pi^2} \right)^{1/3} \Lambda \quad \text{[for SU(2)]},$$

$$\bar{\Lambda} = \left(2 + \frac{\lambda_c^3}{720\pi^2} \right)^{1/3} \Lambda \quad \text{[for SU(3)]}. \tag{6.35}$$

Equations (6.35), in turn, follow from the above-mentioned fact that at the point λ_c, where $e = \infty$ and $g = 0$, the total pressure P is continuous for the thermal transition between deconfining and preconfining phase. No higher loop corrections to the one-loop result for the pressure in the deconfining phase take place at λ_c since there the fluctuations $a_\mu^{1,2}$ [SU(2)] and $a_\mu^{1,2,4,5,6,7}$ [SU(3)] decouple. Equating at $\lambda_c = \frac{\bar{\Lambda}}{\Lambda}\bar{\lambda}_c$ the right-hand side of Eq. (5.63) in Sec. 5.2.2 with the right-hand side of the following expression for the total pressure in the preconfining phase [according to Eq. (6.34) the quantum part $-\Delta V$ is negligible],

$$P(\bar{\lambda}_c) = -\bar{\Lambda}^4 \left[\frac{6\bar{\lambda}_c^4}{(2\pi)^6}\bar{P}(0) + \frac{\bar{\lambda}_c}{2} \right], \tag{6.36}$$

yields the relation in Eqs. (6.35) for SU(2). For SU(3) one needs to equate the right-hand sides of

$$P(\lambda_c) = -\Lambda^4 \left\{ \frac{8\lambda_c^4}{(2\pi)^6}\bar{P}(0) + 2\lambda_c \right\} \tag{6.37}$$

and

$$P(\bar{\lambda}_c) = -\bar{\Lambda}^4 \left\{ \frac{12\bar{\lambda}_c^4}{(2\pi)^6}\bar{P}(0) + \bar{\lambda}_c \right\}. \tag{6.38}$$

Inspecting Eqs. (6.36) and (6.38), we notice that the SU(3) pressure is just twice the SU(2) pressure in the preconfining phase.

In close analogy to the evolution of the effective coupling e in the deconfining phase the evolution of the magnetic coupling g with temperature is described by the following first-order differential equation for both SU(2) and SU(3):

$$\partial_a \bar{\lambda} = -\frac{12\bar{\lambda}^4}{(2\pi)^6} \frac{aD(a)}{1 + \frac{12\bar{\lambda}^3 a^2}{(2\pi)^6}D(a)}. \tag{6.39}$$

Here $a \equiv 2\pi g\bar{\lambda}^{-3/2}$, and the function $D(y)$ is defined in Eq. (3.19) of Sec. 3.1.1. One shows numerically that throughout the preconfining phase the total pressure is ground-state dominated, and thus it is negative. As in the deconfining phase, the invariance of the Legendre transformations between thermodynamic quantities under the applied coarse-graining implies for the effective theory that $\partial_{(aT)}P = 0$, and for both SU(2) and SU(3) the same evolution equation (6.39) follows.

Numerically, the initial condition for the evolution described by Eq. (6.39) is $g(\bar{\lambda}_c) = 0$ for $\bar{\lambda}_c = 8.478$ [SU(2)] and $\bar{\lambda}_c = 7.376$ [SU(3)]. When lowering $\bar{\lambda} < \bar{\lambda}_c$ the magnetic coupling g rises rapidly and runs into a logarithmic pole at $\bar{\lambda}_{c'}$: $g \propto -\log(\bar{\lambda} - \bar{\lambda}_{c'})$. One has $\bar{\lambda}_{c'} = 7.075$ [SU(2)] and $\bar{\lambda}_{c'} = 6.467$ [SU(3)]. Taking the mass m of the dual gauge mode as an order parameter for the dynamical breaking of U(1) [SU(2)] and U(1)^2 [SU(3)] and postulating that $m = K(T_c - T)^\nu$ for $T \lesssim T_c$ (K and ν real constants), one extracts mean-field critical exponents: $\nu = \frac{1}{2}$. The ratio of energy density ρ and T^4 in the preconfining phase is given as

$$\frac{\rho}{T^4}(\bar{\lambda}) = \frac{(2\pi)^4}{\bar{\lambda}^4}\left[\frac{6\bar{\lambda}^4}{(2\pi)^6}\bar{\rho}(a) + \frac{\bar{\lambda}}{2}\right] \quad \text{[for SU(2)]},$$

$$\frac{\rho}{T^4}(\bar{\lambda}) = \frac{(2\pi)^4}{\bar{\lambda}^4}\left[\frac{12\bar{\lambda}^4}{(2\pi)^6}\bar{\rho}(a) + \bar{\lambda}\right] \quad \text{[for SU(3)]}. \qquad (6.40)$$

With a slight abuse of notation we refer in the following to $\rho(\bar{\lambda})$ as the functional dependence of the energy density on temperature $\bar{\lambda}$ in the *preconfining* phase and to $\rho(\lambda)$ as the functional dependence of the energy density on temperature λ in the *deconfining* phase. Thus $\rho(\bar{\lambda})$ and $\rho(\lambda)$ are different functions of their arguments. With this convention, the energy density ρ exhibits a positive jump $\Delta(\bar{\lambda}_c)$ when lowering the temperature. One has

$$\Delta(\bar{\lambda}_c) \equiv \frac{\rho(\lambda_c + 0) - \rho(\bar{\lambda}_c - 0)}{T_c^4} = \frac{4}{3}\frac{\pi^2}{30} \quad \text{[for SU(2)]},$$

$$\Delta(\bar{\lambda}_c) = \frac{8}{3}\frac{\pi^2}{30} \quad \text{[for SU(3)]}. \qquad (6.41)$$

The existence of the gap Δ in energy density signals that the monopole-antimonopole condensate only builds up gradually as temperature falls below λ_c. This is intuitively understandable because the condensation would require the influx of an infinite number of totally screened monopole-antimonopole pairs from infinity which costs energy. Seen alternatively, to facilitate the condensation of additional monopole-antimonopole pairs (stable condensate) by total screening requires an energy-consuming increase of the average caloron/anticaloron holonomy from almost trivial to maximal. The associated (anti)caloron dissociation releases this energy by an increase of the density of (anti)monopoles that are screened to masslessness. Although this process is hard to grasp microscopically, after spatial coarse-graining the critical temperature at which a stable condensate forms (defined by the property that the system is more likely to be preconfining than deconfining, see Sec. 6.2.3) can be determined exactly.

6.2.3 *Supercooling*

Here we present the results of [Hofmann (2007)]. We claim that a stable condensate of monopoles and antimonopoles exists for temperatures $\bar{\lambda}$ in $\bar{\lambda}_{c'} \leq \bar{\lambda} \leq \bar{\lambda}_*$ where $\bar{\lambda}_* < \bar{\lambda}_c$, and $\bar{\lambda}_{c'} = 8.478$ [for SU(2)] is the position of the logarithmic pole of the magnetic coupling g (see Sec. 6.2.2). Here the criterion for the existence of a stable condensate is that the system is more likely to be found in the preconfining than in the deconfining state. Whether this situation takes place is decided by the respective energy densities: If the energy associated with a given spatial volume is smaller in the preconfining as compared to the deconfining state then a larger Boltzmann weight takes place in the monopole-condensing situation.

Let us now discuss the above claim. At $\bar{\lambda}_{c'}$ one has

$$\frac{\rho(\bar{\lambda}_{c'})}{T_{c'}^4} = 8\pi^4 \bar{\lambda}_{c'}^{-3} \quad \text{[for SU(2)]},$$

$$\frac{\rho(\bar{\lambda}_{c'})}{T_{c'}^4} = 16\pi^4 \bar{\lambda}_{c'}^{-3} \quad \text{[for SU(3)]}. \tag{6.42}$$

On the other hand, continuing the energy density of the deconfining phase down to $\lambda_{c'} = \frac{\bar{\Lambda}}{\Lambda}\bar{\lambda}_{c'}$ and using Eqs. (6.35) yields

$$\frac{\rho(\lambda_{c'})}{T_{c'}^4} = \frac{\pi^2}{15} + \frac{32\pi^4}{4 + \frac{\lambda_c}{720\pi^2}}\bar{\lambda}_{c'}^{-3} \quad \text{[for SU(2)]},$$

$$\frac{\rho(\lambda_{c'})}{T_{c'}^4} = 2\frac{\pi^2}{15} + \frac{32\pi^4}{2 + \frac{\lambda_c}{720\pi^2}}\bar{\lambda}_{c'}^{-3} \quad \text{[for SU(3)]}. \qquad (6.43)$$

The second summands in Eqs. (6.43) practically coincide with the expressions in Eqs. (6.42) for $\frac{\rho(\bar{\lambda}_{c'})}{T_{c'}^4}$. Thus we conclude that $\frac{\rho(\bar{\lambda}_{c'})}{T_{c'}^4} < \frac{\rho(\lambda_{c'})}{T_{c'}^4}$ for both SU(2) and SU(3). But according to Eqs. (6.41) we have $\frac{\rho(\bar{\lambda}_c)}{T_{c'}^4} > \frac{\rho(\lambda_c)}{T_c^4}$ for both SU(2) and SU(3). Since both functions $\frac{\rho(\bar{\lambda})}{T^4}$ and $\frac{\rho(\lambda)}{T^4}$ are continuous in the ranges $\bar{\lambda}_{c'} \leq \bar{\lambda} \leq \bar{\lambda}_c$ and $\lambda_{c'} \leq \lambda \leq \lambda_c$, respectively, there is at least one intersection. Numerically, one shows that only a single intersection $\bar{\lambda}_*$ takes place (see Fig. 6.4). For SU(2) one obtains the following values: $\lambda_* = 12.15$ or $\bar{\lambda}_* = 7.428$. Now, cooling the system, a stable condensate starts to take place at $\bar{\lambda}_*$, and the above claim follows.

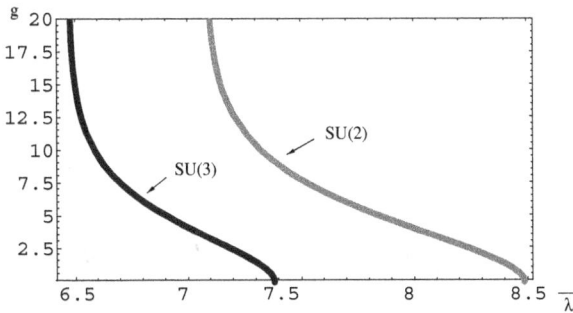

Figure 6.3. Evolution of the magnetic coupling g within the preconfining phase for SU(2) (grey) and SU(3) (black). The dimensionless temperature $\bar{\lambda}$ is defined as $\bar{\lambda} \equiv \frac{2\pi T}{\bar{\Lambda}}$ where the scale $\bar{\Lambda}$ is related to the scale Λ of the deconfining phase by virtue of Eqs. (6.35).

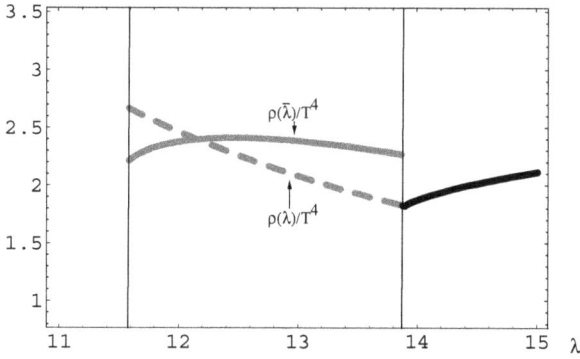

Figure 6.4. The quantities $\frac{\rho(\bar{\lambda})}{T^4}$ (dashed gray line) and $\frac{\rho(\bar{\lambda})}{T^4}$ (solid gray line) in the preconfining phase as functions of λ for SU(2). The intersection point is at $\lambda_* = 12.15$. Vertical lines mark the boundaries of the preconfining phase. The black solid line is associated with $\frac{\rho(\lambda)}{T^4}$ in the deconfining phase.

Let us now investigate whether topological defects within the monopole-antimonopole condensate are spatially resolved within the realm of temperatures $\bar{\lambda}_{c'} \leq \bar{\lambda} \leq \bar{\lambda}_*$ where such a condensate dominates the thermodynamics of the SU(2) Yang–Mills system. The typical, maximal core-size $R_{\text{cor}}(\bar{\lambda})$ of an instable center-vortex loop (see Sec. 6.3), is given as $R_{\text{cor}}(\bar{\lambda}) \sim \frac{1}{m} = \frac{1}{g|\varphi|}$. For $\bar{\lambda}_{c'} \leq \bar{\lambda} \leq \bar{\lambda}_* = 7.428$ we have $g \geq 8.3$ according to Eq. (6.39). Thus $\frac{R_{\text{cor}}(\bar{\lambda})}{|\varphi|^{-1}} \leq 0.12$. As a consequence, collapsing center-vortex loops are not resolved since their core-size is much smaller than the minimal length scale $|\varphi|^{-1}$ of the effective theory for the preconfining phase, and the monopole-antimonopole condensate is spatially homogeneous in the effective theory.

To describe the average effect of tunneling between the two trajectories $\frac{\rho(\bar{\lambda})}{T^4}$ and $\frac{\rho(\lambda)}{T^4}$ for $\bar{\lambda}_* \leq \bar{\lambda} \leq \bar{\lambda}_c$ and $\lambda_* \leq \lambda \leq \lambda_c$, respectively, one may think of the following "droplet" model. Let V and V_{tot} be two three-dimensional volumina such that V is contained in V_{tot}. The thermal probability density $P(V, V_{\text{tot}}, \bar{\lambda})$ for a fraction $\frac{V}{V_{\text{tot}}}$ of space to be filled with condensed magnetic monopoles and antimonopoles is given as

$$P(V, V_{\text{tot}}, \bar{\lambda}) \equiv d(\bar{\lambda})\bar{\lambda}^3 \frac{\exp\left[d(\bar{\lambda})\bar{\lambda}^3(V_{\text{tot}} - V)\right]}{\exp\left[d(\bar{\lambda})\bar{\lambda}^3 V_{\text{tot}}\right] - 1}, \tag{6.44}$$

where $d(\bar{\lambda}) \equiv \Delta(\bar{\lambda})\frac{\bar{\lambda}^3}{(2\pi)^3}$, and Δ is the temperature-dependent (positive) difference between trajectories $\frac{\rho(\bar{\lambda})}{T^4}$ and $\frac{\rho(\bar{\lambda})}{T^4}$. Notice that for $d(\bar{\lambda})\bar{\lambda}^3 V_{\text{tot}} \gg 1$ the probability density $P(V, V_{\text{tot}}, \bar{\lambda})$ ceases to depend on V_{tot}. In the model of Eq. (6.44) and for SU(2) the average polarization number N_p of the U(1) gauge field calculates as

$$\begin{aligned}
N_p(\bar{\lambda}) &= \int_0^{V_{\text{tot}}} dV\, P(V, V_{\text{tot}}, \bar{\lambda}) \left(3\frac{V}{V_{\text{tot}}} + 2\frac{V_{\text{tot}} - V}{V_{\text{tot}}}\right) \\
&= 2 + \int_0^{V_{\text{tot}}} dV\, P(V, V_{\text{tot}}, \bar{\lambda})\frac{V}{V_{\text{tot}}}.
\end{aligned} \tag{6.45}$$

Keeping V_{tot} fixed, this yields $\lim_{\bar{\lambda}\to\bar{\lambda}_*} N_p = \lim_{d\to 0} N_p = \frac{5}{2}$. A similar model for the regime $\bar{\lambda}_{c'} \le \bar{\lambda} \le \bar{\lambda}_*$ shows that N_p increases towards $N_p = 3$ for $\bar{\lambda} \searrow \bar{\lambda}_{c'}$.

6.3 Abrikosov–Nielsen–Olesen (ANO) Vortex Lines and Center-Vortex Loops

We have referred at various places to the concept of a center-vortex loop. Here we would like to be more definite about the properties of these solitonic field configurations. After their discoverers [Abrikosov (1957); Nielsen and Olesen (1973)] topologically stabilized line-like solitons of an Abelian Higgs model are called Abrikosov–Nielsen–Olesen (ANO) vortex lines. These comprise of quanta of magnetic flux determined by their topological charge or winding number. Since we consider Yang–Mills theories free of any external sources there are no isolated charges and sinks for these magnetic flux lines in a situation where point-like magnetic monopoles are condensed. Thus ANO vortex lines must form closed loops. As we shall see, in SU(2) and SU(3) Yang–Mills theory quasistable loops are generated by having nontrivial elements of the magnetic center group act locally on a complex background field, thus the name center-vortex loop. Depending on whether the diameter of the flux core is finite or vanishing, the corresponding soliton rapidly collapses or is quasistable, respectively. In the presence of a stable condensate of magnetic monopoles subject to a finite value of the magnetic coupling g the former situation takes place while quasistable thin center-vortex loops start to form at $T_{c'}$ where $g = \infty$.

6.3.1   *ANO vortex in the Abelian Higgs model*

In this section we investigate key properties of ANO vortex lines in a U(1) Abelian Higgs model. The presentation follows closely the one in [Hofmann (2005)]. Just as the topological defects in the deconfining phase of an SU(2) or SU(3) Yang–Mills theory are screened BPS monopoles, the defects within the monopole-antimonopole condensate of the preconfining phase are screened and closed magnetic flux lines (vortex-loops). Topologically, the existence of stable (unresolvable) magnetic (anti)monopoles in the deconfining phase, where SU(2) or SU(3) is broken to U(1) or U(1)^2 by small-holonomy (anti)calorons, respectively, is guaranteed by the nontrivial homotopy group $\Pi_2(SU(2)/U(1) = S_2) = \mathbf{Z}$ of which only the elements ± 1 are realized.[11] Inside the preconfining phase magnetic vortex loops are associated with flux lines whose cores are composed of magnetic monopoles and antimonopoles. Within a vortex core massive magnetic monopoles and antimonopoles move into opposite directions, and there is a net magnetic current (see Fig. 6.5). At finite values of g the massiveness of (anti)monopoles within the vortex core is a consequence of the associated local restoration of the U(1) [SU(2)] and U(1)^2 [SU(3)] dual gauge symmetry. The magnetic

---

[11]Recall that for the formation of the inert adjoint field ϕ only (anti)calorons of unit topological charge modulus are allowed to contribute which implies that by caloron dissociation only magnetic monopoles of charge modulus one are liberated.

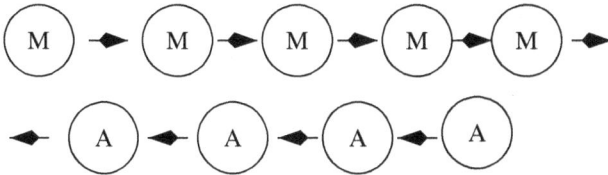

Figure 6.5. Microscopics of the core of a center-vortex-loop segment. Monopoles (M) and antimonopoles (A) of unit magnetic charge modulus move in opposite directions [Del Debbio *et al.* (1997)].

flux, which penetrates a spatial hyperplane perpendicular to the direction of monopole or antimonopole motion, is by Stoke's theorem measured by a line integral $g \oint_C dz_\mu A_\mu^D$ along a circular curve C with infinite radius lying in this hyperplane. Here A_μ^D denotes the (fundamental) gauge field with respect to the dual gauge group U(1) [SU(2)] or either factor in U(1)^2 [SU(3)] generated by the chains of monopoles and antimonopoles. If we choose to evaluate the line integral in a linear Lorentz covariant gauge then we see that the contribution to $dz_\mu A_\mu^D$ of each *moving* monopole or antimonopole is that of a static monopole or antimonopole since the perpendicular part of the gauge field is invariant under boosts along the vortex axis. Thus the state of motion of each monopole or antimonopole is irrelevant for its effect on the total magnetic flux carried by the vortex as long as the net motion of all monopoles or antimonopoles in a given segment defines a definite direction of the vortex axis. In other words, the only quantity that determines the magnetic flux of the vortex line is the *charge* of a monopole.

Topologically argued, ANO vortices take place in the preconfining phase because of a nontrivial homotopy group $\Pi_1(U(1) = S_1) = \mathbf{Z}$. Let us now find an analytic expression for a particular ANO vortex configuration. An SU(2) mesoscopic ANO vortex is governed by the action Eq. (6.29) when neglecting the potential for the field φ. [This potential measures the energy density of the *macroscopic* ground state and thus must be subtracted when discussing the typical energy of a solitonic configuration on a mesoscopic level, that is, for a spatial resolution larger than $|\varphi|$. Notice, however, that the potential of Eq. (6.27) determines the boundary condition at spatial infinity.] The following cylindrically symmetric (with cylinder axis along the x_3 direction) and static ansatz for the gauge field a_μ^D was made in Nielsen and Olesen (1973):

$$a_4^D = 0, \quad a_i^D = \epsilon_{ijk}\hat{r}_j e_k A(r), \tag{6.46}$$

where $\hat{r}$ is a radial unit vector in the $x_1 x_2$-plane, $r = \sqrt{x_1^2 + x_2^2}$ and $\vec{e}$ is a unit vector along the x_3 direction. Writing $\varphi = |\varphi|(r)\exp[i\theta]$, the equations of motion for $|\varphi|(r)$ and $A(r)$ read

$$-\frac{1}{r}\frac{d}{dr}\left(r\frac{d}{dr}|\varphi|\right) + \left(\frac{1}{r} - gA\right)^2 |\varphi| = 0, \tag{6.47}$$

$$-\frac{d}{dr}\left(\frac{1}{r}\frac{d}{dr}(rA)\right) + g|\varphi|^2\left(gA - \frac{1}{r}\right) = 0. \tag{6.48}$$

We keep in mind that $|\varphi|(r \to \infty) = \sqrt{\frac{\bar{\Lambda}^3\beta}{2\pi}}$. Let us now seek solutions to the system of second-order equations (6.47) and (6.48) such that $|\varphi|$ and $A(r)$ satisfy first-order equations. Presuming that

$$\frac{d}{dr}|\varphi| = \left(\frac{1}{r} - gA\right)|\varphi| \tag{6.49}$$

and substituting Eq. (6.49) into Eq. (6.47), we observe that

$$A = -r\frac{d}{dr}A. \tag{6.50}$$

The solution to Eq. (6.50) is

$$A(r) = \frac{\text{const}}{r}. \tag{6.51}$$

Substituting Eq. (6.51) into Eq. (6.48), the constant in Eq. (6.51) is determined to be $\frac{1}{g}$. Equation (6.49) is solved for $r > 0$ by

$$|\varphi|(r) \equiv \sqrt{\frac{\bar{\Lambda}^3\beta}{2\pi}}. \tag{6.52}$$

Thus we have for $r > 0$ found a solution to the system (6.47) and (6.48) which solves the first-order system of (6.49) and (6.50), which possesses one unit of magnetic flux $F_v(r)$, and which exhibits a vanishing vortex core. The magnetic flux $F_v(r)$ evaluates as

$$F_v(r) \equiv \frac{2\pi}{g} = \oint_C dz_\mu \, a_\mu^D , \tag{6.53}$$

where C is a circular curve of radius r embedded in the x_1x_2-plane and centered at $x_1 = x_2 = 0$. The energy density $\rho_v(r)$ of this solution reduces to that of the magnetic[12] field $H(r) = \frac{1}{2\pi r}\frac{d}{dr}F_v(r)$:

$$\rho_v(r) = \frac{1}{2}H^2(r) \equiv 0 \quad (r > 0). \tag{6.54}$$

The pressure $P_v(r)$ is outside the vortex core given as

$$P_v(r) = -\frac{1}{2}\frac{\bar{\Lambda}^3\beta}{2\pi}\frac{1}{g^2r^2} \quad (r > 0). \tag{6.55}$$

Equation (6.55) is the mesoscopic reason for the results of Secs. 6.2.1 and 6.2.2 expressing the dominance of the thermal ground state and thus negativity of the total pressure exerted by the unresolvably vortex-loop "contaminated" monopole-antimonopole condensate and its massive dual gauge-mode excitations. Because of the *negative* pressure in Eq. (6.55) vortex-*loops* start to collapse as soon as they are created at finite coupling g. (The pressure is more negative inside than outside of the vortex-loop.) Notice that in the limit $g \to \infty$ we have $P_v(r) \to 0$: For temperatures below $T_{c'}$ vortex-loops do exist as particle-like, quasistable excitations. Thus a magnetic flux loop acquires the status of a (two-fold degenerate) degree of freedom. Since one can understand the process of flux creation in terms of the local action of a nontrivial magnetic center element of SU(2) or

---

[12]By Stoke's theorem the magnetic field H at $r = 0$ must be proportional to a two-dimensional δ-function. Thus the energy per unit vortex length diverges on the configuration (6.51) and (6.52).

SU(3) onto a scalar background field the name "center-vortex loop" suggests itself.

Let us now discuss more realistic solutions to Eqs. (6.47) and (6.48) than the ones in Eqs. (6.52) and (6.51) which possess infinite energy per vortex length. For finite values of energy per vortex length only approximate analytical solutions to the second-order system (6.47) and (6.48) are known [Nielsen and Olesen (1973)]. Assuming $|\varphi|$ to be constant, which is reasonable when observing the vortex line sufficiently far away from its core, the solution to Eq. (6.48) reads

$$A(r) = \frac{1}{gr} - |\varphi|K_1(g|\varphi|r)$$

$$\rightarrow \frac{1}{gr} - |\varphi|\sqrt{\frac{\pi}{2g|\varphi|r}}\,\exp[-g|\varphi|r] \quad \left(r \gg \frac{1}{g|\varphi|}\right), \quad (6.56)$$

where K_1 is a modified Bessel function of the first kind. Notice that the $1/r$ divergence at $r = 0$ of the solution in Eq. (6.51) is absent in the configuration in Eq. (6.56). Now $|\varphi|$ is not constant inside the vortex core but smoothly approaches zero for $r \rightarrow 0$. So there is a gradient contribution $\frac{E_{\varphi,v}}{2\pi R}$ to the total energy per vortex length $\frac{E_v}{2\pi R}$. Here R denotes the typical radius of a vortex loop. Let us first calculate the magnetic energy per vortex length $\frac{E_{m,v}}{2\pi R}$. One has [Nielsen and Olesen (1973)]

$$\frac{E_{m,v}}{2\pi R} = \frac{1}{2}\int_0^\infty dr\, 2\pi r H^2(r) = \pi|\varphi|^2\int_0^\infty dy\, K_0^2(y)y = \frac{\pi}{2}|\varphi|^2, \quad (6.57)$$

where $|\varphi|$ is given in Eq. (6.52). The gradient contribution $\frac{E_{\varphi,v}}{2\pi R}$ is comparable to $\frac{E_{m,v}}{2\pi R}$. Thus the typical energy E_v of the vortex loop is obtained by multiplying $\frac{E_{m,v}}{2\pi R} + \frac{E_{\varphi,v}}{2\pi R}$ with the typical circumference $2\pi R \sim \frac{1}{g|\varphi|}$ of the loop. We have

$$E_v \sim 2\frac{\pi}{2}|\varphi|^2 \times \frac{1}{g|\varphi|} = \pi\frac{|\varphi|}{g}. \quad (6.58)$$

From Eq. (6.58) we conclude that vortex loops become massless in the limit $g \to \infty$. For $r \gg \frac{1}{g|\varphi|}$ the total pressure $P_v(r)$ of the vortex configuration still is given by Eq. (6.55). Therefore, for finite values of the coupling g, which take place throughout the preconfining phase ($\bar{\lambda}_{c'} < \bar{\lambda} \leq \bar{\lambda}_c$), also realistic vortex loops collapse as soon as they are created.

Problems

6.1. Perform the limit $e \to \infty$ in Eq. (6.5) explicitly.

6.2. Why does the limit $\mu_c \to 0$ in Eq. (6.15) yield the identity $2\zeta(3) = 2\zeta(3)$ where $\zeta(z)$ is Riemann's zeta function?

6.3. What is the first-order (BPS) differential equation for the field φ that is associated with Eq. (6.25)?

6.4. Derive Eqs. (6.35).

6.5. Taking the mass m of the dual gauge mode as an order parameter for the dynamical breaking of U(1) [SU(2)] and U(1)^2 [SU(3)] and postulating that $m = K(T_c - T)^\nu$ for $T \overset{<}{\sim} T_c$, where K and ν are real constants, show analytically that mean-field critical exponents, $\nu = \frac{1}{2}$, take place on the preconfining side of the deconfining–preconfining phase transition. What is the value of K? *Hint:* Investigate the small-mass limit of the evolution equation (6.39) analytically by casting it into a form analogous to Eq. (5.71).

6.6. Show that the jump in energy density across the deconfining-preconfining phase boundary is given as $\Delta(\bar{\lambda}_c) \equiv \frac{\rho(\bar{\lambda}_c+0) - \rho(\bar{\lambda}_c-0)}{T_c^4} = \frac{4}{3}\frac{\pi^2}{30}$ for SU(2) and $\Delta(\bar{\lambda}_c) = \frac{8}{3}\frac{\pi^2}{30}$ for SU(3). *Hint:* Consider Eq. (5.6.4) in the limit $\lambda \searrow \lambda_c$, Eq. (6.40) in the limit $\bar{\lambda} \nearrow \bar{\lambda}_c$, and also Eq. (6.35).

6.7. Construct for SU(2) a droplet model for the regime $\bar{\lambda}_{c'} \leq \bar{\lambda} < \bar{\lambda}_*$ in analogy to that in Eq. (6.44). Create a plot within this regime of the average number of polarization states for the dual gauge mode as a function of $\bar{\lambda}$.

References

Abrikosov, A. A. (1957). On the magnetic properties of superconductors of the second group, *Sov. Phys. JETP*, **5**, 1174.

Andersen, J. O. (1997). The electric screening mass in scalar electrodynamics at high temperature, *Z. Phys. C*, **75**, 147.

Del Debbio, L., Faber, M., Greensite, J., and Olejnik, S. (1997). Center dominance, center vortices, and confinement, *proc. NATO Adv. Res. Workshop on Theor. Phys.* New Developments in Quantum Field Theory in Zakopane, 47.

Giacosa, F. and Hofmann, R. (2007). Nonperturbative screening of the Landau pole, *Phys. Rev. D*, **77**, 065022.

Huang, K. (1963). Statistical Mechanics, John Wiley & Sons, Inc. New York, 1963, Sec. 9.6 and 12.3.

Hofmann, R. (2005). Nonperturbative approach to Yang-Mills thermodynamics, *Int. J. Mod. Phys. A*, **20**, 4123; *Erratum-ibid. A* **21**, 6515.

Hofmann, R. (2007). Yang-Mills thermodynamics, arXiv:0710.0962v3.

Nieto, M. M. (1970). Exact state and fugacity equation for the ideal quantum gases, *J. Math. Phys.*, **11**, 1364.

Nielsen, H. B. and Olesen, P. (1973). Vortexline models for dual strings, *Nucl. Phys. B*, **61**, 45.

Chapter 7
The Confining Phase

The confining phase of an SU(2) or an SU(3) Yang–Mills theory is the one where gauge modes no longer propagate: They are thermo-dynamically not excited because their quasiparticle masses diverge. The onset of the confining phase is characterized by an explosive decay of the preconfining ground state into nearly massless center-vortex loops, that is, ANO vortex loops in the limit of vanishing vortex-core size. By subsequent twisting, energy is passed from this vortex-loop sector of no selfintersections to genuinely massive vortex-loop excitations characterized by finite numbers of selfin-tersections. Disregarding the internal excitability and instability of massive vortex-loops, the process of particle creation builds up an infinite tower of center-vortex loops which are classified accord-ing to their selfintersection number n and (knot-) topology. Each selfintersection contains an isolated magnetic (anti)monopole.[1] The distribution of the sign of charge associated with these magnetic (anti)monopoles[2] and the two-fold directedness of magnetic flux around a given vortex loop provide for additional ways of classi-fication. Even when neglecting spatial translations of rigid vortex loops, the associated hypothetical partition function diverges at any value of temperature because the strong increase of the density of states with energy beats the Boltzmann suppression factor. Thus

---

[1]There is a *factorial* degeneracy of solitons of mass $\sim n\Lambda$.
[2]There is a 2^n-fold degeneracy with respect to the distribution of charge $\pm$.

311

a thermodynamical treatment contradicts itself. One may, however, argue that internal excitations and interactions between distinct solitons enforce a much stronger dependence of the energy on n than the simple model of rigid solitons suggests. To treat this situation, Borel resummation of the analytically continued thermodynamical pressure due to rigid and free solitons is required. Continuation back to the physical region indicates how a departure from thermal behavior emerges as the temperature parameter is increased: The preconfining-confining phase transition truly is nonthermal and of the Hagedorn type [Hagedorn (1965)].

As it seems, interesting physics takes place in the confining phase: a change of the quantum statistics from bosonic to fermionic, a vanishing ground-state energy density and pressure, a mass gap in the excitation spectrum at low temperature, the emergence of order at a distinct, finite noise level in the sector with one selfintersection. Chapter 8 will address these features. Definite applications to physical systems are presently not discussed. However, the idea that neutral and charged leptons are associated with single and one-fold selfintersecting center-vortex loops in the confining phases of SU(2) Yang–Mills theories will surface occasionally.

7.1   Decay of the Preconfining Ground State

7.1.1   *The center-vortex condensate* Φ

In the preconfining phase closed lines of magnetic flux[3] form along regions where the dissociation of large-holonomy calorons/ anticalorons is collectively characterized by a net direction: Inside the cores of vortex loops magnetic monopoles travel in opposite direction to their antimonopoles (see Fig. 6.5). In the preconfining phase vortex loops are unresolved and short-lived because

---

[3]Since there are no isolated magnetic charges in the preconfining phase (monopoles and antimonopoles are condensed) these flux lines cannot terminate in sinks or sources, and thus they form closed loops.

curvature of the vortex line at any point on the vortex induces a motion of this point along the normal to the vortex line thus causing the vortex loop to shrink (collapse). Since no condensate of vortex loops with $n = 0$ can thus form the discrete *magnetic* center symmetries Z_2 [for SU(2)] and Z_3 [for SU(3)] are left unbroken in the preconfining phase. Recall from Sec. 5.1.5 that in the deconfining phase the discrete *electric* center symmetry Z_2 [for SU(2)] and Z_3 [for SU(3)] is induced by singular *temporal* gauge transformations. In the confining phase magnetic center transformations are induced by singular gauge transformations of the dual gauge field a_μ^D along closed spatial contours whose minimal surfaces are pierced by center-vortex loops. The electric center symmetry is restored when temperature is lowered through $T_{c'}$. (The Polyakov-loop expectation vanishes). The condensation of center-vortex loops with $n = 0$ sets in for $T \searrow T_{c'}$ because these solitons become quasistable and massless. Thus they qualify as the constituents of a new ground state which macroscopically is characterized by a complex and scalar field Φ. This condensation introduces a discrete degeneracy with respect to Φ's phase which signals the dynamical breaking of the magnetic center symmetries Z_2 and Z_3.

Let us be more definite about the emergence of the field Φ. The 't Hooft-loop operator $\hat{\Phi}(\vec{x}, C)$ ['t Hooft (1978)] is defined as the exponential of the magnetic flux of the dual gauge field a_μ^D through the minimal surface M_C spanned by an oriented and closed spatial curve C centered at the point $\vec{x}$:

$$\hat{\Phi}(\vec{x}, C) \propto \exp\left[ig \oint_C dz_i\, a_i^D\right]. \tag{7.1}$$

Thus the expectation Φ of $\hat{\Phi}$ changes phase in discrete units under singular gauge transformations of a_μ^D along the contour C. Only those singular gauge transformations are admissible which render $\exp[ig \oint_C dz_i\, a_i^D]$ group elements of SU(2) or SU(3) that are represented by degrees of freedom at $T = T_{c'}$ and below. But, as we have discussed in Sec. 6.3.1, for SU(2) these are two-fold directionally degenerate magnetic fluxes. Thus only those singular gauge transformations, which change Φ by the nontrivial SU(2) center element -1 are admissible. For SU(3) there are two independent two-fold directionally degenerate magnetic fluxes which change Φ by the nontrivial SU(3) center elements $\exp\left[\pm\frac{2\pi}{3}i\right]$. Each center jump of Φ is associated with an extra quantum of magnetic flux due to a center-vortex loop piercing M_C in addition to the multitude of vortex loops which had generated a finite value of Φ to begin with. The process of having an additional vortex pierce M_C proceeds in real time. Thus the *dynamics* associated with Φ phase changes is described by the time depedence of a complex scalar field.[4]

Consider now a spatial circle of infinite radius $S_1^{R=\infty}$ centered at $\vec{x}$. The thermally averaged flux $F_{\pm,0;\text{th}}$ through $M_{S_1^{R=\infty}}$ of a pair of a center-vortex loop and its flux-reversed partner, both at rest, is for vanishing core-size and mass ($\bar{\lambda} \to \bar{\lambda}_{c'}, g \to \infty$) given as

$$\lim_{g \to \infty} F_{\pm,0;\text{th}} = 4\pi \int d^3 p\, \delta^{(3)}\,(\vec{p})\, n_B(\beta_{c'} 2\, E_v(\vec{p}, \bar{\lambda}_{c'}))\, F_{\pm,0} = \begin{cases} 0 \\ \pm\dfrac{\bar{\lambda}_{c'}^{3/2}}{\pi}, \end{cases}$$

(7.2)

where $E_v(0, \bar{\lambda}) \sim \pi\frac{|\varphi(\bar{\lambda})|}{g}$ is the typical mass of a single center-vortex loop at temperature $\bar{\lambda}$ [see Eq. (6.58)]. The case $\lim_{g \to \infty} F_{\pm,0;\text{th}} = 0$ occurs if $M_{S_1^{R=\infty}}$ is pierced an even and an odd number of times by the flux lines of each center-vortex loop. The other two cases take place if $M_{S_1^{R=\infty}}$ is pierced an odd number of times by the flux line of one center-vortex loop and an even number of times by the flux line of the partner. Notice the use of the Bose function n_B for the system

---

[4]Neglecting dynamical aspects of the phase changes, which are, however, essential in understanding the process of relaxing the center-vortex condensate to zero energy density, one could, alternatively describe the SU(2) [SU(3)] situation by one [two independent] real spin variable(s). This would yield an imaginary-time statistical model.

of two center-vortex loops of opposite flux: Even though we believe that each vortex loop should be interpreted as a spin-1/2 fermion (two polarizations also in the case of selfintersections [Reinhardt (2002)]) the system is of total spin zero. It is obvious that in the SU(2) case an identification of $\pm(\bar{\lambda}_{c'}^{3/2}/\pi)$ takes place which is not true for SU(3). Properly normalized, the discrete values in Eq. (7.2) belong to values of the phase of Φ. A phase jump is associated with the creation of a single center-vortex loop. For SU(2) these jumps proceed from 0 to $i\pi$ (vortex) and from $i\pi$ to 0 (flux-reversed vortex). For SU(3) the processes 0 to $\pm i\frac{2\pi}{3}$ and $\pm i\frac{2\pi}{3}$ to 0 create and destroy respectively two distinct species of center-vortex loops. Since the spatial extent of a given center-vortex loop is unresolvable ($g \nearrow \infty$, see discussion in Sec. 6.2.3) and since in the condensate the distance between a center-vortex loop and its flux-reversed partner is zero, the vanishing-curvature situation of $S_1^{R=\infty}$ is saturated at a finite curvature already: $S_1^{R=\infty} \rightarrow S_1^{R<\infty}$.

7.1.2   Real-time relaxation of Φ to zero energy density and pressure

So far we have given topological and thermodynamical arguments on why the onset of condensation of nearly massless center-vortex loops is to take place at $T_{c'}$. On the other hand, it is clear that the decay of the monopole condensate into (twisted, that is, also selfintersecting) center-vortex loops is highly nonthermal and thus cannot be described within the framework of imaginary-time thermodynamics. To elucidate further the process of the decay of the zero-entropy situation of a monopole-antimonopole condensate into a tower of center-vortex-loop species accompanied by a new ground state is the subject of the following discussion.

Let us first argue for the truth of the following statement: The process of decay of the monopole-antimonopole condensate(s) and the formation of the center-vortex condensate are described by real-time dynamics of the scalar and complex order-parameter Φ for the dynamical breakdown of magnetic center symmetry. This dynamics is subject to the following potentials

$$V(\Phi) = \left(\frac{\tilde{\Lambda}^3}{\Phi} - \tilde{\Lambda}\Phi\right)^* \left(\frac{\tilde{\Lambda}^3}{\Phi} - \tilde{\Lambda}\Phi\right) \quad \text{[for SU(2)]}, \qquad (7.3)$$

$$V(\Phi) = \left(\frac{\tilde{\Lambda}^3}{\Phi} - \Phi^2\right)^* \left(\frac{\tilde{\Lambda}^3}{\Phi} - \Phi^2\right) \quad \text{[for SU(3)]}, \qquad (7.4)$$

where $\tilde{\Lambda} \sim 2^{1/3}\,\bar{\Lambda}$ [for SU(2)] and $\tilde{\Lambda} \sim \bar{\Lambda}$ [for SU(3)], and a canonical kinetic term in Φ's effective action. The reasons are as follows. At the onset of center-vortex loop condensation thermal equilibrium is maintained at zero entropy (maximally negative pressure). Periodic BPS-saturated trajectories[5] along Euclidean time, describing the onset of vortex-loop condensation, exist for the potentials in Eqs. (7.3) and (7.4) (see [Hofmann (2000)]). As in the other two phases, this periodicity is due to the pole-term in the "square-root" of V which allows the field Φ to exhibit nontrivial winding number. (The "superpotential" W, defined by the relation $V(\Phi) = |\partial_\Phi W|^2$, has a branch cut, see [Hofmann (2000)].) Furthermore, the potential V needs to satisfy the following requirements: (i) invariance under nontrivial and local [center-element factor in Eq. (7.6) depends on position $\tilde{x}$] magnetic center transformations

$$\Phi \to -\Phi \quad \text{[for SU(2)]} \quad \text{and} \quad \Phi \to \exp\left(\pm\frac{2\pi i}{3}\right)\Phi \quad \text{[for SU(3)]}$$

$$(7.6)$$

---

[5]A canonical kinetic term in Φ's effective action

$$S = \int d^4x \left(\frac{1}{2}(\partial_\mu \Phi)^* \partial^\mu \Phi - \frac{1}{2}V\right) \qquad (7.5)$$

is inherited from the effective action for the field φ describing the monopole-antimonopole condensate of the preconfining phase: At the onset of vortex condensation thermal equilibrium prevails, and the vortex condensate *coincides* with the monopole-antimonopole condensate.

only (no larger symmetry); (ii) dynamical realization of the latter (flux creation, negative tangential curvature for jump-like behavior in real-time corresponding to tunneling processes through the potential barrier along a nearly circular, concentric trajectory in imaginary time); (iii) the minima of V are center-degenerate and at zero energy density (center-vortex loops with $n = 0$ in condensate do not interact and are massless); and (iv) as in the other two phases, a single mass-scale $\tilde{\Lambda}$ enters V. It is easy to check (see also [Hofmann (2005)]), that modulo U(1)-invariant rescalings and up to the addition of terms of the form $\Delta V = \kappa(\tilde{\Lambda}^2 - \tilde{\Lambda}^{-2(m-1)}(\bar{\Phi}\Phi)^m)^{2k}$, $(\kappa > 0, k = 1, 2, 3, \dots, m)$, which only increase the curvature of V at its minima, the potentials in Eqs. (7.3) and (7.4) are uniquely determined by conditions (i) through (iv). Demanding at the onset of the condensation that the (negative) pressure be continuous in the Euclidean formulation, the above quoted relation $\tilde{\Lambda} \sim 2^{1/3} \bar{\Lambda}$ [for SU(2)] and $\tilde{\Lambda} \sim \bar{\Lambda}$ [for SU(3)] follows. Moreover, writing $\Phi = |\Phi| \exp\left[i\frac{\theta}{\Lambda}\right]$, one has

$$\left.\frac{\partial_\theta^2 V(\Phi)}{|\Phi|^2}\right|_{\Phi_{\min}} = \left.\frac{\partial_{|\Phi|}^2 V(\Phi)}{|\Phi|^2}\right|_{\Phi_{\min}} = \begin{cases} 8 & [\text{SU(2)}] \\ 18 & [\text{SU(3)}], \end{cases} \qquad (7.7)$$

where $\Phi_{\min} = \pm\tilde{\Lambda}$ [for SU(2)] and $\Phi_{\min} = \tilde{\Lambda} \exp\left[\frac{2\pi ik}{3}\right]$ ($k = 0, 1, 2$), [for SU(3)]. Since $|\Phi|_{\min}$ is the scale of maximal resolution after Φ has settled into one of its minima we conclude that the field Φ no longer fluctuates.[6] This is consistent with the fact that Φ is a condensate of interaction-free zero-energy configurations which by themselves do not propagate. As a consequence, no spontaneous tunneling to another minimum (flux creation) takes place once the field Φ has settled into $\Phi_{\min}$. In Fig. 7.1 plots of the dimensionless quantity $V/\tilde{\Lambda}^4$ are shown for SU(2) and SU(3) as functions of the normalized center-vortex condensate $\Phi/\tilde{\Lambda}$.

---

[6]The mass $m_{\delta\Phi}$ of possible, propagating excitations, described by small oscillations $\delta\Phi$ about the minima positions $\Phi_{\min}$ of V, is much larger than the resolution $|\Phi|_{\min}$. Thus those excitations cannot be created.

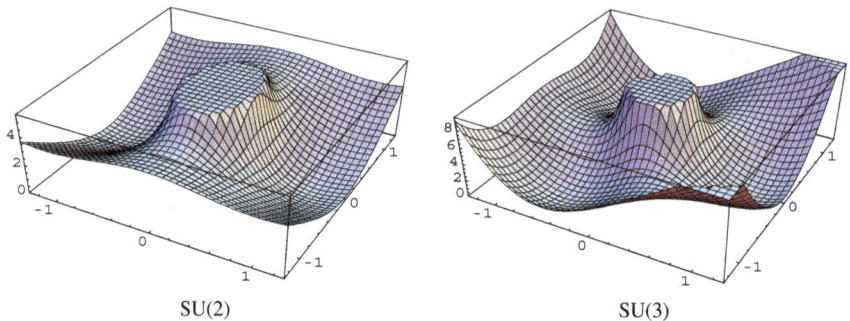

Figure 7.1. The dimensionless potential $V/\tilde{\Lambda}^4$ as a function of the normalized center-vortex condensate $\Phi/\tilde{\Lambda}$. The flat central regions correspond to the onsets of the (cut-off) poles. Notice the regions of negative tangential curvature in between the minima.

7.1.3 *Multiplicity of center-vortex loops with n selfintersections*

Let us now discuss which possibilities exist for the generation of center-vortex loops of nonvanishing selfintersection number n during the relaxation of the field Φ to one of the center-degenerate minima of the potential V. Recall that the attainment of one of these minima is associated with the process of formation of a stable[7] condensate formed of paired (zero-energy, that is, pointlike), magnetic center-vortex loops of $n = 0$ while shaking off energy from the preconfining ground state (monopole-antimonopole condensate) by the ultimate generation of energetic (selfintersecting, $n > 0$) center-vortex loops out of the $n = 0$ sector by twisting. Due to their two-fold directional degeneracy all of these propagating excitations are interpreted as massless or massive spin-1/2 fermions.[8]

---

[7]If the matter content of the universe were to be described by a single SU(2) or SU(3) Yang–Mills theory then field Φ could not move away from such a minimum of V once it has reached it. Since the universe certainly comprises a more complex gauge-group structure one may assign a noise level to this relaxation process which is maintained by other gauge-theory dynamics (see Sec. 7.3) even after Φ has reached a minimum of V.

To keep the discussion void of simple but irritating case differentiations we discuss here the case of SU(2) Yang–Mills theory only. The (fermionic) solitons, generated in the course of the process of turning a monopole-antimononopole condensate into a condensate of massless, pointlike, and thus interaction-free center-vortex loops, are classified mainly by their topology which is related to the number n of selfintersections. At finite mass and thus finite spatial extent of these solitons only contact interactions occur due to the complete decoupling of propagating gauge modes in the confining phase: Collisions of center-vortex loops with $n = 0$ lead to twisting and merging and thus to the creation of selfintersections. These processes convert kinetic energy into mass. The very process of thermalization/virialization proceeds via instable high-mass excitations: Annihilations of oppositely charged intersection points locally elevate the energy density of the field Φ. Subsequently, the field Φ ejects this energy by phase jumps and an increase of its modulus, thereby re-creating virialized center-vortex loops. The following discussion on how this can quantitatively be described in a collective way is borrowed from [Hofmann (2007)].

To start with, let us determine the multiplicity M_n of solitons with n selfintersections and naked mass $n\tilde{\Lambda}$. Again, by "naked mass" we mean that only the masses of the magnetic monopole/antimonopoles residing in the cores of selfintersections (one in each) are added up to describe the mass of the soliton, while possible internal excitations of the soliton are neglected. One has

$$M_n \equiv 2\,N_n\,C_n. \tag{7.8}$$

---

[8]Each selfintersection carries an isolated magnetic monopole at the core of the associated flux eddy whose mass is roughly given by the asymptotic value of its Higgs field, that is, by $\tilde{\Lambda}$. A microscopic approach to the statistics of the stable $n = 1$ sector investigates the behavior under exchange of two such solitons in a gauge where the Dirac strings of the isolated magnetic monopoles in the cores of the selfintersections point in opposite directions.

Here the factors of 2, of N_n, and of C_n stand for the spin-1/2 multiplicity, the number of distinct (knot-)topologies, and multiplicity associated with the distribution of monopole/antimonopole charge over the selfintersections of the soliton, respectively. For C_n one obtains (two possibilities at each selfintersection point)

$$C_n = 2^n. \tag{7.9}$$

In Fig. 7.2 we list soliton (knot-)topologies for $n \leq 3$. It is obvious that N_n represents the number of connected vacuum bubble diagrams with n vertices arising in the perturbative expansion of the effective action of a $\lambda\phi^4$ quantum field theory. Modulo algebraic-in-n factors and for large n the number of such diagrams in a $\lambda\phi^{2\mathcal{M}}$-theory was found by Bender and Wu (1976) to be

$$[n(\mathcal{M} - 1)]! \left[\frac{2\mathcal{M}}{\mathcal{M} - 1} \right]^{\mathcal{M}n} (\mathcal{M} - 1)^n. \tag{7.10}$$

Thus for $\mathcal{M} = 2$ we have

$$N_n \sim n!16^n, \tag{7.11}$$

where the $\sim$ sign signals a dependence modulo a factor of the form $\sum_{l=1}^{L} a_l n^l$ ($L < \infty$ and a_l integer).

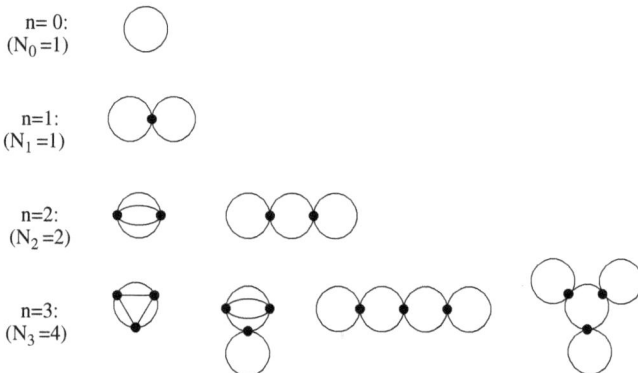

Figure 7.2. A list of the soliton (knot-)topologies for $n \leq 3$. The number N_n is the number of connected vacuum bubbles with n vertices in a $\lambda\phi^4$ quantum field theory.

7.2 Nonthermal Pressure

In the hypothetical case of absolutely stable excitations at arbitrarily large selfintersection number n, which once and forever were thermalized to a given temperature by a single relaxation process of the field Φ to one of the minima of the potential V of Eq. (7.3), there is no ground state contribution to the pressure. Here we are concerned with this hypothetical situation which allows for strong simplifications compared to the direct analytical treatment of the full physical situation and, as we will see below, in principle captures sufficient information to recover the latter collectively. Neglecting the excitability of internal degrees of freedom *within* and the instability *of* a given soliton and disregarding (contact) interactions *between* solitons, the total pressure is represented in terms of an asymptotic series in powers of a dimensionless coupling[9] $\lambda \equiv \exp(-\tilde{\Lambda}/T)$. Notice that λ is strictly smaller than unity for $T \leq \tilde{\Lambda}$. For $T \nearrow \tilde{\Lambda}$ a highly nonthermal situation (Hagedorn transition) occurs. The latter goes with a condensation of the selfintersection points of densely packed center-vortex loops into a new ground state: The monopole-antimonopole condensate of the preconfining phase.

7.2.1 *Naive thermodynamical estimate*

Naively, that is, neglecting internal excitations and the instability of interactions between solitions, the total pressure P at temperature T is represented by an asymptotic series (sum over partial pressures of fermionic spin-1/2 particles associated with the sectors of n selfintersections)

$$P = \sum_{n=0}^{\infty} P_n \equiv T \sum_{n=0}^{\infty} M_n \int_0^{\infty} dp \, \frac{p^2}{2\pi^2} \, \log\left(1 + e^{-\beta\omega_n}\right), \qquad (7.12)$$

---

[9]In Sec. 7.2 we use the symbols λ and $\bar{\lambda}$ with a different meaning than in the other parts of the book. Apologies to the reader.

where $\beta \equiv \frac{1}{T}$, $\omega_n \equiv \sqrt{p^2 + (n\tilde{\Lambda})^2}$, and M_n is given by Eqs. (7.8), (7.9), and (7.11). Explicitly, one has

$$
\begin{aligned}
P_{as} &= \frac{\tilde{\Lambda}^4}{2\pi^2} \hat{\beta}^{-4} \left(\frac{7\pi^4}{180} + \hat{\beta}^3 \sum_{n \geq 1} M_n \int_0^\infty dx\, x^2 \log\left(1 + e^{-\hat{\beta}\sqrt{n^2+x^2}} \right) \right) \\
&\leq \frac{\tilde{\Lambda}^4}{2\pi^2} \hat{\beta}^{-4} \left(\frac{7\pi^4}{180} + \hat{\beta}^3 \sum_{n \geq 1} M_n \int_0^\infty dx\, x^2 \, e^{-\hat{\beta}\sqrt{n^2+x^2}} \right) \\
&= \frac{\tilde{\Lambda}^4}{2\pi^2} \hat{\beta}^{-4} \left(\frac{7\pi^4}{180} + \hat{\beta}^2 \sum_{n \geq 1} M_n \, n^2 \, K_2(n\hat{\beta}) \right) \\
&\sim \frac{\tilde{\Lambda}^4}{2\pi^2} \hat{\beta}^{-4} \left(\frac{7\pi^4}{180} + \sqrt{\frac{\pi}{2}}\, \hat{\beta}^{\frac{3}{2}} \sum_{n \geq 1} M_n \, \lambda^n \, n^{\frac{3}{2}} \right) \\
&\leq \frac{\tilde{\Lambda}^4}{2\pi^2} \hat{\beta}^{-4} \left(\frac{7\pi^4}{180} + \sqrt{2\pi}\, \hat{\beta}^{\frac{3}{2}} \sum_{l=0}^{L} a_l \sum_{n \geq 1} (32\lambda)^n \, n! \, n^{\frac{3}{2}+l} \right).
\end{aligned}
\tag{7.13}
$$

In Eq. (7.13) we have defined $\hat{\beta} \equiv \frac{\tilde{\Lambda}}{T}$, $\lambda \equiv e^{-\hat{\beta}}$, and K_2 denotes the modified Bessel function of the second kind. The first $\leq$ sign holds strictly for the linear truncation of the expansion of the logarithm about unity. The transition from the third to the fourth line in Eq. (7.13) actually appeals to the large-n behavior of $K_2(n\hat{\beta})$, that is, terms of order $(\hat{\beta}n)^{-1}$ have been neglected in the expansion of the nonexponential factor in $K_2(n\hat{\beta})$. This is justified since the sum is dominated by large n [compare with Eqs. (7.8), (7.9), and (7.11)]. The coefficients a_l, which determine the algebraic factor in N_n (see Eq. (7.11)), are unknown at present. From the λ dependence of the sought-after physical pressure it will become clear later that the very concept of thermalization ceases to be useful for $T \nearrow \tilde{\Lambda}$ or $\lambda \nearrow e^{-1}$ due to approach of the Hagedorn transition.

7.2.2   *Borel resummation and analytic continuation*

Our strategy to elute the physical pressure from the asymptotic expansion in Eq. (7.13) is to perform a Borel transformation of this series. Subsequently, we investigate the region of analyticity of the Borel transform. Only for $\lambda < 0$ is the inverse Borel transform real-analytic in λ. However, we can analytically continue our results obtained for $\lambda < 0$ to $\lambda = \mathrm{Re}\,\lambda \pm i0$ ($\mathrm{Re}\,\lambda \geq 0$) thus obtaining a prediction of the theory for the physical real part of the pressure at physical coupling. As we will see below, we obtain by virtue of analyticity an expression for the physical pressure that is unique modulo the jumps of its imaginary part. Since a partition function within the region of convergence never exhibit an imaginary part, our results indicate a departure from thermodynamical equilibrium. Interestingly, this departure grows with an increase of the formal temperature parameter. Our result for the physical pressure, which relies on the naive expression of Eq. (7.13), collectively sums up all internal excitations of n-fold selfintersecting center-vortex loops and accounts for all decay and creation processes of these solitons.

The Borel transformation removes the factors of $n!$ in the coefficients of the power series in Eq. (7.13). One has

$$\bar{P}_{\mathrm{mass}}(\bar{\lambda}) \equiv \sum_{l=0}^{L} a_l \sum_{n \geq 1} \bar{\lambda}^n \, n! \, n^{\frac{3}{2}+l} \xrightarrow{\text{Borel}} B_{\bar{P}_{\mathrm{mass}}}(\bar{\lambda}) \equiv \sum_{l=0}^{L} a_l \sum_{n \geq 1} \bar{\lambda}^n \, n^{\frac{3}{2}+l},$$

$$(7.14)$$

where $\bar{\lambda} \equiv 32\lambda$. A sum over n as in Eq. (7.14) defines the polylogarithm $\mathrm{Li}_s(z)$ for complex numbers s and z with $|z| < 1$ [here $s = -(\frac{3}{2}+l)$ and $z = \bar{\lambda}$]:

$$\mathrm{Li}_{-\left(\frac{3}{2}+l\right)}(\bar{\lambda}) \equiv \sum_{n \geq 1} \bar{\lambda}^n \, n^{\frac{3}{2}+l}. \qquad (7.15)$$

By analytic continuation the function $\mathrm{Li}_{-\left(\frac{3}{2}+l\right)}(\bar{\lambda})$ is defined for a much larger range in $\bar{\lambda}$ than the definition in Eq. (7.15) seems to suggest. In any case, there is a branch cut for positive-real values of $\bar{\lambda}$ and $\bar{\lambda} \geq 1$. One has [Wood (1992)]

$$\mathrm{Im}\,\mathrm{Li}_{-\left(\frac{3}{2}+l\right)}(\bar{\lambda} \pm i0) = \pm\pi\,\frac{(\log(\bar{\lambda}))^{-\left(\frac{5}{2}+l\right)}}{\Gamma\left(-\left(\frac{3}{2}+l\right)\right)} \quad (\bar{\lambda} \geq 1), \qquad (7.16)$$

where $\Gamma(s) \equiv \int_0^\infty dt\, e^{-t}\, t^{s-1}$ denotes the gamma function. Notice that $\Gamma(-(\frac{3}{2}+l))$ is finite and real due to the fractional nature of its negative argument. Notice also that there is a singularity $\propto (\bar{\lambda}-1)^{-\left(\frac{3}{2}+l\right)}$ at $\bar{\lambda} = 1$. To the left (right) of $\bar{\lambda} = 1$ this singularity looks like a real (imaginary), nonintegrable pole which follows from the limit formula [Wood (1992)]

$$\lim_{|\mu|\to 0}\mathrm{Li}_s(e^\mu) = \Gamma(1-s)(-\mu)^{s-1} \quad \left(s = -\left(\frac{3}{2}+l\right) < 1\right). \qquad (7.17)$$

Let us now invert the Borel transformation leading to Eq. (7.14). The inverse Borel transform $\hat{P}_{\mathrm{mass}}(\bar{\lambda})$ of $B_{\bar{P}_{\mathrm{mass}}}(\bar{\lambda})$ is defined as

$$\hat{P}_{\mathrm{mass}}(\bar{\lambda}) \equiv \sum_{l=0}^{L} a_l \hat{P}_l(\bar{\lambda}) \equiv \int_0^\infty dt\, e^{-t}\, B_{\bar{P}_{\mathrm{mass}}}(\bar{\lambda}\,t), \qquad (7.18)$$

where

$$\hat{P}_l(\bar{\lambda}) \equiv \int_0^\infty dt\, e^{-t}\, \mathrm{Li}_{-\left(\frac{3}{2}+l\right)}(\bar{\lambda}\,t). \qquad (7.19)$$

The following integral representation of $\mathrm{Li}_s(z)$ holds for all complex s and z [Lewin (1981)]:

$$\mathrm{Li}_s(z) = \frac{iz}{2}\int_C du\,\frac{(-z)^u}{(1+u)^s\,\sin(\pi u)}, \qquad (7.20)$$

where the path C is along the imaginary axis from $-i\infty$ to $+i\infty$ with an indentation to the left of the origin. Inserting Eq. (7.20) into

Eq. (7.19) for $\bar{\lambda} = -|\bar{\lambda}|$ and interchanging the order of integration, we have

$$\hat{P}_l(\bar{\lambda}) = -i \int_C du \, \frac{(1+u)^{\frac{3}{2}+l}}{1 - e^{-2\pi i u}} \, e^{-\pi i u} \, e^{(1+u)\log(-\bar{\lambda})} \, \Gamma(u+2). \quad (7.21)$$

Since, by Stirling's formula[10] the gamma function $\Gamma(u+2)$ decays exponentially fast for $u \to \pm i\infty$, the integral over u in Eq. (7.21) exists and defines the real-analytic[11] function $\hat{P}_l(\bar{\lambda})$ ($\bar{\lambda} < 0$).

The inverse Borel transformation in Eq. (7.19) can be performed numerically for $l = 0, 1, 2, 3, 4$. In Fig. 7.3 we have plotted the functions $\hat{P}_l(\bar{\lambda})$ ($l = 0, 1, 2, 3, 4$) for $-15 \leq \bar{\lambda} \leq 0$. Modulo powers in u and for $|u| \gg 1$ the integrand in Eq. (7.21) roughly is of the form

$$c \cos(\log(-\bar{\lambda})x) e^{-ax} + s \sin(\log(-\bar{\lambda})x) e^{-ax}$$

$$(c, a \text{ real}, a > 0; s \text{ imaginary}), \quad (7.22)$$

where x denotes the positive-real integration variable. Thus a fit function $\Phi_l(\bar{\lambda})$ of the following form is suggested[12]:

$$\Phi_l(\bar{\lambda}) = \frac{\sum_{r=0}^{R_l} \alpha_{2r+1}(\log(-\gamma_{2r+1}\bar{\lambda}))^{2r+1}}{\sum_{s=0}^{S_l} \beta_{2s}(\log(-\delta_{2s}\bar{\lambda}))^{2s}}, \quad (7.23)$$

where $\gamma_{2r+1}, \delta_{2s}$ are positive-real, and $\alpha_{2r+1}, \beta_{2s}$ are real. Higher (even and odd) powers of $\log(-\bar{\lambda})$ were introduced into the numerator and denominator of Eq. (7.23) (not setting $\delta_{2s} = \gamma_{2r+1} \equiv 1$

---

[10]$\Gamma(z) = \sqrt{2\pi} \, z^{z-\frac{1}{2}} \, e^{-z} \, e^{H(z)}$ where $H(z) \equiv \sum_{n \geq 0}((z+n+1)/2 \log(1 + \frac{1}{z+n}) - 1)$ converges for $z \in \mathbf{C}_-$ and $\lim_{|z| \to \infty} H(z) = 0$ (see [Freitag and Busam (1995)]).

[11]This follows from Eq. (7.19) and the fact that $\text{Im}\left[\text{Li}_{-(\frac{3}{2}+l)}(z)\right] \equiv 0$ for $z \leq 0$.

[12]One has:

$$\int_0^\infty dx \, e^{-ax} \sin(\log(-\bar{\lambda})x) = \frac{a}{a^2 + (\log(-\bar{\lambda}))^2},$$

$$\int_0^\infty dx \, e^{-ax} \cos(\log(-\bar{\lambda})x) = \frac{\log(-\bar{\lambda})}{a^2 + (\log(-\bar{\lambda}))^2} \quad (a > 0).$$

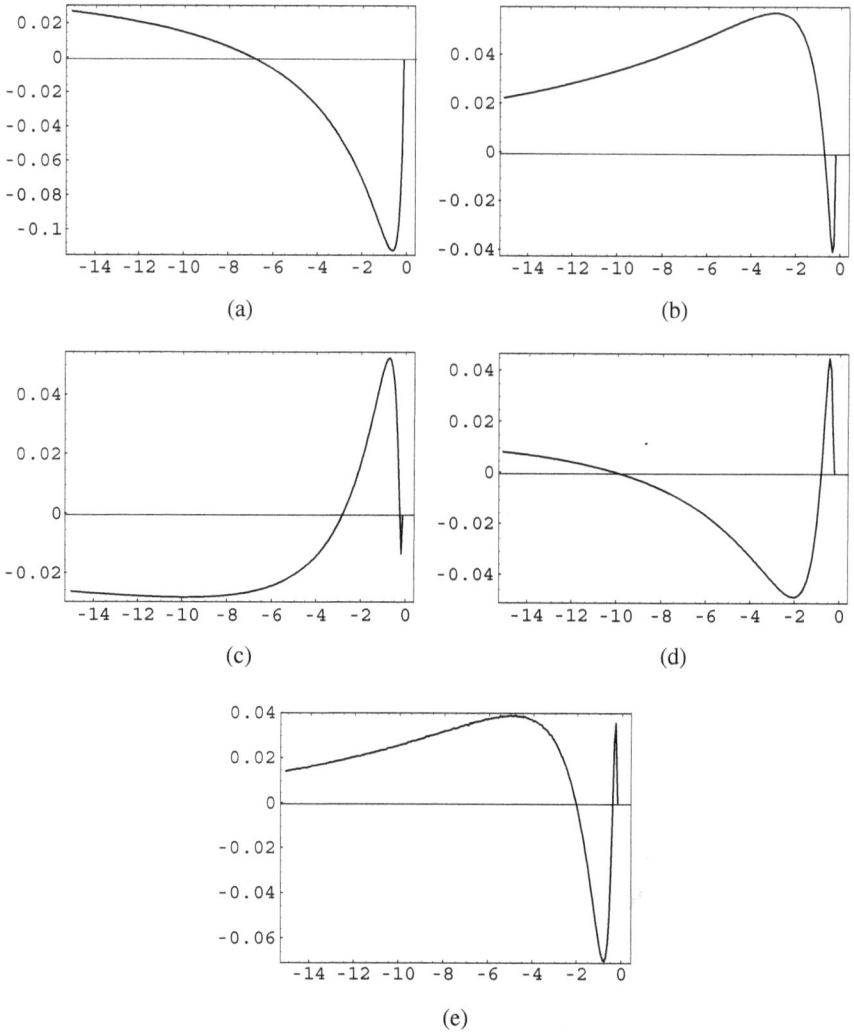

Figure 7.3.   The functions $\hat{P}_l(\bar{\lambda})$ for (a) $l = 0$, (b) $l = 1$, (c) $l = 2$, (d) $l = 3$, (e) $l = 4$ and $-15 \leq \bar{\lambda} \leq 0$. Amusingly, l measures the (increasing) "time" at which snapshots are taken on an "ocean wave" crashing against the "beach" at $\bar{\lambda} = 0$ where $\hat{P}_l \equiv 0$.

and $S_l = R_l + 1 = 1$) because of the contributions to the integral in Eq. (7.21) from moderate values of $|u|$ and the presence of the factor $(1 + u)^{3/2+l}$. One may check that the functions displayed in Fig. 7.3, indeed, fall off logarithmically slowly for $-\bar{\lambda} \to \infty$. Except for the

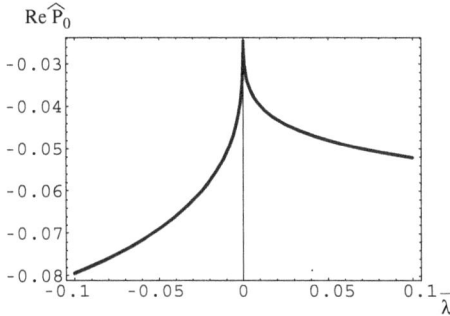

Figure 7.4.   The behavior of the function $\operatorname{Re}\hat{P}_0(\bar{\lambda})$ in the vicinity of $\bar{\lambda} = 0$.

branch cut along the positive-real axis Φ_l is a meromorphic function of $\bar{\lambda}$. As such it can be continued arbitrarily close to the cut.

Let us now consider the case $\bar{\lambda} \geq 0$. The functions $\hat{P}_l(\bar{\lambda})$ exhibit branch cuts along the positive-real axis. This is because of their dependence on powers of $\log(-\bar{\lambda})$ as suggested by Eq. (7.21). At $\bar{\lambda} = 0$ there is a cusp in $\operatorname{Re}\hat{P}_l(\bar{\lambda})$ with $\operatorname{Re}\hat{P}_l(\bar{\lambda} = 0) = 0$ (see Fig. 7.4). The slope of $\operatorname{Re}\hat{P}_l(\bar{\lambda})$ at $\bar{\lambda} = \mp 0$ is $\pm\infty$. Excellent fits reveal the following expressions for Φ_l, $(l = 0, 1, 2, 3, 4, 5)$:

$$\Phi_0(\bar{\lambda}) = 0.0570\frac{\log(-0.154\bar{\lambda})}{1 + 0.220(\log(-0.494\bar{\lambda}))^2},$$

$$\Phi_1(\bar{\lambda}) = \frac{0.0212\log(-10.2\bar{\lambda}) + 0.00142(\log(-0.109\bar{\lambda}))^3}{1 + 0.128(\log(-1.09\bar{\lambda}))^2 + 0.0544(\log(-0.886\bar{\lambda}))^4},$$

$$\Phi_3(\bar{\lambda}) = -0.0544\frac{\log(-0.367\bar{\lambda})}{1 + 0.641(\log(-0.513\bar{\lambda}))^2},$$

$$\Phi_4(\bar{\lambda}) = \frac{-0.0722\log(-1.66\bar{\lambda}) + 0.00780(\log(-1.95\bar{\lambda}))^3}{1 + 0.212(\log(-1.90\bar{\lambda}))^2 + 0.0864(\log(-1.06\bar{\lambda}))^4},$$

$$\Phi_5(\bar{\lambda})$$

$$= \frac{14.7\log(-0.505\bar{\lambda}) - 1.01(\log(-0.917\bar{\lambda}))^3 - 0.0172(\log(-0.0568\bar{\lambda}))^5}{1 + 5.84(\log(-0.0156\bar{\lambda}))^2 + 0.925(\log(-10.3\bar{\lambda}))^4 + 1.68(\log(-0.641\bar{\lambda}))^6}.$$

$$(7.24)$$

Figure 7.5 shows for $\bar{\lambda} \geq 0$ plots of $\operatorname{Re}\Phi_l$, which are continuous across the cut, and $\operatorname{Im}\Phi_l$, whose respective signs are chosen such as to minimize the value of $\bar{\lambda} > 0$ where they first intersect with

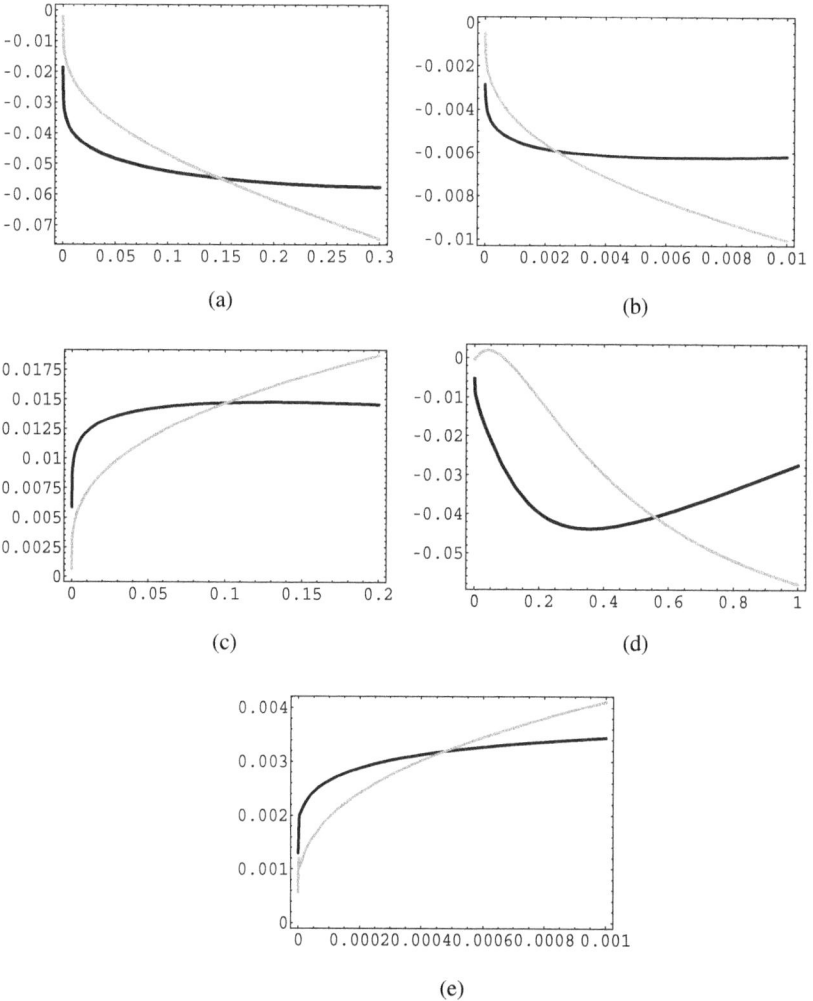

Figure 7.5. The real (black) and the imaginary (gray) part of the functions $\hat{P}_l(\bar{\lambda})$ for (a) $l = 0$, (b) $l = 1$, (c) $l = 2$, (d) $l = 3$, (e) $l = 4$, and positive $\bar{\lambda}$.

Re Φ_l (approach to the cut as $\bar{\lambda} = \mathrm{Re}\,\bar{\lambda} \pm i\,0$). Notice that Re Φ_l and Im Φ_l vanish precisely at $\bar{\lambda} = 0$. Notice also that for $\bar{\lambda} \geq 0$ and in the vicinity of $\bar{\lambda} = 0$ the modulus of Re Φ_l grows much more rapidly than the modulus of Im Φ_l.

We now would like to give an interpretation of these results. Due to the meromorphic nature of the functions Φ_l away from the cut along the positive-real axis and the fact that Re Φ_l is continuous across this cut, we have no choice but to regard, up to a modification involving finitely many terms,[13] the quantity

$$P_{\text{mass}}(\hat{\beta}) \equiv \frac{\tilde{\Lambda}^4}{\sqrt{2}\,\pi^{\frac{3}{2}}\,\hat{\beta}^{\frac{5}{2}}} \text{ Re } \sum_{l=0}^{L} a_l \hat{P}_l(\bar{\lambda}(\hat{\beta})) \qquad (7.25)$$

as a unique prediction for the pressure exerted by the massive modes in the confining phase of an SU(2) Yang–Mills theory [compare with Eq. (7.13)]. We expect this pressure to be positive and rapidly growing for $\bar{\lambda} > 0$ (see [Giacosa, Hofmann, and Schwarz (2006)] where the asymptotic expansion of Eq. (7.13) was used to predict the temperature dependence of the pressure for an "electron" gas). For accurate numerical predictions the integer coefficients a_l would have to be known. In a $\lambda\phi^4$-theory the algebraic dependence of the number of connected vacuum bubble diagrams on n should be extractable from a numerical analysis of the ground state energy of an anharmonic oscillator (see [Bender and Wu (1969, 1971)]).

It is important to realize that at $\hat{\beta} = \infty$ (or $T = 0$ or $\bar{\lambda} = 0$) we have $P_{\text{mass}}(\hat{\beta} = \infty) = 0$. That is, the contribution to the cosmological constant arising from an SU(2) Yang–Mills theory at zero temperature is *nil*. But what about the imaginary part in P_{mass}? As is readily observed from Fig. 7.5 (and is easily proven analytically for all l) the modulus of Re $\Phi_l(\bar{\lambda})$ grows much faster than $|\text{Im } \Phi_l(\bar{\lambda})|$ when increasing $\bar{\lambda}$ starting from $\bar{\lambda} = 0$. Thus for a certain range

---

[13]In Eq. (7.13) we have used the large-n expressions for the coefficients N_n and for $K_2(n\hat{\beta})$. Precise predictions would have to subtract the first few terms from the expression in Eq. (7.25) (evaluated with large-n coefficients) and add terms with realistic multiplicities N_n.

of small values[14] of $\hat{\beta}^{-1}$ the pressure $P_{\text{mass}}(\hat{\beta})$ is dominated by the real part.

The presence of a growing imaginary contamination signals an increasing deviation from thermodynamical behavior.[15] Namely, we interpret the occurrence of a sign-indefinite imaginary part as an indication for the violation of a basic thermodynamical property: spatial homogeneity. That is, the growth of the ratio $R \equiv \text{Im} P_{\text{mass}}/\text{Re} P_{\text{mass}}$ with increasing temperature is a measure for the increasing importance of turbulence-like phenomena in the plasma since imaginary contributions to the pressure lead to localized exponential build-up and collapse of energy density about the equilibrium situation. When the point $R = 1$ is reached the thermodynamical description of the system fails badly; the system then is highly "nervous" and close to the Hagedorn transition.

7.3   Evolving Center-Vortex Loops

Recall that the mass m_D of the dual gauge field diverges at the onset of the decay preconfining monopole-antimonopole condensate, containing collapsing, closed and unresolved magnetic flux lines of finite core-size d, into center-vortex loops. This, in turn, implies that $d \to 0$. As a consequence, center-vortex loops with nonvanishing selfintersection number n become stable solitons in isolation. The region of negative pressure P coincides with the vanishing vortex core, and isolated center-vortex loops become particle-like excitations. As we have seen in Sec. 7.1.3, these solitons are classified

---

[14]For example, $\bar{\lambda} = 0.1, 0.01, 0.0004$ correspond to

$$\hat{\beta}^{-1} = \frac{T}{\Lambda} = -\left(\log\left(\frac{\bar{\lambda}}{32}\right)\right)^{-1} = 0.173, 0.124, 0.0886,$$

respectively.

[15]No well-defined partition function directly generates an imaginary part for the pressure.

essentially according to their (knot-)topology and the distribution of the signs of magnetic charge over all selfintersections.

7.3.1   The case of $n = 0$: Mass gap

The purpose of this section is to investigate the sector with no self-intersection ($n = 0$) in microscopic detail. The presentation relies on [Moosmann and Hofmann (2008a)]. Topologically, there is no reason for the absolute stability of this sector's soliton, and we will argue that on average and as a consequence of a noisy environment a planar center-vortex loop with $n = 0$ shrinks[16] to nothingness within a finite time. Here the role of time is played by a variable measuring the decrease of externally provided resolving power applied to the system. At $T = 0$ this resolving power emerges by interactions of the Yang–Mills theory under investigation with other Yang–Mills theories. For later use we define by "planar" curve or center-vortex loop an embedding of the curve or of the $n = 0$ soliton into a two-dimensional flat and spatial plane. As we shall argue below, the evolution of center-vortex loops in dependence of a decreasing resolving power should be modeled by the curve-shrinking equation [Gage and Hamilton (1986)]. According to the work of [Altschuler and Grayson (1991); Altschuler (1991)] and for low-energy SU(2) Yang–Mills theory the assumption of *planar*, closed curves to accurately describe the late-time behavior of the evolution of center-vortex loops with $n = 0$ is no restriction of generality. This is because (i) any closed nonselfintersecting three-dimensional space curve becomes planar after a finite time of evolving under the curve shrinking equation, and (ii) the formation of a singularity (a loop pinching off to form a cusp) is a planar phenomenon [Altschuler

---

[16]At $n = 0$ there is no selfintersection of flux lines. If the external, average resolving power, associated with a noisy environment, is lower than the Yang–Mills scale then the creation of an intersection point is energetically excluded because one unit of Yang–Mills scale is required to isolate a monopole at the core of an associated flux eddy.

(1991)]. This phenomenon, however, is excluded to occur in low-energy, that is, weakly resolved SU(2) Yang–Mills theory because the energy required to form the intersection point is not available. It is an established fact that the curve-shrinking evolution of planar and closed $n = 0$ curves terminates in round points[17] after a finite time [Gage and Hamilton (1986); Grayson (1987)].

For an isolated SU(2) Yang–Mills theory, the role of the noisy environment, providing for a finite, average resolving power, is played by the sectors with $n > 0$ if the temperature parameter does not vanish. If the SU(2) theory under consideration is part of a world which is subject to additional gauge symmetries, then a portion of the environment arises from interactions with these theories. In any case, a planar center-vortex loop of finite length L and selfintersection number $n = 0$ maintains its small mass by frequent interactions with the environment characterized by the finite resolving power Q_0. The center-vortex loop shrinks towards a round point if the resolving power Q is decreased, and the situation of a round point is reached at some finite resolution $Q_* < Q_0$. Thus the center-vortex loop is unresolvable for $Q < Q_*$. Therefore, all properties that are related to the existence of extended lines of center flux, observable for $Q_0 \geq Q > Q_*$, do not occur for $Q \leq Q_*$, and center-vortex loops with $n = 0$ vanish from the spectrum of confining SU(2) Quantum Yang–Mills theory. Since center-vortex loops with $n > 0$ have a finite mass (roughly speaking positive-integer multiples of the Yang–Mills scale Λ) we observe a gap in the mass spectrum of the theory when probing the system with average resolution $Q \leq Q_*$.

By embedding an $n = 0$ center-vortex loop of finite core-size (preconfining phase) of an isolated SU(2) Yang–Mills theory into a flat two-dimensional surface, a hypothetical observer at some point

---

[17]By "round point" it is meant that the isoperimetric ratio L^2/A, where L denotes the length of the curve and A the enclosed area, approaches the value 4π from above as $L, A \to 0$. That is, in this limit the curve becomes circular.

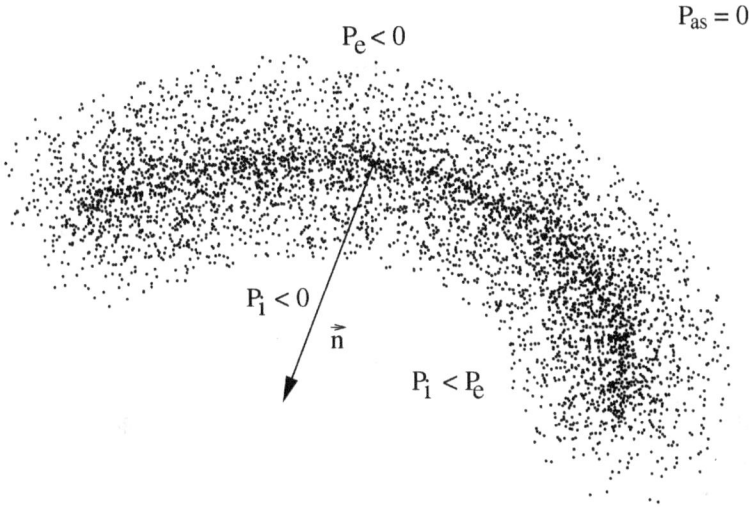

Figure 7.6. Highly space-resolved snapshot of a curve-sector belonging to a center-vortex loop. An increasing density of points represents a measure for an increasing restoration of the magnetic center symmetry towards the core of the flux line. This increase of center-symmetry restoration is accompanied by a local decrease of pressure. That is, the pressure P_i in the region pointed to by the normal vector $\vec{n}$ is more negative than the pressure P_e, thus leading to a motion of this sector along $\vec{n}$.

$\vec{x}_0$ in the plane measures a positive or a negative curvature along a given sector of the vortex line. This means, respectively, more or less negative pressure on the observer's side of this curve sector leading to the sector's motion towards or away from the observer (see Fig. 7.6). The speed of this motion is a monotonic function of curvature. On average, this shrinks the center-vortex loop.

One possibility to describe this situation is by the following flow equation in the (dimensionless) parameter τ

$$\partial_\tau \vec{x} = \frac{1}{\sigma} \partial_s^2 \vec{x}, \tag{7.26}$$

where s is arc length, $\vec{x}$ is a point on the planar center-vortex loop, and σ is a fixed string tension associated with the physics of monopoles and antimonopoles in the unresolved vortex core. Notice that, physically, τ is interpreted as a strictly monotonic decreasing (dimensionless) function of the ratio $\frac{Q}{Q_0}$. After rescalings, $\hat{x} \equiv \sqrt{\sigma}x, \xi = \sqrt{\sigma}s$, Eq. (7.26) assumes the following form

$$\partial_\tau \hat{x}(u, \tau) = \partial_\xi^2 \hat{x} = k(u, \tau)\vec{n}(u, \tau), \tag{7.27}$$

where u is a (dimensionless) curve parameter ($d\xi = |\partial_u \hat{x}| \, du$), $\vec{n}$ the (inward-pointing) Euclidean unit normal, k the scalar curvature, defined as

$$k \equiv |\partial_\xi^2 \hat{x}| = \left| \frac{1}{|\partial_u \hat{x}|} \partial_u \left(\frac{1}{|\partial_u \hat{x}|} \partial_u \hat{x} \right) \right|, \tag{7.28}$$

$|\vec{v}| \equiv \sqrt{\vec{v} \cdot \vec{v}}$, and $\vec{v} \cdot \vec{w}$ denotes the Euclidean scalar product of the vectors $\vec{v}$ and $\vec{w}$ lying in the plane. In the following we resort to a slight abuse of notation by using the same symbol $\hat{x}$ for the functional dependence on u or ξ.

It is worth mentioning that Eq. (7.27) expresses a special case of the local condition that the rate of decrease of the (dimensionless) curve length $L(\tau) = \int_0^{L(\tau)} d\xi = \int_0^{2\pi} du \, |\partial_u \hat{x}(u, \tau)|$ is maximal with respect to a variation of the velocity $\partial_\tau \hat{x}$ of a given point on the curve at fixed $|\partial_\tau \hat{x}|$ [Smith, Broucke, and Francis (2005)]:

$$\frac{dL(\tau)}{d\tau} = - \int_0^{L(\tau)} d\xi \, k\vec{n} \cdot \partial_\tau \hat{x}. \tag{7.29}$$

The one-dimensional "heat" equation (7.27) is well understood mathematically [Gage and Hamilton (1986); Grayson (1987)] and represents the one-dimensional analog of the Ricci equation describing the curvature-homogenization of certain three-manifolds [Perelman (2002, 2003a,b)].

The present section interprets the shrinking of closed planar curves, describing the physical situation of the late-time behavior of center-vortex loops with $n = 0$ within the confining phase of SU(2) or SU(3) Yang–Mills theories, as a Wilsonian renormalization-group evolution governed by an effective action. This effective action is defined purely in geometric terms. In the presence of an environment, represented by the parameter σ, this action possesses a natural decomposition into a conformal and a nonconformal factor. One of our goals is to show that the transition to the conformal limit of vanishing mean curve length really is a critical phenomenon characterized by a mean-field exponent if a suitable parameterization of the effective action is used. To see this, various initial conditions are chosen to generate an ensemble whose partition function is invariant under curve-shrinking. A (second-order) phase transition is characterized by the critical behavior of the coefficient associated with the nonconformal factor in the effective action. That is, in the presence of an environment the (nearly massless) $n = 0$ sector of low-energy, confining SU(2) Yang–Mills dynamics, generated during the downward a Hagedorn transition, practically disappears after a finite time, leading to an asymptotic mass gap.

Let us now provide knowledge on the properties of the shrinking of embedded (nonselfintersecting) curves in a plane [Grayson (1987)]. The properties of the τ-evolution of smooth, embedded, and closed curves $\hat{x}(u, \tau)$ subject to Eq. (7.27) were investigated in [Gage and Hamilton (1986)] for the purely convex case and in [Grayson (1987)] for the general case. The main results of [Grayson (1987)] are that an embedded curve with finitely many points of inflection remains embedded and smooth when evolving under Eq. (7.27) and that such a curve flows to a round point for $\tau \nearrow T$ where $0 < T < \infty$. That is, asymptotically the curve converges (with respect to the C^∞-norm) to a shrinking circle: $\lim_{\tau \to T} L(\tau) = 0$ and $\lim_{\tau \to T} A(\tau) = 0$, A being the (dimensionless) area enclosed by the curve, such that the isoperimetric ratio $\frac{L^2(\tau)}{A(\tau)}$ approaches the value 4π from above. For later use, we present the following two identities (see Lemmas 3.1.2 and 3.1.7 of [Gage and Hamilton (1986)]):

$$\partial_\tau L = -\int_0^L d\xi\, k^2 = -\int_0^{2\pi} du\, |\partial_u \hat{x}| k^2, \qquad (7.30)$$

$$\partial_\tau A = -2\pi. \qquad (7.31)$$

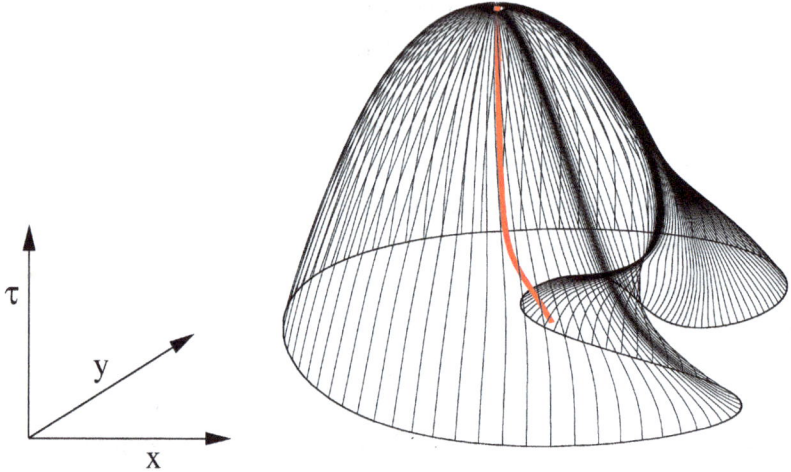

Figure 7.7.   Plot of the evolution of a planar center-vortex loop under Eq. (7.27). The thick central line depicts the graph of the center-vortex loop's "center of mass". The flow is started at $\tau = 0$ and ends at $\tau = T$. The so-called method of lines was employed to solve Eq. (7.27), for an explanation of this method see Sec. 7.3.2.

Setting $A(\tau = 0) \equiv A_0$, the solution to Eq. (7.31) is

$$\frac{A(\tau)}{A_0} = 1 - \frac{2\pi\tau}{A_0}. \qquad (7.32)$$

By virtue of Eq. (7.32) the critical value T is related to A_0 as

$$T = \frac{A_0}{2\pi}. \qquad (7.33)$$

We now wish to interpret curve-shrinking as a Wilsonian renormalization group flow taking place in the $n = 0$ planar center-vortex loop sector: That is, a partition function, defined as a statistical average (subject to a suitably defined weight) over $n = 0$ center-vortex loops, remains invariant under a change of the resolution expressed by the flow parameter τ.

To device a geometric ansatz for the effective action $S = S[\hat{x}(\tau)]$, representable in terms of integrals over local densities in ξ (reparametrization invariance), the following reflections on symmetries are in order.

(i) Scaling symmetry $\hat{x} \rightarrow \lambda \hat{x}, \lambda \in \mathbf{R}_+$: For both $\lambda \rightarrow \infty$ and $\lambda \rightarrow 0$, implying $\lambda L \rightarrow \infty$ and $\lambda L \rightarrow 0$ at fixed L, the action S should be invariant under further finite rescalings (decoupling of the fixed length scale $\sigma^{-1/2}$).

(ii) Euclidean point symmetry of the plane (rotations, translations and reflections about a given axis): Sufficient but not necessary for this is a representation of S in terms of integrals over scalar densities with respect to these symmetries. That is, the action density should be expressible as a series involving products of Euclidean scalar products of $\frac{\partial^n}{\partial \xi^n} \hat{x}, n \in \mathbf{N}_+$, or constancy. However, an exceptional scalar integral over a nonscalar density can be devised: Consider the area A, calculated as

$$A = \left| \frac{1}{2} \int_0^{2\pi} d\xi \, \hat{x} \cdot \vec{n} \right|. \qquad (7.34)$$

Obviously, the density $\hat{x} \cdot \vec{n}$ in Eq. (7.34) is not a scalar under translations.

We now resort to a factorization ansatz for S as

$$S = F_c \times F_{nc}, \qquad (7.35)$$

where in addition to Euclidean point symmetry F_c (F_{nc}) is (is not) invariant under $\hat{x} \rightarrow \lambda \hat{x}$. In principle, infinitely many operators can be defined to contribute to F_c. Since the evolution generates circles for $\tau \nearrow T$ higher derivatives of the curvature k with respect to ξ rapidly converge to zero [Gage and Hamilton (1986)]. We expect this to be true also for Euclidean scalar products involving higher derivatives $\frac{\partial^n}{\partial \xi^n} \hat{x}$. To yield conformally invariant expressions such integrals need to be multiplied by powers of $\sqrt{A}$ and/or L or the inverse of integrals involving lower derivatives. At this stage, we are not capable of constraining the expansion in derivatives by additional physical or mathematical arguments. To be pragmatic, we simply set F_c equal to the isoperimetric ratio:

$$F_c(\tau) \equiv \frac{L(\tau)^2}{A(\tau)}. \qquad (7.36)$$

We conceive the nonconformal factor F_{nc} in S as a formal expansion in inverse powers of L. Since we regard the renormalization group evolution of the effective action as induced by the flow of an ensemble of curves, where the evolution of each member is determined by Eq. (7.27), we allow for an explicit τ dependence of the coefficient c of the lowest nontrivial power $\frac{1}{L}$. In principle, this sums up the contribution to F_{nc} of certain higher-power operators which do not exhibit an explicit τ dependence. Hence we make the following ansatz

$$F_{nc}(\tau) = 1 + \frac{c(\tau)}{L(\tau)}. \qquad (7.37)$$

The initial value $c(\tau = 0)$ is determined from a physical boundary condition such as the mean length $\bar{L}$ at $\tau = 0$ which determines the mean mass $\bar{m}$ of a center-vortex loop as $\bar{m} = \sigma\bar{L}$.

For later use we investigate the behavior of $F_{nc}(\tau)$ as $\tau \nearrow T$ for an ensemble consisting of a single curve only and require the independence of the partition function under changes in τ. Integrating Eq. (7.30) in the vicinity of $\tau = T$ under the boundary condition that $L(\tau = T) = 0$, we have

$$L(\tau) = \sqrt{8\pi}\sqrt{T - \tau}. \qquad (7.38)$$

Since $F_c(\tau \nearrow T) = 4\pi$ independence of the partition function under the flow in τ implies that

$$c(\tau) \propto \sqrt{T - \tau}. \qquad (7.39)$$

That is, F_{nc} approaches constancy for $\tau \nearrow T$ which brings us back to the conformal limit by a finite renormalization of the conformal factor F_c of the action. In this parameterization of S, $c(\tau)$ can thus be regarded as an order parameter for conformal symmetry with mean-field critical exponent.

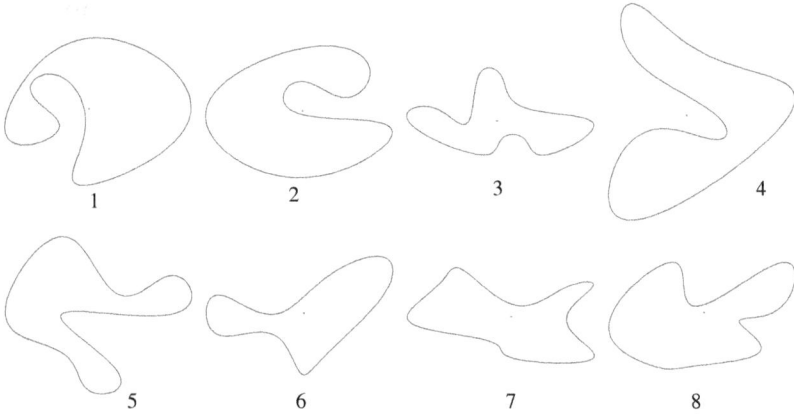

Figure 7.8. Initial curves contributing to the ensembles E_M (see text). Points locate the respective positions of the center of mass.

Let us now numerically investigate the effective action $S[\hat{x}(\tau)]$ resulting from a partition function Z with respect to a nontrivial ensemble E. This partition function is defined as the average

$$Z = \sum_i \exp\left(-S[\hat{x}_i(\tau)]\right) \qquad (7.40)$$

over the ensemble $E = \{\hat{x}_1(u, \tau), \dots\}$. Let us denote by E_M an ensemble consisting of M curves where E_M is obtained from E_{M-1} by adding a new curve $\hat{x}_M(u, \tau)$. In Fig. 7.8 eight initial curves are depicted which in this way generate the ensembles E_M for $M = 1, \dots, 8$.

We are interested in a situation where all curves in E_M shrink to a point at the same value $\tau = T$. Because of Eqs. (7.32) and (7.33) we thus demand that at $\tau = 0$ all curves in E_M initially have the same area A_0. The effective action S in Eq. (7.35) (when associated with the ensemble E_M we will denote it by S_M) is determined by the function $c_M(\tau)$ [compare with Eq. (7.37)]. The flow $c_M(\tau)$ follows from the requirement of τ independence of Z_M:

$$\frac{d}{d\tau} Z_M = 0. \qquad (7.41)$$

This is an implicit, first-order ordinary differential equation for $c_M(\tau)$ which needs to be supplemented with an initial condition $c_{0,M} = c_M(\tau = 0)$. A natural initial condition is to demand that the quantity

$$\bar{L}_M(\tau = 0) \equiv \frac{1}{Z_M(\tau = 0)} \sum_{i=1}^{M} L[\hat{x}_i(\tau = 0)] \exp\left(-S_M[\hat{x}_i(\tau = 0)]\right)$$

$$(7.42)$$

coincide with the mean value $\tilde{L}_M(\tau = 0)$ defined as

$$\tilde{L}_M(\tau = 0) \equiv \frac{1}{M} \sum_{i=1}^{M} L[\hat{x}_i(\tau = 0)]. \qquad (7.43)$$

Setting $\bar{L}_M(\tau = 0) = \tilde{L}_M(\tau = 0)$, a value for $c_{0,M}$ follows. In [Moosmann and Hofmann (2008a)] also a modified factor $F_{nc}(\tau) = 1 + \frac{c(\tau)}{A(\tau)}$ in Eq. (7.35) was considered. In this case the choice of initial condition $\bar{L}_M(\tau = 0) = \tilde{L}_M(\tau = 0)$ leads to $F_{nc}(\tau = 0) \equiv 0$. While the geometric effective action S is profoundly different for such a modification of $F_{nc}(\tau)$ physical results such as the evolution of the variance of center-of-mass position agree remarkably well (see below). That is, S itself is not a physical object. Rather, going from one ansatz for S_M to another describes a particular way of redistributing the weight in the ensemble with no significant impact on the physics. This is in contrast to quantum field theory and conventional statistical mechanics where the action in principle is related to the physical properties of a given member of the ensemble.

For the curves depicted in Fig. 7.8 we make the convention that $A_0 \equiv 2\pi \times 100$. It then follows by Eq. (7.33) that $T = 100$. The dependence $c_M^2(\tau)$ is plotted[18] in Fig. 7.9. According to Fig. 7.9 it appears that the larger the ensemble the closer is $c_M^2(\tau)$ to the evolution of a single circle of initial radius $R = \sqrt{\frac{A_0}{\pi}}$. That is, for

---

[18]Numerically, we employ the so-called "method of lines" to solve Eq. (7.27). For an explanation see Sec. 7.3.2.

$c_M^2(\tau)$

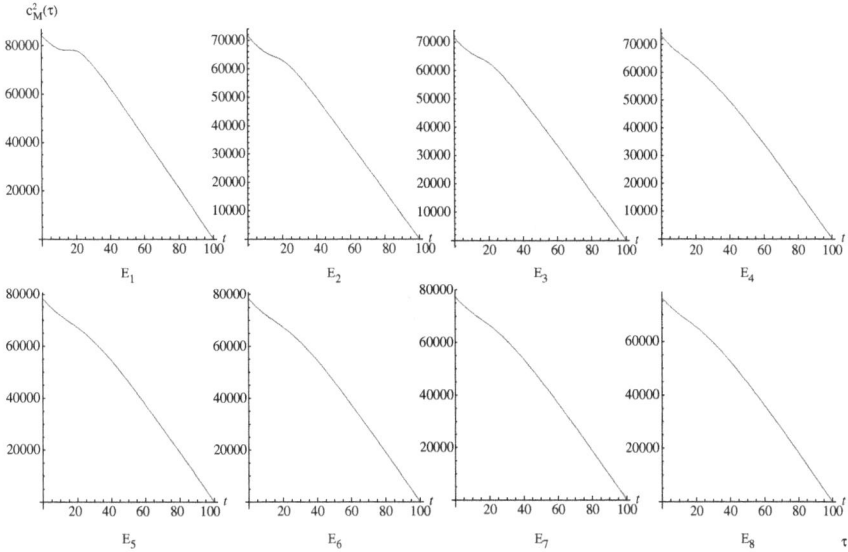

Figure 7.9. The square of the coefficient $c_M(\tau)$ entering the action of Eq. (7.35) by virtue of Eq. (7.37) for various ensemble sizes. Notice the early onset of the linear drop of $c_M^2(\tau)$ and the saturation in M for $M \geq 5$. The slope of $c_M^2(\tau)$ near $\tau = T$ neither depends on $c_{0,M}^2 \equiv c_M^2(\tau = 0)$ nor on the initial choice of $\bar{L}$.

growing M the function $c_M^2(\tau)$ approaches the form

$$c_{as,M}^2(\tau) = k_M(T - \tau), \qquad (7.44)$$

where the slope k_M depends on the strength of deviation from circles of the representatives in the ensemble E_M at $\tau = 0$, that is, on the variance ΔL_M at a given value A_0. Physically speaking, the value of $\tau = 0$ is associated with a certain initial resolution of the measuring device [the strictly monotonic function $\tau(Q)$, Q being a physical scale such as energy or momentum transfer, which expresses the characteristics of the measuring device and the measuring process], the value of A_0 describes the strength of noise associated with the environment (A_0 determines how fast the conformal limit of circular points is reached), and the values of $c_{0,M}$ and k_M [see Eq. (7.44)] are associated with the conditions at which the system to be evolved

is prepared. Notice that this interpretation is valid for the action
$S_M = \frac{L(\tau)^2}{A(\tau)}\left(1 + \frac{c_M(\tau)}{L(\tau)}\right)$ only.

We are now in a position to compute the flow of a more local "observable", namely, the mean center-of-mass (COM) position in a given ensemble and the statistical variance of the COM position. The COM position $\hat{x}_{COM}$ of a given curve $\hat{x}(\xi, \tau)$ is defined as

$$\hat{x}_{COM}(\tau) = \frac{1}{L(\tau)}\int_0^{L(\tau)} d\xi\, \hat{x}(\xi, \tau). \tag{7.45}$$

Because it is physically more interesting, we will below present only results on the statistical variance of the COM position.

Let us assume that at $\tau = 0$ the ensembles E_M that are modified by a translation applied to each representative for a coincidence of its COM position with the origin. Recall that such a modification $E_M \rightarrow E'_M$ does not alter the (effective) action (Euclidean point symmetry). At $\tau = 0$ the statistical variance of the COM position is prepared to be nil, physically corresponding to an infinite resolution applied to the system by the measuring device.

The mean COM position $\bar{\hat{x}}_{COM}$ over ensemble E'_M is defined as

$$\bar{\hat{x}}_{COM}(\tau) \equiv \frac{1}{Z_M}\sum_{i=1}^M \hat{x}_{COM,i}(\tau)\exp\left(-S_M[\hat{x}_i(\tau)]\right). \tag{7.46}$$

The scalar statistical deviation $\Delta_{M,COM}$ of $\bar{\hat{x}}_{COM}$ over the ensemble E'_M is defined as

$$\Delta_{M,COM}(\tau) \equiv \sqrt{\text{var}_{M,COM;x}(\tau) + \text{var}_{M,COM;y}(\tau)}, \tag{7.47}$$

where

$$\text{var}_{M,COM;x} \equiv \frac{1}{Z_M}\sum_{i=1}^M \left(x_{COM,i}(\tau) - \bar{x}_{COM}(\tau)\right)^2 \exp\left(-S_M[\hat{x}_i(\tau)]\right)$$

$$= -\bar{x}^2_{COM}(\tau) + \frac{1}{Z_M}\sum_{i=1}^M x^2_{COM,i}(\tau)\exp\left(-S_M[\hat{x}_i(\tau)]\right) \tag{7.48}$$

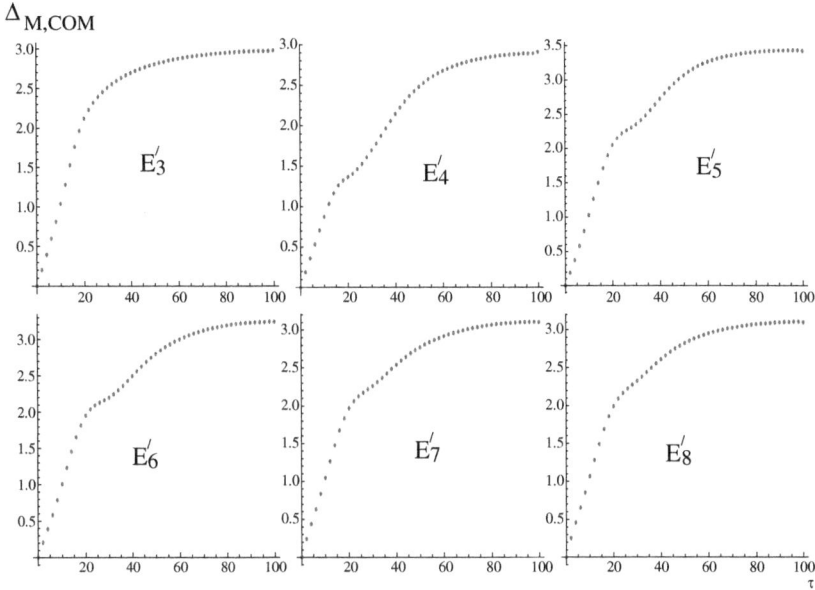

Figure 7.10. Plots of $\Delta_{M,\text{COM}}(\tau)$ for $M = 3, \dots , 8$ when evaluated with the action $S_M = \frac{L(\tau)^2}{A(\tau)}\left(1 + \frac{c_M(\tau)}{L(\tau)}\right)$. Notice the rapid generation of an uncertainty in the COM position under the flow and the saturation of this uncertainty when approaching the conformal limit $\tau \nearrow T$. There is also saturation of the curves with a growing ensemble size. This appears to express that the condensate of center-vortex loops is associated with a definite, intrinsic position uncertainty of its constituents.

and similarly for the coordinate y. In Fig. 7.10 plots of $\Delta_{M,\text{COM}}(\tau)$ are shown when $\Delta_{M,\text{COM}}(\tau)$ is evaluated over the ensembles $E'_3, \dots , E'_8$ with the action

$$S_M = \frac{L(\tau)^2}{A(\tau)}\left(1 + \frac{c_M(\tau)}{L(\tau)}\right)$$

and subject to the initial condition $\bar{L}_M(\tau = 0) = \tilde{L}_M(\tau = 0)$. In Fig. 7.11 the corresponding plots of $\Delta_{M,\text{COM}}(\tau)$ are depicted as obtained with the action

$$S_M = \frac{L(\tau)^2}{A(\tau)}\left(1 + \frac{c_M(\tau)}{A(\tau)}\right)$$

$\Delta_{M,\text{COM}}$

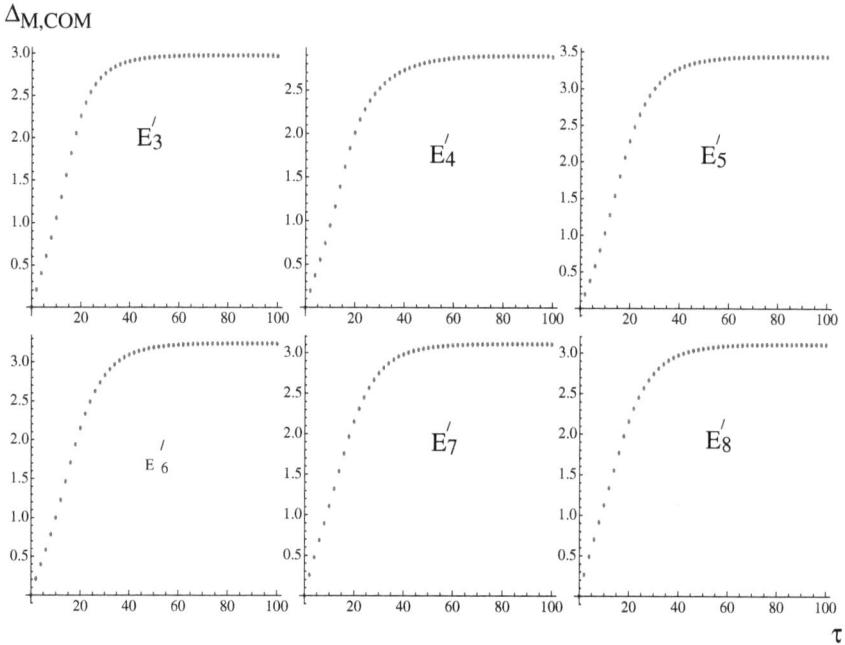

Figure 7.11.   Plots of $\Delta_{M,\text{COM}}(\tau)$ for $M = 3, \ldots , 8$ when evaluated with the action $S_M = \frac{L(\tau)^2}{A(\tau)}\left(1 + \frac{c_M(\tau)}{A(\tau)}\right)$. Notice the qualitative agreement of the curve shape with those displayed in Fig. 7.10. Notice also that the values of $\Delta_{M,\text{COM}}$ agree with those of Fig. 7.10 within the saturation regime.

and subject to the initial condition $\bar{L}_M(\tau = 0) = \tilde{L}_M(\tau = 0)$. In this case, one has $c_M(\tau) = -A(\tau)$ leading to equal weights for each curve in E'_M.

In view of the results obtained so far, we would say that an ensemble of evolving planar center-vortex loops in the $n = 0$ sector qualitatively resembles the quantum mechanics of a free point-particle[19] of mass m in, say, space dimension one. Namely, denoting for the moment (dimensionful) time by τ and the space coordinate by x. An initially localized square of the wave function ψ with $|\psi(\tau = 0, x)|^2 \propto \exp\left[-\frac{x^2}{a_0^2}\right]$, where

---

[19]It is of no relevance at this point whether this particle carries spin or not.

$\Delta x(\tau = 0) = a_0$, according to unitary time-evolution in free-particle quantum mechanics, evolves as $|\psi(\tau, x)|^2 = |\exp[-i\frac{H\tau}{\hbar}]$
$\psi(\tau = 0, x)|^2 \propto \exp[-\frac{(x-\frac{p}{m}\tau)^2}{a^2(\tau)}]$ where $H = \frac{\hat{p}^2}{2m}$ is the free-particle Hamiltonian, p is the overall spatial momentum of the packet, and $a(\tau) \equiv a_0\sqrt{1 + (\frac{\hbar\tau}{ma_0^2})^2}$. Thus, time-evolution in free-particle quantum mechanics and the process of lowering resolution in a statistical system describing planar, nonintersecting center-vortex loops share the property that in both systems the evolution generates out of a small initial position uncertainty (corresponding to a large initial resolution Δp) a larger position uncertainty in the course of the evolution. Possibly, future development will show that interference effects in quantum mechanics can be traced back to the nonlocal nature of center-vortex loops.

Let us now summarize our exploratory findings for the $n = 0$ sector of center-vortex loop excitations in the confining phase of an SU(2) Yang–Mills theory: An attempt has been undertaken to interpret the effects of an environment on two-dimensional planar center-vortex loops as they emerge in the confining phase of an SU(2) Yang–Mills theory, in terms of a Wilsonian renormalization-group flow determined by purely geometric entities. Our (mainly numerical) analysis uses established mathematics on the shrinking of embedded and closed curves in the plane. In the case of non-intersecting center-vortex loops ($n = 0$) the role of the environment is played by all sectors with $n > 0$ in addition to a possible explicit environment. In a particular parameterization of the effective action we observe critical behavior as the limit of round points is approached. That is, planar, non-intersecting center-vortex loops disappear from the spectrum for resolving powers smaller than a critical and finite value. Using this formalism to compute the evolution of the mean values of local observables, such as the center-of-mass position, a behavior is generated that qualitatively resembles the associated unitary evolution in quantum mechanics. We also have found evidence that this situation is practically not altered when changing the ansatz for the effective, geometric action concerning the expansion of its nonconformal factor.

7.3.2   *The case of $n = 1$: Emergence of order*

Let us now discuss the sector of center-vortex loops with $n = 1$ whose statistical behavior under variations of overall resolution is microscopically described by the curve-shrinking equation (7.27). Following [Moosmann and Hofmann (2008b)], we subject their statistics to the same geometrically motivated ansatz for the action S in the partition function Z of Eq. (7.40). Here, the assumption that curve-shrinking for closed curves of the figure-eight type is considered for immersions in a two-dimensional flat plane only [Grayson (1989)] is an essential constraint of generality [Grayson (1987)]. This is because the evolution in three dimensions of center-vortex loops of the figure-eight type may not have a planar limit. Our results for evolution in the plane indicate that spontaneous order emerges at a finite resolution.

This is interesting because there are condensed matter systems, where at low resolution an environment confines valence electrons to spatial planes[20] The standard quantum mechanical point-particle description associates strong correlations with the electrons in such two-dimensional systems at low temperature.

At $n = 1$ we have a point on the curve singled out: the location of the selfintersection point where practically the entire mass of the soliton resides [Hofmann (2007)]. Notice that as long as both wings of center flux are of finite size the position of the intersection point can be shifted at almost no cost of energy. In particular, if the inner angle α between in- and out-going center-flux at the intersection is sufficiently small then a motion of points on the vortex

---

[20]The question of how well a center-vortex loop with $n = 1$ in the confining phase of an SU(2) Yang–Mills theory describes an electron (or a muon or a tau-particle) is not an essential subject of this book.

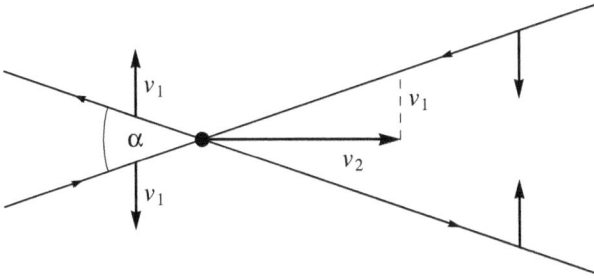

Figure 7.12. Points on the center flux lines moving oppositely on a line perpendicular to the bisecting line of the angle α with velocity modulus v_1. For sufficiently small α the velocity modulus v_2 of the intersection point is superluminal: $v_2 = v_1 \cot \frac{\alpha}{2}$.

line directed perpendicular to the bisecting line of the angle α easily generates a velocity of the intersection point which exceeds the speed of light (see Fig. 7.12). Recall, that the path-integral formulation of quantum mechanics admits such superluminal motion in the sense that the respective trajectories do contribute to transition amplitudes.

The transition from the nonintersecting to the selfintersecting center-vortex loop sector is by twisting nonintersecting curves. The emergence of a localized (anti)monopole in the process is due to its capture by oppositely directed center fluxes in the intersection region, that is, within the "eye of the storm". By a rotation of the left half-plane in Fig. 7.13(a) by an angle of π [see Fig. 7.13(b)], each wing of the center-vortex loops forms a closed flux loop by itself, thereby introducing equally directed center fluxes at the intersection point. This does not allow for an isolation of a single, spinning (anti)monopole in the core of the intersection and thus is topologically equivalent to the untwisted case Fig. 7.13(a). However, another rotation of the left-most half-plane in Fig. 7.13(c) introduces an intermediate loop which by shrinking is capable of isolating a spinning (anti)monopole due to oppositely directed center fluxes. Notice that in the last stage of such a shrinking process (short distances between the cores of the flux lines), where propagating dual gauge modes

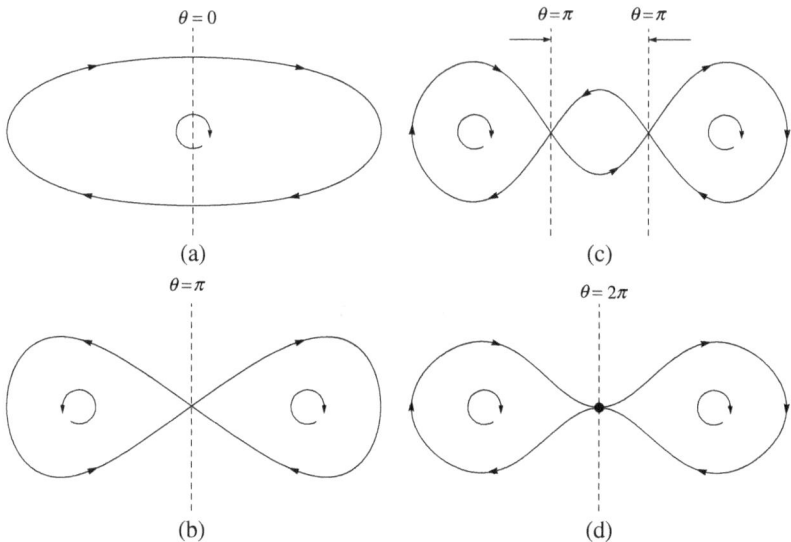

Figure 7.13.   (Topological) transition from the $n = 0$ sector (a), (b), (c) to the $n = 1$ sector (d) by twisting and subsequent capture of a magnetic (anti)monopole in the core of the final intersection. Arrows indicate the direction of center flux.

are available,[21] there is repulsion due to Biot–Savart which needs to be overcome. This necessitates an investment of energy which manifests itself in terms of the mass of the isolated (anti)monopole situated within the "eye of the storm". Alternatively, the emergence of an isolated (anti)monopole is possible by a simple pinching of the untwisted curve, again having to overcome local repulsion in the final stage of this process.

For the analysis performed in the present section we solely regard the situation depicted in Fig. 7.13(d). Thus we no longer distinguish between relative directions of the center flux within two given curve segments: This is not relevant for the evolution microscopically described by the same curve-shrinking flow as applied to

---

[21]On large distances these modes are infinitely massive which is characteristic of the confining phase.

the sector with $n = 0$ [Moosmann and Hofmann (2008a)]. The situation was mathematically analyzed in [Grayson (1989)]. Since the direction of center flux is inessential for the shrinking process we may actually treat this situation in a way as depicted in Fig. 7.13(b) where the curve is defined to be a smooth immersion into the plane with exactly one double point and a total rotation number zero: $\int_0^L d\xi\, k = 0$. Here the (dimensionless) curve length L is given by $L(\tau) = \int_0^{L(\tau)} d\xi = \int_0^{2\pi} du\, |\partial_u \hat{x}(u,\tau)|$. Notice that this is topologically distinct from the case Fig. 7.13(d) where one encounters a nonvanishing rotation number which is not smoothly deformable to zero.

In the $n = 0$ case a smooth, embedded curve shrinks to a round point under the flow for $\tau \nearrow T < \infty$ [Gage and Hamilton (1986); Grayson (1987)]. That is, the isoperimetric ratio approaches 4π from above. The curve in situation Fig. 7.13(d) separates the plane into three disjoint areas two of which are finite and denoted by A_1 and A_2. We understand T as the finite value of τ where either A_1 or A_2 or both vanish. Thus at $\tau = T$, a physical singularity is encountered which terminates the flow.

Recall that in the $n = 0$ case the rate of area change is a constant, $\frac{dA}{d\tau} = -2\pi$. This is no longer true for $n = 1$. Rather, we have that

$$A_1(\tau) - A_2(\tau) = \text{const.} \qquad (7.49)$$

Also, for $n = 1$ we have in comparison to the $n = 0$ case the relaxed constraint that $-4\pi \leq \frac{d(A_1+A_2)}{d\tau} \leq -2\pi$. In contrast to the $n = 0$ case the isoperimetric ratio for the $n = 1$ case is bounded for $\tau \nearrow T$ if and only if $A_1 \neq A_2$. Notice that the case $A_1 = A_2$ is extremely fine-tuned physically.

In analogy to the case $n = 0$ we now wish to interpret curve-shrinking as a Wilsonian renormalization-group flow: A partition function, defined as a statistical average (according to a suitably defined weight) over $n = 1$ center-vortex loops, is to be left invariant under a decrease of the resolution determined by the flow parameter τ. Reasoning as for the $n = 0$ sector (see Sec. 7.3.1), we again resort to the ansatz of Eq. (7.35) for the action $S[\hat{x}(\tau)]$ determining the weight in the partition function of the system [see Eq. (7.40)]. The two factors in Eq. (7.35) are defined in Eqs. (7.36) and (7.37).

Let us numerically investigate the effective action $S[\hat{x}(\tau)]$ resulting from the partition function Z of Eq. (7.40) with respect to a nontrivial ensemble $E = \{\hat{x}_1, \dots\}$. More definitely, let E_M be an ensemble consisting of M curves. Here E_M is obtained from E_{M-1} by adding a new curve $\hat{x}_M(u, \tau)$. The effective action S in Eq. (7.35) (when associated with the ensemble E_M we will denote it by S_M) is determined by the function $c_M(\tau)$ [compare with Eq. (7.37)] whose flow follows from the requirement of the τ independence of Z_M:

$$\frac{d}{d\tau} Z_M = 0. \tag{7.50}$$

Just like Eq. (7.41), this is an implicit, first-order ordinary differential equation for $c(\tau)$ which needs to be supplemented with an initial condition $c_{0,M} = c_M(\tau = 0)$. Again, we demand that the $\bar{L}_M(\tau = 0)$ of Eq. (7.42) coincide with $\tilde{L}_M(\tau = 0)$ of Eq. (7.43). As a consequence, a value for $c_{0,M}$ follows. Like in the case $n = 0$ we also consider the modified factor

$$F_{nc}(\tau) = 1 + \frac{c(\tau)}{A(\tau)} \tag{7.51}$$

in Eq. (7.35). While the ansatz for the geometric effective action is profoundly different for such a modification of $F_{nc}(\tau)$, physical results such as the evolution of the variance of the intersection point will turn out to agree remarkably well (see below).

We normalize all curves to possess the same initial area $A_0 = A_{0,1} + A_{0,2}$ and, since we are now interested in the position of the intersection point [location of (anti)monopole], we have applied a translation to each curve in the ensembles E_M such that all intersections initially coincide with the origin. Since the critical value T of the flow parameter τ varies from curve to curve we order the members of the maximal-size ensemble $E_{M=16}$ into subensembles $E_{M<16}$ such that $T_{i=1} \geq T_{i=2} \geq \cdots \geq T_M$. The ensembles E_M obtained in this way are referred to as T-ordered. We also have performed all

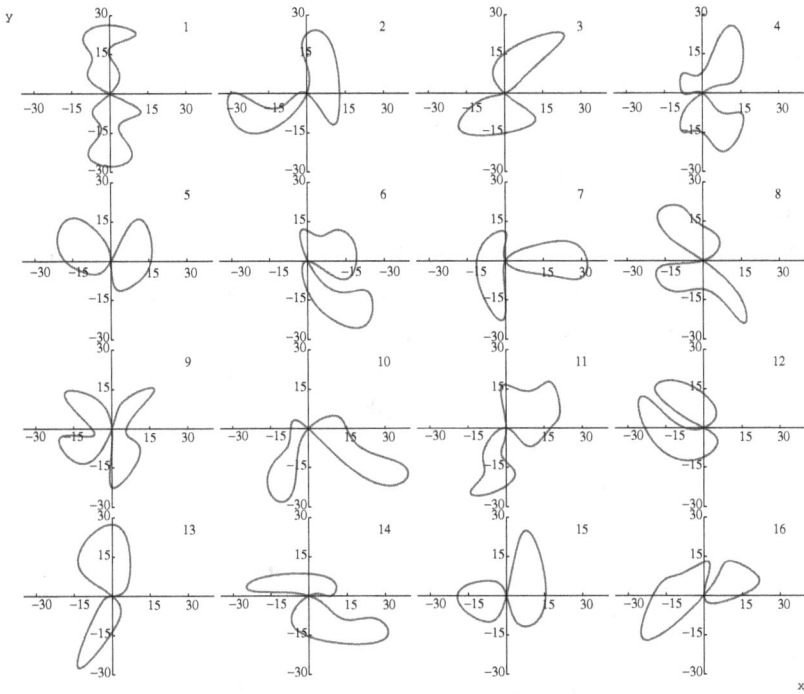

Figure 7.14. Initial curves $\hat{x}_i(u, \tau = 0)$ contributing to the ensemble $E_{M=16}$. The intersection points coincide with the origin, and all curves have the same area $200\,\pi$. By definition $E_{M=16}$ is T-ordered.

simulations with ensembles $E'_{M<16}$ whose members are picked randomly from $E_{M=16}$ and have obtained strikingly similar results for ensemble averages of "observables" using $E_{M<16}$ and $E'_{M<16}$ for the τ-evolution to the left of $\min\{T_i | \hat{x}_i \in E'_{M<16}\}$.

The maximal-size ensemble $E_{M=16}$ at $\tau = 0$ is depicted in Fig. 7.14 with the universal choice $A_0 = 200\,\pi$. The curves in Fig. 7.14 are arranged in a T-ordered way. We have $T_{i=1} = 65 \geq T_{i=2} \geq \cdots \geq T_{M=16} = 43$. In Fig. 7.15 the evolution of an initial curve under curve-shrinking is shown as observed from two points of view. The flow is started at $\tau = 0$ and stopped at a value of τ shortly below T. In Fig. 7.16 the flow of the intersection points, corresponding to the initial curves depicted in Fig. 7.14, is shown. The search for solutions to the second-order partial differential Eq. (7.27) subject to periodic boundary conditions in the curve parameter, $\hat{x}(u = 0, \tau = 0) = \hat{x}(u = 2\pi, \tau = 0)$, and for the initial conditions $\hat{x}(u, \tau = 0)$ depicted in Fig. 7.14 was performed numerically using the method of lines. That is, the partial differential equation was discretized on a uniform grid in the parameter u yielding a semi-discrete problem in terms of a system of ordinary differential equations in τ. Figure 7.15 indicates why this technique is called method of lines. As one can also see from Fig. 7.15, a set of discrete points on the curve, although remaining equidistant in u, may evolve under the flow such that the spatial distances between adjacent neighboring points fall below the numerical precision. Numerically, the flow then encounters a ficti-tious singularity (not to be confused with the earlier-mentioned sin-gularities). To recognize such a situation automatically, Eq. (7.49) was exploited: The evolution was stopped as soon as a sizable deviation occured from what Eq. (7.49) predicts. The configuration obtained at this value of τ was fitted in such a way that a new discretization in u yielded well-separated points to re-start the methods of lines. Equation (7.49) was also used as an indicator for the final, nonficti-tious singularity at T where A_1 or A_2 or both vanish.

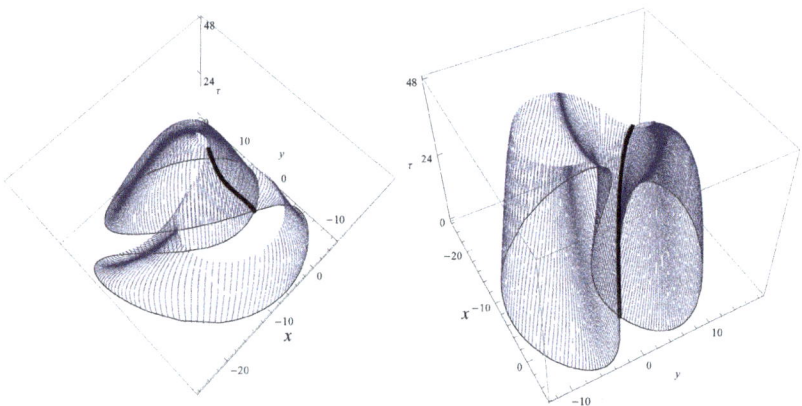

Figure 7.15. Plot of the evolution of an $n = 1$ center-vortex loop (curve 12 of Fig. 7.14) under Eq. (7.27). The thick central line indicates the trajectory of the intersection point which coincides with the origin at $\tau = 0$.

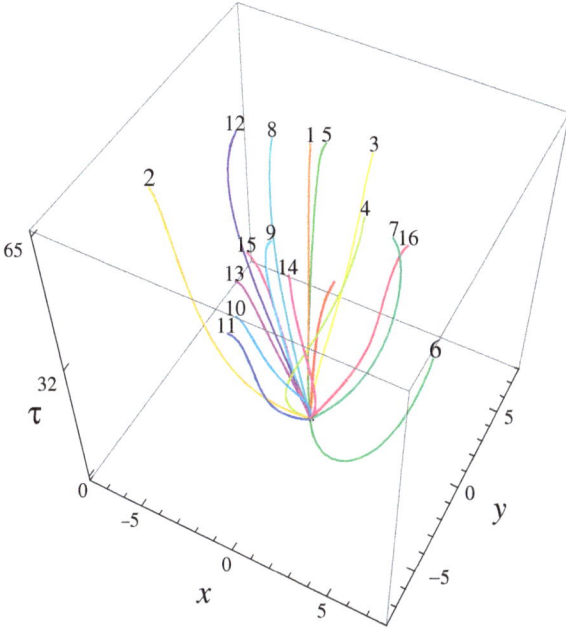

Figure 7.16. Flow of the intersection points for the initial curves depicted in Fig. 7.14.

For all ensembles E_M the τ dependence of the coefficient c_M in Eq. (7.37) roughly behaves like $\propto \sqrt{T_M - \tau}$ where T_M is the weakly ensemble-dependent minimal value of $\{T_i\}$. For the modified action $S_M = \frac{L(t)^2}{A(t)}\left(1 + \frac{c_M(t)}{A(t)}\right)$ the coefficient c_M is well-approximated by a linear function $\propto T_M - \tau$. For T-ordered ensembles the results for c_M for the actions subject to Eqs. (7.37) and (7.51) are shown in Figs. 7.17 and 7.18, respectively. The results for the ensembles E'_M do not differ sizably from those presented in Figs. 7.17 and 7.18.

The mean position $\bar{\hat{x}}_{\text{int}}$ of the intersection point over the ensemble E_M is defined as

$$\bar{\hat{x}}_{\text{int}}(\tau) \equiv \frac{1}{Z_M} \sum_{i=1}^{M} \hat{x}_{\text{int},i}(\tau) \exp\left(-S_M[\hat{x}_i(\tau)]\right), \qquad (7.52)$$

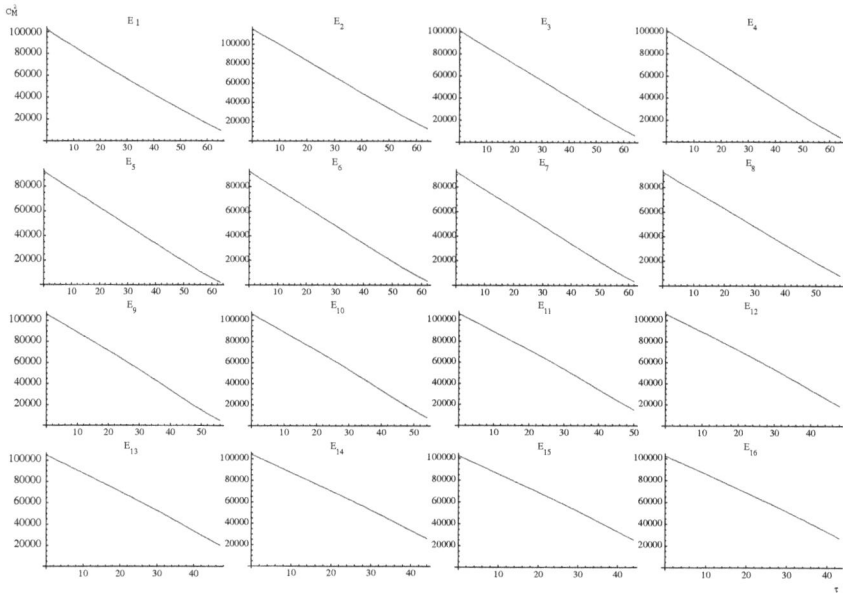

Figure 7.17.   The squares of the coefficients $c_M(\tau)$ entering the ansatz for action of Eq. (7.35) specializing to Eq. (7.37) for T-ordered ensembles up to $M = 16$.

where $\hat{x}_{\text{int},i}(\tau)$ is the intersection point of curve $\hat{x}_i$ at evolution parameter value τ. The scalar statistical deviation $\Delta_{M,\text{int}}$ of $\bar{\hat{x}}_{\text{int}}$ over the ensemble E_M is defined as

$$\Delta_{M,\text{int}}(\tau) \equiv \sqrt{\text{var}_{M,\text{int};x}(\tau) + \text{var}_{M,\text{int};y}(\tau)}, \qquad (7.53)$$

where

$$\text{var}_{M,\text{int};x} \equiv \frac{1}{Z_M} \sum_{i=1}^{M} (x_{\text{int},i}(\tau) - \bar{x}_{\text{int}}(\tau))^2 \exp\left(-S_M[\hat{x}_i(\tau)]\right)$$

$$= -\bar{x}_{\text{int}}^2(\tau) + \frac{1}{Z_M} \sum_{i=1}^{M} x_{\text{int},i}^2(\tau) \exp\left(-S_M[\hat{x}_i(\tau)]\right) \quad (7.54)$$

and similarly for the coordinate y. In Fig. 7.19 plots of $\Delta_{M,\text{int}}(\tau)$ are shown when evaluated over the ensembles $E_1, \ldots, E_{16}$ subject to the

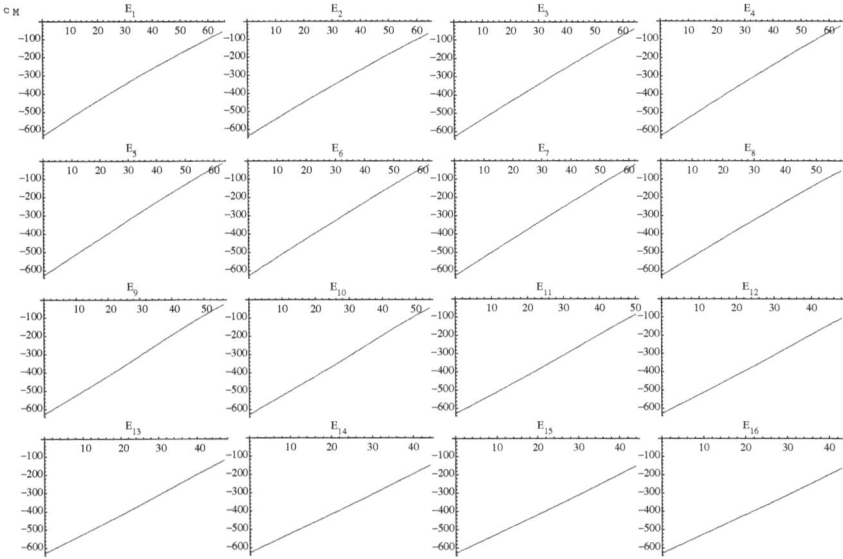

Figure 7.18. The coefficient $c_M(\tau)$ entering the ansatz for the effective action of Eq. (7.35) specializing to Eq. (7.51) for T-ordered ensembles up to $M = 16$.

action

$$S_M = \frac{L(\tau)^2}{A(\tau)}\left(1 + \frac{c_M(\tau)}{L(\tau)}\right)$$

and the initial condition $\bar{L}_M(\tau = 0) = \tilde{L}_M(\tau = 0)$. In Fig. 7.20 the corresponding plots of $\Delta_{M,\text{int}}(\tau)$ are depicted as obtained with the action

$$S_M = \frac{L(\tau)^2}{A(\tau)}\left(1 + \frac{c_M(\tau)}{A(\tau)}\right)$$

and subject to the initial condition $\bar{L}_M(\tau = 0) = \tilde{L}_M(\tau = 0)$. Relaxing the constraint of T-ordering ($E_M \to E_M'$) does not entail a qualitative change of the results.

The results presented in Figs. 7.19 and 7.20 are quite unexpected since within the $n = 0$ sector the variance of the COM saturates rapidly to finite values (see Figs. 7.10 and 7.11). In contrast, for the $n = 1$ sector the variance of the location of selfintersection initially

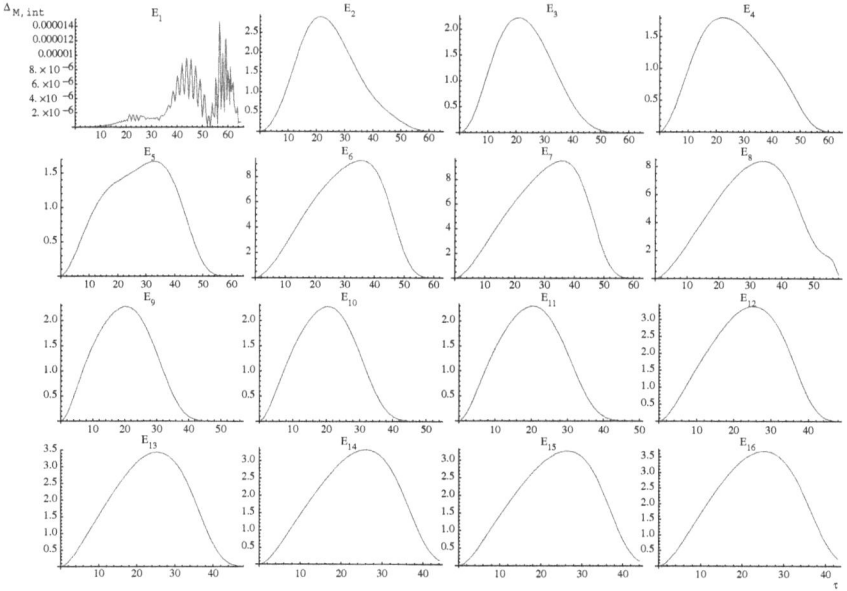

Figure 7.19. Plots of $\Delta_{M,\text{int}}(\tau)$ for the T-ordered ensembles E_M with $M = 1, \ldots, 16$. The ansatz $S_M = \frac{L(\tau)^2}{A(\tau)}\left(1 + \frac{c_M(\tau)}{L(\tau)}\right)$ for the geometric action was employed. The result for E_1, which theoretically should be identically nil, represents a measure for the numerical accuracy (compare with the results for $E_{M>1}$).

increases, reaches a maximum, and decreases to zero at a *finite* value of τ. Thus we conclude that within the $n = 1$ sector a particular member of E_M (or E'_M) is singled out by its weight being close to unity, starting at a finite value of τ, and the contribution of all other members of the ensemble then is severely suppressed in the partition function (emergence of order). This is a remarkable consequence of the fact that all members of the ensemble, though explicitly void of interactions,[22] are correlated via their common environment. Mathematically, such a correlation is a consequence of the renormalization-group requirement of Eq. (7.50).

_____

[22]The $n = 1$ partition functions consider the contribution of single $n = 1$ particle states only.

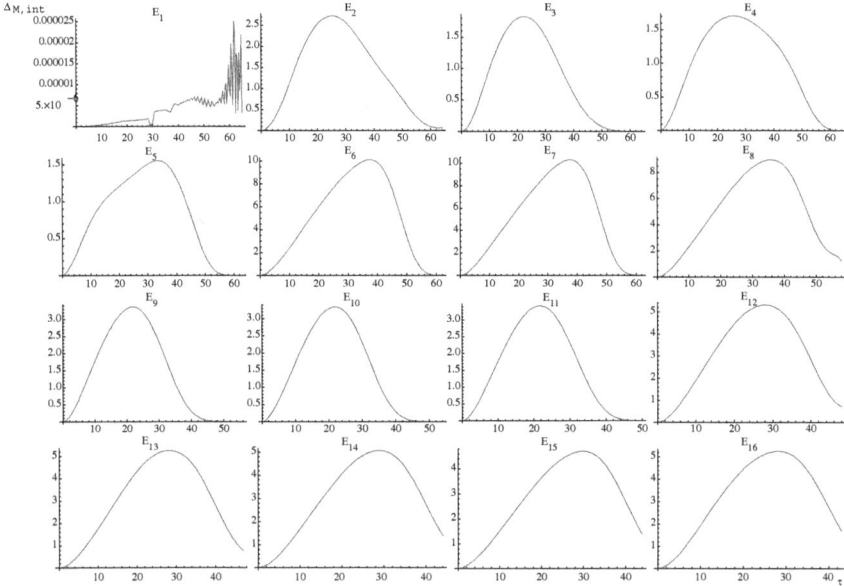

Figure 7.20. Plots of $\Delta_{M,\text{int}}(\tau)$ for the T-ordered ensembles E_M with $M = 1, \ldots, 16$. The ansatz $S_M = \frac{L(\tau)^2}{A(\tau)}\left(1 + \frac{c_M(\tau)}{A(\tau)}\right)$ for the geometric action was employed. The result for E_1, which theoretically should be identically nil, represents a measure for the numerical accuracy (compare with the results for $E_{M>1}$).

We may readily confirm the emergence of order in the $n = 1$ sector by evaluating the entropy Σ_M. The τ dependence of entropy Σ_M is defined as

$$\Sigma_M(\tau) \equiv \log Z_M + \frac{1}{Z_M} \sum_{i=1}^{M} \exp\left(-S_M[\hat{x}_i(\tau)]\right) S_M[\hat{x}_i(\tau)], \qquad (7.55)$$

where $S_M[\hat{x}_i(\tau)]$ is given by Eq. (7.35). In Fig. 7.21 plots are shown for $\Sigma_M(\tau)$ (when evaluated with the action $S_M = \frac{L(\tau)^2}{A(\tau)}\left(1 + \frac{c_M(\tau)}{L(\tau)}\right)$ for T-ordered ensembles of size $M = 1, \ldots, 16$. These graphs look much like the ones generated using the action $S_M = \frac{L(\tau)^2}{A(\tau)}\left(1 + \frac{c_M(\tau)}{A(\tau)}\right)$. Notice the continuous approach of zero at finite values of τ. This, again, implies that order emerges in the system when decreasing the resolution.

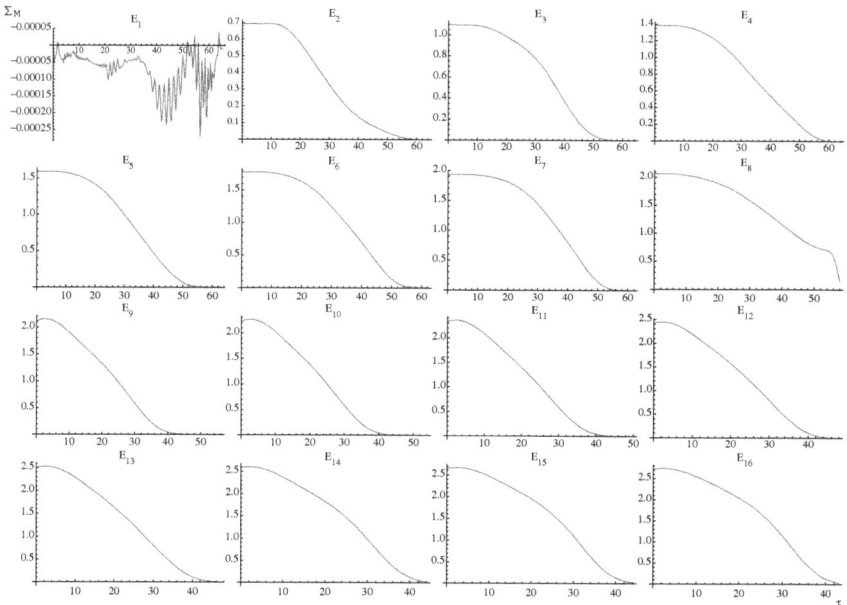

Figure 7.21. Flow of the entropies Σ_M for T-ordered ensembles of size $M = 1, \ldots, 16$ when evaluated with the action $S_M = \frac{L(\tau)^2}{A(\tau)} \left(1 + \frac{c_M(\tau)}{L(\tau)}\right)$. The situation does not change qualitatively if the action $S_M = \frac{L(\tau)^2}{A(\tau)} \left(1 + \frac{c_M(\tau)}{A(\tau)}\right)$ is used.

Let us summarize our results. As a consequence of the curve-shrinking flow and renormalization-group invariance we observe the unexpected result that a statistical ensemble of $n = 1$ curves, immersed in a two-dimensional plane, starts to create order below some finite resolution in the sense that, starting from finite values of τ, only a particular member of the ensemble survives the evolution. That is, the entropy attributed to the ensemble is practically zero for sufficiently large values of τ. For the location of the selfintersection (an (anti)monopole), which may be interpreted as the charge of an electron (or muon or tau-lepton) in physics applications, this means that no dissipation of energy, as provided by the environment, can be mediated by the (anti)monopole if the resolution falls below a critical, *finite* value. The validity of this remarkable result depends on the two-dimensionality of space and the fact that we consider the sector with $n = 1$. One may thus speculate that the reason why electrons in ion arsenide systems appear highly ordered below a finite, critical temperature (high-T_c superconductivity) is their extendedness (one-fold selfintersecting center-vortex loops in a confining-phase SU(2) Yang–Mills theory) into the two-dimensional plane.

Problems

7.1. Exhibit why the potentials defined in Eqs. (7.3) and (7.4) are modulo rescalings by U(1)-invariant functions of Φ and $\bar{\Phi}$ and up to addition of terms of the form $\Delta V = \kappa(\tilde{\Lambda}^2 - \tilde{\Lambda}^{-2(m-1)}(\bar{\Phi}\Phi)^m)^{2k}$ $(\kappa > 0, k = 1, 2, \ldots, m)$, uniquely determined by conditions (i) through (iv) of Sec. 7.1.2.

7.2. Argue why N_n in Eq. (7.8) represents the number of connected vacuum bubble diagrams with n vertices in a perturbatively accessed $\lambda\phi^4$ quantum field theory.

References

Altschuler, S. J. and Grayson, M. A. (1991). Shortening space curves and flow through singularities, *IMA Preprint Series*, **823**.

Altschuler, S. J. (1991). Singularities of the curve shrinking flow for space curves, *J. Differential Geometry*, **34**, 491.

Bender, C. M. and Wu, T. T. (1969). Anharmonic oscillator, *Phys. Rev.*, **184**, 1231.

Bender, C. M. and Wu, T. T. (1971). Large order behavior of perturbation theory, *Phys. Rev. Lett.*, **27**, 461.

Bender, C. M. and Wu, T. T. (1976). Statistical analysis of Feynman diagrams, *Phys. Rev. Lett.*, **37**, 117.

Freitag, E. and Busam, R. (1995). *Funktionentheorie* (Springer), p. 204, Theorem 1.14.

Gage, M. and Hamilton, R. S. (1986). The heat equation shrinking convex plane curves, *J. Differential Geometry*, **23**, 69.

Giacosa, F., Hofmann, R., and Schwarz, M. (2006). Explosives Z pinch, *Mod. Phys. Lett. A*, **21**, 2709.

Grayson, M. A. (1987). The heat equation shrinks embedded plane curves to round points, *J. Differential Geometry*, **26**, 285.

Grayson, M. A. (1989). The shape of a figure-eight under the curve shortening flow, *Invent. Math.*, **96**, 177.

Hagedorn, R. (1965). Statistical thermodynamics of strong interactions at high energies, *Nuovo Cim. Suppl.*, **3**, 147.

Hofmann, R. (2000). BPS saturated vacua interpolation along one compact dimension, *Phys. Rev. D*, **62**, 065012.

Hofmann, R. (2005). Yang–Mills thermodynamics: The confining phase, arXiv: hep-th/0508212.

Hofmann, R. (2007). Yang–Mills thermodynamics at low temperature, *Mod. Phys. Lett. A*, **22**, 2657.

't Hooft, G. (1978). On the phase transition towards permanent quark confinement, *Nucl. Phys. B*, **138**, 1.

Lewin, L. (1981). *Polylogarithms and Associated Functions* (Elsevier North Holland), p. 236, Eq. (7.193).

Moosmann, J. and Hofmann, R. (2008a). Evolving center-vortex loops, *ISRN Math. Phys.*, **2012**, Article ID 236783.

Moosmann, J. and Hofmann, R. (2008b). Center-vortex loops with one selfintersection, *ISRN Math. Phys.*, **2012**, Article ID 601749.

Perelman, G. (2002). The entropy formula for the Ricci flow and its geometric applications, arXiv: math/0211159.

Perelman, G. (2003a). Ricci flow with surgery on three-manifolds, arXiv: math/0303109.

Perelman, G. (2003b). Finite extinction time for the solutions to the Ricci flow on certain three-manifolds, arXiv: math/0307245.

Reinhardt, H. (2002). Topology of center-vortex loops, *Nucl. Phys. B*, **628**, 133.

Smith, S., Broucke, M. E., and Francis, B. A. (2005), *Proc. 44th IEEE Conference on Decision and Control, and European Control Conference*, Seville, Spain, Dec. 12–15.

Wood, D. (1992). Technical report, University of Kent Computing Laboratory, **15-92**, June, University of Kent, Canterbury, UK.

PART 2

Applications

Chapter 8

The Approach of Thermal Lattice Gauge Theory

The present chapter compares results for thermodynamical quantities, obtained by the essentially analytic continuum approach to deconfining and preconfining SU(2) and SU(3) Yang–Mills thermodynamics discussed in Chapters 5 and 6, with results generated by an alternative approach: lattice gauge theory. For a comprehensive introduction to the well-developed method of lattice gauge theory the reader is referred to [Rothe (2005)].

The lattice-gauge theory approach to four-dimensional Yang–Mills theory, though strongly dependent on numerical simulation, is widely considered the only first-principle method capable of extracting quantitative information on thermodynamical quantities that depend sensitively on the so-called soft sector. In some cases lattice gauge theory indeed is highly successful in this respect. As we shall argue however, the point of view that practiced lattice gauge-theory represents a universally precise way of addressing Yang–Mills theory nonperturbatively appears to be too optimistic. Namely, the most severe limitation of lattice gauge theory in describing the continuum physics of thermalized Yang–Mills theory is its formulation on a discrete spacetime lattice with Euclidean signature and lattice constant $a > 0$.

On one hand, the finite resolution imposed by a certain choice of a requires the lattice action to be a coarse-grained version of the continuum Yang–Mills action. In actual simulations, the coarse-graining process often is for pragmatic reasons assumed to be dominated by perturbation theory. This implies the use of a naive discretization of the continuum action [Wilson (1974)] with all resolution dependence carried by the coupling constant running perturbatively[1] (Wilson action). Such a form of the lattice action may be justifiable in the simulation of observables, which do not sensitively depend on nonperturbative infrared physics, but leads to incorrect results due to the omission of topologically nontrivial field configurations in the coarse-graining process. So-called perfect lattice actions, which work their way around this problem, are expensive to simulate (see [Niedermeyer (1999); Hasenfratz (1998)] for a general discussions), and therefore are rarely used.

On the other hand, the assumption of a Euclidean signature on the lattice by definition excludes a proper description of nonthermal effects. But, according to Chapter 7, the decay process of the monopole-antimonopole condensate into center-vortex loops and the subsequent formation of a condensate of massless center-vortex loops are highly nonthermal.

Finally, lattices are of finite spatial extent. In particular, in the vicinity of (second-order) phase transitions, where correlation lengths diverge due to the existence of massless excitations, these strong correlations are captured only partially by manageable lattice sizes.

So considering lattice gauge theory as a convenient physics-simulating tool is dangerous; this attitude has a potential to be misleading. However, to employ the lattice as a means to learn about the role played by topologically implied extended degrees of freedom (see [Belavin *et al.* (1975); 't Hooft (1974); Polyakov (1974); Abrikosov (1957); Nielsen and Olesen (1973)]) in the emergence of important collective phenomena is highly valuable and indeed has motivated much of the work presented in Chapters 5–7.

---

[1]No higher-dimensional operators creep up in such a coarse-graining owing to the perturbative renormalizability of the theory (see discussion in Sec. 2.1).

8.1   Pressure, Energy Density, and Entropy Density

Without explaining the lattice gauge theory method [Rothe (2005)] we compare in this section the results for thermodynamical quantities with those obtained in the effective theories of Chapters 5 and 6. We strongly adhere to the presentation of [Hofmann (2005)]. At the level of 1% accuracy it is sufficient to work with one-loop precision when discussing thermodynamical quantities in the effective theory for the deconfining phase.[2]

SU(2) case

In the deconfining phase the ratio of pressure P and T^4 is given as

$$\frac{P}{T^4} = -\frac{(2\pi)^4}{\lambda^4}\left[\frac{2\lambda^4}{(2\pi)^6}(2\bar{P}(0) + 6\bar{P}(2a)) + 2\lambda\right], \qquad (8.1)$$

where the function $\bar{P}(a)$ and the dimensionless mass parameter a are defined in Eqs. (5.66) and (5.67), respectively. In the preconfining phase we have

$$\frac{P}{T^4} = -\frac{(2\pi)^4}{\bar{\lambda}^4}\left[\frac{6\bar{\lambda}^4}{(2\pi)^6}\bar{P}(a) + \frac{\bar{\lambda}}{2}\right]. \qquad (8.2)$$

In the deconfining phase (compare with Sec. 5.2.2), the ratio of energy density ρ and T^4 is given as

$$\frac{\rho}{T^4} = \frac{(2\pi)^4}{\lambda^4}\left[\frac{2\lambda^4}{(2\pi)^6}(2\bar{\rho}(0) + 6\bar{\rho}(2a)) + 2\lambda\right], \qquad (8.3)$$

where the function $\bar{\rho}(a)$ is defined as

$$\bar{\rho}(a) \equiv \int_0^\infty dx\, x^2 \frac{\sqrt{x^2 + a^2}}{\exp(\sqrt{x^2 + a^2}) - 1}. \qquad (8.4)$$

---

[2]Recall that in the preconfining phase thermodynamical quantities are one-loop exact.

In the preconfining phase (compare with Sec. 6.2.2), we have

$$\frac{\rho}{T^4} = \frac{(2\pi)^4}{\bar{\lambda}^4}\left[\frac{6\bar{\lambda}^4}{(2\pi)^6}\bar{\rho}(a) + \frac{\bar{\lambda}}{2}\right]. \tag{8.5}$$

The ratio of entropy density s and T^3 is given as

$$\frac{s}{T^3} = \frac{1}{T^4}(\rho + P). \tag{8.6}$$

Because the ground-state contributions of ρ and P cancel in $\frac{s}{T^3}$ this quantity is not as infrared-sensitive as, e.g., $\frac{\rho}{T^4}$ or $\frac{P}{T^4}$: Lattice gauge-theory simulations using the Wilson action, thus ignoring the long-distance correlating effects of topological, nonperturbative field configurations in the process of coarse-graining the continuum action, are in a position to measure the entropy density at low temperatures within reasonable precision. Recall that topological configurations give rise to a departure from the naive perturbative form invariance of the Yang–Mills action — its closest analog in lattice gauge theory being the Wilson action — under the renormalization flow towards lower resolution.

| SU(3) case |

In the deconfining phase (compare with Sec. 5.2.2), we have

$$\frac{P}{T^4} = -\frac{(2\pi)^4}{\lambda^4}\left[\frac{2\lambda^4}{(2\pi)^6}(4\bar{P}(0) + 3(4\bar{P}(a) + 2\bar{P}(2a))) + 2\lambda\right] \tag{8.7}$$

and

$$\frac{\rho}{T^4} = \frac{(2\pi)^4}{\lambda^4}\left[\frac{2\lambda^4}{(2\pi)^6}(4\bar{\rho}(0) + 3(4\bar{\rho}(a) + 2\bar{\rho}(2a))) + 2\lambda\right]. \tag{8.8}$$

In the preconfining phase (compare with Sec. 6.2.2), we have

$$\frac{P}{T^4} = -\frac{(2\pi)^4}{\bar{\lambda}^4}\left[\frac{12\bar{\lambda}^4}{(2\pi)^6}\bar{P}(a) + \bar{\lambda}\right] \tag{8.9}$$

P/T⁴

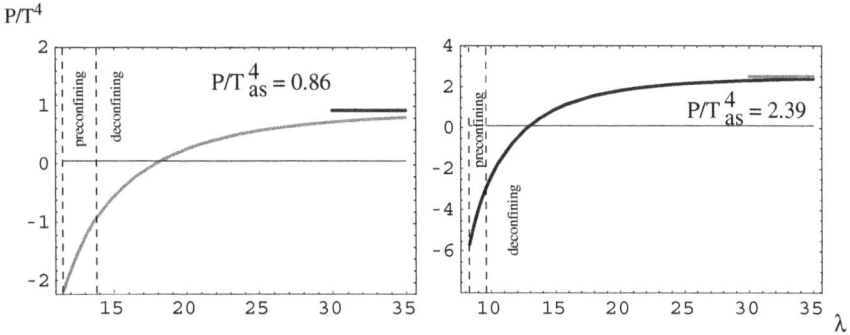

Figure 8.1. $\frac{P}{T^4}$ as a function of temperature for SU(2) (left panel) and SU(3) (right panel). The thick horizontal lines to the right indicate the respective asymptotic values (large λ); the dashed vertical lines are the phase boundaries.

and

$$\frac{\rho}{T^4} = \frac{(2\pi)^4}{\bar{\lambda}^4}\left[\frac{12\bar{\lambda}^4}{(2\pi)^6}\bar{\rho}(a) + \bar{\lambda}\right]. \tag{8.10}$$

The ratio of entropy density s and T^3 is given in Eq. (8.6), where now the SU(3) expressions for P and ρ are used. In converting $\bar{\lambda}$ to λ relations (6.35) are applied.

The result for $\frac{P}{T^4}$ is plotted in Fig. 8.1 as a function of temperature throughout the deconfining and preconfining phases where thermal equilibrium is intact. Figure 8.2 depicts SU(3) lattice gauge theory results. Notice that the pressure is negative in the deconfining phase close to λ_c and even more so in the preconfining phase where the ground state strongly dominates the thermodynamics of infrared-sensitive quantities. Notice also the negative pressure in Fig. 8.2 as obtained close to the phase transition. Here the differential method is used in the lattice gauge theory simulation. (For a discussion of differential versus integral method see Sec. 8.2 below.) Comparing Figs. 8.1 and 8.2, we conclude that the finite-size constraints of realistic lattices have a severe effect on the obtained values for the pressure shortly above the deconfining-preconfining transition, and even more so below this transition.

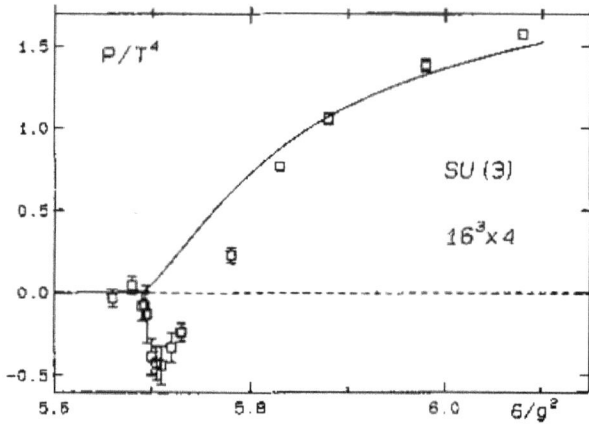

Figure 8.2. $\frac{P}{T^4}$ as a function of temperature for SU(3) as obtained on a ($16^3 \times$ 4) lattice using the differential method with a universal two-loop perturbative β function [Brown *et al.* (1988);Deng (1988)] and using the integral method (solid line) [Engels *et al.* (1990)]. The figure is taken from [Engels *et al.* (1990)] (kind permission by Elsevier under licence number 2538190971886).

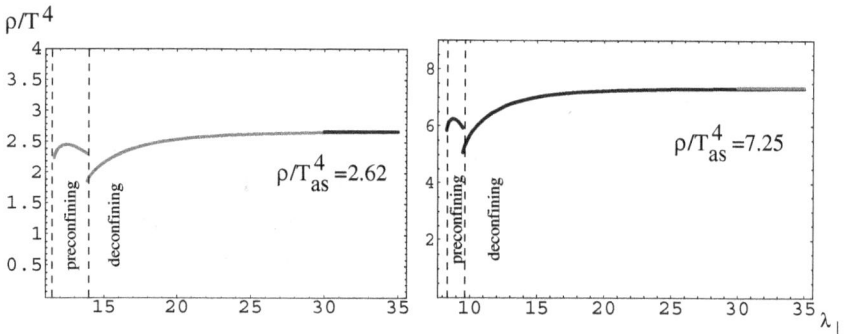

Figure 8.3. $\frac{\rho}{T^4}$ as a function of temperature for SU(2) (left panel) and SU(3) (right panel). The thick horizontal lines to the right indicate the respective asymptotic values (large λ), the dashed vertical lines are the phase boundaries.

The result for $\frac{\rho}{T^4}$ as a function of temperature throughout the deconfining and preconfining phases is shown in Fig. 8.3. Figure 8.4 depicts an SU(2) lattice result [Engels *et al.* (1982)]. Notice in Fig. 8.3 (small) discontinuities at λ_c. Their occurrence is explained by the

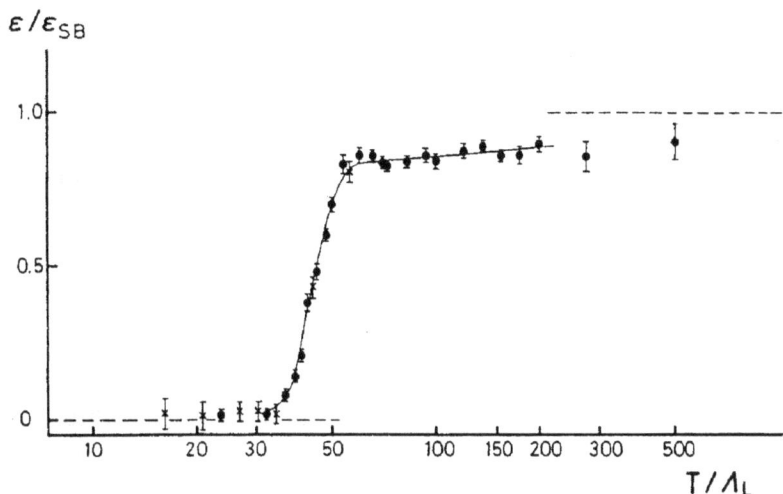

Figure 8.4. $\frac{p}{T^4}$ as obtained from the SU(2) lattice simulation in [Engels *et al.* (1982)] (kind permission by Elsevier under licence number 2538200058910).

fact that by crossing the deconfining–preconfining phase boundary the system needs to generate an extra polarization for each dual gauge mode compared to the two polarizations of a tree-level massless mode. Extra polarizations are extra fluctuating degrees of freedom which increase the energy density on the preconfining side of the phase boundary. The situation is somewhat peculiar: On one hand, the order parameter m (mass of the dual gauge mode in the preconfining phase) for the dynamical breaking of the dual gauge groups $U(1)_D$ [SU(2)] and $U(1)_D^2$ [SU(3)] is continuous. On the other hand, there is a small amount of latent heat being released across the *preconfining–deconfining* transition. (That is, by heating up the system starting in the preconfining phase.) Such a situation can be resolved by considering tunneling[3] between the deconfining trajectory, continued into the preconfining phase (supercooling), and the preconfining trajectory (see Sec. 6.2.3). Again, the energy density is

---

[3]In this way, the order parameter m fluctuates at a given value of T, and no latent heat is discontinuously released.

dominated by the ground state contribution in the deconfining phase close to the deconfining–preconfining transition and even more so in the preconfining phase. Notice also that $\rho = -P$ at the point $T_{c'}$, where the system starts to condense center-vortex loops at the pre-confining phase boundary, that the preconfining phase is narrower for SU(3) than it is for SU(2), and that $\frac{\rho}{T^4}$ dips in a much steeper way at the deconfining–preconfining transition for SU(3) than for SU(2).

The result for the interaction measure $\frac{\Delta}{T^4} \equiv \frac{(\rho - 3P)}{T^4}$ is shown in Fig. 8.5. Figures 8.6 and 8.7(b) are lattice results. Notice the rapid approach to the free-gas limit in Fig. 8.5 and the large values of $\frac{\Delta}{T^4}$ in the preconfining phase. Interestingly, there is a small bump to the left of the phase boundary in Fig. 8.6.

The result for the ratio of the specific heat per unit volume $c_V \equiv \frac{d\rho}{dT}$ and T^3 is shown in Fig. 8.8 (Fig. 8.9 is an SU(2) lattice result [Engels *et al.* (1982)].) The quantity $\frac{c_V}{T^3}$ peaks both at the deconfining–preconfining and the preconfining–confining transition. The finite peak at the former phase boundary is in agreement with the deconfining–preconfining transition being essentially second-order. Moreover, we have $\frac{c_V}{T^3}|_{T_c;SU(3)} \sim 3\frac{c_V}{T^3}|_{T_c;SU(2)}$. This explains why lattice simulations prefer to identify the confining transition, which we

Δ/T^4

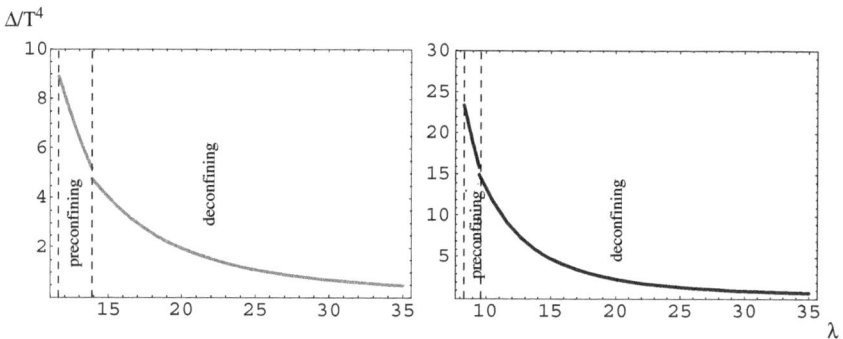

Figure 8.5. The interaction measure $\frac{\Delta}{T^4}$ as a function of temperature for SU(2) (left panel) and SU(3) (right panel). The asymptotic value in both cases is $\frac{\Delta}{T^4} = 0$; the dashed vertical lines are the phase boundaries.

$(\varepsilon - 3P)/T^4$

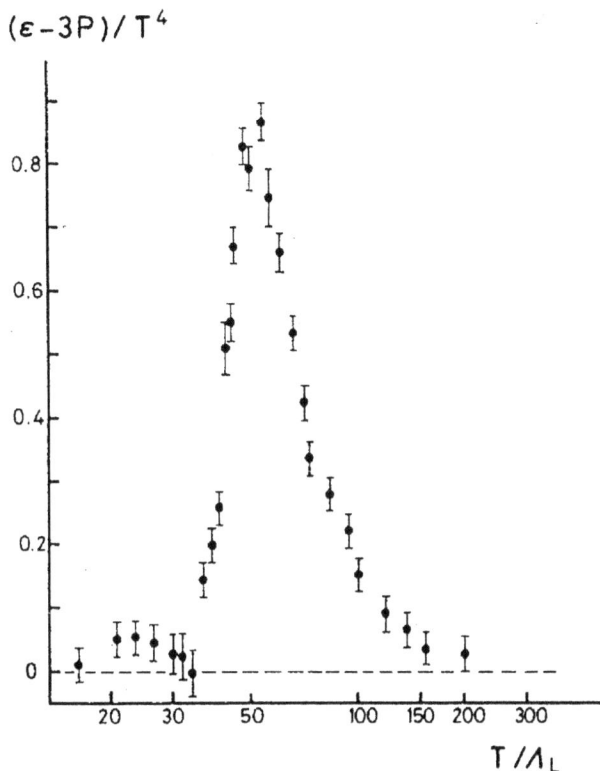

Figure 8.6. $\frac{\Delta}{T^4}$ as obtained from the SU(2) lattice simulation in [Engels *et al.* (1982)] (kind permission by Elsevier under licence number 2538200058910).

expose here as the deconfining–preconfining transition, as weakly first-order for SU(3).

The result for $\frac{s}{T^3}$ is shown in Fig. 8.10. Figure 8.11 depicts a lattice result for SU(3) obtained with the differential method [Brown *et al.* (1988)]. The entropy density *s* is a measure for the "mobility" of gauge modes. Notice the jump of s/T^3 which, again, is explained by the additional polarization of the dual gauge mode in the preconfining phase. Notice also that $\frac{s}{T^3}$ vanishes at the point $T_{c'}$ where the system starts to condense center-vortex loops. At this point dual gauge modes are infinitely heavy: The thermodynamics is completely determined by the ground state. The numerical agreement

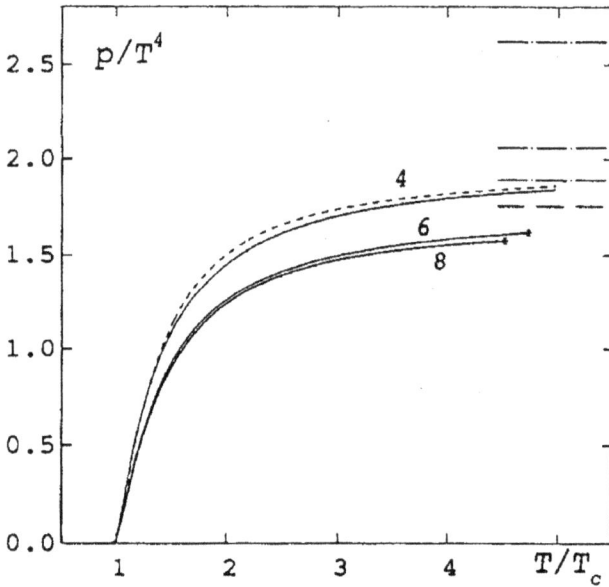

(a)

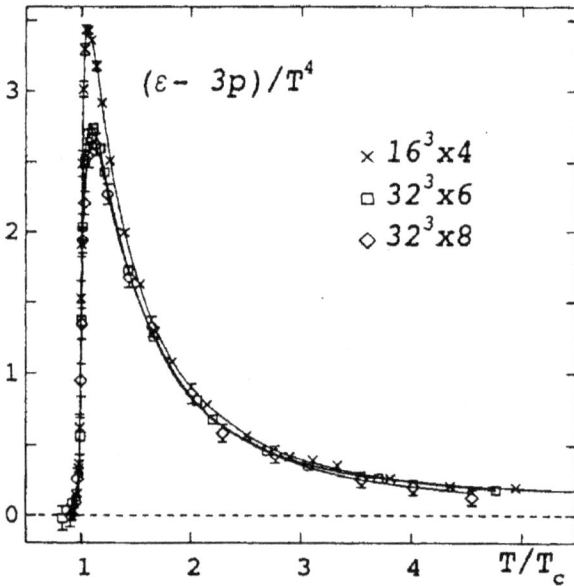

(b)

Figure 8.7. $\frac{P}{T^4}$ and $\frac{\Delta}{T^4}$ as obtained from the SU(3) lattice simulation in [Boyd *et al.* (1995, 1996)] (kind permission by Elsevier under licence number 2538200730276).

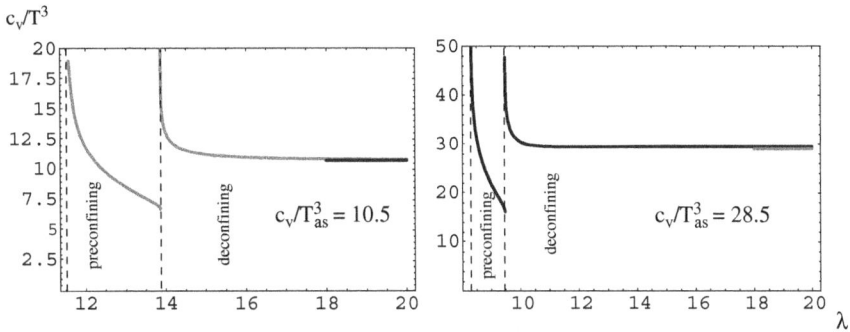

Figure 8.8. $\frac{c_v}{T^3}$ as a function of temperature for SU(2) (left panel) and SU(3) (right panel). The horizontal lines signal the respective asymptotic values; the dashed vertical lines are the phase boundaries.

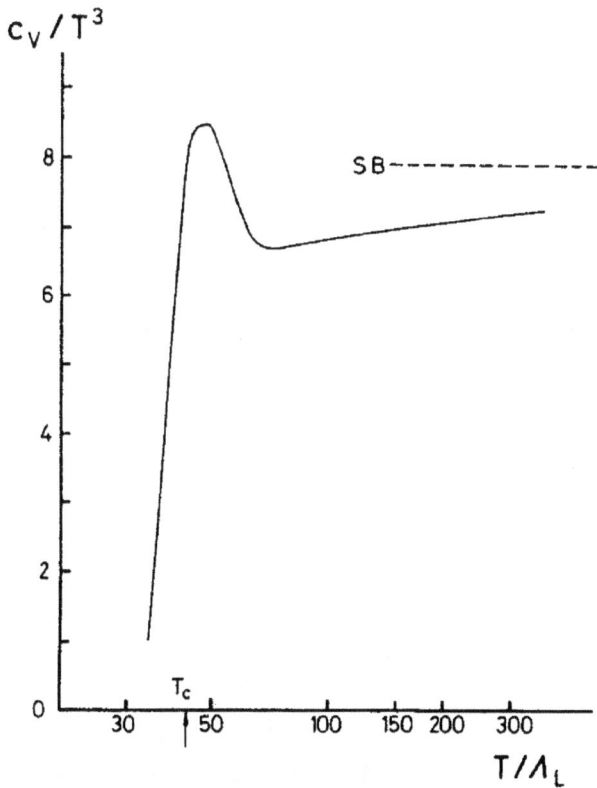

Figure 8.9. $\frac{c_v}{T^3}$ as obtained from the SU(2) lattice simulation in [Engels *et al.* (1982)] (kind permission by Elsevier under licence number 2538200058910).

s/T^3

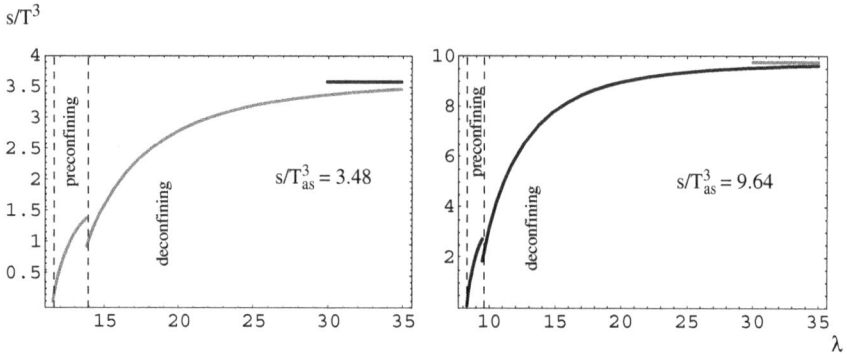

Figure 8.10. $\frac{s}{T^3}$ as a function of temperature for SU(2) (left panel) and SU(3) (right panel). The thick horizontal lines to the right signal the respective asymptotic values.

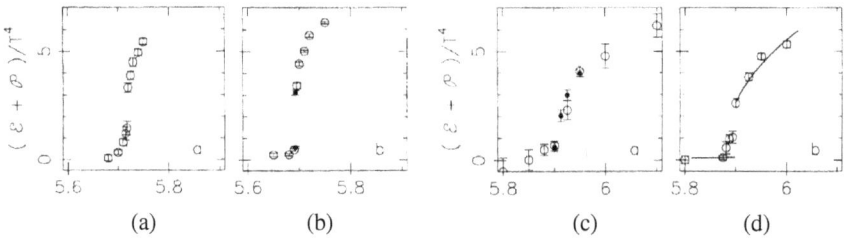

Figure 8.11. $\frac{s}{T^3}$ as a function of β obtained in SU(3) lattice gauge theory using the differential method and a perturbative beta function [Deng (1988)]. The simulations were performed on lattices of size (a) $16^3 \times 4$, (b) $24^3 \times 4$, (c) $16^3 \times 6$ (open circles) and $20^3 \times 6$ (closed circles), and (d) $24^3 \times 6$. Using the $(24^3 \times 6)$-lattice, the critical value of β is between 5.8875 and 5.90.

between the lattice result (d) (largest lattice) in Fig. 8.11 and the SU(3) result in Fig. 8.10 is striking. The two data points to the left of the jump in (d) indicate that an ambiguity exists for the value of $\frac{s}{T^3}$ very close to the transition. The jump itself corresponds to the large slope of $\frac{s}{T^3}$ on the deconfining side of the phase boundary in Fig. 8.10. The observed agreement is explained by the small sensitivity of the quantity $\frac{s}{T^3}$ on the ground-state physics making the finite-volume lattice simulation (differential method) reliable close to the deconfining–preconfining transition.

8.2 Differential versus Integral Method: Thermodynamical Quantities

What are the reasons for the qualitative difference between the pressure results obtained in the lattice gauge-theory simulations of [Boyd *et al.* (1995, 1996)] using the integral method and those of [Brown *et al.* (1988); Deng (1988)] using the differential method? While the differential method is based on the definition

$$P = T\frac{\partial \ln Z}{\partial V},$$ (8.11)

which is proper for a lattice of *finite* volume V, the integral method assumes the thermodynamical limit $V \to \infty$ from the start. In this limit one has

$$P = T\frac{\ln Z}{V},$$ (8.12)

and thus the pressure equals minus the free energy density. In Eqs. (8.11) and (8.12) Z denotes the partition function.

The official reason for the introduction of the integral method (see for example [Engels *et al.* (1990)]) was to avoid the use of the imprecisely known β function in the strong-coupling regime of the theory.[4] A prescription involving the subtraction of the zero-temperature result (symmetric lattice) is used in [Boyd *et al.*

---

[4]For experts: Based on the definition in Eq. (8.11), the β function multiplies the sum of spatial and time plaquette averages in the expression for the pressure. When using the definition in Eq. (8.12), the derivative of the pressure with respect to the bare coupling $\bar{\beta}$ [$\bar{\beta} = \frac{6}{\bar{g}^2}$ for SU(3)] can be expressed as an expectation over minus the sum of spatial and timelike plaquettes without the beta function prefactor. Thus the pressure is, up to an unknown integration constant, obtained in terms of an integral of a sum of plaquette averages over β. The integration constant is chosen in such a way that the pressure vanishes at a temperature well below T_c. Instead of only integrating over minus the sum of spatial and timelike plaquette expectations, an extra term was added to the *integrand* [Boyd *et al.* (1995, 1996)] to assure that the pressure vanishes at $T = 0$. The added term equals

(1995, 1996)] which does not follow from the definition in Eq. (8.12). Moreover, the assumption that $P = 0$ for $T \leq 0.8 T_c$ or so is a strong bias.

The results for $P(T)$, obtained when using the integral method, show a rather large dependence on the spatial size and the time extent N_τ of the lattice [Boyd *et al.* (1995, 1996)]. We believe that this reflects the considerable deviation from the assumed thermodynamical limit for realistic lattice sizes. The problem was addressed in [Engels, Karsch, and Scheideler (1999)] where a correction factor r was introduced to relate P, obtained with the integral method, to P, obtained with the differential method. For a given value of N_τ the factor r was determined from the pressure ratio at $\bar{g} = 0$! Subsequently, this value of r was used at *finite* coupling $\bar{g}$ to extract the spatial anisotropy coefficient c_σ (essentially the β function) by demanding the equality of the pressure obtained with the integral and the differential method. In doing so, again, a zero-temperature subtraction was introduced. It may be questioned whether simply a correction factor r accounts for finite-size effects and whether it is justified to determine r in the limit of noninteracting gauge modes. In addition, it seems that the imprecise knowledge of the β-function, which contains information about fluctuations in the ultraviolet, is much less of a problem for a lattice simulation of the pressure than the missing infrared physics (finite lattice size, use of Wilson action).

Using the universal part of the two-loop perturbative beta function in the differential method, negative values for the pressure were obtained for T close to T_c in [Deng (1988)]. Moreover, rapid approaches of ρ and P to their respective free-gas limits were observed. This is in qualitative (but not quantitative) agreement with the results of the continuum theory (see Figs. 8.1 and 8.3).

---

twice the plaquette expectation taken on a symmetric lattice (the expectation at $T = 0$).

8.3 Analytical Aspects of Thermal Lattice Gauge Theory

In this section we briefly discuss the important role of lattice gauge theory in providing solid knowledge about the microscopics of quantum Yang–Mills theory.

8.3.1 *Identification of relevant field configurations*

Driven by insights [Bogomolnyi (1976); Prasad and Sommerfield (1975); 't Hooft (1974); Polyakov (1974); Abrikosov (1957); Nielsen and Olesen (1973); Harrington and Shepard (1978); Lee and Lu (1998); Kraan and Van Baal (1998 a, b)] about analytically explored low-action nonperturbative field configurations of Euclidean Yang–Mills theory, the subsequent investigation of their properties and their potential role in causing the activation of certain order parameters [Mandelstam (1975); 't Hooft (1982)] for quark confinement, numerous investigations were performed in lattice gauge theory to study the dynamical behavior of these field configurations in dependence of temperature.

In particular, the identification of calorons and the role of an overall change in holonomy during a phase transition [Bruckmann *et al.* (2005); Ilgenfritz *et al.* (2005, 2006); Garcia Perez (1999); Diakonov *et al.* (2004)], the extraction of magnetic monopole currents in maximal Abelian gauge and their influence on the Wilson loop (an order parameter for confinement) [Bornyakov *et al.* (2002); Ichie and Suganuma (1999)], and the particulars of center-vortex percolation and condensation, together with an investigation of the topological charge attributed to their intersections, see for example [Faber, Greensite, and Olejnik (2001); Langfeld, Reinhardt, and Schafke (2001); Gattnar *et al.* (2005)], was performed. Most of these studies draw their interpretations from an insightful interplay between the analytical understanding of the continuum physics of isolated field configurations in the realm of the semiclassical approximation[5] and well-aimed questions concerning their collective behavior. Often these questions can be answered conclusively by lattice simulations. In this way solid puzzle pieces in our understanding of nonperturbative gauge theory dynamics were collected over the years. Much of the intuition leading to the results of Chapters 5 through 7 is due to the results of this type of lattice-based research.

8.3.2   *Lattice manifestation of ϕ and spatial Wilson loop*

An important question is how lattice simulations could possibly measure directly the effects of the adjoint scalar field ϕ of Chapter 5. Since the emergence of this field selfconsistently is tied to the maximal resolution $|\phi|$ and since the lattice spacing a is associated with a much higher spatial resolution (perturbative scaling regime) the field ϕ is over-resolved and thus does not play a *direct* role in lattice simulations. It should be possible, however, to perform the average of the nonlocal operator tr $\int d^3x\, t^a F_{\mu\nu}(\tau,\vec{0})\ \{(\tau,\vec{0}),(\tau,\vec{x})\}F_{\mu\nu}(\tau,\vec{x})\{(\tau,\vec{x}),(\tau,\vec{0})\}$ of (5.3) in Chapter 5 over cooled lattice configurations[6] after singular gauge fixing (see Sec. 4.3.1). As a result, the sine-function dependence on τ of Fig. 5.3 should be observed. This, however, is a nontrivial and cumbersome simulation.

Let us now re-address the spatial Wilson loop. Recall that the effective theory for the deconfining phase in Sec. 5.4.4 yields a *perimeter* law while lattice gauge theory based on the Wilson action implies that this quantity obeys an *area* law at high temperatures [Borgs

---

[5]Integration of quadratic fluctuations about a given gauge-field configuration usually taken to be a minimum of the classical action (for the case of charge-one instantons and calorons see ['t Hooft (1976)] and [Diakonov *et al.* (2004)], respectively).

[6]The integration over $\vec{x}$ needs to be performed over the entire spatial extent of the lattice.

(1985)]. Here we would like to point out that lattice gauge simulations of the spatial Wilson loop employing the *Wilson* action are prone to miss certain aspects of the monopoles physics within the deconfining phase. As mentioned above, in lattice simulations the typical spatial resolution — the inverse spatial lattice spacing a — even for temperatures only a few times T_c is considerably larger [Manousakis and Polonyi (1987); Bali *et al.* (1993)] then the natural resolution[7] $|\phi|$ of the effective action associated with continuum and infinite volume Yang–Mills thermodynamics. Due to the contribution of topologically nontrivial field configurations to the partition function a renormalization-group evolved perfect lattice action certainly is more complicated than the Wilson action obtained from the continuum Yang–Mills action by naive discretization: Using the Wilson action, the perturbative scaling regime for the fundamental coupling hardly makes any reference to bulk properties of the highly nonperturbative ground state physics [Giacosa and Hofmann (2007)]. A perfect lattice action, however, is extremely expensive to compute [Luscher and Weisz (1985); Hasenfratz and Niedermayer (1994)] at finite temperature and thus is practically never used.

Appealing to the Wilson action, the introduction of a finite spatial lattice spacing a actually acts physically (which it should not) in separating monopoles from antimonopoles. In other words, the lattice partition function is *not* invariant under a change in scale of resolution $a \to a' \neq a$ at high temperatures since the only explicit resolution dependence of the Wilson action is encoded in a perturbative running of the fundamental gauge coupling. As a result, a net magnetic flux is measured through the spatial contour in lattice simulations. Thus, we suspect that these lattice computations are not sufficiently adapted to describe the subtle effects attributed to the dissociation of large holonomy (anti)calorons as they take

---

[7]By "natural" we mean that at maximal resolution $|\phi|$ the effective action density for deconfining Yang–Mills thermodynamics is of the simple form of Eq. (5.42).

place in infinite-volume Yang–Mills thermodynamics on a spacetime continuum.

8.3.3 *Where is the preconfining phase?*

Why is the preconfining phase, predicted to exist in the continuum approach of Chapters 5 and 6, undetected on the lattice in the simulations of order parameters like the Polyakov or the 't Hooft loop?

A first answer to this question relates to the finite spatial extent of the lattice cutting off long-range correlations that do occur as a result of screened and stable magnetic monopoles and anti-monopoles becoming massless as $T \searrow T_c$. Due to finite lattice sizes the deconfining phase, when simulated by lattice gauge theory, is continued towards temperatures lower than the value of T_c predicted by the continuum approach. Recall that the extent of the preconfining phase is small: $(T_c - T_{c'})/T_c \sim 0.17$ for SU(2). The fact that the formation of a stable monopole-antimonopole condensate takes place at a temperature, which is even below T_c (see Sec. 6.2.3), makes the effective region, where the dual gauge mode on average propagates with close to three polarizations, yet smaller. In lattice simulations, this provokes that the occurence of a much more severe condensation phenomenon (the condensation of center-vortex loops [de Forcrand and D' Elia (1999)]) is detected without the resolution of the preceeding step of stable monopole-antimonopole condensation.

A second answer relates to the use of the Wilson as compared to the perfect lattice action which close to the phase transition, where nonperturbatively generated operators composed of fundamental gauge fields should play an important role, is likely to be an unreliable approximation.

The third answer relates to the nonapplicability of Euclidean thermodynamics, as presumed by any thermal lattice simulation, for $T \searrow T_{c'}$. In a simulation on a Euclidean lattice nonequilibrium physics, which necessarily takes place across the Hagedorn transition discussed in Sec. 7.2, is projected onto a presumed equilibrium situation. To the author it is questionable whether and in what sense this projection captures the important aspects of the (downward) nonthermal transition and the subsequent relaxation of non-intersecting center-vortex loops into a condensate of massless particles (see also discussion in [Scheffler *et al.* (2008)]).

References

Abrikosov, A. A. (1957). On the magnetic properties of superconductors of the second group, *Sov. Phys. JETP*, **5**, 1174.

Bali, G. S., *et al.* (1993). The Spatial string tension in the deconfined phase of the (3+1)-dimensional SU(2) gauge theory, *Phys. Rev. Lett.*, **71**, 3059.

Belavin, A. A., *et al.* (1975). Pseudoparticle solution of the Yang–Mills equations, *Phys. Lett. B.*, **59**, 85.

Bogomolnyi, E. B. (1976). The stability of classical solutions, *Sov. J. Nucl. Phys.*, **24**, 449.

Borgs, C. (1985). Area law for spatial Wilson loops in high-temperature lattice gauge theory, *Nucl. Phys. B*, **261**, 455.

Bornyakov, V. G., *et al.* (2002). Anatomy of the lattice magnetic monopoles, *Phys. Lett. B*, **537**, 291.

Boyd, G., *et al.* (1995). Equation of state for the SU(3) gauge theory, *Phys. Rev. Lett.*, **75**, 4169.

Boyd, G. *et al.* (1996). Thermodynamics of SU(3) lattice gauge theory, *Nucl. Phys. B*, **469**, 419.

Brown, F. R., *et al.* (1988). Nature of the deconfining phase transition in SU(3) lattice gauge theory, *Phys. Rev. Lett.*, **61**, 2058.

Bruckmann, F., *et al.* (2005). Calorons with nontrivial holonomy on and off the lattice, *Nucl. Phys. Proc. Suppl.*, **140**, 635.

Deng, Y. (1988). The energy density and pressure in Su(3) lattice gauge theory at finite temperature, *Battavia 1988, Proc. Lattice 1988*, p. 334.

de Forcrand, P. and D'Elia, M. (1999). On the relevance of center vortices to QCD, *Phys. Rev. Lett.*, **82**, 4582.

Diakonov, D., *et al.* (2004). Quantum weights of dyons and of instantons, *Phys. Rev. D*, **70**, 036003.

Engels, J., *et al.* (1982). Gauge field thermodynamics for the SU(2) Yang–Mills system, *Nucl. Phys. B*, **205**, 545.

Engels, J., et al. (1990). Nonperturbative thermodynamics of SU(N) gauge theories, *Phys. Lett. B*, **252**, 625.

Engels, J., Karsch, F., and Scheideler, T. (1999). Determination of the anisotropy coefficients for SU(3) gauge actions from the integral and matching methods, *Nucl. Phys. B*, **564**, 302.

Faber, M., Greensite, J., and Olejnik, S. (2001). Direct Laplacian center gauge, *JHEP*, **0111**, 053.

Garcia Perez, M. (1999). Calorons on the lattice: A new perspective, *JHEP*, **9906**, 001.

Gattnar, J., et al. (2005). Center vortices and Dirac eigenmodes in SU(2) lattice gauge theory, *Nucl. Phys. B*, **716**, 105.

Giacosa, F. and Hofmann, R. (2007). Nonperturbative screening of the Landau pole, *Phys. Rev. D*, **77**, 065022, (2008).

Harrington, B. J. and Shepard, H. K. (1978). Periodic euclidean solutions and the finite-temperature Yang–Mills gas, *Phys. Rev. D*, **17**, 2122.

Hasenfratz, P. and Niedermayer, F. (1994). Perfect lattice action for asymptotically free theories, *Nucl. Phys. B*, **414**, 785.

Hasenfratz, P. (1998). Prospects for perfect actions, *Nucl. Phys. B — Proc. Suppl.*, **63**, 53.

Hofmann, R. (2005). Nonperturbative approach to Yang–Mills thermodynamics, *Int. J. Mod. Phys. A*, **20**, 4123; Erratum-*ibid. A*, **21**, 6515 (2006).

Ichie, H. and Suganuma, H. (1999). Abelian dominance for confinement and random phase property of off diagonal gluons in the maximally Abelian gauge, *Nucl. Phys. B.* **548**, 365.

Ilgenfritz, E.-M., et al. (2005). The monopole content of topological clusters: Have the KvB calorons been found? *Phys. Rev. D*, **71**, 034505.

Ilgenfritz, E.-M., et al. (2006). Calorons and monopoles from smeared SU(2) lattice fields at nonzero temperature, *Phys. Rev. D*, **73**, 094509.

Kraan, T. C. and Van Baal, P. (1998a). Exact T-duality between calorons and Taub-NUT spaces, *Phys. Lett. B*, **428**, 268.

Kraan, T. C. and Van Baal, P. (1998b). Periodic instantons with non-trivial holonomy, *Nucl. Phys. B*, **533**, 627.

Langfeld, K., Reinhardt, H., and Schafke, A. (2001). Center vortex properties in the Laplacian center gauge of SU(2) Yang–Mills theory, *Phys. Lett. B*, **504**, 338.

Lee, K. and Lu, C. (1998). SU(2) calorons and magnetic monopoles, *Phys. Rev. D*, **58**, 025011-1.

Luscher, M. and Weisz, P. (1985). Computation of the action for on-shell improved lattice gauge theories at weak coupling, *Phys. Lett. B*, **158**, 250.

Mandelstam, S. (1975). Vortices and quark confinement in nonabelian gauge theories, *Phys. Lett. B*, **53**, 476.

Manousakis, E. and Polonyi, J. (1987). Nonperturbative length scale in high temperature QCD, *Phys. Rev. Lett.*, **58**, 847.

Niedermayer, F. (1999). Perfect lattice actions, *Nucl. Phys. B — Proc. Suppl.*, **60**, 257.

Nielsen, H. B. and Olesen, P. (1973). Vortexline models for dual strings, *Nucl. Phys. B*, **61**, 45.

Polyakov, A. M. (1974). Particle spectrum in the quantum field theory, *JETP Lett.*, **20**, 194.

Prasad, M. K. and Sommerfield, C. M. (1975). An exact solution for the 't Hooft monopole and the Julia–Zee dyon, *Phys. Rev. Lett.*, **35**, 760.

Rothe, H. J. (2005). *Lattice Gauge Theories: An Introduction*, World Sci. Lect. Notes Phys., **74**, 1–605 (World Scientific publishing, 2005).

Scheffler, S., Hofmann, R., and Stamatescu, I.-O. (2008). Scalar field theory with a non-standard potential, *Phys. Rev. D*, **77**, 065015.

't Hooft, G. (1974). Magnetic monopoles in unified gauge theories, *Nucl. Phys. B.*, **79**, 276.

't Hooft, G. (1976). Computation of the quantum effects due to a four-dimensional pseudoparticle, *Phys. Rev. D*, **14**, 3432; Erratum-*ibid. Phys. Rev. D*, **18**, 2199, (1978).

't Hooft, G. (1982). The topological mechanism for permanent quark confinement in a nonabelian gauge theory, *Phys. Scripta*, **25**, 133.

Wilson, K. G. (1974). Confinement of quarks, *Phys. Rev. D*, **10**, 2445.

Chapter 9

Black-Body Anomaly

9.1 Introduction

In this chapter we discuss a potential application of deconfining and preconfining SU(2) Yang–Mills thermodynamics to a physical system: the low-temperature photon gas. Recall from Chapter 5 that the deconfining phase of an SU(2) Yang–Mills theory, after an appropriate spatial coarse-graining, possesses two vector modes that are tree-level massive and one tree-level massless "photon". Recall also that interactions between these excitations are very weak due to strong constraints on momentum transfers between and on the off-shellness of these modes. Due to a decoupling of the massive modes the photon experiences no interactions at all at the critical temperature T_c for the deconfining–preconfining transition. The effect of interactions on thermal propagation, which are described by the photon's polarization tensor, peaks at about $2\,T_c$ and decays in a power-like way for temperature increasing beyond this point (see Sec. 5.4.7.5).

Since, apart from (important) modifications at very low frequencies, the Cosmic Microwave Background (CMB) at its present temperature of $T_{\mathrm{CMB}} \sim 2.73\,\mathrm{K}$ exhibits a perfect black-body shape, one is tempted to identify the photon of deconfining SU(2) Yang–Mills thermodynamics[1] with the physical photon and to set $T_{\mathrm{CMB}} = T_c$.

---

[1]Since only one photon species is observed in nature and gauge-symmetry breaking induced by an adjoint Higgs field (deconfining phase) is as SU(2)→U(1) the gauge group SU(2) is the only candidate for a fundamental, non-Abelian gauge symmetry underlying photon physics.

In stating this, a number of questions arise: (i) Why would we enlarge the U(1) gauge principle of quantum electrodynamics (QED) which in the absence of electric charges implies that the photon is a massless and free particle? (ii) If we consider the photon to be the tree-level massless mode of an SU(2) theory in its deconfining phase, how large are the deviations from its U(1) thermal propagation properties as induced by non-Abelian effects and are these deviations reconcilable with observation/experiment? (iii) If we can assure that SU(2)-induced deviations are obervationally/experimentally admissible anomalies of the photon physics described by U(1), can we identify physical systems to track down this anomalous behavior in nature? (iv) What are the implications of an SU(2) gauge principle describing photon propagation for the other fundamental interactions so successfully described by the present Standard Model (SM) of particle physics? Recall that in the SM pointlike fermions are charged according to the fundamental representation of the gauge group $U(1)_Y \times SU(2)_W \times SU(3)_C$. (v) What are possible implications of the SU(2) thermal ground state, composed of interacting, topologically charged field configurations, for the former and for the present state of expansion and the spatial geometry of our universe?

QED was independently developed by [Feynman (1949a,b, 1950)], [Schwinger (1948a,b)], and [Tomonaga (1946)] based on contributions by Dirac, Pauli, Wigner, Jordan, Heisenberg, Fermi, Weisskopf, Nordsieck, Bethe, Bloch, Dyson, and others. QED as a U(1) gauge theory is a quantum field theory describing the interaction between electrically charged point particles and photons very successfully up to energy–momentum transfers of about 100 GeV [Straessner (2010)] by virtue of perturbation theory. A major success of QED is its prediction of a radiatively induced correction to the gyromagnetic factor[2] of the Dirac electron which up to the seventh

---

[2]The gyromagnetic factor multiplicatively renormalizes the tree-level relation between the electron spin and its magnetic dipole moment.

decimal is in agreement with experiment! However, in view of question (i) we point out two problems that are inherent to QED.

Problem 1 (QED): This problem concerns the strength of interaction at large energy–momentum transfers. The invariance of infinitely resolved (bare) n-point functions under finite resolution changes (see Sec. 2.3.2) predicts for QED that, perturbatively, the fine structure constant $\alpha = \frac{Q^2}{4\pi}$ (Q the charge of the electron) rises with an increase of energy–momentum transfer. This *asymptotic slavery* is associated with a positive beta function, compare with the opposite situation in SU(N) Yang–Mills theory (see Sec. 2.3.3). In other words, a finite value of the coupling α at infinite resolution implies vanishing coupling at any finite resolution. That is, if QED is viewed as a fundamental theory then, as a consequence of the Abelian gauge principle, the minimal, local interaction between the photon and the electron implies that no interaction takes place at all at finite energy–momentum transfer! This dilemma prompted I. Pomeranchuk and L. D. Landau to disbelief in quantum field theory.

One possible solution could be to proclaim QED an effective theory, valid only up to some finite resolution, such that the continuum limit (infinite resolution), defining the theory via its classical action, should not be taken literally. This, however, does not instruct the practitioner in how to overcome the technical gap between local definition and application at finite resolution since the *maximal* value of the latter is unknown *a priori*. As a matter of principle, such a gap can only be bridged if, in approaching the limit of infinite resolution, a theory becomes asymptotically free. But this is not the case for QED.

In non-Abelian theories, however, asymptotic freedom sets in if the number of fermion species charged fundamentally under the gauge group is sufficiently small (for calculations of the Yang–Mills beta function in covariant gauge (see ['t Hooft (1973); Politzer (1973); Gross and Wilczek (1973)]; for an earlier calculation in Coulomb gauge see [Khriplovich (1969, 1970)]). Thus, superficially seen, another possibility to avoid the problem pointed out by Pomeranchuk and Landau is to suppose that its U(1) gauge group is the result of a dynamical gauge symmetry breaking with the fundamental gauge group being non-Abelian. On closer inspection, however, a fundamental SU(2) gauge symmetry underlying photon physics actually would just imply a healthy perturbative running of the fundamental SU(2) gauge coupling g but would not address the problem of the running of the electron-photon coupling α because the electron is charged with respect to the *effective* U(1) gauge symmetry. In other words, since in SU(2) thermodynamics the gauge symmetry breaking to U(1) takes place at any temperature larger than the critical temperature T_c of the preconfining–deconfining transition the old problem of asymptotic slavery would persist unless the electron (and other charged leptons[3]) would experience drastic deviations from point-particle behavior at extremely high energy–momentum transfer and would cease to exist above some temperature T_e (T_μ etc.). Such a situation would arise if leptons were extended solitons (non-intersecting and one-fold self-intersecting center-vortex loops belonging to a product of SU(2) gauge groups see [Giacosa and Hofmann (2007)]). To summarize, the problem of asymptotic slavery in QED is not resolved by postulating the U(1) gauge group of electromagnetism to be an effective manifestation of a fundamental SU(2) gauge symmetry. It could be resolved, however, if electrically charged, leptons would turn out to be low-temperature, solitonic field configurations in non-Abelian, fundamental gauge dynamics.

Problem 2 (QED): This problem addresses the perturbative ground-state estimate in QED. Considering the electron[4] to be described by a Dirac field, the idea of a negatively charged sea of

---

[3]We consciously avoid here a discussion of composite, electrically charged particles such as the proton or the pion (see, however, [Hofmann (2005)]).
[4]From now on we consider the electron as a generic representative for the other charged lepton species.

negative-energy solutions is the basis for the impressively quantitative description of electron-positron pair creation. On the other hand, perturbative quantum fluctuations of the electron and photon fields would generate a ground-state energy density ρ_{QED}^{gs} comparable to the fourth power of the cutoff: $\rho_{QED}^{gs} \sim Q_{QED}^4$ where Q_{QED} is the scale of maximal energy–momentum transfer in QED. Taking the LEP data as a confirmation of QED up to $Q \sim 100\,\text{GeV}$, we would then conclude that $\rho_{QED}^{gs} > 10^{44}\,\text{eV}^4$ modulo the effects of the Dirac sea. Comparing this with the measured value of vacuum energy density of about $10^{-12}\,\text{eV}^4$ [Perlmutter *et al.* (1998);Riess *et al.* (1998)], we conclude that there could be a discrepancy of about 56 orders of magnitude between the value of dark energy density and ρ_{QED}^{gs} alone! This possibility[5] by itself should be sufficient motivation for searching a description beyond QED maintaining the latter's successes in describing electron-photon interactions. For the photon-physics part of QED the small nonperturbative SU(2) ground-state contribution is a first hint: Quantum corrections to the nonperturbative ground-state estimate $\rho^{gs} = 4\pi\Lambda^3 T$ in the deconfining phase of an SU(2) Yang–Mills theory are negligible (see Sec. 5.2.1). Setting T equal to the present temperature of the Cosmic Microwave Background (CMB), $T \sim 10^{-4}\,\text{eV}$, ρ^{gs} can be made smaller than the observed amount $10^{-12}\,\text{eV}^4$ of dark energy by an appropriate choice of the Yang–Mills scale Λ. Therefore, an unacceptably large contribution to the rate of accelerated cosmological expansion can be avoided in an SU(2) theory for photon physics.

Let us now turn to a brief discussion of question (ii). In Sec. 5.4.7.5 the (anti)screening effects in the thermal propagation of the tree-level massless, transverse mode were computed and are

---

[5]The negative energy density of the Dirac sea would have to be extremely fine-tuned (depth of the sea) to cancel down to a value of $10^{-12}\,\text{eV}^4$ the positive energy density arising from quantum fluctuations. Such a situation is hardly acceptable.

summarized in terms of a modified dispersion law: $p_0^2 = \vec{p}^2 + G(T, |\vec{p}|)$ where function G is positive (negative) at small (large) momentum modulus $|\vec{p}|$. Function G emerges by a thermal tadpole loop involving tree-level heavy gauge modes, and it would vanish in a pure U(1) gas where photons are free particles. We will study G's effect on the black-body spectra of energy density and radiance in Secs. 9.3 and 9.5.1, and, compared to the conventional theory, we will observe small deviations at small frequencies ω if temperature is larger but comparable to T_c. More specifically, detectable deviations are related to the spectral gap $0 \leq \omega \leq \omega^*$ where ω^* decays as $\frac{1}{\sqrt{T}}$ for $T \gg T_c$ [Ludescher and Hofmann (2008)]: According to this prediction, thermalized photons do not propagate if their frequencies ω lie in the range $0 \leq \omega \leq \omega^*$. Very closely above T_c, however, thermal photons propagate in a perfectly conventional way because the tree-level heavy modes, which are responsible for the radiative emergence of G, start to decouple by their common mass approaching a logarithmic pole. For T slightly below T_c low-frequency photons start to acquire a Meissner mass m_γ due to monopole-antimonopole condensation, and evanescence of modes with $\omega < m_\gamma$ leads to a distortion of the conventional Rayleigh–Jeans spectrum [Hofmann (2009)] (see Sec. 9.4.3). In summary, deviations in the thermal propagation of photons induced by SU(2) effects are exceptions to the rule implied by the conventional U(1) theory. These exceptions occur in terms of spectral anomalies at low frequencies for temperatures slightly below and above T_c, and they imply the possibility of an observational determination of T_c.

Turning to question (iii), sensible answers depend on the acceptable value of T_c. Let us thus identify constraints on T_c and subsequently nominate interesting (astro)physical and cosmological systems potentially exhibiting SU(2)-induced anomalies. On one hand, the fact that, modulo possible deviations at very small frequencies, the Cosmic Background Explorer (COBE) has observed a perfect black-body spectrum of the CMB [Boggess *et al.* (1992)], excludes the possibility for T_c to be below the present CMB temperature of about

2.73 K $\sim 10^{-4}$ eV. On the other hand, there seem to be sizable distortions of the CMB black-body spectrum at very low frequencies, that is, below 10 GHz [see the discussion in Fixsen *et al.* (2009)]. The observed spectral excess can not be explained by thermal emission from diffuse sources and certainly is not related to an SU(2)-induced radiative modification of the photon dispersion law [Schwarz, Hofmann, and Giacosa (2007); Ludescher and Hofmann (2008)]. (Recall that radiative modifications peak at about $2 T_c$ and that they yield *no spectral power* at all (gap) while the CMB data at very low frequencies indicate an *excess* in the spectral power [Fixsen *et al.* (2009)].) Because this excess efficiently is explained by the evanescence of low-frequency modes due to the onset of monopole-antimonopole condensation we thus are led to conclude that the critical temperature T_c of an SU(2) Yang–Mills theory describing photon propagation is slightly higher than the present temperature of the CMB. With this information at hand, the following phenomena are interesting to investigate in view of SU(2)-induced deviations from conventional photon physics: (a) cold, dilute, and old inner-galactic hydrogen clouds of line temperature between 5 and 10 K whose unusually large content of *atomic hydrogen* was pointed out in [Knee and Brunt (2001)] (see Sec. 9.4.2 and [Keller, Hofmann, and Giacosa (2007)]), (b) the orientation and magnitude of the CMB dipole [Fixsen *et al.* (1994)], (c) the anomalous suppression of the CMB temperature-temperature correlation at angles larger than 60 degrees, (d) the alignment of low lying multipoles (see Sec. 10.2 and [Szopa and Hofmann (2008); Ludescher and Hofmann (2009); Copi *et al.* (2010)] for a corresponding analysis of the data of the Wilkinson Microwave Anisotropy Probe (WMAP)), and (e) the discrepancy between the low redshift for (instantaneous) re-ioniztion of the Universe, as extracted from a spectral analysis of the light from early quasars, and the high redshift attributed to this process when deduced from the angular power spectrum of the CMB temperature-temperature correlation function (WMAP and Planck satellite, see Sec. 10.4).

Question (iv) addresses modifications, implied by SU(2)-based photon physics, of the description of other interactions contained in our present SM of particle physics. Concerning electroweak symmetry breaking, the SM's Higgs sector would be excluded by a constructible primordial nucleosynthesis: Postulating SU(2)-based photon physics, a balance between the Hubble expansion rate and the rate for electroweak proton-neutron conversion at the freeze-out temperature $T_{\rm fr} \sim 1\,{\rm MeV}$ can only then be maintained if the Fermi coupling G_F were about 12% larger than its zero-temperature value [Schwarz, Hofmann, and Giacosa (2007)]. This is because cosmological expansion would be driven by six additional relativistic degrees of freedom (three polarizations for each of the two nearly massless tree-level heavy gauge modes). A rise of G_F by 12% when increasing T from zero to $1\,{\rm MeV}$ is, however, impossible in the SM.

Since the cascade of dynamical gauge symmetry breaking

$$SU(2) \rightarrow U(1) \rightarrow Z_2^{\rm mag} \rightarrow 1$$

of an SU(2) theory, cooled through its two phase transitions, provides for decoupling gauge modes, no external Higgs mechanism is required to explain the low-energy locality of the weak interactions. The point of view that the electron and its neutrino are associated with the two (quasi)stable low-temperature excitations in the confining phase of an SU(2) Yang–Mills theory (non-intersecting and one-fold selfintersecting center-vortex loops, Sec. 7.3) assures that no ground state-energy density is generated from this sector (see Sec. 7.1.2). Moreover, the positional uncertainty of a pointlike electron in quantum mechanics would then be interpreted in terms of shifts in the intersection point of the associated center-vortex loop [Moosmann and Hofmann (2008)]. It is natural to extend this idea to the other lepton families of the Standard Model implying the existence of very weak, and very-very weak interactions. Thus one should assign indices to each of nature's potential, fundamental SU(2) gauge-group factor as follows: SU(2)$_{\rm CMB}$ for photon physics; SU(2)$_e$ for the electron, its neutrino, and the weak interactions; $SU(2)_\mu$ for the muon, its neutrino, and the very weak interactions; and $SU(2)_\tau$ for the tau-lepton, its neutrino, and the very-very weak interactions. The associated Yang–Mills scales are $\Lambda_e \sim m_e \sim 0.5\,\text{MeV}$, $\Lambda_\mu \sim m_\mu \sim 100\,\text{MeV}$, and $\Lambda_\tau \sim m_\tau \sim 1.8\,\text{GeV}$, respectively. Interactions mediated by the gauge modes of a given $SU(2)$ theory are felt universally by the excitations of the other $SU(2)$ theories through maximal mixing. Finally, for a fermion of a given $SU(2)$ theory not to be confined by the ground state of another $SU(2)$ theory, all these theories must have the same electric-magnetic parity, for otherwise the flux emanating from an isolated charge (intersection point of a center-vortex loop) to its isolated anticharge would be squeezed into a tube by the condensate of center-vortex loops of the electric-magnetically dual gauge theory. As a consequence, a confining potential would emerge. This, however, is not observed to occur in the leptonic sector.

The idea, that leptons of a given family emerge upon lowering the temperature of the associated $SU(2)$ Yang–Mills theory across its Hagedorn transition (see Sec. 7.2), would indeed imply QED and the weak, very weak, and very-very weak interactions to be effective theories in the sense that their matter sectors (confining phases) dissolve into asymptotically free gauge modes at sufficiently high temperature (or energy–momentum transfer in a local vertex generated by nonthermal scattering).

Question (v) is on the implications of pure $SU(2)$ and $SU(3)$ Yang–Mills theories for cosmology, that is, for the former and present states of expansion of our universe and of its spatial geometry. Since all $SU(2)$ and $SU(3)$ Yang–Mills theories other than $SU(2)_{\text{CMB}}$ are in their confining phase at $T_{\text{CMB}} \sim 2.73$ K the only contribution to dark energy would arise from $SU(2)_{\text{CMB}}$. But the energy density of the thermal ground state of $SU(2)_{\text{CMB}}$ only amounts to about 0.36% of the cosmologically inferred density of dark energy [Hofmann (2005)]! This and the fact that the violation of parity and charge-conjugation (CP) symmetry by the weak interactions is not explained by the above-mentioned product of $SU(2)$

and SU(3) Yang–Mills theories subject to maximal mixing and could be resolved by invoking the axial anomaly [Adler (1969); Adler and Bardeen (1969); Bell and Jackiw (1977); 't Hooft (1976); Fujikawa (1979, 1980)]. Namely, additional energy density would be produced through a quantum-anomaly induced potential for a would-be Goldstone field A of a spontaneously broken, global $U(1)_A$ symmetry. This potential would be representing topologically nontrivial gauge-field configurations of $SU(2)_{CMB}$ which invoke an anomaly in the conservation of the axial current ['t Hooft (1976)]. At the Planck scale, the latter would be embodied by massless fermions whose chiral symmetry is broken by gravitational torsion [Giacosa, Hofmann and Neubert (2008)]. This would imply a Planckian modulus for the field A: $|A| \sim 10^{19}$ GeV. This and $\Lambda_{CMB} \sim 10^{-4}$ eV together would explain the observed amount of cosmological dark energy [Frieman *et al.* (1995); Giacosa and Hofmann (2007)] (see Sec. 10.3).

9.2   The Cosmic Microwave Background (CMB)

9.2.1   *Historical remarks and results of CMB explorations*

Since the CMB is the prime physical system to which the postulate on photon propagation being fundamentally ruled by an SU(2) rather than a U(1) gauge principle is of relevance, we will briefly sketch how its existence increasingly became known and how subsequent explorations of its properties revealed details about the physics of its origin.

The discussion on the existence of a thermal relic radiation was started in 1941 by A. McKellar in the context of an analysis of interstellar absorption lines. Throughout the 1940s and 1950s Dicke, Gamow, Alpher, Herman, Le Roux, and Shmaonov continued this discussion with predictions of the present CMB temperature ranging from 2.3 to 50 K. A clear statement of the CMB's detectibility was made by Doroshkevich and Novikov (1964). The definite discovery of the CMB by Penzias and Wilson took place in 1965 [Penzias and Wilson (1965)]. These authors reported on an isotropically persistent excess antenna temperature of about 3.5 K as measured by their Dicke radiometer precision instrument which was planned for use in radio astronomy and satellite communication experiments. Thus the discovery of the CMB was not as accidental as it may seem in hindsight: This discovery could be made because of the dedicated development of an optimized microwave antenna in association with a delicate radiometer measuring the *difference* signal to an internal load. At feeble CMB radiances, such a radiometer was needed to cancel out its own noise. Also, the investigators' awareness of the above-mentioned discussion, stretching over a quarter of a century, surely was useful.

Starting in the early 1970s, the mounting evidence for the thermal, largely isotropic nature of the CMB increasingly was accepted as the signature for Big-Bang cosmology possibly including a very early epoch of inflation. It soon was realized that within Big-Bang cosmology small relative inhomogeneities in the matter density of the universe of order 10^{-4} to 10^{-5} would cause an imprint onto the CMB in terms of (primary) temperature anisotropies of a similar magnitude [Peebles and Yu (1970); Zeldovich (1972); Doroshkevich, Zeldovich, and Sunyaev (1977)]. Competing Russian and American satellite-based experiments, RELIKT-1 and COBE-FIRAS, confirmed the black-body spectral shape of the CMB down to a wave number of $2 \, \text{cm}^{-1}$ with data collection taking place throughout the 1980s and early 1990s, respectively. In 1992 data analysis from both experiments led to the discovery of primary temperature anisotropies. In addition, a dipole anisotropy, which in [Peebles and Wilkinson (1968)] was attributed to the relativistic Doppler effect invoked by a motion of our local group of galaxies at a speed of $627 \pm 22 \, \text{km/s}$ relative to the CMB rest frame in the direction of galactic longitude $l = 276 \pm 3°$ and latitude $b = 3 \pm 3°$, was discovered in the first-year data taken by the Differential Microwave Radiometer aboard of COBE [Kogut *et al.* (1993)]. Throughout the 1990s a series of ground- and balloon-based experiments quantified CMB anisotropies on smaller angular scales with the goal to measure the first acoustic peak. The latter is related to an oscillation of the plasma at the surface of last scattering (conventionally associated with a redshift of $z = 1089$; see, however, Sec. 10.4) which could not be resolved by COBE. In 2000 the BOOMERanG mission spotted a maximum of power at an angular scale of about one degree. Together with complementary cosmological data these results implied that the universe is spatially flat. Over the next three years the Degree Angular Scale Interferometer (DASI) detected polarization of the CMB, and the Cosmic Background Imager (CBI) recorded the first E-mode polarization spectrum. In 2001, a second American CMB space mission — the Wilkinson Microwave Anisotropy Probe (WMAP) — was launched by NASA. This satellite orbits around the second Earth–Sun Lagrangian point (L2) to perform precise measurements of anisotropies over the full sky. WMAP provided detailed information on the angular power spectrum of various correlation functions at angular scales below $1°$. The extracted constraints on cosmological parameters are consistent with cosmological inflation, and rather unexpected signatures were seen at large angles in the temperature-temperature correlation function. The Planck Surveyor, launched in May 2009 by ESA, also orbits L2 to measure the temperature distribution and polarization of the CMB on even smaller angular scales, and decisive results on the physics of CMB decoupling and the possible influence of gravitational waves were expected from this mission and, indeed, have been delivered. The frequency bands used by both missions, WMAP and Planck, are similar at the low frequency end (22 GHz and 30 GHz, respectively). The very low frequency regime (below 3 GHz), where strong deviations of the CMB line temperature from the baseline temperature are observed (see [Fixsen *et al.* (2009)] and references therein), are thus not addressed by WMAP and the Planck mission. A re-launch of the Absolute Radiometer for Cosmology, Astrophysics, and Diffuse Emission (ARCADE), dedicated to balloon-borne measurements of line temperature at very low frequencies, is, however, planned to enhance the dataset in this regime [Kogut (2010)].

9.3 SU(2)$_{CMB}$ and Thermal Photon Propagation

The possibility that the U(1) gauge symmetry, underlying the propagation of photons in the present SM of particle physics, emerges from a fundamental SU(2) Yang–Mills theory as a result of dynamical gauge symmetry breaking is intriguing. Recall that in a thermal version of pure and deconfining SU(2) gauge theory this symmetry breaking is induced by the effective thermal ground-state estimate. The latter occurs upon a spatial coarse-graining over spatially overlapping calorons of topological charge modulus one (see Secs. 5.1.3 and 5.1.4). The fact that below the critical temperature T_c photon mass is generated by the Meissner effect arising from a condensate of (electric) monopoles[6] seems to influence the CMB at very low frequencies (see [Fixsen *et al.* (2009)] and Sec. 9.4 below). This would fix T_c to be very closely above the present CMB baseline temperature of about 2.73 K. The purpose of the present section is to investigate the low-frequency modifications in low-temperature black-body spectra as predicted by deconfining SU(2) Yang–Mills thermodynamics.[7]

9.3.1 *Modified dispersion law and spectral energy density*

The modification of the dispersion law for thermalized photons — provided we postulate that in unitary gauge they are the tree-level massless gauge modes in the deconfining phase of SU(2)$_{CMB}$ — was computed in Sec. 5.4.7.5 in terms of the T and $|\vec{p}|$ dependence of the

---

[6]An electric-magnetically dual interpretation of the SU(2) field strength needs to be made: After partial gauge-symmetry breaking an electric U(1) field is a magnetic U(1) field and vice versa. In particular, a magnetic monopole with respect to U(1) $\subset$ SU(2) is an electric charge in the real world, see Sec. 5.3.

[7]Except for the process of the photon gas attaining a temperature by thermalization with the black-body walls no explicit interaction with electrically charged SM matter and thus no mixing of U(1)$_Y$ and SU(2)$_W$ needs to be considered, but see Secs. 9.6.1 and 9.6.2.

(anti)screening function G [compare with Eq. (5.187)]. The calculation was performed in natural units, $\hbar = k_B = c = 1$.

The spectral energy density $I(\omega)$ of a gas of free (transverse) photons in thermal equilibrium with a heat reservoir is given by the number of modes per volume, available within a frequency interval, times the average (thermal) energy per mode. One has

$$
\begin{aligned}
d\rho = I(\omega)d\omega &= 2\,\frac{d^3p}{(2\pi)^3} \times \frac{\omega}{e^{\frac{\omega}{T}} - 1} \\
&= \frac{1}{\pi^2}\frac{\omega}{e^{\frac{\omega}{T}} - 1}p^2(\omega)\frac{dp}{d\omega}d\omega \\
\Rightarrow I(\omega) &= \frac{1}{\pi^2}\frac{\omega}{e^{\frac{\omega}{T}} - 1}p^2(\omega)\frac{dp}{d\omega},
\end{aligned}
\tag{9.1}
$$

where $p \equiv |\vec{p}|$. The occurrence of the factors $p^2(\omega)$ and $\frac{dp}{d\omega}$ in Eq. (9.1) signals the need for the specification of a dispersion law $\omega = \omega(p)$. In a U(1) theory this dispersion simply is given as $\omega^2 = p^2$. For the photons of deconfining SU(2) Yang–Mills thermodynamics the dispersion law is substantially altered at low frequencies and temperatures, an effect which is described by the (anti)screening function[8] G computed in Sec. 5.4.7.5. Namely, one has

$$
p_0^2 - p^2 \equiv \omega^2 - p^2 = G(p, T).
\tag{9.2}
$$

Therefore, we obtain from Eq. (9.1)

$$
I_{U(1)}(\omega) \to I_{SU(2)}(\omega) = I_{U(1)}(\omega) \times \frac{\left(\omega - \frac{1}{2}\frac{d}{d\omega}G\right)\sqrt{\omega^2 - G}}{\omega^2}\,\theta(\omega - \omega^*),
\tag{9.3}
$$

where $\omega^* \equiv \sqrt{G(p = 0, T)}$,

$$
I_{U(1)}(\omega) \equiv \frac{1}{\pi^2}\frac{\omega^3}{\exp[\frac{\omega}{T}] - 1},
\tag{9.4}
$$

---

[8]Microscopically, this screening or antiscreening takes place by monopole-antimonopole creation or scattering of photons off monopoles and anti-monopoles, respectively.

I/T³

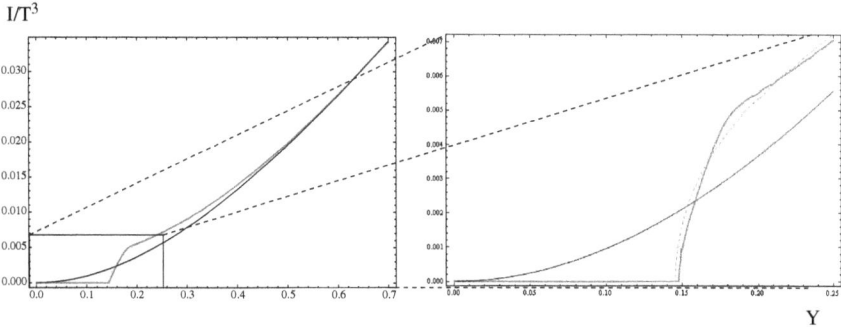

Y

Figure 9.1.   Plots of the dimensionless spectral energy density $\frac{I}{T^3}$ as a result of the one-loop selfconsistent calculation (solid gray curves) and for the approximation $p_\mu p^\mu = 0$ (dashed gray curves) at $T = 2\,T_c$. Here $Y \equiv \frac{\omega}{T}$. Solid black curves depict the conventional Planck spectrum. Notice the excellent agreement with the result of the approximate calculation.

and $\theta(x)$ is the Heaviside step function: $\theta(x) = 0$ for $x < 0, \theta(x) = 1/2$ for $x = 0$, and $\theta(x) = 1$ for $x > 0$. In Fig. 9.1 a comparison of the modification of low-frequency black-body spectral energy densities obtained in the fully selfconsistent calculation of Sec. 5.4.7.5 and in the approximation $p_\mu p^\mu = 0$ (or $\omega \equiv \pm p$) of Sec. 5.4.7.3 is depicted for the temperature $T = 2\,T_c$ as a function of $Y \equiv \frac{\omega}{T}$. Notice the excellent agreement of the approximate and the result of the one-loop selfconsistent resummation. In Fig. 9.2 comparisons between the approximate result and the result of the one-loop selfconsistent resummation are made for temperatures $T = 3\,T_c$ and $T = 4\,T_c$. Notice that, compared to the former, the latter approaches the Planck spectrum faster with rising temperature.

As we shall see in Sec. 9.5.1, only the regime of total screening (spectral gap) $0 \le Y \le Y^*$ is in principle detectable by bolometry or by radiometry (absorption of power by an antenna[9]). Recall that the

---

[9]The total energy density, that is, the integral $\int_0^\infty d\omega\, I_{U(1)}(\omega)$ is in principle detectable gravitationally. As we will see in Sec. 9.4, the Yang–Mills scale of SU(2)$_{CMB}$ is of order 10^{-4} eV, and thus *local* gravitational effects of this theory merely are of academic interest.

I/T³

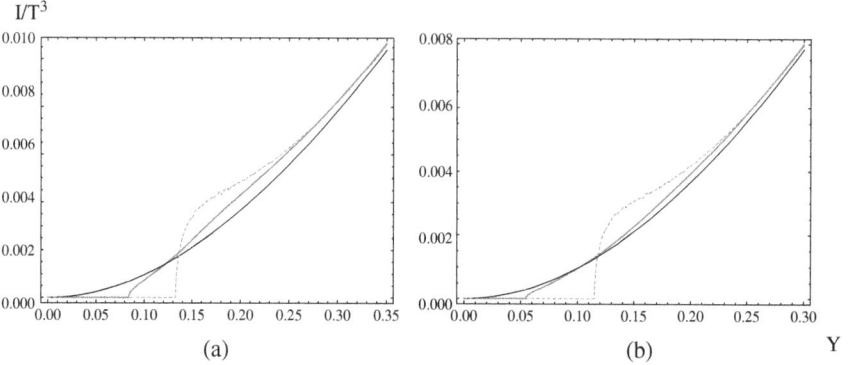

(a)

(b)

Y

Figure 9.2.   Plots of the results for the dimensionless spectral energy density $\frac{I}{T^3}$ obtained in the one-loop selfconsistent calculation (solid gray curves) and for the approximation $p_\mu p^\mu = 0$ (dashed gray curves) for (a) $T = 3\,T_c$ and (b) $T = 4\,T_c$. Here $Y \equiv \frac{\omega}{T}$. Solid black curves depict the conventional Planck spectrum. Notice the much faster approach with rising temperature of the Planck spectrum in the one-loop resummation result as compared to the result of the approximate calculation.

spectral gap $\omega^* = T\,Y$ was numerically found to decay as $T^{-1/2}$ for $T \gg T_c$ (see Fig. 5.30 and text below Eq. (5.205)).

9.4   Determination of T_c

In this section we pursue the question how nature tells us what the value of T_c is if we assume that the tree-level massless gauge mode in the deconfining phase of SU(2)$_{\text{CMB}}$ is the propagating, thermalized photon. Again, all calculations in this section are performed in natural units $\hbar = k_B = c = 1$.

9.4.1   *Very weak evidence for $T_c \sim 2.73$ K: Extragalactic magnetic fields*

Let us first sketch the observational situation. The magnitude of extragalactic magnetic fields, which are believed to be responsible for the seeding of interstellar magnetic fields (see [Kronberg (1994)] and references therein), can be bounded from above via measurements of the Faraday rotation in the polarized radio emission from distant quasars and/or distortions of the spectrum and polarization properties of the CMB [Barrow, Ferreira, and Silk (1997)]. These measurements imply upper limits on the primordial field strengths of the order of 10^{-8} to 10^{-9} Gauss. Galactic magnetic fields are several orders of magnitude stronger, and thus amplification of the primordial seed fields must occur somehow in galaxies. Several amplification mechanisms are under discussion which are adapted to the nature of the magnetic field observed (differently correlated length scales in spiral as compared to elliptical galaxies and galaxy clusters). On the other hand, an analysis of the opening angle of secondary cascade gamma rays induced by initial gamma rays of energy larger than 1 TeV, sets a lower limit of 3×10^{-16} Gauss on extragalactic magnetic fields [Neronov and Vovk (2010)]. Finally, primordial seed fields cannot be explained by magnetohydrodynamics (MHD) [Widrow (2002)]. Namely, while MHD processes can stretch, twist and amplify a given magnetic field no new field can be generated where it did not already exist. As of present there appears to be no consensus on how primordial seed fields emerge.

It is tempting to connect the nature of the highly conducting, preconfining thermal ground state of SU(2)$_{\mathrm{CMB}}$ below T_c with the occurrence of these seed fields. For a rough comparison of orders of magnitudes one may set the energy density $\rho_{\mathrm{CMB}}^{\mathrm{gs}} = 4\pi \Lambda_{\mathrm{CMB}}^3 T = 2\frac{(2\pi)^4}{(13.87)^3} T_c^4 \frac{T}{T_c}$ of the (deconfining) thermal ground state of SU(2)$_{\mathrm{CMB}}$ equal to the energy density $\rho_B = \frac{1}{2\mu_0} \vec{B}^2$ (here: SI units) of a static magnetic-field $\vec{B}$. By inserting the lower bound of 3×10^{-16} Gauss [Neronov and Vovk (2010)], one obtains at $T \sim T_c$ a lower bound of $\sim 2.35 \times 10^{-9}$ eV for T_c. To work with the energy density $\rho_{\mathrm{CMB}}^{\mathrm{gs}}$ of the deconfining ground state and to assume $T \sim T_c$ is justified by the ground-state dominance in this regime and by the narrowness of the preconfining phase (see Sec. 6.2.2). Inserting the upper bound of 10^{-9} Gauss at $T \sim T_c$, one arrives at an upper bound of $\sim 4.3 \times 10^{-6}$ eV for T_c. So in both cases the present CMB temperature T_{CMB} of 2.35×10^{-4} eV would be much higher. The existence of primordial magnetic seed fields being associated with the highly conducting, preconfining ground state of SU(2)$_{\text{CMB}}$ would, however, require that $T_c \geq T_{\text{CMB}}$. Still, primordial seed fields can be attributed to SU(2)$_{\text{CMB}}$ provided that (in contrast to the above estimate) only a tiny fraction of the ground state energy density $\rho_{\text{CMB}}^{\text{gs}}$ is actually converted to the energy density ρ_B of the magnetic field. Because of the steeply rising magnetic coupling g in the preconfining phase (see Fig. 6.3 in Sec. 6.2.2), this would only be possible if T_{CMB} is below T_c by a very small amount. For practical purposes one would thus set $T_c = T_{\text{CMB}}$. We stress that this reasoning provides only very weak evidence for $T_c = T_{\text{CMB}}$ since we had to assume that primordial seed fields exclusively are due to the ground state of SU(2)$_{\text{CMB}}$ turning preconfining at present.

Alternatively, one may associate primordial seed fields with longitudinal photon modes in the deconfining phase which describe the propagation of longitudinal magnetic fields in the real world (see end of Sec. 5.4.7.6 [Falquez, Hofmann and Baumbach (2011)]). In this case the only constraint on T_c is that during the (dynamo-mechanism driven) generation of the galactic field out of "primordial" seed fields the actual temperature of the CMB must be larger than T_c. Since the bulk of galaxies has formed well before redshift nine or smaller this gives a weak upper bound on T_c of about 10 times the present temperature T_{CMB} of the CMB. But for T_c considerably larger than T_{CMB} we would not be in a position to detect the CMB at present: Its photons would then be thermodynamically decoupled degrees of freedom. Therefore, we can tighten the above bound to $T_c \sim T_{\text{CMB}}$ which is what we had obtained already by associating preconfining ground-state effects with the emergence of "primordial" seed fields.

9.4.2   Weak evidence for $T_c \sim 2.73$ K: Cold clouds of dilute atomic hydrogen within the Milky Way

Let us now turn to an astrophysical system which implies stronger evidence that $T_c \sim 2.725$ K. Specifically, we look at the gradual metamorphosis of cold ($T = 5-10$ K) extended objects such as the hydrogen cloud GSH139-03-69 observed in between spiral arms of the outer Milky Way [Knee and Brunt (2001)]. The object possesses an estimated age of about 50 million years, exhibiting a brightness temperature T_B of $T_B \sim 20$ K with cold regions of $T_B \sim 5-10$ K and an atomic number density of ~ 1.5 cm^{-3}. The puzzle about this and similar structures relates to their high content of atomic hydrogen in view of their unexpectedly old age. Namely, numerical simulations of the cloud evolution subject to standard interatomic forces suggest a much lower time-scale of less than 10 million years (see [Goldsmith, Li, and Krčo (2007); Meyer and Lauroesch (2006); Redfield and Linsky (2007)] and references therein) for the generation of a substantial fraction of H_2 molecules.

A quantitative estimate of the increased stability of atomic hydrogen clouds due to the effects of SU(2)$_{CMB}$ in thermalized photon propagation is beyond the scope of the present book. However, in Sec. 10.1 we compute the two-point correlator of photonic energy density at $T = 5-10$ K. Assuming that $T_c = 2.725$ K, this correlator is suppressed by up to a factor of two at the typical interatomic distances in the hydrogen cloud GSH139-03-69. This implies that atomic interactions, leading to the formation of molecules, are suppressed, leaving, in turn, a higher content of atomic hydrogen than predicted by the standard theory.

To make the situation more explicit the Coulomb potential $V(r)$ ($r \equiv |\vec{x}|$) of a heavy point charge and for photons being described by a pure U(1) as well as an SU(2) gauge theory was computed in [Keller, Hofmann, and Giacosa (2007)]. In a U(1) theory the Coulomb potential is given as

$$V_{U(1)}(r) = \frac{1}{(2\pi)^3} \int d^3 p \, \frac{e^{-i\vec{p}\cdot\vec{x}}}{\vec{p}^2} = \frac{1}{2\pi^2} \int_0^\infty dp \, \frac{\sin pr}{pr}$$

$$= \frac{1}{2\pi^2 r} \int_0^\infty d\xi \, \frac{\sin \xi}{\xi} = \frac{1}{4\pi r}. \tag{9.5}$$

Going from U(1) to SU(2), we take into account the resummed one-loop polarization of Sec. 5.4.7.5 by letting $\vec{p}^2 \to \vec{p}^2 + G$ in the denominator of the integrand in Eq. (9.5). Although the function G is known for on-shell photons with $\omega^2 = \vec{p}^2 + G$ only this recipe should work well at sufficiently large distances since the off-shellness of the photon then is sufficiently small. Thus for SU(2) we approximately[10] have

$$V_{\text{SU(2)}}(r) = \frac{1}{(2\pi^3)} \int d^3p \, \frac{e^{-i\vec{p}\cdot\vec{x}}}{\vec{p}^2 + G(T, |\vec{p}|, \Lambda)}$$

$$= \frac{1}{2\pi^2 r} \int_0^{\infty} dp \, \frac{p}{p^2 + G(T, p, \Lambda)} \sin pr. \qquad (9.6)$$

The integral in Eq. (9.6) needs to be performed numerically. In Fig. 9.3 both potentials, $V_{\text{U(1)}}(r)$ and $V_{\text{SU(2)CMB}}(r)$, as well as their ratio, are plotted as functions of r. The computed suppression of the Coulomb potential should have an analog in the dipole-dipole interaction responsible for the formation of hydrogen molecules from hydrogen atoms in the standard theory. Such a suppression, however, can only be relevant to cold (T $\sim$ 5–10 K) and dilute hydrogen clouds within the Milky Way [Knee and Brunt (2001)] if $T_c \sim T_{\text{CMB}}$.

9.4.3 *Evidence for $T_c \sim 2.73\,K$: CMB at very low frequencies*

Let us now discuss an observational result which provides evidence that T_c of SU(2)$_{\text{CMB}}$ is just slightly lower than the CMB baseline temperature $T_0 = T_{\text{CMB}} = 2.725\,\text{K}$.

---

[10]A point charge immersed into the plasma locally distorts the latter's ground state, and, strictly speaking, the theory to describe this distortion in a fundamental way still needs to be worked out. Yet the result of a measurement of the screening of the potential sufficiently far away from the location of the charge should have a description in terms of unadulterated SU(2) Yang–Mills thermodynamics.

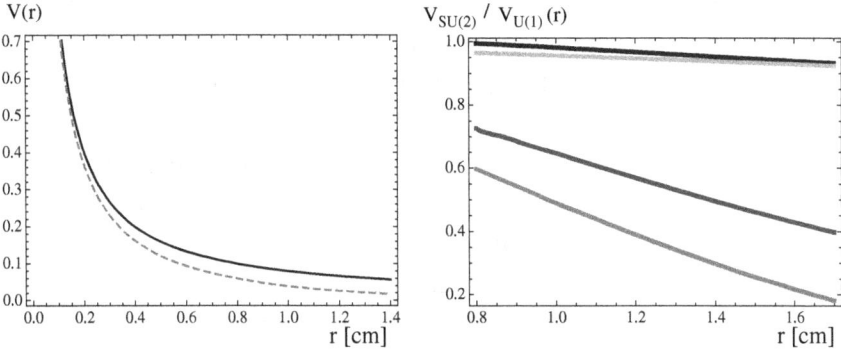

Figure 9.3. Left panel: Plot of the potentials $V_{U(1)}(r)$ (solid line) and $V_{SU(2)CMB}(r)$ (dashed line) as a function of r at $T = 2.5\,T_{CMB}$. Right panel: Plot of the ratio $V_{SU(2)CMB}(r)/V_{U(1)}(r)$ as a function of r. The temperature is set to $T = 1.5\,T_c$ (black), $T = 2.0\,T_c$ (dark gray), $T = 2.5\,T_c$ (gray), and $T = 3.0\,T_c$ (light gray). Notice how close to unity the curve at $T = 3.0\,T_c$ is as compared to the curve at $T = 2.5\,T_c$ in the right-hand side panel.

9.4.3.1 Arcade 2 and earlier radio frequency surveys of the CMB

Arcade 2 is a balloone-borne instrument to observe the CMB at very low frequencies (bands at $\nu = 3, 8, 10, 30$, and $90\,\text{GHz}$). On 22 July 2006, Arcade 2 flew at an altitude of $37\,\text{km}$ achieving a 8.4% sky coverage. The mission of this flight was an alternating observation of a calibrator black-body emitter and the sky. Here the adjustment of the calibrator's temperature was such that the two signals were nulling one another after an appropriate subtraction of foreground radiation. As a result, the temperature of the calibrator as a function of frequency (CMB line temperature) was obtained. The interesting feature in this data was the detection of a clear excess (statistically significant at the level of five standard deviations) of the CMB line temperature $T(\nu)$ at 3 and $8\,\text{GHz}$ as compared to the CMB baseline temperature T_0. This excess is expressed by a fit to following power-law [Fixsen *et al.* (2009)]

$$T(\nu) = T_0 + T_R \left(\frac{\nu}{\nu_0}\right)^\beta. \tag{9.7}$$

The Arcade 2 data are represented as: $T_0 = 2.725\,\mathrm{K}$ (within errors FIRAS's CMB baseline temperature [Mather *et al.* (1994)] obtained by a fit to the CMB spectrum at frequencies ranging from 60 GHz to 600 GHz), $\nu_0 = 1\,\mathrm{GHz}$, $T_R = 1.19 \pm 0.14\,\mathrm{K}$, and $\beta = -2.62 \pm 0.04$. The claim by the Arcade 2 collaboration that this spectacular low-frequency deviation from a perfect black-body spectrum ($T(\nu) \equiv$ const.) is not an artifact of galactic foreground subtraction, is unlikely related to an average effect of distant point sources, and that these results naturally continue earlier radio frequency data [Roger *et al.* (1999); Maeda *et al.* (1999); Haslam *et al.* (1981); Reich and Reich (1986)] appears to rest on firm ground. For example, a CMB line temperature of $21200 \pm 5125\,\mathrm{K}$ at 22 MHz was observed in [Roger *et al.* (1999)]. We take this observational situation as a motivation for an unconventional, SU(2)$_{\mathrm{CMB}}$-based explanation of Eq. (9.7) which, at the same time, determines the value of T_c.

9.4.3.2 *Evanescent CMB photons at low frequencies*

Let us now connect the observed excess in CMB intensity at low frequencies with the physics of the deconfining–preconfining phase-transition under which the photon acquires a Meissner mass by its direct coupling to the superconducting thermal ground state inside the preconfining phase. The presentation closely follows [Hofmann (2009)].

As we have seen in Sec. 6.2.2, the tree-level massless mode acquires a Meissner mass $m_\gamma = g|\varphi|$ deep inside the preconfining phase of SU(2) Yang–Mills thermodynamics. Here g is the dual gauge coupling which vanishes at $T = T_c$ and rises rapidly (critical exponent $\frac{1}{2}$) when T falls below T_c. Moreover, the modulus $|\varphi| = \sqrt{\frac{\bar{\Lambda}^3}{2\pi T}}$ is part of the description of the monopole condensate, parameterized by the preconfining manifestation $\bar{\Lambda}$ of the Yang–Mills scale. As discussed in Sec. 9.4.1, on astrophysical length scales, the superconducting, preconfining ground state enforcing this Meissner mass may presently contribute to extragalactic magnetic fields which, in turn, may seed stronger magnetic fields within galaxies. It is important to stress that m_γ is induced and calculable in a situation of thermal equilibrium for $T < T_c$ and that it vanishes in the deconfining phase. Again, in the latter phase the photon is precisely massless modulo mild (anti)screening effects, which peak at $T \sim 2\,T_c$, reflecting the fact that a subgroup U(1) of the underlying SU(2) gauge symmetry is respected by the deconfining ground state [Hofmann (2005)].

The fact that m_γ is a Meissner mass implies the evanescence of photons of circular frequency $\omega < m_\gamma$. (That the wavelike aspect of thermal photon propagation dominates the low-frequency part of the CMB black-body spectrum, where ARCADE 2 and predecessor experiments detect a significant excess of line temperature, is guaranteed by the results of Sec. 9.6.2.) This, however, is *not* what happens in the deconfining phase [Ludescher and Hofmann (2008)]. There, by a coupling to effective, massive vector modes, the prohibition of photon propagation at low temperatures $T > T_c$ and frequencies [Schwarz, Hofmann, and Giacosa (2007); Ludescher and Hofmann (2008)] is energetically balanced by the creation of nonrelativistic and screened monopoles and antimonopoles (see Sec. 5.5). Thus energy leaves the photon sector to re-appear in terms of (anti)monopole mass in the deconfining phase, and no evanescent photon fields are generated at low frequencies. We have seen in Sec. 5.4.7.5 that at $T = T_c$ no spectral distortions compared to the conventional Planck spectrum arise because the tree-level heavy vector modes are thermodynamically decoupled.

On the preconfining side of the phase boundary Meissner massive photons do not propagate for $\omega < m_\gamma$. They create a spectral intensity attributed to an *evanescent* photon field, fluctuating with random phase and amplitude at a given spatial point, and the corresponding modes are no longer thermalized. Evanescent photons collectively carry the energy density $\Delta\rho(T_c)$ that formerly massless CMB photons have lost due to their interaction with the new ground state (superconductor [Hofmann (2005)]). Because of an upward jump $\Delta\rho$ in energy density as temperature is lowered through T_c (discontinuous increase of number of photon polarizations states from two to three) the new ground state is not available everywhere in space (supercooling) but rather starts to nucleate in small regions with the probability of a certain bubble size determined by $\Delta\rho$ (see Sec. 6.2.3). Very shortly below T_c, where the portion of superconducting versus normally conducting volume in space is small, it is thus reasonable to assume that photon mass, understood as an average over space, does not yet imply a change in the number of polarization states.

Due to their nonpropagating nature frequencies belonging to the evanescent, nonthermal, and random modes of the photon field are distributed according to a Gaussian of width m_γ. This Gaussian is normalized to $\Delta\rho(T_c)$, and we approximately have

$$\Delta\rho(T_c) = \int_0^\infty d\omega (I_{\gamma,\mathrm{dec}} - I_{\gamma,\mathrm{prec}})|_{T=T_c}, \qquad (9.8)$$

where

$$I_{\gamma,\mathrm{dec}} = \frac{1}{\pi^2} \frac{\omega^3}{\exp(\omega/T) - 1} \quad \text{and}$$

$$I_{\gamma,\mathrm{prec}} = \frac{1}{\pi^2} \frac{\sqrt{\omega^2 - m_\gamma^2}\,\omega^2}{\exp(\omega/T) - 1} \theta(\omega - m_\gamma). \qquad (9.9)$$

Here $\theta(x)$ is the Heaviside step function. Introducing the dimensionless photon mass $\mu_\gamma \equiv \frac{m_\gamma}{T_c}$ yields

$$\Delta\rho = \frac{T_c^4}{\pi^2} \left(\frac{\mu_\gamma^3}{3} + F(\mu_\gamma) \right) \quad \text{where}$$

$$F(\mu_\gamma) \equiv \int_{\mu_\gamma}^\infty dy \, \frac{y^2}{e^y - 1} \left(y - \sqrt{y^2 - \mu_\gamma^2} \right). \qquad (9.10)$$

For the CMB spectral intensity at T slightly below T_c, we thus have

$$I_\gamma = 2\frac{\Delta\rho}{\sqrt{2\pi}\,m_\gamma}\exp\left(-\frac{\omega^2}{2m_\gamma^2}\right) + \theta(\omega - m_\gamma)\frac{1}{\pi^2}\frac{\sqrt{\omega^2 - m_\gamma^2}\,\omega^2}{\exp(\omega/T_c) - 1}. \qquad (9.11)$$

Since $\omega/T_c \ll 1$ ($\nu \le 3.4\,\mathrm{GHz}$ corresponds to $\omega \le 21.5\,\mathrm{GHz}$ and for line temperature in units of circular frequency one has $T(\nu) \ge T_c = 356\,\mathrm{GHz}$) we are deep inside the Rayleigh–Jeans regime. Thus for calibrator photons, which are precisely massless (see below), we may write

$$I_{\gamma,\mathrm{dec}} = \frac{\omega^2 T}{\pi^2}. \qquad (9.12)$$

Let us again explain the physics underlying Eqs. (9.12) and (9.11). As already mentioned, we assume that the CMB temperature is just slightly below T_c. This introduces a tiny coupling to the SU(2) pre-confining ground state which endows low-frequency photons with a Meissner mass m_γ if they have propagated for a sufficiently long time above this ground state. (The correlation length of the latter at T_c is of the order of 1 km [Hofmann (2005)].) This is certainly true of CMB photons. As a consequence, modes with $\omega < m_\gamma$ become evanescent, thus nonthermal, and are spectrally distributed in frequency according to the first term in Eq. (9.11). For $\omega > m_\gamma$ CMB photons do propagate albeit with a suppression in spectral energy density as compared to the ideal Planck spectrum. Recall that m_γ is understood as an average over space and that we assume that the number of polarizations is still very close to two. Because meeting this assumption may be less accurate at small *propagating* frequencies the spectral model of Eq. (9.11) is not to be taken literally there. The respective spectral *integral*, however, is.

A calibrator photon, on the other hand, is fresh in that the distance between emission at the black-body wall and absorption at the radiometer is just a small multiple of its wavelength. For sufficiently small coupling g (or for T sufficiently close to but below T_c) this short propagation path is therefore insufficient to generate a mass m_γ even at low frequencies. As a consequence, none of the calibrator

modes is forced into evanescence. That the thermal ground state of
$SU(2)_{CMB}$ sustains wave- and not particle-like excitations within the
low-frequency regime of the present CMB introduced in Sec. 9.4.3.1
and therefore is subject to the very concept of evanescence is discussed in Sec. 9.6.2. To summarize, CMB frequencies approximately
obey the spectral distribution I_γ [see Eq. (9.11)], while low-frequency
calibrator photons are distributed according to $I_{\gamma,\text{dec}}$ [see Eq. (9.12)].
From now on we set T_c equal to the CMB baseline temperature
$T_c = T_0 = 2.725\,\text{K}$ ($T_c = 356.76\,\text{GHz}$ when expressed in terms of
a circular frequency).

Let us now determine m_γ from radio frequency survey data
of the CMB. The essence of Aracde 2's and earlier radio frequency
surveys' experimental philosophy is to null at a given frequency the
CMB signal by that of a calibrator black-body. At the low frequencies
considered there is practically no difference between antenna and
thermodynamical temperature [Fixsen *et al.* (2009)]. Thus the observationally imposed condition for the extraction of a line temperature
$T(\nu)$ is:

$$I_\gamma = I_{\gamma,\text{dec}}. \tag{9.13}$$

Setting $m_\gamma = 0.1\,\text{GHz}$, the corresponding spectral situation is
depicted in Fig. 9.4. For the extraction of m_γ from the data let us
introduce the following two dimensionless quantities

$$y \equiv \frac{\omega}{T_c}, \quad \tau \equiv \frac{T}{T_c}. \tag{9.14}$$

With these definitions and appealing to Eqs. (9.10)–(9.12), Eq. (9.13)
is re-cast as

$$\tau = \sqrt{\frac{2}{\pi}}\, y^{-2}\, \exp\left(-\frac{y^2}{2\mu_\gamma^2}\right)\left(\frac{\mu_\gamma^2}{3} + \frac{F(\mu_\gamma)}{\mu_\gamma}\right) + \theta(y - \mu_\gamma)\frac{\sqrt{y^2 - \mu_\gamma^2}}{e^y - 1}. \tag{9.15}$$

The following table lists out our results for m_γ, extracted from
the data using Eq. (9.15), in units of ordinary (not circular)
frequency ν:

Source	ν[GHz]	T[K]	μ_γ	m_γ[GHz]
Roger	0.022	21200 ± 5125	$0.001821^{+0.000423}_{-0.000419}$	$0.1034^{+0.0240}_{-0.0238}$
Maeda	0.045	4355 ± 520	$0.001704^{+0.000169}_{-0.000166}$	$0.0968^{+0.0095}_{-0.0095}$
Haslam	0.408	16.24 ± 3.4	$0.003611^{+0.000152}_{-0.000325}$	$0.205^{+0.0086}_{-0.0185}$
Reich	1.42	3.213 ± 0.53	$0.0093^{+0.0007}_{-0.00153}$	$0.528^{+0.0397}_{-0.0869}$
Arcade2	3.20	2.792 ± 0.010	$0.0211^{+0.0001}_{-0.0001}$	$1.198^{+0.0057}_{-0.0057}$
Arcade2	3.41	2.771 ± 0.009	$0.02253^{+0.0001}_{-0.0001}$	$1.279^{+0.0057}_{-0.0057}$

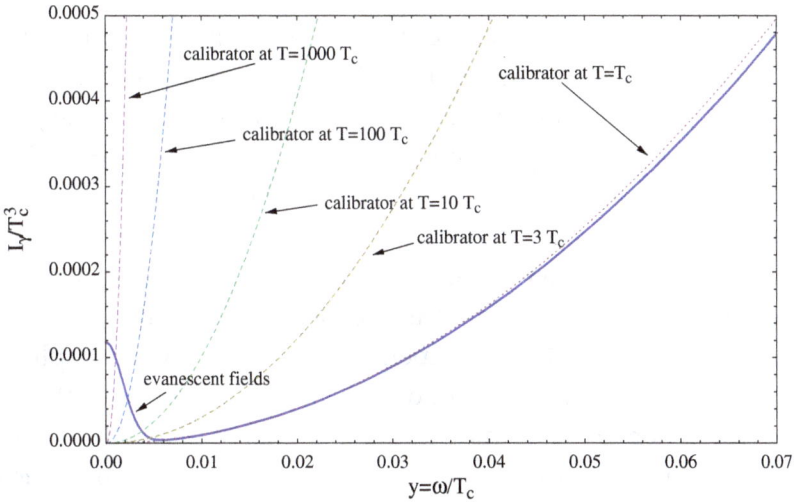

Figure 9.4. The normalized spectral intensities of CMB modes (thick line) at $T = T_c = 2.725$ K and $m_\gamma = 0.1$ GHz (in units of ordinary frequency) and of calibrator modes (dotted and dashed lines) at various temperatures. A null experiment asks for an intersection of the former with a representative of the one-parameter (T) family of the latter-type curves at a given frequency. Since for $y \to 0$ the Gaussian becomes stationary one has in this limit $T(y) = \text{const.} \times y^{-2}$, that is, the asymptotic spectral index reads $\beta_{as} = -2$. With the low-frequency data presently available one has $\beta \sim -2.6$ [Fixsen *et al.* (2009)].

Notice the good agreement of the values for m_γ as extracted from the data of [Roger *et al.* (1999); Maeda *et al.* (1999)] where $\nu < m_\gamma$. The other data [Haslam *et al.* (1981); Reich and Reich (1986); Fixsen *et al.* (2009)] yield $\nu > m_\gamma$ which is in the regime where we do not expect the spectral model for CMB photons to be good (average number of polarizations depends nontrivially on frequency). Still, the value of m_γ obtained from [Haslam *et al.* (1981)] is only twice as large as that arising from the data of [Roger *et al.* (1999)] or [Maeda *et al.* (1999)] at a frequency which is, respectively, twenty or ten times larger!

Let us now discuss whether a Meissner mass of $\sim 100\,\text{MHz}$ is compatible with other experimental/observational facts. Is such a scenario not ruled out by experiments, such as radar versus laser ranging to the moon and the limits on the photon mass obtained by terrestial Coulomb-law measurements or the measurement of the magnetic fields of astrophysical objects (see [Nakamura *et al.* (2010)])? At this point we need to discuss what it really means to have the thermalized photon field (at a temperature $T_0 = 2.725\,\text{K}$) acquire a Meissner mass: Whether or not the propagation properties of the photon are altered as compared to conventional wisdom sensitively depends on the temperature of the thermal ensemble it belongs to, and it depends on its frequency. To be above the thermal noise of the CMB any experiment trying to detect a photon mass must work with local photon-field energy densities that are by many orders of magnitude larger than that of the CMB. This can either be done directly by looking for deviations in electrostatic or magnetostatic field configurations as compared to the predictions of the standard theory or indirectly by searching for modified dispersion laws in propagating photon fields. In both cases the associated field modes usually are not thermalized. The existence of a correlation between an electric potential gradient and a temperature gradient in solid state systems is known for a long time (thermoelectric power, Seebeck effect). We have shown in Sec. 5.6, that the deconfining SU(2) ground state with its abundance of short- and long-lived charge carriers acts as a medium which implies such a correlation. Even though a static background field or a laser beam or radar does not describe a homogeneous thermodynamical setting, one may for a rough argument appeal to an adiabatic approximation setting the experimentally measured energy density equal to that of thermal (deconfining) $SU(2)_{CMB}$, to deduce the local temperature this energy density would correspond to were the system actually thermalized. In any experimental circumstance searching for a universal (by assumption independent of temperature) photon mass, this would yield a temperature far above $T_0 = 2.725\,K$. But it was shown in [Ludescher and Hofmann (2008)] how rapidly the thermalized $SU(2)$ photon approaches $U(1)$ behavior with increasing temperature by a power-like decrease of the modulus of its screening function. For example, the dimensionless spectral gap ω^*/T in black-body spectra, defining the center of the spectral region where non-Abelian effects are most pronounced, decays as $T^{-3/2}$. Thus experimental set-ups designed to obtain photon-mass bounds roughly correspond to temperatures where the photon behaves in a rather Abelian way explaining the very low mass bounds obtained. That is, for the photon to exhibit measurable deviations of its dispersion law compared to the $U(1)$ theory it must belong to a *thermal* bath at temperatures ranging from just below T_0 (Meissner mass) up to about $10\,K$ (momentum-dependent screening mass).

What about the physics just around T_0? Is there a possibility that principle laws of thermodynamics are not obeyed? For example, consider the following set-up. Two black bodies, one at T_1 just below T_0, the other at T_2 just above T_0, are immersed into a photon gas which is exactly at temperature T_0. Photons exchanged by the two black bodies are restricted to frequencies below $m_\gamma \sim 100\,MHz$. Would then not black body 1 transfer energy to black body 2 due to its larger spectral intensity below m_γ — in contradiction to the second law of thermodynamics? The answer is no because the photons of black body 1, supporting this bump in the spectrum, are *evanescent*. Thus they cannot propagate out of black body 1's cavity. Also, if $T_0 < T_1 < T_2$

and both T_1 and T_2 are not too far above T_0 then the rapid rise with temperature of the spectral intensity (in the Rayleigh–Jeans regime linearly) would assure that black body 1 warms up at the expense of black body 2 at temperatures not far above T_0 in spite of the spectral modifications due to screening and antiscreening at small frequencies.

Since we may not trust our spectral model for CMB modes locally if $\nu > m_\gamma$ (both expressed as ordinary frequencies) it is not surprising that considerable deviations occur for the extracted values of m_γ compared to the low-frequency situation. The *integral* of the spectral model, which enters into the normalization $\Delta\rho$ of half the Gaussian in Eq. (9.11), however, is a quantity that is robust against local changes of the spectrum. Thus we are inclined to trust our result $m_\gamma \sim 0.1\,\mathrm{GHz}$ extracted at low frequencies. Based on the present work two predictions, arising from the SU(2)$_{\mathrm{CMB}}$ postulate, can be made: First, since the low-frequency data on line temperatures are efficiently explained by this theory being very little below the critical temperature of its deconfining–preconfining phase boundary one has $T_c = 2.725\,\mathrm{K}$. This allows for the prediction of a sizable anomaly in the low-frequency part of the thermal spectral intensity at higher, absolutely given temperatures, say at $T = 2\,T_c \sim 5.4\,\mathrm{K}$ [Schwarz, Hofmann, and Giacosa (2007); Ludescher and Hofmann (2008)]. Second, we predict that the spectral index β for the line temperature $T(\nu)$, measured by nulling the CMB signal by a black-body reference, approaches $\beta_{\mathrm{as}} = -2$ for $\nu \searrow 0$. The tendency of an increase of β when extracting the parameters of Eq. (9.7) from low-frequency as compared to higher-frequency data sets is nicely seen in Table 5 of [Fixsen *et al.* (2009)].

The here-presented argument that the CMB should be on the verge of undergoing a phase transition at its present baseline temperature $T_0 = T_{\mathrm{CMB}} = 2.725\,\mathrm{K}$ towards a phase, where its thermal photon modes become massive, implies consequences for particle physics (recall discussion in Sec. 9.1).

9.4.3.3 Thermodynamical decoupling: Coexistence of several temperatures

Since in between the stars of a galaxy or within Earth's atmosphere the CMB is not the only system comprising of thermalized photons within a specified spatial volume one may wonder whether a coexistence of several temperatures is possible in one and the same SU(2)$_{CMB}$ theory. By coexistence we mean that the distinct photon gases and their respective thermal ground states do no influence one another. In a pure U(1) theory such a coexistence of various temperatures is a feature granted by the facts that photons do not interact among themselves and no thermal ground state emerges.

Let $\{T_1 < \cdots < T_i < \cdots < T_n, n > 1\}$ be a set of temperatures associated with n thermal photon gases. Interactions of the photons are unmeasurably suppressed if T_1 is sufficiently larger than T_{CMB} and well below $\Lambda_e \sim 0.5\,\mathrm{MeV}$: Each gas then essentially consists of free photons. The question of how large a difference $\Delta T_i \equiv T_{i+1} - T_i$ is required to suppress coupling effects between the two respective thermal systems to below a given bound has not been investigated on a quantitative level yet. Since in the thermal spectrum all non-Abelian effects die off in a power-like fashion with rising temperature we do not expect any measurable effects that would turn out to be problematic for the postulate SU(2)$_{CMB}$. As for the ground-state physics the following situation would arise for sufficiently large $\Delta T_i > 0$: Stable monopoles belonging to T_i are of a larger core size than those belonging to T_{i+1} [see Eq. (4.50) of Sec. 4.2.2 or Eq. (4.100) of Sec. 4.3.2]. The latter monopoles are, however, denser within the thermal ground state (compare with Sec. 5.5.2). Thus T_i monopoles would only see a spatial average of the charges of T_{i+1} monopoles [see Eq. (4.50) of Sec. 4.2.2]. For sufficiently large ΔT_i the average charge density is small causing T_{i+1} monopoles to be ignored by the physics of T_i monopoles. On the other hand, a given small T_{i+1} monopole is surrounded by the smeared charge of a T_i monopole. Thus the local activity of the T_{i+1} monopole in emitting and absorbing T_{i+1} photons is not influenced by the presence of the

T_i monopole. Based on perturbative techniques (linear response) a more quantitative analysis of this situation should be possible.

9.5   Laboratory Experiment on Black-Body Anomaly

Our predictions for the modified black-body spectrum that rely on the screening function G, which is reliably computed on the one-loop level, so far were concerning the spectral energy density. This quantity is, however, not directly accessible experimentally.[11]

For a bolometric and radiometric detection of low-frequency modifications of black-body spectra at temperatures ranging from, say, 5–20 K, the relevant quantity is the spectral radiance L. In conventional U(1) theory both quantities, spectral energy density I and radiance L, are proportional to one another: $L(v, T) = \frac{c}{4\pi} \times I(v, T)$ where c denotes the velocity of light in vacuum, and v is frequency [Grum and Becherer (1979)].

In a deconfining SU(2) plasma, however, the factor c in front of $I(v)$ must be replaced by the group velocity $v_g \equiv \partial_{|\vec{p}|} E$ where $E = hv$ is the photon's energy in dependence of its spatial momentum modulus $|\vec{p}|$ and of temperature T. Recall that this dependence is introduced by the SU(2) (anti)screening function G, see Secs. 5.4.7.5.

The purpose of this section is an investigation of the characteristics of the radiance spectrum at low temperature $T > T_c$ and frequency as well as a discussion of observable effects induced by SU(2)$_{CMB}$. We follow closely the presentation in [Falquez, Hofmann, and Baumbach (2010)].

9.5.1   *Bolometry and radiometry of SU(2) photons*

Here we derive the photonic radiance of an SU(2) Yang–Mills theory subject to a modified dispersion relation at low temperatures and

---

[11]Academically, it is in principle accessible through gravitational interactions. Moreover, in Sec. 10.2 we intepret the contribution to the energy density of the CMB due to the black-body anomaly as a source for the emergence of large-angle temperature anisotropies.

frequencies [Ludescher and Hofmann (2008)] in comparison with the conventional U(1) Planck spectrum. We work in SI units throughout Sec. 9.5. In these units the spectral energy density $I_{SU(2)}(\nu, T)$ reads:

$$I_{SU(2)}(\nu, T) = \frac{8\pi}{h^2} \frac{\omega}{e^{\frac{\hbar\omega}{kT}} - 1} \vec{p}^2(\omega) \frac{d|\vec{p}|}{d\omega}, \qquad (9.16)$$

where $(\hbar\omega)^2 = (cp_0)^2$ and

$$p_0^2 - \vec{p}^2 = G(|\vec{p}|, T). \qquad (9.17)$$

In a deconfining SU(2) plasma the factor c in the relation $L(\nu, T) = \frac{c}{4\pi} \times I(\nu, T)$ is replaced by the photon's group velocity v_g defined as

$$v_g \equiv \partial_{|\vec{p}|} E = \hbar\, \partial_{|\vec{p}|}\, \omega = \partial_{|\vec{k}|}\, \omega, \qquad (9.18)$$

where $E = h\nu = cp_0$ is photonic energy, and $\vec{k}$ is the wave-number vector related to momentum $\vec{p}$ as $\vec{p} = \hbar\vec{k}$. Thus, in calculating the spectral radiance $L(\nu, T)$ for SU(2) photons, the factor $\frac{d|\vec{p}|}{d\omega}$ drops out, and we obtain

$$L_{SU(2)}(\nu, T) = \frac{2h}{c^2} \frac{\nu^3}{e^{\frac{h\nu}{kT}} - 1} \times \left(1 - \frac{c^2 G}{(h\nu)^2}\right) \theta\left(\nu - \nu^*\right)$$

$$= L_{U(1)}(\nu, T) \times \left(1 - \frac{c^2 G}{(h\nu)^2}\right) \theta(\nu - \nu^*), \qquad (9.19)$$

where $L_{U(1)}$ is the Planckian spectral radiance.

To make contact with the real world the critical temperature T_c for the deconfining–preconfining phase-transition, which is the only free parameter of the thermalized SU(2) quantum Yang–Mills theory, must be determined experimentally. In Sec. 9.4 we have given observational reasons why T_c should coincide with the baseline temperature $T_0 = 2.725\,K$ of the present cosmic microwave background. After this observationally inferred fix of T_c the modified radiance spectra for SU(2)$_{CMB}$ photons may be calculated for various physical temperatures. In Figs. 9.5–9.7 we show results for $T = 5.4, 8.0$, and $11.0\,K$, respectively. Both the calculated SU(2)$_{CMB}$ radiance and the

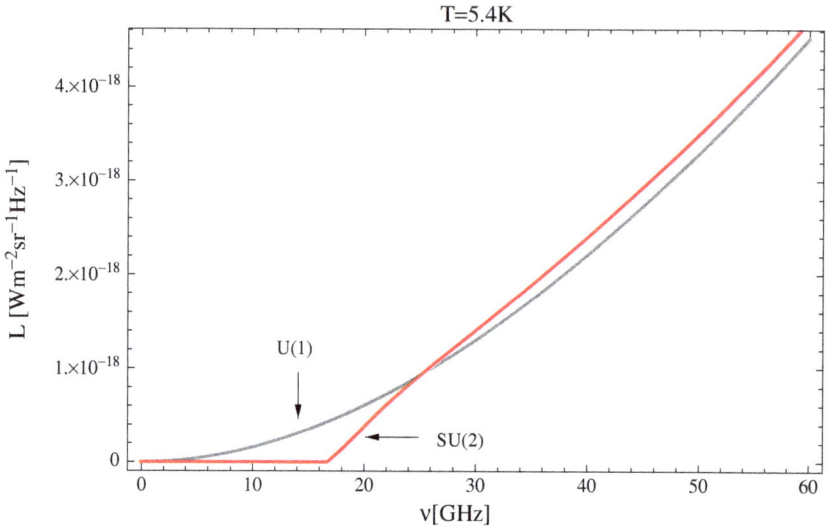

Figure 9.5.   Comparison between the black-body spectral radiances at $T = 5.4\,\text{K}$, associated with $SU(2)_{CMB}$ (red) and conventional $U(1)$ theory (gray).

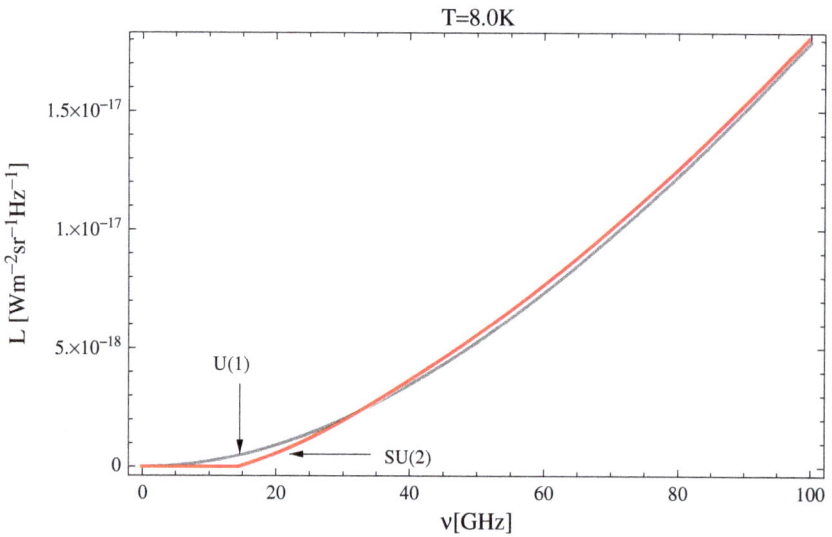

Figure 9.6.   Comparison between the black-body spectral radiances at $T = 8.0\,\text{K}$, associated with $SU(2)_{CMB}$ (red) and conventional $U(1)$ theory (gray).

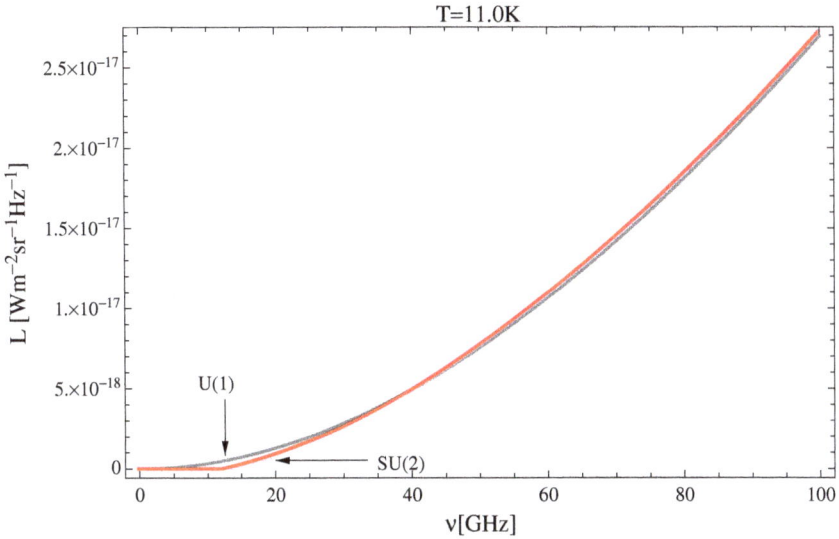

Figure 9.7. Comparison between the black-body spectral radiances at $T = 11.0\,\text{K}$, associated with SU(2)$_{CMB}$ (red) and conventional U(1) theory (gray).

conventional U(1) Planck spectrum is depicted. Notice the regime of total screening (suppression of spectral radiance down to zero) and the cross-over to a regime of slight antiscreening (excess of spectral radiance) in all cases. Figure 9.8 shows the difference in spectral radiance $\Delta L(v)$, defined as

$$\Delta L(v, T) = L_{\text{SU(2)}}(v, T) - L_{\text{U(1)}}(v, T) \qquad (9.20)$$

at $T = 5.4\,\text{K}$.

A type of bolometric experiment can be conceived as follows. Let the aperture of an isolated low-temperature U(1) black-body cavity at[12] temperature T_1 and that of an SU(2) black-body cavity of identical characteristics (volume, geometry, emissivity, aperture) at temperature T_2 face each other, and let them exchange radiant energy. Have the SU(2) black-body be linked to a large heat reservoir

---

[12]We shall discuss in Sec. 9.5.2 how such a U(1) black-body cavity can be prepared in an actual experiment.

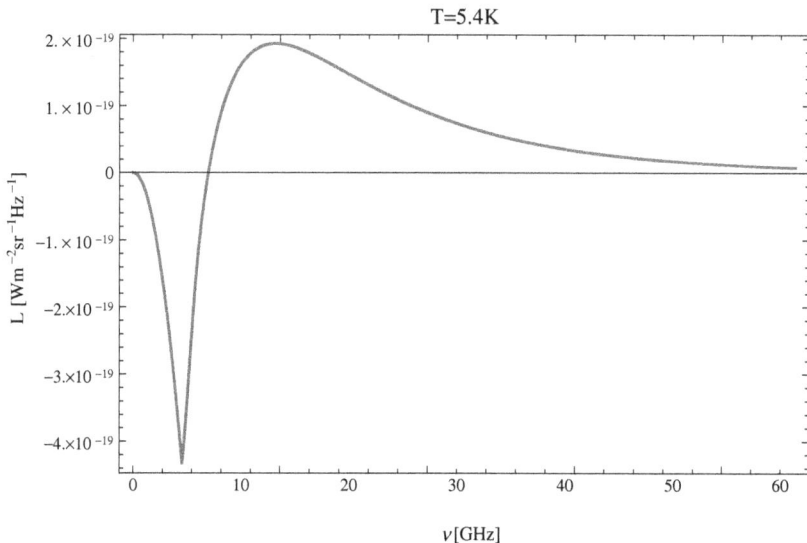

Figure 9.8.   The difference in spectral radiance $\Delta L(\nu, T)$ between an SU(2)$_{\text{CMB}}$ and a conventional U(1) theory at $T = 5.4\,\text{K}$ as a function of (small) frequency ν.

to keep its wall temperature T_2 constant. Furthemore, switch in a bandwidth filter within the common aperture of the U(1) and SU(2) black-body cavities. This band-width filter is tuned to the region of the SU(2) black-body gap: It absorbs photons above frequency ν^* no matter which cavity they come from.[13] U(1) photons within the SU(2) spectral gap are absorbed by the SU(2) thermal ground state [Ludescher *et al.* (2009)]: They create stable but unresolvable monopole-antimonopole pairs. Thus a small amount of energy per time flows across the common aperture from the U(1) cavity to the

---

[13]From the three effective SU(2) gauge modes only the tree-level massless mode (the photon) interacts with electric charges and so can be absorbed by the material in the bandwidth filter. If a propagating photon emerges from the SU(2) cavity then the absorbed power per frequency interval is identical to that of a propagating photon stemming from the U(1) cavity because the additional factor $\dfrac{1}{1 - \frac{c^2 G}{(h\nu)^2}}$ in the SU(2) spectral radiance [see Eq. (9.19)], is canceled [see discussion below Eq. (9.28)].

SU(2) cavity, and zero radiation temperature within the U(1) cavity will asymptotically be reached (radiation refrigeration).

For $T_1 = T_2$ and blocking off the gap region $0 \leq \nu \leq \nu^*(T_1)$ by a complementary bandwidth filter, the regime of propagating photon frequencies would not allow for any heat exchange since $L_{SU(2)}(\nu, T_2) d\Omega_{SU(2)}$ and $L_{U(1)}(\nu, T_1 = T_2) d\Omega_{U(1)}$ do match (see Eq. (9.22)). Here $d\Omega_{SU(2)}$ and $d\Omega_{U(1)}$ are solid-angle elements defined on a sphere centered at a point within the common aperture.

By Snell's law [Krall, Trivelpiece, and Gross (1973)] we have

$$\frac{d\Omega_{SU(2)}}{d\Omega_{U(1)}} = \frac{v_{ph}^2}{c^2}, \tag{9.21}$$

where the phase velocity v_{ph} is defined as $v_{ph} \equiv \frac{\omega}{k} = \frac{1}{\sqrt{1 - \frac{c^2 G}{(h\nu)^2}}}$. There-

fore we obtain

$$L_{SU(2)}(\nu, T_2) \frac{d\Omega_{SU(2)}}{d\Omega_{U(1)}} = L_{U(1)}(\nu, T_1 = T_2). \tag{9.22}$$

A potentially interesting quantity for experiments is the *radiance U(1) line temperature $T_P(\nu)$*. The quantity $T_P(\nu)$ is the temperature a conventional U(1) black-body must possess in order to reach the following (bolometric) equilibrium condition [Krall, Trivelpiece, and Gross (1973)] (compare with Eq. (9.22)):

$$L_{U(1)} = L_{SU(2)} \times \frac{1}{1 - \frac{c^2 G}{(h\nu)^2}}. \tag{9.23}$$

One has

$$T_P(L_{SU(2)}(\nu, T)) \equiv \frac{h\nu}{k} \frac{1}{\ln\left[\frac{2h}{c^2}\left(1 - \frac{c^2 G}{(h\nu)^2}\right)\frac{\nu^3}{L_{SU(2)}(\nu,T)} + 1\right]}. \tag{9.24}$$

Again, on the right-hand side of Eq. (9.23) the factor $(1 - \frac{c^2 G}{(h\nu)^2})$ cancels $\left(1 - \frac{c^2 G}{(h\nu)^2}\right)$ in $L_{SU(2)}$ (see Eq. (9.19)). Recall that apart from the factor $\theta(\nu - \nu^*)$ the factor $\left(1 - \frac{c^2 G}{(h\nu)2}\right)$ distinguishes SU(2) from U(1) spectral radiance. Therefore T_P is the same as the SU(2) wall temperature T_2 except for the regime of total screening $0 \leq \nu \leq \nu^*$ where $T_P \equiv 0$. That is, no heat is effectively exchanged by frequencies above the spectral gap according to the bolometric equilibrium condition (9.23), and within the spectral gap no U(1) photons enter into the balance of Eq. (9.24).

For completeness let us give some characterization of the factor

$$\left(1 - \frac{c^2 G}{(h\nu)^2}\right)\theta(\nu - \nu^*)$$

which converts U(1) to SU(2) spectral radiance and comprises of the characteristic frequencies ν^*, ν_c, ν_M which are implicitly defined as follows:

$$|\vec{p}|(\nu^*) = 0,$$

$$G(\nu_c, T) = 0, \quad \text{and}$$

$$\frac{G(\nu_M, T)}{\nu_M^2} = \min\left\{\frac{G(\nu, T)}{\nu^2}\right\}.$$

(9.25)

The points ν^*, ν_c, and ν_M describe the onset of total screening (no photon propagation), the cross-over between screening and antiscreening ($G = 0$), and maximal antiscreening ($G < 0$), respectively. For $T > 8\,\text{K}$ the critical points ν_c, ν_M and ν^* were numerically fitted to a power law in T. One obtains the following results:

$$\frac{\nu_c(T)}{\text{GHz}} = 1.83\left(\frac{T}{\text{K}}\right)^{1.12} + 13.48,$$

$$\frac{\nu_M(T)}{\text{GHz}} = 3.45\left(\frac{T}{\text{K}}\right)^{1.08} + 12.90,$$

(9.26)

$$\frac{\nu^*(T)}{\text{GHz}} = 42.70\left(\frac{T}{\text{K}}\right)^{-0.53} + 0.21.$$

Figure 9.9 shows calculated points overlaid with the fitted curves.

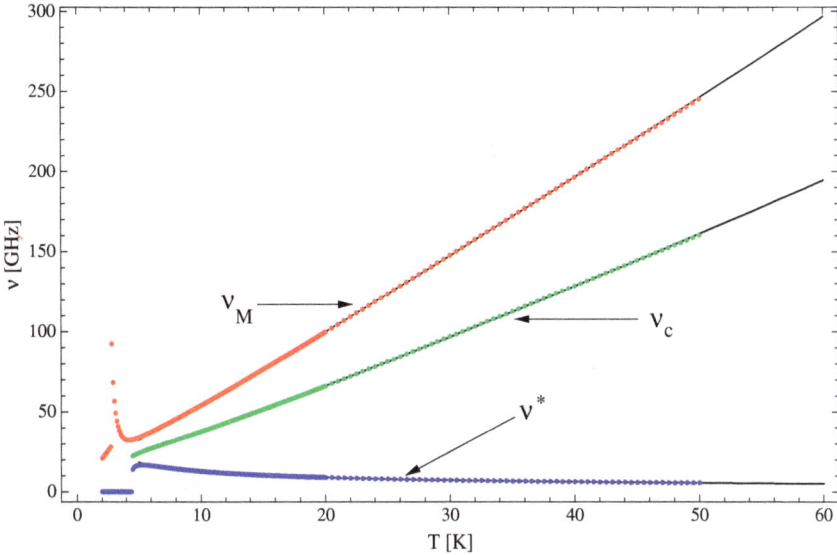

Figure 9.9.   Plots of the T dependence of the characteristic points of $L_{SU(2)}(T, \nu)$.

We now consider a radiometric approach by placing an antenna inside the SU(2) plasma. Independently of (lossless) propagation properties one arrives at the following expression for the power $P_{\Delta\nu}(\nu)$ within band-width $\Delta\nu$ absorbed by the antenna [Grum and Becherer (1979)]:

$$P_{\Delta\nu}(\nu) = \theta(\nu - \nu^*) \int_{\nu}^{\nu+\Delta\nu} d\nu' \, \frac{h\nu'}{\exp\left[\frac{h\nu'}{kT}\right] - 1}. \qquad (9.27)$$

For $h\nu \ll kT$, which is certainly the case for the spectral range we are interested in, Eq. (9.27) simplifies as

$$P_{\Delta\nu}(\nu) \approx \theta(\nu - \nu^*) \, \Delta\nu kT. \qquad (9.28)$$

Apart from the θ-function prefactor in Eqs. (9.27) and (9.28) the expressions are identical to the U(1) situation since a factor $|\vec{k}|^2$ in $L_{SU(2)}(\nu, T)$ is canceled by a factor $|\vec{k}|^{-2}$ (see [Grum and Becherer (1979)]). Only those radiometric measurements inside the predicted frequency regime of *total* screening may therefore be able to test the postulate SU(2)$_{\text{CMB}}$.

9.5.2   *Preparation of a U(1) black-body cavity at low temperatures*

The concept of a U(1) line temperature relies on one's ability to prepare a black-body source in such a way that it emits according to Planck's radiation law even at small temperatures where we expect modifications. In the single-photon counting experiment of [Tada *et al.* (2006)] the (two-step laser) excitation of the $111_{s_{1/2}}$ Rydberg state of ^{85}Rb atoms towards the $111_{p_{3/2}}$ state by absorption of thermal photons of frequency 2527 MHz, prepared within a tunable cavity, can be exploited to learn about the temperature dependence of the mean photon number $\bar{n}(T)$ at this frequency. The measurement was performed for temperatures T ranging from 67 mK up to 1 K. As discussed in [Tada *et al.* (2006)], no deviation of $\bar{n}(T)$ from the U(1) expected Bose–Einstein distribution was observed. In this context, it is important to note that a static, electric stray field of $|\vec{E}| \sim 25$ mV/cm was present in the cavity during the experiment. Since a homogeneous, static electric field $\vec{E}$ of energy density well above that of the thermal ground state effectively decouples the photon from the ground state via a suppression of scattering off the massive vector modes, a black-body cavity, whose volume is transcended by $\vec{E}$, effectively acts as a U(1) emitter.

Let us explain this in more detail. A static electric field can cause a sizable distortion of the thermal SU(2) ground state which is sufficient to render the system effectively U(1). Namely, the (stable and unresolvable) electric monopoles residing in the thermal ground state are accelerated by the external field in a parallel or antiparallel way, and thus acquire kinetic energy which they subsequently disperse by collisions. This increases the energy density of the thermal ground state to an effective temperature largely disparate to the temperature of the radiation which is kept in thermal equilibrium with the cavity walls. Photons thus are liberated from any ground-state-induced effect, and their dispersion law becomes trivial, that is, of the convential U(1) type. This effect may be used to test whether thermalized but otherwise uncorrupted photon propagation is due to thermal SU(2) gauge dynamics.

Let us be more quantitative. The energy density ρ_E of the external electric field is given as

$$\rho_E = \frac{\epsilon_0}{2} \vec{E}^2, \tag{9.29}$$

where in SI units $\epsilon_0 = 8.8542 \, 10^{-12} \, \text{J}/(\text{Vm}^2)$. The energy density of the SU(2) thermal ground state ρ_{gs}, in SI units, is given as

$$\rho_{gs} = 4\pi \, \Lambda_{CMB}^3 \frac{k_B}{(\hbar c)^3} T, \tag{9.30}$$

where $\Lambda_{CMB} = 2\pi \frac{k_B T_c}{\lambda_c}$, $T_c = 2.725 \, \text{K}$, $\lambda_c = 13.87$ [Hofmann (2005)], $k_B = 1.3807 \times 10^{-23} \, \text{J/K}$, $c = 2.9979 \times 10^8 \, \text{m/s}$, and $\hbar = \frac{6.6261}{2\pi} 10^{-34} \, \text{Js}$. Setting $\rho_E = \rho_{gs}$ at $|\vec{E}| \sim 25 \, \text{mV/cm}$ [Tada et al. (2006)], we obtain an effective ground state temperature of about $10^3 \, \text{K}$. Therefore, the energy density of the external electric field roughly is three orders of magnitude larger than that of the SU(2) thermal ground state at 1 K. Thus we may safely assume a decoupling of the thermal ground-state physics from the physics of thermal excitations in the experiment of [Tada et al. (2006)]. In general, we may reason as follows: To not distort the SU(2) ground state physics sizably an admissible field strength should obey the condition that ρ_E be less than ρ_{gs}. For example, demanding that $\rho_E \sim 0.1 \, \rho_{gs}$ at $T = 5.4 \, \text{K}$ implies an electric field strength of about $|\vec{E}| \sim 0.2 \, \text{mV/cm}$. Thus to produce a U(1) black-body cavity at low temperature one should work with a much larger value of $|\vec{E}|$. A more refined treatment of the effects induced on the thermal ground state by external fields, considered as small perturbations to undistorted SU(2) Yang–Mills thermodynamics, should be performed within the realm of linear-reponse theory. We leave this to future investigation.

To summarize the results of the present section, we have investigated the experimental consequences of the assumption that, fundamentally, photon propagation is described by an SU(2) rather than a U(1) gauge principle (postulate SU(2)$_{\text{CMB}}$). We have considered bolometric and radiometric methods, and we have shown that the only region, where differences to the conventional theory can be predicted, is associated with the spectral gap $0 \leq \nu \leq \nu^*(T)$ with the T dependence of ν^* given in Eq. (9.26) (total screening). We also have elucidated how a U(1) black-body cavity can be prepared at low temperature by virtue of an effective decoupling its thermal ground state from its photonic excitations by the application of a static, homogeneous electric field of sufficient strength.

9.5.3   Wave-guide loads to detect the black-body anomaly?

Here we briefly comment on the feasibility of employing wave-guide loads to search for the black-body anomaly at low temperatures and frequencies. The advantage of a set-up, whose principle functional components are described in Fig. 9.10 (see also [Penzias (1964)]), would be a significant reduction in cost if the radiometer viewing a cold absorber of temperature, say 5 K, could itself be at room temperature [Kogut (2010)]. In such an experiment, the effects of the radiometer noise temperature are canceled by the radiometer alternatingly observing (Dicke switch) two identical wave-guide loads A and B, with the time-integral over the differential signal being nulled. To test for SU(2) effects a static electric field is applied across A or across B or across both.

The problem with such a set-up is the following: While the radiometer sees a 5-K radiation from the absorber within its bandwidth the absorber sees 300-K radiation from the radiometer. Thus at any point within the wave guide and within any given frequency band the angular distribution of radiance is highly anisotropic. This,

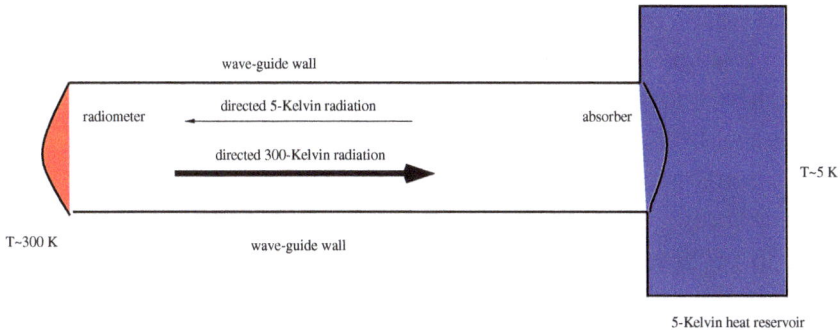

Figure 9.10. Principle set-up for an experiment using a wave-guide load. The absorber, which is immersed into a heat reservoir of temperature, say, 5 K, is observed by a radiometer at room temperature (~ 300 K). The wave-guide walls are practically ideally reflecting and do not emit extra radiation. While the radiometer sees a 5-K radiation from the absorber within its bandwidth the absorber sees 300-K radiation from the radiometer. Thus, at any point inside the wave-guide the isotropy of radiation is strongly violated in contradiction with the assumptions made for the derivation of the black-body anomaly.

however, violates one of the basic assumptions made in the construction of the thermal SU(2) ground state (see Sec. 5.1.3). We conclude that in a wave-guide experiment, the absorber as well as the radiometer must be at an identical low temperature of, say, 5 K. While this renders the set-up more expensive a cold comfort is that the requirements on the temperature stability of the load are less severe than for the set-up with room-temperature radiometer due to reduced integration times.

9.6   Thermal Ground State and Particle-Wave Duality of the Photon

In this final section we ask the question whether the concept of a deconfining thermal ground state of pure SU(2) Yang–Mills theory(ies), as derived in Sec. 5.1, can be extended to obtain a more fundamental description of wave-like, nonthermal propagation of electromagnetic fields than that provided by classical electromagnetism. As it turns out, the peripheries of HS (anti)calorons, which constitute the thermal ground state, provide densities of static electric and magnetic dipole moments invoked by an electromagnetic disturbance (response to radiation source being, say, an oscillating charge). This disturbance's field-energy density equals that of the thermal ground state at a certain (fictitious) temperature. The corresponding vacuum parameters of classical electromagnetism, electric permittivity ϵ_o and magnetic permeability μ_0, defined in terms of the ratios of the respective dipole densities to their respective exciting field strengths, turn out to be temperature independent. The condition that (anti)caloron centers are not sampled by the wavelength of the probe field results in a temperature independent inequality stating that, in the rest frame of the radiation source, the ratio of the square of the intensity to wavelength is bounded from above by the 9th power of the Yang–Mills scale [Grandou and Hofmann (2015)]. Since the Yang–Mills scale of SU(2)$_{\mathrm{CMB}}$ is very low, $\Lambda_{\mathrm{CMB}} = 2\pi \frac{k_B T_c}{\lambda_c} \sim 1.0638 \times 10^{-4}\,$eV (see Sec. 9.4.3), this would imply that most of the waves of sizable intensities throughout the accessible electromagnetic spectrum, as experimentally generated/manipulated/detected and observationally probed, would be excluded by this inequality. The way out is to postulate a mixing in a product of at least two SU(2) Yang–Mills gauge groups whose Yang–Mills scales are hierarchically distinct: SU(2)$_{\mathrm{CMB}} \times$ SU(2)$_e$ with $\Lambda_e \sim m_e \sim 5 \times 10^5\,$eV ($m_e$ the mass of the electron). As in [Hofmann (2015)] we will discuss what the options of such a (dynamical) mixing are towards explaining the particle-wave duality of the photon in terms of a sampling of the central and the peripheral regions of (anti) calorons by an electromagnetic disturbance.

9.6.1 *Propagation of electromagnetic plane waves*

Let us first discuss which aspects of Harrington-Shepard (HS) (anti)calorons, see Sec. 4.3.1, are physically interpretable in a 3 + 1 dimensional Minkowskian spacetime. Again, if not stated otherwise we work in supernatural units, $\hbar = c = k_B = 1$. HS (anti)calorons are singular at $\tau = r = 0$ ($r \equiv \sqrt{\vec{x}^2}$), and we have argued in Sec. 5.3 that, in spite of the fact that the (regular) action density in a Euclidean spacetime region containing this point is τ dependent and thus does not have meaning after analytic continuation to Minkowskian space-time signature (Wick rotation), the action of the HS (anti)caloron, given as

$$S_{C,A} = \frac{8\pi^2}{g^2} \int_{S_\delta^3} d\Sigma_\mu K_\mu = \frac{8\pi^2}{g^2}, \qquad (9.31)$$

is physically interpretable. Here S_δ^3 is a three-sphere of radius δ, centered at $\tau = r = 0$, K_μ denotes the Chern-Simons current evaluated on an HS (anti)caloron, and g is the coupling constant of the fundamental theory, which, as discussed in Sec. 5.3, should be set equal to $g = e = \sqrt{8}\pi/\sqrt{\hbar}$ for HS (anti)calorons of scale parameter $\rho \sim |\phi|^{-1}$ relevant in forming the thermal ground state. Thus $S_{C,A} = \hbar$ in the limit $\delta \to 0$. This limit, however, is invariant under Wick rotation. On the other hand, we have seen in Sec. 4.3.1.2 that for $s \equiv \pi \frac{\rho^2}{\beta} \gg r \gg \beta$ and for $r \gg s \gg \beta$ the field strength of an HS (anti)caloron is static and as such also invariant under Wick rotation. Therefore, both the HS (anti)caloron action as well as its spatially peripheral field strength are, in principle, observables in Minkowskian spacetime. We remark at this point that the condition $s \gg \beta$, which is required for Eqs. (4.76) and (4.77) of Sec. 4.3.1.2 to be valid, is always satisfied for $\rho \sim |\phi|^{-1}$ (deconfining phase). Namely, one has

$$\frac{s}{\beta} = \pi \left(\frac{\rho}{\beta}\right)^2 = \pi \left(\frac{\lambda^{3/2}}{2\pi}\right)^2 = \frac{\lambda^3}{4\pi} \geq 212.3, \qquad (9.32)$$

where $\lambda \equiv \frac{2\pi T}{\Lambda} \geq \lambda_c = 13.87$.

In the following we would like to investigate in what sense the vacuum parameters of classical electrodynamics, namely the electric permittivity ϵ_0 and the magnetic permeability μ_0, can be reduced to the physics of the static, non-Abelian, and (anti)selfdual monopole and dipole configurations represented by HS (anti)calorons in the regimes $\beta \ll r \ll s$ and $r \gg s \gg \beta$, respectively.

We start with the case $r \gg s$. In order to not affect spatial homogeneity on scales comparable to or smaller than s the electromagnetic field, which propagates through the deconfining thermal ground state in the absence of explicit electric charges, is considered a plane wave of wave length l much larger than s. Such a field effectively sees a density of selfdual dipoles, see Eq. (4.77) and (4.78) of Sec. 4.3.1.2. Because they are given by $p_i^a = s\delta_i^a$ their dipole moments align along the direction of the exciting electric or magnetic field both in space and in the SU(2) algebra su(2). Also, since p_i^a is given in terms of the modulus $|\phi|$ of the *inert* field ϕ and $T = \beta^{-1}$ this dipole moment is invariant under the process of having HS (anti)calorons spatially approach (and overlap) one another. Note that at this stage the definition of what is to be viewed as an Abelian direction in su(2) is a global gauge convention such that all spatial orientations of the dipole moment p_i^a are *a priori* thinkable. That is, dynamical Abelian projection of the non-Abelian situation in Eq. (4.77) is owed to the Abelian and dipole aligning nature of the exciting, massless field. This massless field exists because of the adjoint Higgs mechanism invoked by the inert field ϕ which, in turn, is attributed to HS (anti)caloron centers, $r \ll s$ (see Secs. 5.1.3, 5.1.4, and 5.1.5 and Eq. (9.34) below). In Sec. 9.6.2 we will discuss why in almost all instances of electromagnetic wave propagation the massive vector modes (non-Abelian directions in the SU(2) algebra) of the deconfining phase of an SU(2) Yang–Mills theory are completely decoupled. We thus can be content with investigating the Abelian, massless direction of the SU(2) algebra only.

Per spherical spatial coarse-graining volume V_{cg} of radius $|\phi|^{-1} = \rho = \sqrt{\frac{\Lambda^3}{2\pi T}}$ with

$$V_{cg} = \frac{4}{3}\pi |\phi|^{-3}, \tag{9.33}$$

the center of a selfdual HS caloron or the center of an antiselfdual HS anticaloron reside. Note the large hierachy between s (the minimal

spatial distance to the center of a (anti)caloron, which allows to iden-
tify the static, (anti)selfdual dipole) and the radius of the sphere $|\phi|^{-1}$
defining V_{cg},

$$\frac{s}{|\phi|^{-1}} = \frac{1}{2}\lambda^{3/2} \geq 25.83 \left(\frac{\lambda}{\lambda_c}\right)^{3/2}. \tag{9.34}$$

If the exciting field is electric then it sees twice the electric dipole p_i^a
(cancellation of magnetic dipole between caloron and anticaloron),
if it is magnetic it sees twice the magnetic dipole p_i^a (cancellation
of electric dipole between caloron and anticaloron, $\vec{E} = -\vec{B}$ $\leftrightarrow$
$-\vec{E} = \vec{B}$). To be definite, let us discuss the electric case in detail,
characterized by an exciting Abelian field $\vec{E}_e$. The modulus of the
according dipole density $\vec{D}_e || \vec{E}_e$ is given as

$$|\vec{D}_e| = \frac{2s}{V_{cg}} = \frac{3}{4\pi}\Lambda^2\lambda_c^{1/2}\left(\frac{\lambda}{\lambda_c}\right)^{1/2}. \tag{9.35}$$

In classical electromagnetism the relation between the fields $\vec{E}_e$ and
$\vec{D}_e$ is

$$\vec{D}_e = \epsilon_0\vec{E}_e, \tag{9.36}$$

where

$$\epsilon_0 = 5.52703 \times 10^7 \frac{Q}{\mathrm{V\,m}} \tag{9.37}$$

is the electric permittivity of the vacuum, and $Q = 1.602 \times 10^{-19}\,\mathrm{A\,s}$
denotes the electron charge (unit of elementary charge), now both
in SI units.

According to electromagnetism the energy density ρ_{EM} carried
by an external electromagnetic wave with $|\vec{E}_e| = |\vec{B}_e|$ is

$$\rho_{\mathrm{EM}} = \frac{1}{2}(\epsilon_0\vec{E}_e^2 + \frac{1}{\mu_0}\vec{B}_e^2) = \frac{1}{2}(\epsilon_0 + \frac{1}{\mu_0})\vec{E}_e^2. \tag{9.38}$$

In natural units we have $\epsilon_0\mu_0 = 1/c^2 = 1$, and therefore $\mu_0 = 1/\epsilon_0$.
(To assume $\epsilon_0\mu_0 = 1$ just represents a short cut, it would have come
out automatically if we had treated the magnetic case explicitly.)

Thus

$$\rho_{\text{EM}} = \epsilon_0 \vec{E}_e^2 \,. \tag{9.39}$$

The $\vec{E}_e$-field dependence of ρ_{EM} is converted into a fictitious temperature dependence by demanding that the temperature of the thermal ground state of SU(2)$_{\text{CMB}}$ adjusts itself such as to accomodate ρ_{EM},

$$\rho_{\text{EM}} = 4\pi\Lambda^3 T \quad \Leftrightarrow \quad |\vec{E}_e| = \Lambda^2 \sqrt{2\frac{\lambda_c}{\epsilon_0} \left(\frac{\lambda}{\lambda_c}\right)^{1/2}} \,. \tag{9.40}$$

Eq. (9.40) generalizes the thermal situation of ground-state energy density, where ground-state thermalization is induced by a thermal ensemble of excitations, to the case where the thermal ensemble is missing but the probe field induces a fictitious temperature and energy density to the (vibrating) ground state. Combining Eqs. (9.35), (9.36), and (9.40), and introducing the ratio ξ between the non-Abelian monopole charge Q' in the dipole and the (Abelian) electron charge[14] Q, we obtain

$$\epsilon_0[Q(\text{V m})^{-1}] = \frac{3}{\sqrt{32\pi}} \left(\frac{\Lambda[\text{m}^{-1}]}{\Lambda[\text{eV}]}\right)^{1/2} \xi Q \sqrt{\epsilon_0[Q(\text{V m})^{-1}]} \quad \Leftrightarrow$$

$$\epsilon_0[Q(\text{V m})^{-1}] = \frac{9}{32\pi^2} \frac{\Lambda[\text{m}^{-1}]}{\Lambda[\text{eV}]} (\xi Q)^2 \,. \tag{9.41}$$

Notice that ϵ_0 does not exhibit any temperature dependence and thus no dependence on the field strength $\vec{E}_e$. It is a constant. In particular, ϵ_0 does not relate to the state of fictitious ground-state thermalization which would associate to the rest frame of a local heat bath.

To produce the measured value for ϵ_0 as in Eq. (9.37) the ratio ξ in Eq. (9.41) is required to be

$$\xi \equiv \frac{Q'}{Q} = 19.56 \,. \tag{9.42}$$

---

[14]In natural units, the actual charge of the monopole constituents within the (anti)selfdual dipole is $1/g$ where g is the undetermined fundamental gauge coupling. This is absorbed into ξ.

Thus, compared to the electron charge, the charge unit associated with a (anti)selfdual non-Abelian dipole, residing in the thermal ground state, is gigantic.

Discussing μ_0, we could have proceeded in complete analogy to the case of ϵ_0. (In this case μ_0^{-1} would define the ratio between the modulus of the magnetic dipole density and the magnetic flux density $|\vec{B}_e|$.) Here, however, the comparison between non-Abelian magnetic charge and an elementary, magnetic, and Abelian charge is not facilitated since the latter does not exist in electrodynamics.

Let us now ask the question what the condition that the wavelength l of the electromagnetic disturbance considered in this section is much larger than s implies when invoking SU(2)$_{\text{CMB}}$. One has

$$l \gg \frac{\lambda_c^2}{2\Lambda_{\text{CMB}}}\left(\frac{\lambda}{\lambda_c}\right)^2 = 1.1254\,\text{m}\left(\frac{T}{2.726\,\text{K}}\right)^2. \tag{9.43}$$

We now turn to the case $r \ll s$. To rely on the presence of the inert adjoint scalar field ϕ of the effective theory, r needs to be larger than the spatial coarse-graining scale $|\phi|^{-1} = \frac{1}{2\pi}\lambda_c^{3/2}\left(\frac{\lambda}{\lambda_c}\right)^{3/2}\beta \geq 8.22\,\beta$. Within the according regime $|\phi|^{-1} \leq r \ll s$ of spatial distances from the caloron center at $(\tau = 0, \vec{x} = 0)$ an electromagnetic wave of wave length l sees the selfdual field of a static, non-Abelian monopole of electric and magnetic charge as in Eq. (4.76) of Sec. 4.3.1.2 which is centered at $\vec{x} = 0$. A selfdual Abelian field strength $E_i = B_i$ of this monopole is obtained [Goddard and Olive (1978)] as

$$E_i = B_i = \frac{\phi^a}{|\phi|}E_i^a = \frac{\phi^a}{|\phi|}B_i^a \tag{9.44}$$

with the field ϕ gauged from unitary gauge $\phi^a = 2|\phi|\delta^{a3}$ into "hedgehog" gauge $\phi^a = 2|\phi|\hat{x}^a$. The according gauge transformation is give in terms of the group element $\Omega \equiv \cos\frac{1}{2}\psi - i\hat{k}\cdot\vec{\sigma}\sin\frac{1}{2}\psi$ where σ_i, $(i = 1,2,3)$, are the Pauli matrices, $\hat{k} \equiv \frac{\hat{e}_3\times\hat{x}}{\sin\theta}$, $\hat{e}_3$ is the third vector of an orthonormal basis of space, $\theta \equiv \angle(\hat{e}_3,\hat{x})$, and $\psi = \theta$ for $0 \leq \theta \leq \pi - \epsilon$, which smoothly drops to zero at $\theta = \pi$, and the limit $\epsilon \to 0$ is understood [Goddard and Olive (1978)]. For the monopole field E_i to be normalized to charge $-2Q'$ one[15] thus has

$$E_i = B_i = -\frac{2Q'}{4\pi\epsilon_0}\frac{\hat{x}_i}{r^2} = -\frac{2Q'\mu_0}{4\pi}\frac{\hat{x}_i}{r^2}. \tag{9.45}$$

The electric or magnetic poles of Eq. (9.45) should independently react by harmonic and linear acceleration to the presence of an external electric or magnetic field $\vec{E}_e$ or $\vec{B}_e$, respectively, forming a monochromatic electromagnetic wave of frequency $\omega = \frac{2\pi}{T}$. At $\vec{x} = 0$ one has

$$\vec{E}_e = \vec{E}_0 \sin(\omega t), \tag{9.46}$$

and readily derives (as in Thomson scattering) that the induced dipole moment $\vec{p}$, say, for the electric case, is given as

$$\vec{p} = -\frac{\vec{E}_e(2\,Q')^2}{m\omega^2}. \tag{9.47}$$

Interestingly, by virtue of Eq. (9.45) the squared charge of the pole, $(2\,Q')^2$, cancels out in $\vec{p}$ because its mass m carries an identical factor (only the electric (magnetic) monopole is linearly and harmonically accelerated by the external electric (magnetic) field $\vec{E}_e$ ($\vec{B}_e$) and hence m carries electric (magnetic) field energy only):

$$m = \frac{1}{2}\epsilon_0\,4\pi \int_{|\phi|^{-1}}^{\infty} dr\, r^2\, E_i E_i$$
$$= \frac{1}{8\pi\epsilon_0}(2\,Q')^2 \int_{|\phi|^{-1}}^{\infty} \frac{dr}{r^2} = \frac{1}{8\pi\epsilon_0}(2\,Q')^2|\phi| \;\Rightarrow$$
$$\vec{p} = -\frac{8\pi\epsilon_0\vec{E}_e}{|\phi|\omega^2}. \tag{9.48}$$

Again, the volume V_{cg}, which underlies the dipole moment $\vec{p}$ by containing a caloron or an anticaloron center, is given by Eq. (9.33),

---

[15]The factor two in front of the monopole charge Q' is due to a contribution to the monopole field strength of the anticaloron identical to that of the caloron.

and we have

$$|\vec{D}_e| = \frac{|\vec{p}|}{V_{cg}} = 6\,\epsilon_0 \frac{|\vec{E}_e||\phi|^2}{\omega^2}\,, \tag{9.49}$$

and therefore

$$\epsilon_0 \equiv \frac{|\vec{D}_e|}{|\vec{E}_e|} = 6\,\epsilon_0 \frac{|\phi|^2}{\omega^2}\,. \tag{9.50}$$

In Eq. (9.50) also the vacuum permittivity ϵ_0 cancels out, and we are left with the condition

$$\omega = \sqrt{6}\,|\phi| \quad \Leftrightarrow \quad l = \sqrt{\frac{2}{3}}\pi|\phi|^{-1} = \sqrt{\frac{2}{3}}\pi\Lambda^{-1}\lambda_c^{1/2}\left(\frac{\lambda}{\lambda_c}\right)^{1/2}, \tag{9.51}$$

where temperature T (or λ), again, is set by the local field strengths of the electromagnetic probe according to Eqs. (9.38) and (9.40). Let us see whether the second of Eqs. (9.51) is consistent with $|\phi|^{-1} \le r = l \ll s$. The former inequality is selfevident, and the latter follows from

$$\frac{s}{l} = \sqrt{\frac{3}{8}}\frac{\lambda_c^{3/2}}{\pi}\left(\frac{\lambda}{\lambda_c}\right)^{3/2} = 10.069\left(\frac{\lambda}{\lambda_c}\right)^{3/2}. \tag{9.52}$$

By setting $\lambda = \lambda_c$ we obtain from Eqs. (9.51) a minimal wavelength

$$l_{min} = \sqrt{\frac{2}{3}}\pi\Lambda^{-1}\lambda_c^{1/2} = 0.112\,\text{m}. \tag{9.53}$$

This wavelength is about a factor of ten smaller than the lowest possible value expressed by Eq. (9.43).

Eqs. (9.43) and (9.40) indicate that an uncertainty-like relation between field $|\vec{E}_e|$ strength and wave length l takes place as follows

$$|\vec{E}_e|^4 l^{-1} \ll \frac{8\Lambda^9}{\epsilon_0^2}\,. \tag{9.54}$$

Therefore, the larger the probe intensity the longer its wave length is required to be in order to be supported by thermal ground-state physics. Since there is no temperature dependence in (9.54) this statement should be regarded valid independently of the thermo-dynamical condition that $\lambda \geq \lambda_c$. As we shall see in Sec. 9.6.2 a violation of this thermodynamical condition is indeed required for a description of the propagation of electromagnetic waves in non-thermal situations subject to an SU(2) Yang–Mills theory of scale $\Lambda_e \sim m_e \sim 0.5\,\text{MeV}$. As a consequence of the fictitious tempera-ture parameter being much lower then $T_c \sim m_e$, one expects the decoupling (absence) of the massive vector modes of this theory in phenomena that appeal to the long-range, wave-like propagation of the electromagnetic field. Here we conclude that the thermal ground state of SU(2)$_{\text{CMB}}$ supports the propagation of a nonthermal probe of energy density (or intensity) $\epsilon_0 |\vec{E}_e|^2$ in terms of Harrington-Shepard (anti)calorons (trivial holonomy) if the probe's wave length l is suf-ficiently large. To address the nonthermal propagation of shorter wavelength and/or higher intensities, see (9.54), additional, mix-ing SU(2) gauge factors of hierarchically larger Yang–Mills scales have to be postulated. At present, however, it is not clear how the effectiveness of the successful Standard Model of particle physics in describing electroweak processes can be achieved in terms of such a more fundamental framework of pure Yang–Mills dynamics.

9.6.2 *Photons re-visited*

Let us now be more specific on how condition (9.54) for a single Yang–Mills theory of scale Λ can be evaded when constructing a luminiferous, Poincaré invariant aether for the entire observed electromagnetic spectrum. In deriving the field equations of classi-cal electrodynamics Maxwell assumed that electromagnetic distur-bances are those of a medium required for them to propagate in anal-ogy to distortions of the stationary flow of a fluid in hydrodynamics [Maxwell (1873)]. If existent then such a medium needs to be of a peculiar nature, however, since it fails to associate with a preferred rest frame, a feat first demonstrated by Michelson and Morley and re-confirmed many times ever since: the speed of light c is a constant of nature and as such does not depend on the observer's state of motion relative to a source. Consequences of this experimental fact, expressed by the group of Lorentz transformations linking inertial frames, are laid out by Special Relativity and have been vindicated by countless experiments. In classical electrodynamics, constancy of the phase velocity c of electromagnetic waves is implied by the constancy of ϵ_0 and μ_0 – the electric permittivity and magnetic permability of free space. On the other hand, experience associates a particle-like or quantum nature to the carrier of the electromagnetic force – the photon – whose energy $E = 2\pi\hbar\nu$ and momentum modulus $p = \frac{2\pi}{c}\nu$ are independent of the intensity of the monochromatic electromagnetic wave it associates with, but proportional to frequency ν. From now on we use (super)natural units $c = \hbar = k_B = \epsilon_0 = \mu_0 = 1$, k_B indicating Boltzmann's constant.

The purpose of this last section of the chapter is to propose a framework to address the particle-wave duality of the photon, based on the extension of the gauge group of electromagnetism U(1) to a product of mixing SU(2) groups belonging to pure Yang–Mills theories. As we will argue, such a setting promises to reconcile the seemingly paradoxical aspects of electromagnetic disturbances in terms of a nontrivial vacuum structure. This vacuum is structured according to different aspects of the thermal ground state of an SU(2) Yang–Mills theory, depending on whether (anti)caloron centers or peripheries are probed by the excitation, see Sec. 9.6.1.

While field ϕ represents spatially densely packed (anti)caloron centers the effective gauge field $a_\mu^{\rm gs}$ represents the collective effect of (anti)caloron overlap, accompanied by transient holonomy changes [Nahm (1980–1983);Kraan and Van Baal I (1998);Kraan and Van Baal II (1998); Lee and Lu (1983); Diakonov et al. (2004)], as facilitated by their static peripheries.

For a given caloron, the peripheral field strength is that of a self-dual, static dipole. A density of such polarized dipoles represents the response of the vacuum to any externally provided disturbance, say, an oscillating electric charge. On the other hand, field ϕ breaks the

SU(2) gauge symmetry of the underlying, classical Euclidean Yang–Mills action down to U(1) which means that only one of the three directions of the SU(2) algebra su(2) is massless, the (large) mass of the other two directions being fixed by (low-temperature) ambient thermodynamics in a large bulk volume. A natural question to ask is under what conditions the associated electric and magnetic dipole densities of the thermal ground state can be regarded a medium induced by, at the same time supporting, wave-like propagation of the massless mode. By promoting a_μ^{gs} to a monochromatic electro-magnetic wave $a_\mu^{a=3}$, associated with the massless su(2) direction [16], and by identifying its mean Minkowskian energy density with ρ^{gs}, one arrives at (see also Eq. (9.40))

$$|\vec{E}_e| = \Lambda^2 \sqrt{2\lambda}, \tag{9.55}$$

where $|\vec{E}_e|$ represents the mean electric field-strength modulus of $a_\mu^{a=3}$. Eq. (9.55) together with the condition that wavelength l must not resolve the interior of an (anti)caloron, $l \gg s(\lambda)$, imply the following T independent statement

$$|\vec{E}_e|^4 l^{-1} = |\vec{E}_e|^4 \nu \ll 8\Lambda^9. \tag{9.56}$$

The Yang–Mills scale Λ thus determines the maximum of intensity at a given frequency ν and vice versa commensurate with wave-like

---

[16] In SU(2)$_{CMB}$ massive modes $a_\mu^{a=1,2}$ interact with the massless mode $a_\mu^{a=3}$ by tiny radiative corrections only at a thermodynamical temperature bounded from below by that of the present CMB, see Sec. 9.4.3, and thus can be ignored in the effective Yang–Mills equation $D^\mu G_{\mu\nu} = 2ie[\phi, D_\nu\phi]$ when discussing the propagation of $a_\mu^{a=3}$. This equation thus reduces to the vacuum Maxwell equation $\partial^\mu F_{\mu\nu}^3 = 0$ with $F_{\mu\nu} = \partial_\mu a_\nu^{a=3} - \partial_\nu a_\mu^{a=3}$ which, indeed, is solved by a plane wave. The latter is subject to undetermined normalisation, frequency, and phase. Note that, as is the case for a_μ^{gs}, the Euclidean, time averaged energy density $\text{tr}\frac{1}{2}(\vec{E}_e^2 - \vec{B}_e^2)$ of $a_\mu^{a=3}$ vanishes such that solely the potential $\text{tr}V(\phi) = \rho^{gs}$ in the effective action determines the mean energy density of such a plane wave, see Sec. 9.6.1.

propagation. Although derived from the two T dependent relations $l \gg s$ with

$$s(\lambda) = \frac{1}{2}\lambda^2\Lambda^{-1} \qquad (9.57)$$

and (9.55) condition (9.56) should be regarded universally valid: In a nonthermal situation, ν is not required to satisfy any additional constraint as implied by the existence of a critical thermodynamical temperature $\lambda_c = 13.87$ for the deconfining-preconfining phase transition.

For SU(2)$_{\mathrm{CMB}}$, the Yang–Mills theory proposed to underly all experimentally investigated thermal photon gases including the Cosmic Microwave Background (CMB), see Sec. 9.4, one has $\Lambda_{\mathrm{CMB}} \sim 10^{-4}\,\mathrm{eV}$. According to (9.56) the energy density $|\vec{E}_e|^2$ is bounded by $|\vec{E}_e|^2 \ll \sqrt{8\frac{\Lambda_{\mathrm{CMB}}^9}{\nu}}$. For $\nu = 10^6\,\mathrm{Hz}$ (radio frequency) one obtains $|\vec{E}_e|^2 \ll 2.3 \times 10^{-21}\,\mathrm{J\,cm^{-3}}$. For a comparison, the U(1) energy density of the CMB at this frequency, measured with a spectral band width of $\Delta\nu = 10^4\,\mathrm{Hz}$, is $8\pi T\nu^2\Delta\nu = 3.51 \times 10^{-37}\,\mathrm{J\,cm^{-3}}$. Thus, such a radio wave would represent a signal discernible from the thermal noise of the CMB if the latter where described by a conventional U(1) theory. Higher-frequency monochromatic waves are bounded by energy densities reduced by a factor $1/\sqrt{\frac{\nu}{10^6\,\mathrm{Hz}}}$, and it is clear that condition (9.56) is violated for a wealth of phenomena, attributed to the propagation of electromagnetic waves, when setting $\Lambda = \Lambda_{\mathrm{CMB}}$.

The way out is to postulate the existence of additional SU(2) factors with Yang–Mills scales hierarchically larger than Λ_{CMB} which, thermodynamically seen, are in confining phases under ambient conditions such that massive modes do not propagate. One could consider $\Lambda = \Lambda_e \sim 5 \times 10^5\,\mathrm{eV}\sim m_e$, m_e denoting the electron mass. Then (9.56) no longer is in contradiction with experience: propagation of high-intensity and high-frequency massless waves is accomodated by the large value of the Yang–Mills scale Λ_e.

The process of thermalisation in SU(2)$_{\mathrm{CMB}}$ towards black-body radiation, which is surrounded by a cavity wall providing emitting and absorbing electrons, would then proceed as follows. At a thermodynamical wall temperature T with $\Lambda_e > T \gg T_{CMB} = 2.725\,K=$ $\frac{13.87}{2\pi}\Lambda_{CMB}$ radiation emitted by the wall electrons satisfies (9.56) with $\Lambda = \Lambda_e$. This radiation represents classical waves in SU(2)$_e$. A priori, their spectral energy density thus is given by the Rayleigh-Jeans law

$$u_{RJ} = 8\pi T \nu^2 = \frac{2}{\pi} T^3 x^2 \quad (x \equiv 2\pi\nu/T), \qquad (9.58)$$

which expresses an obvious and well-known ultraviolet catastrophe. The latter, however, does not take place if classical SU(2)$_e$ waves excite photons from the thermal ground state of SU(2)$_{CMB}$. Namely, setting $\Lambda = \Lambda_{CMB}$ in a thermal situation, the condition that wavelength l must be larger than s (see Eq. (9.57)) for wave-like propagation amounts to

$$l = \frac{2\pi}{xT} \gg s = \frac{2\pi^2 T^2}{\Lambda_{CMB}^3} \quad \Leftrightarrow \quad x \ll \frac{1}{\pi}\left(\frac{\Lambda_{CMB}}{T}\right)^3. \qquad (9.59)$$

Hence, condition (9.59) is violated at extremely small frequencies already, that is, for $x > \frac{1}{\pi}\left(\frac{\Lambda_{CMB}}{T}\right)^3$. For such frequencies the quantum of action, localised in thus probed (anti)caloron centers, participates in the thermodynamics of fluctuations by indeterministic materialisations of quanta of energy and momentum $2\pi\nu$. In assuming that their numbers are suppressed by associated Boltzmann weights the Bose-Einstein distribution function $n_B(x) = 1/(e^x - 1)$ is implied, corresponding to a spectral energy density

$$u_{Planck} = \frac{2}{\pi} T^3 \frac{x^3}{e^x - 1}. \qquad (9.60)$$

Function u_{Planck} peaks at $x = 2.82$, is normalisable to $\frac{\pi^2}{15} T^4$ (Stefan-Boltzmann law), and is bounded from above by u_{RJ}. This provides for an energetic reason why the "rotation" from SU(2)$_e$ to SU(2)$_{CMB}$ is invoked in the emergence of black-body radiation. Setting the critical, thermodynamical temperature T_c for the deconfining-preconfining phase transition in SU(2)$_{CMB}$ equal to

$T_c = T_0 = 2.725$ K, see Sec. 9.4.3, one obtains for the right-hand side of (9.59) a value of $x = \frac{1}{\pi} \left(\frac{\Lambda_{CMB}}{T} \right)^3$ which corresponds to 1.68 GHz. This supports the proposal in Sec. 9.4.3 that the CMB radio excess indeed is attributed to evanescent $SU(2)_{CMB}$ *waves*.

To view photons as thermal excitations of the thermal ground state in $SU(2)_{CMB}$ would relate to the photoelectric effect in the following way. In $SU(2)_e$ an incident monochromatic wave of frequency, say, $\sim 10^{14}$ Hz, drives the dissipation of radiation-field energy within a thin surface layer of a bulk metal or semiconductor (the classical skin effect with skin depth, e.g., in copper, of a few nanometers at $\nu \sim 10^{14}$ Hz) such that an equilibrium between energy entry into this surface layer and heat flow towards the bulk is established. Such an equilibrium is characterised by a thermodynamical temperature $T \ll \Lambda_e = m_e$. By the above argument, this local thermal environment, however, is subject to $SU(2)_{CMB}$. Since in $SU(2)_{CMB}$ $s(T)$ in Eq. (9.57) is much larger than $l = \nu^{-1}$ the excitations of the thermal ground state are photons. (Unequality (9.56) is violated.) That is, the incoming $SU(2)_e$ wave of intensity $|\vec{E}_e|^2$ is not supported within such a thermal surface layer: by probing the interior of $SU(2)_{CMB}$ (anti)calorons it "decays" into photons of energy and momentum $2\pi\nu$. With a finite probability [Mandel, Sudarshan, and Wolf (1964)], such a quantum of energy and momentum is transferred to a layer electron which, in turn, is expelled from the material thus becoming a photoelectron. Modulo a material dependent work function (a function of material parameters such as skin depth, electric and heat conductivity, etc.), the maximal kinetic energy of photoelectrons thus is given by $2\pi\nu$ while their flux is proportional to the wave's intensity (energy conservation after the above described dynamical equilibrium is established).

The thermal ground state of $SU(2)_e$ ($\Lambda = \Lambda_e = m_e$), on the other hand, could explain on a deeper level the transition from Thomson (T) scattering of a classical wave to Compton (C) scattering of a photon off an electron, which is well described by Quantum Electrodynamics. The associated total cross section [Klein and Nishina

(1929)] — an intensity independent quantity – is given as

$$
\begin{aligned}
\sigma_C &= \frac{3}{4}\sigma_T \left[\frac{1+x}{x^3} \left(\frac{2x(1+x)}{1+2x} - \log(1+2x) \right) \right. \\
&\qquad \left. + \frac{1}{2x} \log(1+2x) - \frac{1+3x}{(1+2x)^2} \right] \\
&= \sigma_T [1 - 2x + \frac{26}{5}x^2 + O(x^3)], \quad (x \equiv v/m_e).
\end{aligned}
\tag{9.61}
$$

As x rises to order unity, an increasingly strong violation of (9.56) takes place [17]. Here the scattering off an isolated electron does not generate any local, thermal equilibrium but the incoming beam itself is of an increasingly corpuscular structure as x grows.

References

Adler, S. L. (1969). Axial vector vertex in spinor electrodynamics, *Phys. Rev.*, **177**, 2426.

Adler, S. L. and Bardeen, W. A. (1969). Absence of higher order corrections in the anomalous axial vector divergence equation, *Phys. Rev.*, **182**, 1517.

Barrow, J. D., Ferreira, P. G., and Silk, J. (1997). Constraints on primordial magnetic field, *Phys. Rev. Lett.*, **78**, 3610.

Bell, J. S. and Jackiw, R. (1969). A PCAC puzzle: $\pi^0 \to \gamma\gamma$ in the sigma model, *Nuovo Cim. A*, **60**, 47.

Boggess, N. W., *et al.* (1992). The COBE mission — its design and performance two years after launch, *Astrophys. J.*, **397**, 420.

Copi, C. J., *et al.* (2010). Large angle anomalies in the CMB, *Adv. Astron.*, **2010**, 847541.

Diakonov, D., *et al.* (2004). Quantum weights of dyons and of instantons, *Phys. Rev. D*, **70**, 036003.

Doroshkevich, A. G. and Novikov, I. D. (1964). Mean density of radiation in the metagalaxy and certain problems in relativistic cosmology, *Soviet Physics Doklady*, **9**, 111.

---

[17]Since σ_C is defined as an intensity independent quantity the level of violation of (9.56) is to be inferred at constant $|\vec{E}_e|^4$. In a nonthermal situation $\Lambda_e = m_e$ is the only mass scale in SU(2)$_e$, and thus it is natural to set $|\vec{E}_e|^4 = m_e^8$. As a consequence, (9.56) requires $v \ll 8 m_e$ to obtain a beam void of particle-like aspects.

Doroshkevich, A. G., Zeldovich, Y. B., and Sunyaev, R. A. (1977). Fluctuations of the microwave background radiation in the adiabatic and entropic theories of galaxy formation, in *The Large Scale Structure of the Universe, Proceedings of the Symposium*, pp. 393.

Falquez, C., Hofmann, R., and Baumbach, T. (2010). Modification of black-body radiance at low temperatures and frequencies, *Ann. d. Phys.*, **522**, 904.

Falquez, C., Hofmann, R., and Baumbach, T. (2011). Charge-density waves in deconfining SU(2) Yang–Mills thermodynamics, arXiv:1106.1353 [hep-th].

Feynman, R. P. (1949a). Space-time approach to quantum electrodynamics, *Phys. Rev.*, **76**, 769.

Feynman, R. P. (1949b). The theory of positrons, *Phys. Rev.*, **76**, 749.

Feynman, R. P. (1950). Mathematical formulation of the quantum theory of electromagnetic interaction, *Phys. Rev.*, **80**, 440.

Fixsen, D. J., *et al.* (1994). Cosmic microwave background dipole spectrum measured by the COBE FIRAS instrument, *Astrophys. J.*, **420**, 445.

Fixsen, D. J., *et al.* (2009). ARCADE 2 measurement of the extra-galactic sky temperature at 3-90 GHz, arXiv:0901.0555.

Frieman, J. A., *et al.* (1995). Cosmology with ultralight pseudo Nambu–Goldstone bosons, *Phys. Rev. Lett.*, **75**, 2077.

Fujikawa, K. (1979). Path integral measure for gauge invariant fermion theories, *Phys. Rev. Lett.*, **42**, 1195.

Fujikawa, K. (1980). Path integral for gauge theories with fermions, *Phys. Rev. D*, **21**, 2848; Erratum-*ibid. Phys. Rev. D*, **22**, 1499.

Giacosa, F. and Hofmann, R. (2007). A Planck-scale axion and SU(2) Yang–Mills dynamics: Present acceleration and the fate of the photon, *Eur. Phys. J. C*, **50**, 635.

Giacosa, F., Hofmann, R., and Neubert, M. (2008). A model for the very early Universe, *JHEP*, **0802**, 077.

Goddard, P. and Olive, D. I. (1978). Magnetic monopoles in gauge field theories, *Rep. Prog. Phys..* **41**, 1357.

Goldsmith, P., Li, D., and Krčo, M. (2007). The transition from atomic to molecular hydrogen in interstellar clouds: 21 cm signature of the evolution of cold atomic hydrogen in dense clouds, *Astrophys. J.*, **654**, 273.

Grandou, T. and Hofmann, R. (2015). Thermal ground state and nonthermal probes, *Adv. Math. Phys.*, **2015**, Article ID 197197.

Gross, D. J. and Wilczek, F. (1973). Ultraviolet behavior of nonabelian gauge theories, *Phys. Rev. Lett.*, **30**, 1343.

Grum, F. and Becherer, R. J. (1979). *Optical Radiation Measurements. Volume 1: Radiometry* (Academic Press).

Haslam, C. G. T., *et al.* (1981). A 408 MHz all-sky continuum survey, *Astron. & Astrophys.*, **100**, 209.

Hofmann, R. (2005). *Nonperturbative approach to Yang–Mills thermodynamics*, *Int. J. Mod. Phys. A*, **20**, 4123; Erratum- *ibid. A*, **21**, 6515, (2006).

Hofmann, R. (2009). Low-frequency line temperatures of the CMB, *Ann. d. Phys.*, **18**, 634.

Hofmann, R. (2015). Electromagnetic waves and photons, arXiv:1508.02270.

Keller, J., Hofmann, R., and Giacosa, F. (2007). Correlation of energy density in deconfining SU(2) Yang–Mills thermodynamics, *Int. J. Mod. Phys. A*, **23**, 5181, (2008).

Khriplovich, I. B. (1969). Green's functions in theories with non-abelian gauge group, *Yad. Fiz.*, **10**, 409.

Khriplovich, I. B. (1970). Green's functions in theories with non-abelian gauge group, *Sov. J. Nucl. Phys.*, **10**, 235.

Klein, O. and Y. Nishina (1929), Über die Streuung von Strahlung durch freie Elektronen nach der neuen relativistischen Quantenmechanik nach Dirac, *Zeitschr. f. Physik*, **52**, 853.

Knee, L. B. G. and Brunt, C. M. (2001). A massive cloud of cold atomic hydrogen in the outer Galaxy, *Nature*, **412**, 308.

Kogut, A., *et al.* (1993). Dipole anisotropy in the COBE differential microwave radiometers first-year sky maps, *Astrophys. J.*, **419**, 1.

Kogut, A. (2010). Private communication.

Kronberg, P. P. (1994). Extragalactic magnetic fields, *Rept. Prog. Phys.*, **57**, 325.

Kraan, T. C. and Van Baal, P. (1998a). Exact T-duality between calorons and Taub-NUT spaces, *Phys. Lett. B*, **428**, 268.

Kraan, T. C. and Van Baal, P. (1998b). Periodic instantons with non-trivial holonomy, *Nucl. Phys. B*, **533**, 627.

Krall, N. A., Trivelpiece, A. W., and Gross, R. A. (1973). Principles of plasma physics, *American J. Phys.*, **41**, 1380.

Lee, K. and Lu, C. (1983). SU(2) calorons and magnetic monopoles, *Phys. Rev. D*, **58**, 025011-1.

Ludescher, J. and Hofmann, R. (2008). Thermal photon dispersion law and modified black-body spectra, *Ann. d. Phys.*, **18**, 271, (2009).

Ludescher, J. and Hofmann, R. (2009). CMB dipole revisited, arXiv:0902.3898 [hep-th].

Ludescher, J., *et al.* (2009). Spatial Wilson loop in continuum, deconfining SU(2) Yang–Mills thermodynamics, *Ann. d. Phys.*, **19**, 102.

Mather, J. C. *et al.* (1994). Measurement of the cosmic microwave background spectrum by the COBE FIRAS instrument, *Astrophys. J.*, **420**, 439.

Maeda, K., *et al.* (1999). A 45-MHz continuum survey of the northern hemisphere, *Astron. & Astrophys. Suppl. Ser.*, **140**, 145.

Mandel, L., Sudarshan, E. C. G., and Wolf, E. (1964), Theory of photoelectric detection of light fluctuations, *Proc. Phys. Soc.*, **84**, 435.

Maxwell, J. C. (1873), A treatise on electricity and magnetism, vol. I, *Clarendon Press (Oxford)*.

Maxwell, J. C. (1873), A treatise on electricity and magnetism, vol. II, *Clarendon Press (Oxford)*.

Meyer, D. M. and Lauroesch, J. T. (2006). A cold nearby cloud inside the local bubble, *Astrophys. J.*, **650**, L67.

Moosmann, J. and Hofmann, R. (2008). Center-vortex loops with one selfintersection, *ISRN Math. Phys.*, 2012, Article ID 601749.

Nahm, W. (1980). A simple formalism for the BPS monopole, *Phys. Lett. B*, **90**, 413.

Nahm, W. (1981). All self-dual multimonopoles for arbitrary gauge groups, CERN preprint TH-3172.

Nahm, W. (1982). *The Construction of all Self-dual Multimonopoles by the ADHM Method* in *Monopoles in Quantum Field Theory*, ed. N. Craigie *et al.* (World Scientific, Singapore), p. 87.

Nahm, W. (1983). Self-dual monopoles and calorons in *Trieste Group Theor. Method 1983*, p. 189.

Nakamura, K., *et al.* (2010). The review of particle physics, *J. Phys. G*, **37**, 075021.

Neronov, A. and Vovk, I. (2010). Evidence for strong extragalactic magnetic fields from Fermi observations of TeV blazars, *Science*, **328**, 73.

Penzias, A. A. (1964). Helium-cooled reference noise source in a 4-kMc waveguide, *Rev. Scient. Instrum.*, **36**, 68.

Penzias, A. and Wilson, R. W. (1965). A measurement of excess antenna temperature at 4080 Mc/s, *Astrophys. J.*, **142**, 419.

Peebles, P. J. E. and Yu, J. T. (1970). Primeval adiabatic perturbation in an expanding universe, *Astrophys. J.*, **162**, 815.

Peebles, P. J. E. and Wilkinson, T. (1968). Comment on the anisotropy of the primeval fireball, *Phys. Rev.*, **174**, 2168.

Perlmutter, S., *et al.* (1998). Measurements of Omega and Lambda from 42 high redshift supernovae, *Astrophys. J.*, **517**, 565.

Politzer, H. D. (1973). Reliable perturbative results for strong interactions? *Phys. Rev. Lett.*, **30**, 1346.

Reich, P. and Reich, W. (1986). A radio continuum survey of the northern sky at 1420 MHz. II, *Astron. & Astrophys. Suppl. Ser.*, **63**, 205.

Redfield, S. and Linsky, J. L. (2007). The structure of the local interstellar medium IV: dynamics, morphology, physical properties, and implications of cloud-cloud interactions, arxiv:0709.4480 [astro-ph].

Riess, A. G., *et al.* (1998). Observational evidence from supernovae for an accelerating Universe and a cosmological constant, *Astronom. J.*, **116**, 1009.

Roger, R. S., *et al.* (1999). The radio emission from the Galaxy at 22 MHz, *Astron. & Astrophys. Suppl. Ser.*, **137**, 7.

Schwinger, J. (1948a). On quantum-electrodynamics and the magnetic moment of the electron, *Phys. Rev.*, **73**, 416.

Schwinger, J. (1948b). Quantum electrodynamics. I. A covariant formulation, *Phys. Rev.*, **74**, 1439.

Schwarz, M., Hofmann, R., and Giacosa, F. (2007). Gap in the black-body spectrum at low temperature, *JHEP*, **0702**, 091.

Straessner, A. (2010). *Gauge Boson Production at LEP*, Springer Tracts in Modern Physics, Vol. 235, p. 55.

Szopa, M. and Hofmann, R. (2008). A model for CMB anisotropies on large angular scales, *JCAP*, **03**, 001.

Tada, M., *et al.* (2006). Single-photon detection of microwave blackbody radiations in a low-temperature resonant-cavity with high Rydberg atoms, *Phys. Lett. A*, **349**, 488.

't Hooft, G. Unpublished.

't Hooft, G. (1976). Symmetry breaking through Bell–Jackiw anomalies, *Phys. Rev. Lett.*, **37**, 8.

Tomonaga, S. (1946). On a relativistically invariant formulation of the quantum theory of wave fields, *Progr. Theor. Phys.*, **1**, 27.

Widrow, L. M. (2002). Origin of galactic and extragalactic magnetic fields, *Rev. Mod. Phys.*, **74**, 775.

Zeldovich, Y. B. (1972). A hypothesis, unifying the structure and the entropy of the Universe, *Mon. Not. Roy. Astron. Soc.*, **160**, 1P.

Chapter 10

Astrophysical and Cosmological Implications of SU(2)$_\text{CMB}$

In the previous chapter we have provided reasons to seriously consider thermal photon gases based on SU(2)$_\text{CMB}$. The present chapter investigates implications of this postulate for the stability of galactic low-temperature and low-density atomic-hydrogen clouds [Knee and Brunt (2001)]. Moreover, the large-angle domain of the CMB [Copi *et al.* (2010)], the emergence of dark energy due to topologically nontrivial field configurations of SU(2)$_\text{CMB}$ interacting with a Planck-scale axion field [Frieman *et al.* (1995);Giacosa and Hofmann (2007)], and the fate of the CMB on cosmological time-scales [Giacosa and Hofmann (2007)] are addressed. Next, we discuss implications of the nontrivial equation of state of an SU(2)$_\text{CMB}$ gas of thermal relic photons – the CMB – for the temperature-redshift relation and how this seems to explain the discrepancy between the redshift of (instantaneous) re-ionization as extracted from high-redshift quasar spectra and from fits to the angular power spectrum of the CMB temperature-temperature correlation function [Hofmann (2015)]. Finally, we speculate about cosmic neutrinos receiving their mass through interaction with the CMB and investigate the influence of this postulate on the CMB temperature-redshift relation and on the cosmological model at high redshift.

10.1 Cold and Dilute Clouds of Atomic Hydrogen

The purpose of this section is to compute the two-point correlator of the canonical energy density Θ_{00} of photons in a thermalized gas. This correlation function is a measure for the occurrence of energy transport within such a gas which is of relevance when investigating why at low densities and radiation temperatures immersed hydrogen atoms act like the constituents of an ideal atomic gas (no interaction between atoms). The presentation given below follows [Keller, Hofmann, and Giacosa (2008)] very closely.

Apart from astrophysical considerations, the computation of the two-point correlator of Θ_{00} in a thermalized gas is technically interesting. Namely, in the deconfining phase of SU(2) Yang–Mills thermodynamics this quantity is accurately calculable in terms of the one-loop photon polarization tensor. In the present section we present results in a condensed way to warrant a reasonably efficient flow of the arguments. For calculational details we refer the reader to [Keller (2008)].

10.1.1 *Two-point correlation of energy density in thermal U(1) gauge theory*

Here we compute the two-point correlation of the canonical energy density in a pure, thermalized U(1) gauge theory. Our results will serve as a benchmark for the more involved calculation of SU(2) Yang–Mills theory.

10.1.1.1 *General strategy*

The two-point correlation of the energy density is computed by letting derivative operators, associated with the structure of the energy–momentum tensor $\Theta_{\mu\nu}$, act on the real-time propagator of the U(1) gauge field.

The traceless and symmetric (Belinfante) energy–momentum tensor of a pure U(1) gauge theory is given as

$$\Theta_{\mu\nu} = -F_\mu^\lambda F_{\nu\lambda} + \frac{1}{4}g_{\mu\nu}F^{\kappa\lambda}F_{\kappa\lambda}, \tag{10.1}$$

where $F_{\mu\nu} \equiv \partial_\mu A_\nu - \partial_\nu A_\mu$, and A_μ denotes the U(1) gauge field.

We are interested in computing the connected correlation function $\langle\Theta_{00}(x)\Theta_{00}(y)\rangle$ in four-dimensional Minkowskian spacetime ($g_{00} = 1$). This is done by applying Wick's theorem to express $\langle\Theta_{00}(x)\Theta_{00}(y)\rangle$ in terms of the propagator $D_{\mu\nu}$ of the field A_μ. In Coulomb gauge and momentum space the latter is given as

$$D_{\mu\nu}(p, T) = D_{\mu\nu}^{vac}(p, T) + D_{\mu\nu}^{th}(p, T)$$

$$= -P_{\mu\nu}^T(p)\frac{i}{p^2 + i\epsilon} + i\frac{u_\mu u_\nu}{\vec{p}^2} - P_{\mu\nu}^T(p)2\pi\delta(p^2)n_B(\beta|p_0|), \tag{10.2}$$

where $u_\mu = (1, 0, 0, 0)$, $\beta \equiv \frac{1}{T}$,

$$P_{00}^T(p) \equiv P_{0i}^T(p) = P_{i0}^T(p) = 0,$$

$$P_{ij}^T(p) \equiv \delta_{ij} - \frac{p_i p_j}{\vec{p}^2}, \tag{10.3}$$

and $n_B(x) \equiv \frac{1}{e^x-1}$. By virtue of Eq. (10.1) one obtains

$$\langle\Theta_{00}(x)\Theta_{00}(y)\rangle = 2\langle\partial_{x^0}A^\lambda(x)\partial_{y^0}A^\tau(y)\rangle\langle\partial_{x^0}A_\lambda(x)\partial_{y^0}A_\tau(y)\rangle$$

$$- g_{00}\langle\partial_{x^0}A^\lambda(x)\partial_{y^\sigma}A^\tau(y)\rangle\langle\partial_{x^0}A_\lambda(x)\partial_{y_\sigma}A_\tau(y)\rangle$$

$$+ g_{00}\langle\partial_{x^0}A^\lambda(x)\partial_{y_\sigma}A^\tau(y)\rangle\langle\partial_{x^0}A_\lambda(x)\partial_{y^\tau}A_\sigma(y)\rangle$$

$$- g_{00}\langle\partial_{x_\kappa}A^\lambda(x)\partial_{y^0}A^\tau(y)\rangle\langle\partial_{x^\kappa}A_\lambda(x)\partial_{y^0}A_\tau(y)\rangle$$

$$+ g_{00}\langle\partial_{x_\kappa}A^\lambda(x)\partial_{y^0}A^\tau(y)\rangle\langle\partial_{x^\lambda}A_\kappa(x)\partial_{y^0}A_\tau(y)\rangle$$

$$+ \frac{g_{00}^2}{2}\langle\partial_{x_\kappa}A^\lambda(x)\partial_{y_\sigma}A^\tau(y)\rangle\langle\partial_{x^\kappa}A_\lambda(x)\partial_{y^\sigma}A_\tau(y)\rangle$$

$$- \frac{g_{00}^2}{2}\langle\partial_{x_\kappa}A^\lambda(x)\partial_{y_\sigma}A^\tau(y)\rangle\langle\partial_{x^\kappa}A_\lambda(x)\partial_{y^\tau}A_\sigma(y)\rangle$$

$$-\frac{g_{00}^2}{2}\langle\partial_{x_\kappa}A^\lambda(x)\partial_{y_\sigma}A^\tau(y)\rangle\langle\partial_{x^\lambda}A_\kappa(x)\partial_{y^\sigma}A_\tau(y)\rangle$$

$$+\frac{g_{00}^2}{2}\langle\partial_{x_\kappa}A^\lambda(x)\partial_{y_\tau}A^\sigma(y)\rangle\langle\partial_{x^\lambda}A_\kappa(x)\partial_{y^\sigma}A_\tau(y)\rangle$$

$$+2\langle\partial_{x^0}A^0(x)\partial_{y^0}A^0(y)\rangle\langle\partial_{x^0}A_0(x)\partial_{y^0}A_0(y)\rangle$$

$$-2\langle\partial_{x^0}A^0(x)\partial_{y^\tau}A^0(y)\rangle\langle\partial_{x^0}A_0(x)\partial_{y_\tau}A_0(y)\rangle$$

$$+2\langle\partial_{x^\tau}A^0(x)\partial_{y^\sigma}A^0(y)\rangle\langle\partial_{x_\tau}A_0(x)\partial_{y_\sigma}A_0(y)\rangle$$

$$-2g_{00}\langle\partial_{x^0}A^0(x)\partial_{y^0}A^0(y)\rangle\langle\partial_{x^0}A_0(x)\partial_{y^0}A_0(y)\rangle$$

$$+4g_{00}\langle\partial_{x^0}A^0(x)\partial_{y^\tau}A^0(y)\rangle\langle\partial_{x^0}A_0(x)\partial_{y_\tau}A_0(y)\rangle$$

$$-2g_{00}\langle\partial_{x^\tau}A^0(x)\partial_{y^\sigma}A^0(y)\rangle\langle\partial_{x_\tau}A_0(x)\partial_{y_\sigma}A_0(y)\rangle$$

$$+\frac{g_{00}^2}{2}\langle\partial_{x^0}A^0(x)\partial_{y^0}A^0(y)\rangle\langle\partial_{x^0}A_0(x)\partial_{y^0}A_0(y)\rangle$$

$$-g_{00}^2\langle\partial_{x^0}A^0(x)\partial_{y^\tau}A^0(y)\rangle\langle\partial_{x^0}A_0(x)\partial_{y_\tau}A_0(y)\rangle$$

$$+\frac{g_{00}^2}{2}\langle\partial_{x^\tau}A^0(x)\partial_{y^\sigma}A^0(y)\rangle\langle\partial_{x_\tau}A_0(x)\partial_{y_\sigma}A_0(y)\rangle.$$

$$(10.4)$$

10.1.1.2   *Decomposition into thermal and vacuum parts*

In evaluating the expression in Eq. (10.4) the derivative operators are taken out of the expectation, and the momentum-space expression of (10.2) is used for the propagator of the U(1) gauge field A^μ. At this stage ambiguities related to the various possibilities of time-ordering in the two-point function of the gauge field A_μ cancel out. The last nine lines in Eq. (10.4) arise from the term $\propto u_\mu u_\nu$ in the propagator of Eq. (10.2). The expression Eq. (10.4) (see Fig. 10.1) separates into purely thermal, purely vacuum, and mixed contributions. Here we only consider the purely thermal and the purely vacuum cases.

Using

$$P^{T\lambda\tau}(p)P^T_{\lambda\tau}(k) = 1 + \frac{(\vec{p}\cdot\vec{k})^2}{\vec{p}^2\vec{k}^2},$$

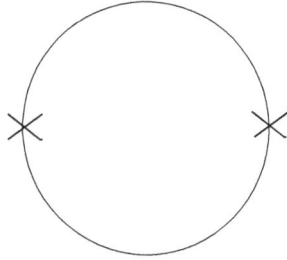

Figure 10.1. Feynman diagram for the correlator $\langle \Theta_{00}(x)\Theta_{00}(y)\rangle$ in a pure U(1) gauge theory. Crosses denote the insertion of the local, composite operator Θ_{00} at x and y.

$$P^T_{\lambda\sigma}(p)k^\lambda k^\sigma = \vec{k}^2 - \frac{(\vec{p}\cdot\vec{k})^2}{\vec{p}^2},$$

$$P^{T\kappa\tau}(p)P^T_{\kappa\sigma}(k)p^\sigma k_\tau = -(\vec{p}\cdot\vec{k})\left(\frac{(\vec{p}\cdot\vec{k})^2}{\vec{p}^2\vec{k}^2}-1\right), \tag{10.5}$$

the purely thermal contribution $\langle \Theta_{00}(x)\Theta_{00}(y)\rangle^{\mathrm{th}}$ (performing the trivial integration over the zero-components of the momenta) reads

$$\langle \Theta_{00}(\vec{x})\Theta_{00}(\vec{y})\rangle^{\mathrm{th}}$$

$$= \left(\int \frac{d^3p}{(2\pi)^3}|\vec{p}|n_B(\beta|\vec{p}|)e^{i\vec{p}\vec{z}}\right)^2$$

$$+ \int \frac{d^3p}{(2\pi)^3}\int \frac{d^3k}{(2\pi)^3}\left(\frac{\vec{p}\vec{k}}{|\vec{p}||\vec{k}|}\right)^2|\vec{p}||\vec{k}|n_B(\beta|\vec{p}|)n_B(\beta|\vec{k}|)e^{i\vec{p}\vec{z}}e^{i\vec{k}\vec{z}},$$

$$\tag{10.6}$$

where $\vec{z} \equiv \vec{x} - \vec{y}$. In deriving Eq. (10.6) it was assumed that $x^0 = y^0$ which amounts to neglecting oscillatory terms in $x^0 - y^0$. This prescription should yield a measure for the time-averaged energy transport between points $\vec{x}$ and $\vec{y}$ and is technically much easier to handle than the case $x^0 \neq y^0$.

Inserting the vacuum part of the propagator in Eq. (10.2) into Eq. (10.4), performing similar contractions as in Eq. (10.5), and rotating to Euclidean signature $(p_0, k_0 \rightarrow ip_0, ik_0, x^0, y^0 \rightarrow -ix^0, -iy^0,$ $g_{\mu\nu} \rightarrow -\delta_{\mu\nu})$, we obtain

$$\langle \Theta_{00}(x)\Theta_{00}(y) \rangle^{\text{vac}}$$

$$= \frac{9}{2}\left(\int \frac{d^4p}{(2\pi)^4} p_0^2 \frac{e^{ip\zeta}}{p^2}\right)^2 + \frac{1}{2}\left(\int \frac{d^4p}{(2\pi)^4} |\vec{p}|^2 \frac{e^{ip\zeta}}{p^2}\right)^2$$

$$+ 6 \int \frac{d^4p}{(2\pi)^4} \int \frac{d^4k}{(2\pi)^4} \left(\frac{\vec{p}\vec{k}}{|\vec{p}||\vec{k}|}\right) p_0 k_0 |\vec{p}||\vec{k}| \frac{e^{ip\zeta}}{p^2} \frac{e^{ik\zeta}}{k^2}$$

$$+ \frac{1}{2} \int \frac{d^4p}{(2\pi)^4} \int \frac{d^4k}{(2\pi)^4} \left(\frac{\vec{p}\vec{k}}{|\vec{p}||\vec{k}|}\right)^2 \left(9p_0^2 k_0^2 + |\vec{p}|^2|\vec{k}|^2\right) \frac{e^{ip\zeta}}{p^2} \frac{e^{ik\zeta}}{k^2}$$

$$+ 2\left(\int \frac{d^4p}{(2\pi)^4} p_0^2 \frac{e^{ip\zeta}}{\vec{p}^2}\right)^2$$

$$+ 2 \int \frac{d^4p}{(2\pi)^4} \int \frac{d^4k}{(2\pi)^4} \left(\frac{\vec{p}\vec{k}}{|\vec{p}||\vec{k}|}\right) p_0 k_0 |\vec{p}||\vec{k}| \frac{e^{ip\zeta}}{\vec{p}^2} \frac{e^{ik\zeta}}{\vec{k}^2}$$

$$+ \frac{9}{2} \int \frac{d^4p}{(2\pi)^4} \int \frac{d^4k}{(2\pi)^4} \left(\frac{\vec{p}\vec{k}}{|\vec{p}||\vec{k}|}\right)^2 |\vec{p}|^2|\vec{k}|^2 \frac{e^{ip\zeta}}{\vec{p}^2} \frac{e^{ik\zeta}}{\vec{k}^2}, \tag{10.7}$$

where $\zeta = x - y$ and $p\zeta \equiv p_\mu \zeta_\mu$, $k\zeta \equiv k_\mu \zeta_\mu$, $p^2 \equiv p_\mu p_\mu$, and $k^2 \equiv k_\mu k_\mu$. The last three lines in Eq. (10.7) arise from the term $\propto u_\mu u_\nu$ in the propagator of Eq. (10.2).

10.1.1.3   *Results*

To evaluate the integrals in Eqs. (10.6) and (10.7) we introduce rescaled momenta $\tilde{p}_\mu \equiv \beta p_\mu$ and $\tilde{k}_\mu \equiv \beta k_\mu$. The integrals in Eqs. (10.6) and (10.7) are then expressed in terms of three-dimensional and four-dimensional spherical coordinates, respectively, and the integration over azimuthal angles is performed in a straightforward way in both cases. As a result, the integrals over the remaining variables factorize for each term in Eqs. (10.6) and (10.7).

In case of $\langle\Theta_{00}(\mathbf{x})\Theta_{00}(\mathbf{y})\rangle^{\text{th}}$ we choose $\tilde{z}$ to point into the spatial three-direction. We make use of the spherical expansion of the exponential,

$$e^{i|\vec{\tilde{q}}||\tilde{z}|\cos\theta} = \sum_{l=0}^{\infty} i^l(2l+1)j_l(|\vec{\tilde{q}}||\tilde{z}|)P_l(\cos\theta), \qquad (10.8)$$

where $\theta \equiv \angle(\vec{\tilde{q}},\tilde{z})$, $\tilde{z} \equiv \frac{z}{\beta}$, $\vec{\tilde{q}}$ can be $\vec{\tilde{p}}$ or $\vec{\tilde{k}}$, and j_l denotes a spherical Bessel function. In Eq. (10.6) we express polynomial factors in $\cos\theta$ in terms of linear combinations of Legendre polynomials $P_l(\cos\theta)$. Exploiting the following orthonormality relation

$$\frac{2m+1}{2}\int_{-1}^{+1} dx P_n(x)P_m(x) = \delta_{mn} \quad (m,n \text{ integer}) \qquad (10.9)$$

and integrating in Eq. (10.6) over the polar angle θ, we arrive at

$$\langle\Theta_{00}(\vec{\tilde{x}})\Theta_{00}(\vec{\tilde{y}})\rangle^{\text{th}} = \frac{1}{(2\pi)^6\beta^8}\left[\frac{64\pi^2}{3}\left(\int_0^{\infty} d|\vec{\tilde{p}}|\frac{|\vec{\tilde{p}}|^3}{e^{|\vec{\tilde{p}}|}-1}j_0(|\tilde{z}||\vec{\tilde{p}}|)\right)^2\right.$$

$$\left. +\frac{32\pi^2}{3}\left(\int_0^{\infty} d|\vec{\tilde{p}}|\frac{|\vec{\tilde{p}}|^3}{e^{|\vec{\tilde{p}}|}-1}j_2(|\tilde{z}||\vec{\tilde{p}}|)\right)^2\right]. \qquad (10.10)$$

Performing the integration over $|\vec{\tilde{p}}|$, we have [Gradshteyn and Ryzhik (1965)]

$$\langle\Theta_{00}(\vec{\tilde{x}})\Theta_{00}(\vec{\tilde{y}})\rangle^{\text{th}}$$

$$= \frac{1}{(2\pi)^6\beta^8}\left[\frac{64\pi^2}{3}\left(\frac{1}{|\tilde{z}|^4} - \frac{\pi^3\coth(\pi|\tilde{z}|)\operatorname{cosech}^2(\pi|\tilde{z}|)}{|\tilde{z}|}\right)^2 + \frac{32\pi^2}{3}\times\right.$$

$$\left.\left(\frac{-8+\pi|\tilde{z}|\left(3\coth(\pi|\tilde{z}|)+\pi|\tilde{z}|(3+2\pi|\tilde{z}|\coth(\pi|\tilde{z}|))\operatorname{cosech}^2(\pi|\tilde{z}|)\right)}{2|\tilde{z}|^4}\right)^2\right].$$

$$(10.11)$$

In case of $\langle\Theta_{00}(x)\Theta_{00}(y)\rangle^{\text{vac}}$ we again restrict ourselves to $x_0 = y_0$ to be able to compare with $\langle\Theta_{00}(\vec{x})\Theta_{00}(\vec{y})\rangle^{\text{th}}$. Again, $\vec{z}$ is taken to point into the three-direction. In analogy to the case of $\langle\Theta_{00}(\vec{x})\Theta_{00}(\vec{y})\rangle^{\text{th}}$ the combination of four-dimensional integrals over p and k in Eq. (10.7) factorizes upon azimuthal integration. We make use of the four-dimensionally adapted spherical expansion of the exponential as

$$e^{i|\tilde{q}||\tilde{z}|\sin\psi\cos\theta} = \sum_{l=0}^{\infty} i^l(2l+1)j_l(|\tilde{q}||\tilde{z}|\sin\psi)P_l(\cos\theta), \qquad (10.12)$$

where ψ is the second polar angle in four dimensions, and $\tilde{q}$ can be $\tilde{p}$ or $\tilde{k}$. Expressing polynomial factors in $\cos\theta$ in terms of linear combinations of Legendre polynomials $P_l(\cos\theta)$, exploiting their orthonormality relation (10.9), and upon integration over the first polar angle θ, we arrive at

$$\langle\Theta_{00}(\vec{x})\Theta_{00}(\vec{y})\rangle^{\text{vac}}$$

$$= \frac{1}{(2\pi)^8\beta^8}\left[96\pi^2\left(\int_0^\infty d|\tilde{p}||\tilde{p}|^3\int_0^\pi d\psi\,\sin^2\psi\cos^2\psi\,j_0(|\tilde{p}||\tilde{z}|\sin\psi)\right)^2\right.$$

$$+48\pi^2\left(\int_0^\infty d|\tilde{p}||\tilde{p}|^3\int_0^\pi d\psi\,\sin^2\psi\cos^2\psi\,j_2(|\tilde{p}||\tilde{z}|\sin\psi)\right)^2$$

$$+96\pi^2\left(\int_0^\infty d|\tilde{p}||\tilde{p}|^3\int_0^\pi d\psi\,\sin^3\psi\cos\psi\,j_1(|\tilde{p}||\tilde{z}|\sin\psi)\right)^2$$

$$+\frac{32\pi^2}{3}\left(\int_0^\infty d|\tilde{p}||\tilde{p}|^3\int_0^\pi d\psi\,\sin^4\psi\,j_0(|\tilde{p}||\tilde{z}|\sin\psi)\right)^2$$

$$+\frac{16\pi^2}{3}\left(\int_0^\infty d|\tilde{p}||\tilde{p}|^3\int_0^\pi d\psi\,\sin^4\psi\,j_2(|\tilde{p}||\tilde{z}|\sin\psi)\right)^2$$

$$+ 32\pi^2 \left(\int_0^\infty d|\tilde{p}| |\tilde{p}|^3 \int_0^\pi d\psi \, \cos^2 \psi \, j_0(|\tilde{p}||\tilde{z}| \sin \psi) \right)^2$$

$$+ 32\pi^2 \left(\int_0^\infty d|\tilde{p}| |\tilde{p}|^3 \int_0^\pi d\psi \, \sin \psi \cos \psi \, j_1(|\tilde{p}||\tilde{z}| \sin \psi) \right)^2$$

$$+ 24\pi^2 \left(\int_0^\infty d|\tilde{p}| |\tilde{p}|^3 \int_0^\pi d\psi \, \sin^2 \psi \, j_0(|\tilde{p}||\tilde{z}| \sin \psi) \right)^2$$

$$\left. + 48\pi^2 \left(\int_0^\infty d|\tilde{p}| |\tilde{p}|^3 \int_0^\pi d\psi \, \sin^2 \psi \, j_2(|\tilde{p}||\tilde{z}| \sin \psi) \right)^2 \right],$$

$$(10.13)$$

where now $|\tilde{p}| \equiv \sqrt{\tilde{p}_0^2 + \tilde{p}_1^2 + \tilde{p}_2^2 + \tilde{p}_3^2}$, and j_0, j_1, j_2 are spherical Bessel functions. In Eq. (10.13) the last four lines arise from the term $\propto u_\mu u_\nu$ in the propagator, see Eq. (10.2). They vanish. Physically, this is explained by the fact that no energy transfer between points $\vec{x}$ and $\vec{y}$ is mediated by the Coulomb part of the photon propagator. Notice that upon performing the integration over ψ the terms in the third and seventh lines vanish for symmetry reasons. The final result is

$$\langle \Theta_{00}(\vec{x}) \Theta_{00}(\vec{y}) \rangle^{\text{vac}}$$

$$= \frac{1}{(2\pi)^8 \beta^8} \cdot 96\pi^2 \cdot \left(\frac{2\pi}{|\tilde{z}|^4} \right)^2 + \frac{1}{(2\pi)^8 \beta^8} \cdot 48\pi^2 \cdot \left(\frac{-8\pi}{|\tilde{z}|^4} \right)^2$$

$$+ \frac{1}{(2\pi)^8 \beta^8} \cdot \frac{32\pi^2}{3} \cdot \left(\frac{-2\pi}{|\tilde{z}|^4} \right)^2 + \frac{1}{(2\pi)^8 \beta^8} \cdot \frac{16\pi^2}{3} \cdot \left(\frac{8\pi}{|\tilde{z}|^4} \right)^2$$

$$= \frac{15}{\pi^4 \beta^8 |\tilde{z}|^8} \sim \frac{0.15399}{|\tilde{z}|^8}. \qquad (10.14)$$

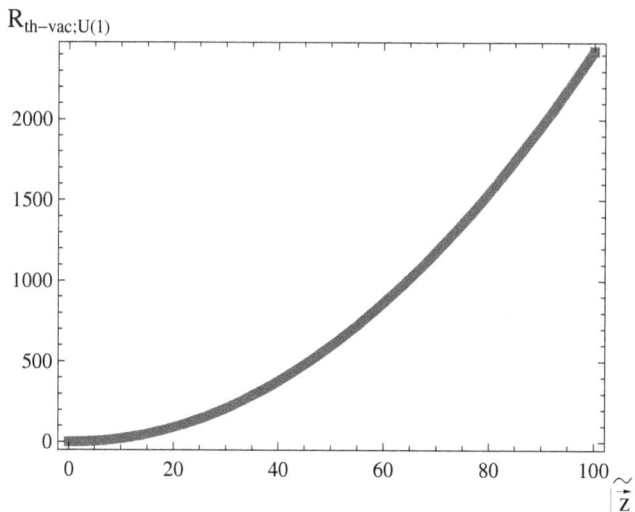

Figure 10.2. The ratio $R_{\text{th}-\text{vac};U(1)} \equiv \frac{\langle \Theta_{00}(\vec{x})\Theta_{00}(\vec{y})\rangle^{\text{th}}}{\langle \Theta_{00}(\vec{x})\Theta_{00}(\vec{y})\rangle^{\text{vac}}}$ as a function of $|\vec{\tilde{z}}|$ for a thermalized pure U(1) gauge theory.

The result of Eq. (10.14) can be checked by a position-space calculation in Feynman gauge (see Sec. 2.3.1), where the vacuum part of the propagator is given as

$$D_{\mu\nu}^{\text{vac}}(x) = \frac{1}{4\pi^2 x^2} g_{\mu\nu}. \tag{10.15}$$

In Fig. 10.2 the ratio $R_{\text{th}-\text{vac};U(1)} \equiv \frac{\langle \Theta_{00}(\vec{x})\Theta_{00}(\vec{y})\rangle^{\text{th}}}{\langle \Theta_{00}(\vec{x})\Theta_{00}(\vec{y})\rangle^{\text{vac}}}$ is shown as a function of $|\vec{\tilde{z}}|$. At $T = 5.5, 8.2, 10.9\,\text{K}$ a distance $|\vec{z}|$ of 1 cm corresponds to $|\vec{\tilde{z}}| \sim 24, 36, 48$, respectively. Thus the thermal part of $\langle \Theta_{00}(\vec{x})\Theta_{00}(\vec{y})\rangle$ dominates the vacuum part by at least a factor of hundred in these examples.

10.1.2   *The case of deconfining thermal SU(2) gauge theory*

Here we consider the massless mode surviving the dynamical gauge symmetry breaking SU(2)→U(1) upon spatial coarse-graining in the deconfining phase of SU(2) Yang–Mills thermodynamics (see Chapter 5). As we have seen, the effective theory for this phase describes spatially coarse-grained calorons of trivial holonomy and of unit topological charge modulus in terms of the inert, adjoint Higgs field ϕ (see Sec. 5.1). The field strength $F_{\mu\nu} = \partial_\mu A_\nu - \partial_\nu A_\mu$ of the Abelian theory then can be replaced by the 't Hooft tensor $\mathcal{F}_{\mu\nu}$ to define the SU(2) gauge-invariant energy–momentum tensor according to Eq. (10.1) (compare with Sec. 4.2.2 where the role of ϕ is played by A_4 in the Euclidean SU(2) Yang–Mills theory). Owing to the presence of the field ϕ, one has

$$\mathcal{F}_{\mu\nu} \equiv \frac{1}{|\phi|}\phi_a G^a_{\mu\nu} - \frac{1}{e|\phi|^3}\epsilon^{abc}\phi_a(D_\mu\phi)_b(D_\nu\phi)_c, \qquad (10.16)$$

where $G^a_{\mu\nu}$ is the SU(2) field strength of topologically trivial, coarse-grained fluctuations, D_μ denotes the adjoint covariant derivative, and e is the effective gauge coupling. Obviously, the quantity defined by the right-hand side of Eq. (10.16) is SU(2) gauge-invariant. Notice that in unitary gauge $\phi_a = \delta_{a3}|\phi|$ the 't Hooft tensor $\mathcal{F}_{\mu\nu}$ reduces to the Abelian field strength $F_{\mu\nu}$ defined on the massless gauge field $A_\mu \equiv a^3_\mu$.

Our goal is to obtain a measure for the average energy transfer between points $\vec{x}$ and $\vec{y}$, as mediated by a^3_μ, subjecting it to interactions with the two massive excitations. This energy transfer is characterized by the two-point correlator $\langle\Theta_{00}(\mathbf{x})\Theta_{00}(\mathbf{y})\rangle^{\text{th}}$ where Θ_{00} is now calculated as in Eq. (10.1), replacing A_μ by a^3_μ.

10.1.2.1   *Radiative modification of dispersion law for massless mode*

In Secs. 5.4.7.3 and 5.4.7.5, the transverse part of the one-loop polarization tensor $\Pi_{\mu\nu}$ for the massless, transverse mode a^3_μ was computed assuming $(p^2 = 0)$ and selfconsistently $(p^2 = G)$, respectively. As a result, a modification of the dispersion law

$$p_0^2 = \vec{p}^2 \rightarrow p_0^2 = \vec{p}^2 + G(|\vec{p}|, T, \Lambda) \qquad (10.17)$$

is obtained where Λ denotes the Yang–Mills scale related to the critical temperature T_c of the deconfining–preconfining phase-transition as $T_c = \frac{\lambda_c}{2\pi}\Lambda = \frac{13.87}{2\pi}\Lambda$. Here we account for radiative corrections in

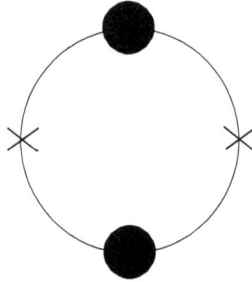

Figure 10.3. Feynman diagram for the correlator $\langle\Theta_{00}(x)\Theta_{00}(y)\rangle$ on the tree-level massless mode in a thermalized, deconfining SU(2) gauge theory including resummed radiative corrections to lowest order. Blobs signal a selfconsistent resummation of the transverse part of the one-loop polarization, crosses denote the insertion of the composite operator Θ_{00}.

the correlator $\langle\Theta_{00}(\vec{x})\Theta_{00}(\vec{y})\rangle^{\text{th}}$ in terms of a resummation of the one-loop screening function G of the massless, transverse mode only (see Fig. 10.3).

10.1.2.2   *Thermal part of two-point correlation*

After integration over p_0 and k_0 the thermal part $\langle\Theta_{00}(\mathbf{x})\Theta_{00}(\mathbf{y})\rangle^{\text{th}}$ in analogy to Eq. (10.6) calculates as

$$\langle\Theta_{00}(\vec{x})\Theta_{00}(\vec{y})\rangle^{\text{th}}$$

$$= \frac{1}{2}\left(\int\frac{d^3p}{(2\pi)^3}\sqrt{\vec{p}^2+G}\,n_B(\beta\sqrt{\vec{p}^2+G})e^{i\vec{p}(\vec{x}-\vec{y})}\right)^2$$

$$+ \frac{1}{2}\left(\int\frac{d^3p}{(2\pi)^3}\frac{\vec{p}^2}{\sqrt{\vec{p}^2+G}}n_B(\beta\sqrt{\vec{p}^2+G})e^{i\vec{p}(\vec{x}-\vec{y})}\right)^2$$

$$+ \frac{1}{2}\int\frac{d^3p}{(2\pi)^3}\int\frac{d^3k}{(2\pi)^3}\left(\frac{\vec{p}\vec{k}}{|\vec{p}||\vec{k}|}\right)^2\sqrt{\vec{p}^2+G}\sqrt{\vec{k}^2+G}$$

$$\times n_B(\beta\sqrt{\vec{p}^2+G})n_B(\beta\sqrt{\vec{k}^2+G})e^{i\vec{p}(\vec{x}-\vec{y})}e^{i\vec{k}(\vec{x}-\vec{y})}$$

$$+ \frac{1}{2} \int \frac{d^3p}{(2\pi)^3} \int \frac{d^3k}{(2\pi)^3} \left(\frac{\vec{p}\vec{k}}{|\vec{p}||\vec{k}|} \right)^2 \frac{\vec{p}^2}{\sqrt{\vec{p}^2 + G}} \frac{\vec{k}^2}{\sqrt{\vec{k}^2 + G}}$$

$$\times n_B(\beta\sqrt{\vec{p}^2 + G}) n_B(\beta\sqrt{\vec{k}^2 + G}) e^{i\vec{p}(\vec{x}-\vec{y})} e^{i\vec{k}(\vec{x}-\vec{y})}.$$

$$(10.18)$$

As in the U(1) case we introduce spherical coordinates and evaluate the integrals over the angles analytically. Upon performing the azimuthal integrations for each summand in Eq. (10.18) the respective expression reduces to a square of an integral over the modulus $|\vec{p}|$ of spatial momentum. Employing the modified dispersion law of Eq. (10.17), the integral over the modulus of spatial momentum is replaced in favor of an integral over frequency $\omega = p_0$. Those values of ω, which yield an imaginary modulus of the spatial momentum due to strong screening, are excluded from the domain of integration. This regime of strong screening is in the range $0 < \omega < \omega^*$, where ω^* is defined as

$$(\omega^*)^2 = G(|\vec{p}| = 0, T, \Lambda) \qquad (10.19)$$

(see Sec. 5.4.7.5). Finally, we introduce dimensionless frequency $\tilde{\omega} \equiv \beta\omega$ and the dimensionless screening function $\tilde{G} \equiv \beta^2 G$ to arrive at

$$\langle \Theta_{00}(\vec{x})\Theta_{00}(\vec{y}) \rangle^{\text{th}}$$

$$= \frac{1}{(2\pi)^6 \beta^8} \left[\frac{32\pi^2}{3} \left(\int\limits_{\tilde{\omega}^* \leq \tilde{\omega}} d\tilde{\omega} \left(\tilde{\omega} - \frac{1}{2}\frac{d\tilde{G}}{d\tilde{\omega}} \right) \tilde{\omega}\sqrt{\tilde{\omega}^2 - \tilde{G}} \frac{j_0(\sqrt{\tilde{\omega}^2 - \tilde{G}}|\vec{z}|)}{e^{\tilde{\omega}} - 1} \right)^2 \right.$$

$$+ \frac{32\pi^2}{3} \left(\int\limits_{\tilde{\omega}^* \leq \tilde{\omega}} d\tilde{\omega} \left(\tilde{\omega} - \frac{1}{2}\frac{d\tilde{G}}{d\tilde{\omega}} \right) \frac{(\sqrt{\tilde{\omega}^2 - \tilde{G}})^3}{\tilde{\omega}} \frac{j_0(\sqrt{\tilde{\omega}^2 - \tilde{G}}|\vec{z}|)}{e^{\tilde{\omega}} - 1} \right)^2$$

$$+ \frac{16\pi^2}{3} \left(\int\limits_{\tilde{\omega}^* \leq \tilde{\omega}} d\tilde{\omega} \left(\tilde{\omega} - \frac{1}{2}\frac{d\tilde{G}}{d\tilde{\omega}} \right) \tilde{\omega}\sqrt{\tilde{\omega}^2 - \tilde{G}} \frac{j_2(\sqrt{\tilde{\omega}^2 - \tilde{G}}|\vec{z}|)}{e^{\tilde{\omega}} - 1} \right)^2$$

$$+ \frac{16\pi^2}{3} \left(\int\limits_{\tilde{\omega}^* \le \tilde{\omega}} d\tilde{\omega} \left(\tilde{\omega} - \frac{1}{2} \frac{d\tilde{G}}{d\tilde{\omega}} \right) \frac{(\sqrt{\tilde{\omega}^2 - \tilde{G}})^3}{\tilde{\omega}} \frac{j_2(\sqrt{\tilde{\omega}^2 - \tilde{G}}|\tilde{z}|)}{e^{\tilde{\omega}} - 1} \right)^2 \right].$$

$$(10.20)$$

10.1.2.3   *Estimate for vacuum part of two-point correlation*

Here we would like to obtain an order-of-magnitude estimate for the vacuum part of the two-point correlation of Θ_{00} for the massless mode. For pragmatic reasons, we ignore the modification of the dispersion law in Eq. (10.17). In evaluating the vacuum part, the difference compared to the U(1) case is a restriction of the Euclidean four-momentum p as $p^2 \le |\phi|^2$ due to the existence of a scale of maximal resolution $|\phi|$ in the effective theory of deconfining SU(2) Yang–Mills thermodynamics. Again, we will see that $\langle \Theta_{00}(\vec{x})\Theta_{00}(\vec{y})\rangle^{\text{vac}}$ is a negligible correction to $\langle \Theta_{00}(\vec{x})\Theta_{00}(\vec{y})\rangle^{\text{th}}$ for physically interesting distances.

In case of $\langle \Theta_{00}(x)\Theta_{00}(y)\rangle^{\text{vac}}$ we again set $x_0 = y_0$ to be able to compare with $\langle \Theta_{00}(\mathbf{x})\Theta_{00}(\mathbf{y})\rangle^{\text{th}}$. Introducing $\tilde{\phi} \equiv \beta\phi$, proceeding in a way analogous to the derivation of Eq. (10.13), and performing the ψ- and $|\tilde{p}|$-integrations, we arrive at

$$\langle \Theta_{00}(\vec{x})\Theta_{00}(\vec{y})\rangle^{\text{vac}}$$

$$\sim \frac{1}{(2\pi)^8 \beta^8} \left[96\pi^2 \cdot \frac{\pi^2}{|\tilde{z}|^8} \left(2J_0(|\tilde{z}||\tilde{\phi}|) + |\tilde{z}||\tilde{\phi}|J_1(|\tilde{z}||\tilde{\phi}|) - 2 \right)^2 \right.$$

$$+ 48\pi^2 \cdot \frac{\pi^2}{4|\tilde{z}|^8} \left(2(8J_0(|\tilde{z}||\tilde{\phi}|) + |\tilde{z}||\tilde{\phi}|J_1(|\tilde{z}||\tilde{\phi}|) - 8) + 3|\tilde{z}|^2|\tilde{\phi}|^2 \right.$$

$$\left. \times {}_1F_2 \left(\frac{1}{2}; \frac{3}{2}, 2; -\frac{|\tilde{z}|^2|\tilde{\phi}|^2}{4} \right) \right)^2$$

$$+ \frac{32\pi^2}{3} \cdot \frac{\pi^2}{|\tilde{z}|^8} \left((|\tilde{z}|^2|\tilde{\phi}|^2 - 2)J_0(|\tilde{z}||\tilde{\phi}|) - 3|\tilde{z}||\tilde{\phi}|J_1(|\tilde{z}||\tilde{\phi}|) + 2 \right)^2$$

$$+ \frac{16\pi^2}{3} \cdot \frac{\pi^2}{|\tilde{z}|^8} \left((|\tilde{z}|^2|\tilde{\phi}|^2 - 8)J_0(|\tilde{z}||\tilde{\phi}|) - 6|\tilde{z}||\tilde{\phi}|J_1(|\tilde{z}||\tilde{\phi}|) + 8 \right)^2$$

$$+ 32\pi^2 \cdot \frac{\pi^2|\tilde{\phi}|^8}{64} \left({}_1F_2\left(\frac{1}{2}; \frac{3}{2}, 3; -\frac{|\tilde{z}|^2|\tilde{\phi}|^2}{4}\right) \right)^2$$

$$+ 24\pi^2 \cdot \frac{\pi^2|\tilde{\phi}|^4}{|\tilde{z}|^4} \left(J_2(|\tilde{z}||\tilde{\phi}|) \right)^2$$

$$+ 48\pi^2 \cdot \frac{\pi^2|\tilde{\phi}|^2}{4|\tilde{z}|^6} \left(6J_1(|\tilde{z}||\tilde{\phi}|) + 2|\tilde{z}||\tilde{\phi}|J_2(|\tilde{z}||\tilde{\phi}|) - 3|\tilde{z}||\tilde{\phi}| \right.$$

$$\left. \times {}_1F_2\left(\frac{1}{2}; \frac{3}{2}, 2; -\frac{|\tilde{z}|^2|\tilde{\phi}|^2}{4}\right) \right)^2 \Bigg], \tag{10.21}$$

where ${}_1F_2$ is a hypergeometric function, and J_0, J_1, J_2 are Bessel functions of the first kind (conventions as in [Gradshteyn and Ryzhik (1965)]).

10.1.2.4 *Numerical results*

Our goal is to compare the result of Eq. (10.21) for $\langle \Theta_{00}(\vec{x})\Theta_{00}(\vec{y})\rangle^{vac}$ with $\langle \Theta_{00}(\vec{x})\Theta_{00}(\vec{y})\rangle^{th}$ of Eq. (10.20). The integrals expressing the latter are evaluated numerically.

To make contact with SU(2)$_{CMB}$, whose Yang–Mills scale is $\Lambda = 1.065 \times 10^{-4}$ eV (compare with Sec. 9.4.3), we relate at a given temperature T the dimensionless distance $|\tilde{z}|$ to the physical distance in centimeters. We define

$$R_{th-vac;SU(2)}(|\tilde{z}|) \equiv \frac{\langle \Theta_{00}(\vec{x})\Theta_{00}(\vec{y})\rangle^{th}}{\langle \Theta_{00}(\vec{x})\Theta_{00}(\vec{y})\rangle^{vac}}. \tag{10.22}$$

In Fig. 10.4 the quantity $R_{th-vac;SU(2)}(|\tilde{z}|)$ is plotted as a function of distance for various temperatures and for SU(2)$_{CMB}$. Notice the strong dominance of the thermal part for the allowed distances to the right of the intersection with the dashed line. This line represents the fact that the thermal ground state in deconfining SU(2)

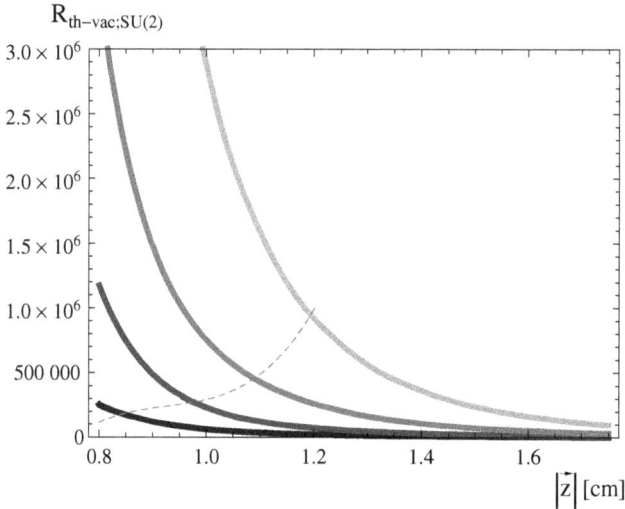

Figure 10.4. The function $R_{th-vac;SU(2)}(|\vec{z}|)$ for SU(2)$_{CMB}$ ($T_c = 2.725$ K) plotted for various temperatures: black curve ($T = 1.5T_c$), dark gray curve ($T = 2.0T_c$), gray curve ($T = 2.5T_c$), and light gray curve ($T = 3.0T_c$). The dashed line separates distances smaller from those larger than $|\phi|^{-1}$.

Yang–Mills thermodynamics constrains quantum fluctuations of the massless mode to be softer than the scale $|\phi|$. Owing to its trivial ground state, no such constraint exists for the thermodynamics of a U(1) gauge theory. Thus the quantum fluctuations in a U(1) theory dominate thermal fluctuations for sufficiently small distances and at a given temperature (see Fig. 10.2).

By virtue of Figs. 10.2 and 10.4 we may neglect the vacuum contributions to $\langle\Theta_{00}(\vec{x})\Theta_{00}(\vec{y})\rangle$ for U(1) and SU(2)$_{CMB}$ in what follows.

An interesting question is how the correlation of the photon energy density in SU(2)$_{CMB}$ compares to that of the pure U(1) case. In Fig. 10.5 the ratio $R_{th;SU(2)-th;U(1)}(|\vec{z}|)$, defined as

$$R_{th;SU(2)-th;U(1)}(|\vec{z}|) \equiv \frac{\langle\Theta_{00}(\vec{x})\Theta_{00}(\vec{y})\rangle^{th,SU(2)}}{\langle\Theta_{00}(\vec{x})\Theta_{00}(\vec{y})\rangle^{th,U(1)}}, \tag{10.23}$$

is plotted for various temperatures as a function of distance $|\vec{z}|$. Notice the suppression of the correlation for SU(2)$_{CMB}$ as compared

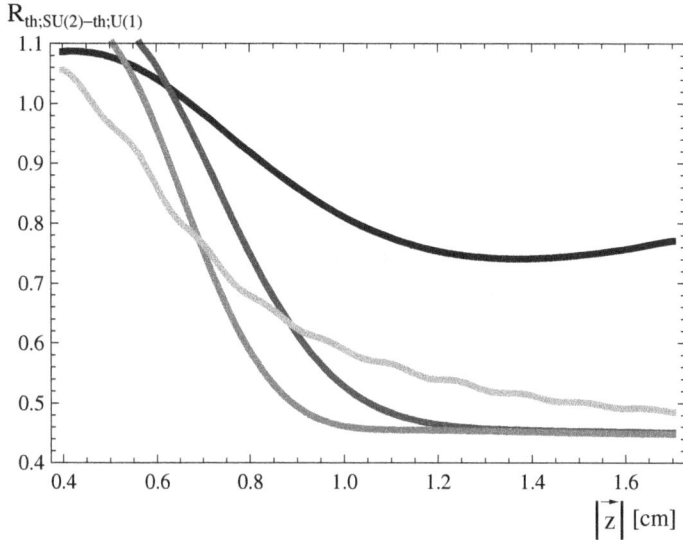

Figure 10.5. The function $R_{th,SU(2)-th,U(1)}(|\vec{z}|)$, as defined in Eq. (10.23), for SU(2)$_{CMB}$ ($T_c = 2.725$ K) and for various temperatures: black curve ($T = 1.5T_c$), dark gray ($T = 2.0T_c$), gray ($T = 2.5T_c$), light gray ($T = 3.0T_c$). Notice the regime of antiscreening for small, unresolved $|\vec{z}|$, and the screening effects of SU(2)$_{CMB}$ for large $|\vec{z}|$.

to the pure U(1) case for the resolved distances. Since $\langle \Theta_{00}(\vec{x})\Theta_{00}(\vec{y}) \rangle$ is a measure for the mediation of the electromagnetic interaction between microscopic objects emitting and absorbing photons, we conclude that this interaction is severly suppressed on distances ~ 1 cm and larger for temperatures ranging between 5 and 10 K, say.

10.1.3 Stability of clouds of atomic hydrogen in the Milky Way

The results of Sec. 10.1.2 could improve our understanding of the gradual metamorphosis of cold ($T = 5$–10 K) astrophysical objects such as the hydrogen cloud GSH139-03-69 observed in between spiral arms of the outer Milky Way [Knee and Brunt (2001)]. We refer the reader to [Goldsmith, Li, and Krčo (2007); Meyer and Lauroesch

(2006); Redfield and Linsky (2007)] for simulational aspects and the observation of other clouds systems with similar characteristics whose main features we have described in Sec. 9.4.2. The main puzzle concerning cloud evolution is the unexpectedly large age at a high content of atomic hydrogen.

Figure 10.5 clearly expresses that the equal-time thermal correlation of energy density in the photon gas described by SU(2)$_{\text{CMB}}$ at temperatures in the range $T = 5$–$10\,\text{K}$ is, as compared to the conventional U(1) theory, suppressed by up to a factor of two at distances of about 1 cm. But such values for (21-cm line) temperature and interatomic distances are typical within the hydrogen cloud GSH139-03-69. As a consequence, atomic interactions, leading to the formation of molecules, should be strongly suppressed. It would be interesting to see how the simulation of the cloud evolution, taking into account the suppression effects shown in Fig. 10.5, would increase the estimate of the content of atomic hydrogen at a given age as compared to the conventional theory. Such an investigation, however, is well beyond the scope of the present book.

Let us summarize the results of Sec. 10.1. We have computed the two-point correlation of the canonical energy density of photons both in the conventional and postulated cases of U(1) and SU(2)$_{\text{CMB}}$ [Hofmann (2005); Giacosa and Hofmann (2007)] respectively. In the real-time formalism of finite-temperature field theory and resumming the polarization tensor for the massless, transverse mode (photon) in the SU(2)$_{\text{CMB}}$ case [Schwarz, Hofmann, and Giacosa (2007); Ludescher and Hofmann (2009)], this correlation was investigated in terms of its thermal and vacuum parts. We have observed that for both U(1) and SU(2)$_{\text{CMB}}$ the vacuum parts are strongly suppressed as compared to the thermal parts for distances $\sim 1\,\text{cm}$ and temperatures of 5–10 K. The most important result of Sec. 10.1 is a strong suppression of the thermal part of the SU(2)$_{\text{CMB}}$-correlator as compared to the thermal part of the U(1)-correlator at spatial distances matching the interatomic distances of hydrogen atoms within unexpectedly stable and cold ($T \sim 5$–$10\,\text{K}$) cloud structures in between the spiral arms of the Milky Way. Thus the mean energy transport and hence the atomic interactions are hamstrung in such structures possibly explaining their unexpectedly large age at a high content of a atomic hydrogen.

10.2   Large-angle Anomalies of the CMB

Based on the statistical analysis of WMAP data (mostly the internal linear combination (ILC) map but see also [Tegmark, de Oliveira-Costa, and Hamilton (2003)]) alignments of the largest angular modes of CMB anisotropy with each other and with the geometry and direction of motion of the Solar System as well as an unusually low angular two-point correlation at the largest angular scales (see Fig. 10.6) were reported in [de Oliveira-Costa et al. (2004); Tegmark, de Oliveira-Costa, and Hamilton (2003); Copi, Huterer, and Starkman (2004)] (see also [Copi et al. (2010)] and references therein). These results contradict the usual assumption of a statistically isotropic sky in analyzing the observed (galactic) foreground- as well as dipole- and monopole-subtracted maps of CMB temperature.

In this section we derive and apply an evolution equation for temperature fluctuations of the CMB which may provide a framework to discuss the above-quoted observational results in view of the black-body anomaly due to SU(2)$_{CMB}$ (see Sec. 9.3). We start by performing a match to the situation of a perfect fluid which enables us to interpret temperature as a scalar field subject to an adiabatic approximation. This situation is not unlike the one of a thermalized condensed matter system where the evolution of temperature inhomogeneities is described by a heat equation. Here temperature plays the role of a scalar field under rotations. Subsequently, we perturb the perfect-fluid situation to describe a dynamic temperature evolution. Finally, we derive the linearized evolution equation for temperature fluctuations sourced by the anomaly in the black-body spectrum of energy density (see Sec. 9.3). We also briefly discuss an alternative,

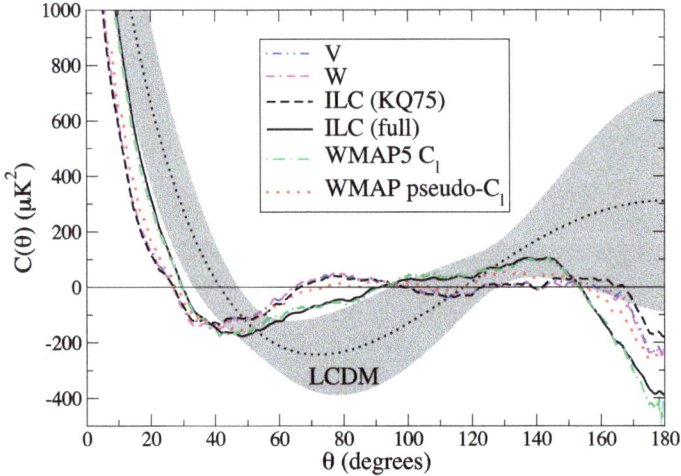

Figure 10.6. The two-point, directly computed angular correlation function $C(\theta)$ from the foreground- as well as dipole- and monopole-subtracted CMB temperature map for two different bands (from WMAP five-year-data, no assumption of statistical isotropy in decomposing cut-sky data into spherical harmonics). Also shown are $C(\theta)$ as directly computed from the ILC map with and without the same cut-sky mask, $C(\theta)$ from a statistically isotropic sky with a best-fit ΛCDM (cosmological constant plus nonrelativistic dark matter) comology (dotted curve) together with 68% cosmic-variance error bars, and $C(\theta)$ as reconstructed from the angular power C_l in the WMAP5 map assuming statistical isotropy. It is obvious that cut-sky maps produce $C(\theta)$ that are consistent with zero for $\theta \geq 60°$ in contrast to the results of the full-sky maps. This indicates violation of statistical isotropy at large angles. Figure adopted from [Copi *et al.* (2010)] with kind permission by D. Schwarz.

purely thermodynamical approach to the emergence of a temperature sink on cosmological scales and late times which makes the nature of the dynamical breaking of statistical isotropy, driven by the integral black-body anomaly, more explicit. The presentation of his section is closely related to [Ludescher and Hofmann (2009); Szopa and Hofmann (2008)], and [Hofmann (2013)].

10.2.1 *Temperature as a scalar field*

Let us now argue why it is possible to regard temperature as a scalar field being a function of cosmological coordinates. We start by considering the energy–momentum tensor $T_{\mu\nu}$ of a perfect fluid whose energy density ρ and pressure p are functions of temperature T:

$$T_{\mu\nu} = (\rho + p)u_\mu u_\nu - pg_{\mu\nu}. \tag{10.24}$$

Here u_μ is the four-velocity of a fluid segment (local rest frame of the heat bath), and the signature of the metric tensor $g_{\mu\nu}$ is $(1, -1, -1, -1)$. We now seek a Lagrangian density $\mathcal{L}$ which produces the right-hand side of Eq. (10.24) upon using the definition for gravitationally consistent energy–momentum:

$$T_{\mu\nu} = \frac{1}{\sqrt{-g}} \frac{\delta \mathcal{L}}{\delta g^{\mu\nu}}, \tag{10.25}$$

where $\det g_{\mu\nu} \equiv g$. For a static, perfect fluid the only two nonscalar covariants available to construct $\mathcal{L}$ are u_μ and $g_{\mu\nu}$. Thus we make the ansatz

$$\mathcal{L} = \sqrt{-g}\left(\alpha u_\mu u_\nu g^{\mu\nu} + \beta\right) \tag{10.26}$$

with scalar parameters α and β to be determined such that the perfect-fluid form in Eq. (10.24) emerges when employing Eq. (10.25). Notice that in varying the Lagrangian density in Eq. (10.26) after $g^{\mu\nu}$ the four-velocity u_μ is kept fixed. The connection between u_μ and $g^{\mu\nu}$ is made subsequently by virtue of Einstein's equations.

Comparing the coefficients in front of the two independent tensor structures in Eq. (10.24) yields:

$$\alpha = \rho + p, \quad \beta = -\rho - 3p. \tag{10.27}$$

Thus the Lagrangian density in Eq. (10.26) becomes

$$\mathcal{L} = \sqrt{-g}\left(\underbrace{(\rho + p)\, u_\mu u_\nu g^{\mu\nu}}_{=1} - \rho - 3p \right) = -2\sqrt{-g}p. \tag{10.28}$$

Specializing to a conventional photon gas with equation of state $\rho = 3p$, we have

$$\mathcal{L} = -\frac{2}{3}\sqrt{-g}\rho. \tag{10.29}$$

Since $\rho \propto T^4$ it follows that temperature itself acquires the status of a scalar field within the static, perfect-fluid situation.

10.2.2   Temperature fluctuations and background cosmology

The Lagrangian density in Eq. (10.29) represents a potential for the scalar field T. We would like to go beyond this adiabatic approximation by allowing for the contributions of derivatives acting on δT where:

$$T = \bar{T}(t) + \delta T(t, \vec{x}). \tag{10.30}$$

Since $\bar{T}(t)$ is a homogeneous, scalar field corresponding to the limit of noninteracting photons the deviation from this limit $\delta T(t, \vec{x})$ is also associated with a scalar field.[1] The deviation $\delta \rho$ of the energy density due to the black-body anomaly induces the fluctuation δT about the mean temperature $\bar{T}$. The latter is redshifted by the evolution of the cosmological background. We consider the conventional black-body part $\bar{\rho}(\bar{T}) \equiv \rho_{U(1)} = \frac{\pi^2}{45}\bar{T}^4$ as a fluid which, among other contributions (cold dark matter and dark energy ΛCDM), sources spatially flat Friedmann–Lemaître–Robertson–Walker (background) cosmology[2]:

$$ds^2 = g_{\mu\nu}dx^\mu dx^\nu \equiv dt^2 - a^2(t)d\vec{x}^2, \tag{10.31}$$

where $a(t)$ denotes the scale factor.

---

[1]Notice that the field T is a classical field and thus does not admit a particle interpretation of its fluctuations.
[2]We are only interested in redshifts z up to $z = 30$. This justifies the assumption of a spatially flat universe driven by ΛCDM.

The usual factor of $\sqrt{-g}$ in the action density is expressed in terms of $\bar{T}$ and $\bar{T}_0 \equiv T_c = T_{\text{CMB}}$ by virtue of the identity

$$\frac{a(t)}{a_0} = \frac{\bar{T}_0}{\bar{T}} \quad \Rightarrow \quad \sqrt{-g} = \left(\frac{\bar{T}_0}{\bar{T}}\right)^3 = a^3 \quad (a_0 \equiv 1). \tag{10.32}$$

In Eq. (10.32) and in the remainder of Sec. 10.2 a subscript '0' refers to today's value of the corresponding quantity.

Leaving the limit of noninteracting photons, the action for $\delta T(t, \vec{x})$ is a series involving scalar combinations of arbitrarily high powers of covariant and thus also ordinary derivatives ∂_μ. The mass scale l^{-1}, which determines the relevance of the n-th power of ∂_μ in this expansion, is, however, determined by the mass of a screened monopole: $l^{-1} \sim \frac{4\pi}{e}\pi T$ where $e \sim \sqrt{8\pi}$. Thus l^{-1} is comparable to temperature itself. Derivatives, however, measure variations of temperature on cosmological scales and thus, for counting purposes, the n-th power of a derivative (n even) can be replaced by the n-th power of the Hubble parameter H. For the regime of redshift we are interested in ($z \leq 30$ or so) the ratio $\frac{H}{T}$ is extremely small,[3] and thus a truncation at $n = 2$ of the expansion of the action into powers of derivatives is justified.

We do not yet have a theoretical prescription at our disposal on how to fix the coefficient in front of this kinetic term ($n = 2$) for δT. Being pragmatic, we allow here for a dimensionless coefficient k whose numerical value needs to be determined observationally. Using Eq. (10.32), the part of the Lagrangian density describing the emergence of CMB temperature anisotropies due to the black-body anomaly being activated at low z is given as

$$\sqrt{-g}\mathcal{L}_{\text{CMB}} = \left(\frac{\bar{T}_0}{\bar{T}}\right)^3 \left(k\partial_\mu \delta T \partial^\mu \delta T - \delta\rho(T)\right), \tag{10.33}$$

---

[3]In a ΛCDM model we have $\frac{H}{T} \sim 10^{-33}$ at $z = 1$.

where $\delta\rho(T) \equiv \rho_{SU(2)CMB} - \rho_{U(1)}$. Let us now define a function $\hat{\rho}(T, T_0)$ as

$$\delta\rho = T_0^2 \hat{\rho}. \tag{10.34}$$

Varying the action associated with Eq. (10.33) with respect to $\delta T = T - \bar{T}$ and linearizing the resulting equation of motion yields:

$$\partial_{\tilde{\mu}}\partial^{\tilde{\mu}}\delta T - \frac{3}{\bar{T}}\partial_\tau \bar{T}\partial_\tau \delta T + \frac{\bar{T}_0^2}{kH_0^2}\left[\frac{1}{2}\frac{d^2\hat{\rho}}{dT^2}\bigg|_{T=\bar{T}}\delta T + \frac{1}{2}\frac{d\hat{\rho}}{dT}\bigg|_{T=\bar{T}}\right] = 0. \tag{10.35}$$

In deriving Eq. (10.35), the coordinate transformation

$$\tilde{x}_0 \equiv \tau = H_0 t, \quad \tilde{x}_i = \frac{da}{dt}x_i, \quad (i = 1, 2, 3) \tag{10.36}$$

was performed. Notice the extremely large factor $(\bar{T}_0/H_0)^2 \sim 10^{60}$ in front of the square brackets in Eq. (10.35). This factor arises because we chose to measure time τ in units of the age of the universe, distances from the origin $\tilde{x}_i$ in units of the actual horizon size $H^{-1} = a/\frac{da}{dt}$ (as long as $|\tilde{x}_i|$ is sufficiently smaller than unity), and temperature in units of $\bar{T}_0 = 2.35 \times 10^{-4}$ eV.

By assuming spherical symmetry in the fluctuation δT, which is relevant for an analysis of the cosmic dipole, Eq. (10.35) simplifies as:

$$0 = \partial_\tau \partial_\tau \delta T - \left(\frac{da}{ad\tau}\right)^2\left[\partial_\sigma \partial_\sigma \delta T + \frac{2}{\sigma}\partial_\sigma \delta T\right] - \frac{3}{\bar{T}}\partial_\tau \bar{T}\partial_\tau \delta T$$

$$+ \frac{\bar{T}_0^2}{kH_0^2}\left[\frac{1}{2}\frac{d^2\hat{\rho}}{dT^2}\bigg|_{T=\bar{T}}\delta T + \frac{1}{2}\frac{d\hat{\rho}}{dT}\bigg|_{T=\bar{T}}\right]. \tag{10.37}$$

In Eq. (10.37) we have introduced $\sigma \equiv \sqrt{\tilde{x}_1^2 + \tilde{x}_2^2 + \tilde{x}_3^2}$.

Equation (10.37) has the form of a two-dimensional wave equation supplemented with additional terms arising on one hand due to the time dependence of the cosmological background ($-\frac{3}{\bar{T}}\partial_\tau \bar{T}\partial_\tau \delta T$) and on the other hand due to the presence of the black-body anomaly: The term $\frac{1}{2}\frac{\bar{T}_0^2}{kH_0^2}\frac{d^2\hat{\rho}}{dT^2}\Big|_{T=\bar{T}}\,\delta T$ will be referred to as "restoring term", and the term $\frac{1}{2}\frac{\bar{T}_0^2}{kH_0^2}\frac{d\hat{\rho}}{dT}\Big|_{T=\bar{T}}$ will be referred to as "source term" in the following.

Let us now provide information on the simple ΛCDM model for the background cosmology[4] which fits the data best [Riess et al. (1998); Perlmutter et al. (1998)]. We assume a spatially flat universe subject to the following Friedmann equation

$$\left(\frac{\dot{a}}{a}\right)^2 = H_0{}^2\left(\frac{\Omega_m}{a^3}+\Omega_\Lambda\right),\tag{10.38}$$

where $\Omega_m=0.24$ and $\Omega_\Lambda=1-\Omega_m=0.76$ (fit obtained from WMAP three-year data [Spergel et al. (2007)]) are the cold dark matter mass density and the dark energy density, respectively, both given in units of the critical density. H_0 is today's value of the Hubble parameter, and $\dot{a}\equiv\frac{da}{dt}$. The solution to Eq. (10.38) is

$$a(t) = \left(\frac{\Omega_m}{\Omega_\Lambda}\right)^{1/3}\left[\sinh\frac{3\sqrt{\Omega_\Lambda}}{2}H_0t\right]^{2/3},\tag{10.39}$$

where H_0 is connected to t_0 (present age of the universe) as

$$H_0t_0 = \frac{1}{3\sqrt{1-\Omega_m}}\ln\frac{2-\Omega_m+2\sqrt{1-\Omega_m}}{\Omega_m}.\tag{10.40}$$

Again, we use the convention that $a_0\equiv a(t_0)=1$. The time dependence of the mean temperature $\bar{T}$ then follows from Eqs. (10.32) and (10.39). We also remind the reader of the relation between scale factor a and redshift z:

$$a(z) = \frac{1}{1+z}\quad(a_0\equiv a(z=0)=1).\tag{10.41}$$

---

[4]The contribution $\bar{\rho}(\bar{T})$, that is, of radiation, to the total energy density of the universe safely is negligible for $z\leq 30$.

10.2.3   Dynamic component in the cosmic dipole: Numerical analysis

The main concern of this section is to relate a discrepancy between the velocity relative to the CMB rest frame of our Local Group of galaxies, gravitationally inferred [Erdogdu et al. (2006)] on one hand and kinematically deduced from the CMB dipole [Peebles and Wilkinson (1968)] (relativistic Doppler effect) on the other hand, to a dynamic CMB dipole component. This dynamic component would be generated by evolving the CMB temperature fluctuation δT through the integrated spectral black-body anomaly of SU(2)$_{CMB}$ according to Eq. (10.37). The main contribution to δT is induced at redshift $z \sim 1$.

10.2.3.1   Principle remarks and boundary conditions

To identify a dynamic component in the CMB *dipole*, which observationally is known to dominate the higher multipoles by at least two orders of magnitude [Kogut et al. (1993); Mather et al. (1994)], the associated solution to Eq. (10.35) must locally exhibit a singled-out direction. This implies spherical symmetry about the center of an initial inhomogeneity: By the source term in Eq. (10.37) the built-up of a three-dimensional radial profile[5] δT is driven.

Let us give two additional reasons why a spherical profile δT is the only option for a dynamic component of the CMB dipole driven by the black-body anomaly. First, a superposition of two solutions associated with two distinct and isolated initial inhomogeneities contained within one another's horizon is not a solution of Eq. (10.35)

---

[5]Simulating the full equation (10.35), that is, not assuming spherical symmetry subject to boundary conditions stated below, one observes that within $\sim 1\%$ deviation the solutions of Eqs. (10.35) and (10.37) coincide. This implies that the source term due to the black-body anomaly generates a spherical profile δT of minimum $\sim 10^{-3}\bar{T}$ during the cosmological evolution. This process smoothens out preexisting fluctuations on large angular scale explaining their suppression compared to small-scale fluctuations.

due to the presence of the spatially homogeneous source term. Second, the superposition of isolated initial inhomogeneities contained within one another's horizon would evolve to populate higher multipoles of strength comparable to that of the dipole. This, however, is ruled out by observation [Kogut *et al.* (1993); Mather *et al.* (1994)]. For a dynamic component to the CMB dipole spherical symmetry of the fluctuation δT is thus inevitable within the horizon of the initial inhomogeneity.

On a three-dimensional radial profile δT *almost* every observer perceives the modulus of the dynamic CMB dipole component. Moreover, this modulus is nearly independent of the observer's position. That is, the mean radial gradient approximately serves to define a direction that is singled out except at the center of the (as we shall see) sink-like configuration δT. This exception, however, occurs with vanishing likelihood geometrically (see Fig. 10.7).

The modulus of the dynamic component $\vec{D}_{dyn}$, as it would be perceived by an observer situated a radial distance σ_0 away from the center of the bump, is defined as follows[6]

$$|\vec{D}_{dyn}| \equiv \int_{\sigma_0}^{1} d\xi \delta T(z=0,\xi) - \int_{\sigma_0-1}^{\sigma_0} d\xi \delta T(z=0,\xi). \qquad (10.42)$$

The upper limit in the first integral arises from the fact that for $\sigma \geq 1$ the nonexistence of a causal connection to the center of the bump forbids the build-up of a profile. The definition in Eq. (10.42) states that $|\vec{D}_{dyn}|$ is roughly given by the mean gradient of δT.

Observationally, the coefficient k in Eq. (10.37) is determined such that the mean gradient in $\delta T(z = 0, \sigma)$ coincides with the

---

[6]Equation (10.42) arises as follows: Looking into (opposite to) the direction of the gradient, a surplus (deficit) of photons stemming from the hotter (colder) tail (central region) of the profile δT is detected by the observer. This allows to define a mean temperature. The amplitude of the dipole then is half the difference between the temperature into and opposite to the direction of the gradient. These temperatures are obtained by performing a radial average over δT within the horizon of the observer.

discrepancy in determining the CMB dipole from the gravitational measurement of the Local-Group velocity [Erdogdu *et al.* (2006)] by virtue of the relativistic Doppler effect [Peebles and Wilkinson (1968)] and by a direct measurement of CMB anisotropy [Kogut *et al.* (1993)]. The velocity of the Local Group $\vec{v}_{LG,dir}$ can be estimated from the gravitational impact on it by all those galaxies contained in successively enlarged concentric, spherical shells and by observing saturation for $zc \geq 6000\,\mathrm{kms}^{-1}$ [Erdogdu *et al.* (2006)] where c is the velocity of light. It is found that $|\vec{v}_{LG,dir}| \sim 400\,\mathrm{kms}^{-1}$ with a typical error of $\sim 50\,\mathrm{kms}^{-1}$ [Erdogdu *et al.* (2006)]. On the other hand, the conventional understanding of the CMB dipole as a purely kinematic effect [Peebles and Wilkinson (1968)], which implies a velocity of the solar system of $\vec{v}_{SS} = (369 \pm 2)\mathrm{kms}^{-1}$ [Hinshaw *et al.* (2006)], plus knowledge of the relative velocity $\vec{v}_{LG-SS}$ between the solar system and the Local Group allows to deduce the velocity of the Local Group[7] as $|\vec{v}_{LG,dedu}| \sim 619\ \mathrm{kms}^{-1}$ [Kogut *et al.* (1993)]. In this way the angle $\delta \equiv \angle \vec{v}_{LG,dedu}, \vec{v}_{LG,dir}$ is extracted as $\delta = (13 \pm 7)^{o}$ [Erdogdu *et al.* (2006)]. As a consequence, a deficit velocity $\vec{v}_{dyn} = \vec{v}_{LG,dedu} - \vec{v}_{LG,dir}$ arises. Again, we propose that this apparent deficit is due to the black-body anomaly of SU(2)$_{CMB}$ driving the build-up of the three-dimensional spherical profile δT. For later use we state that the dipole $\vec{D}$, which emerges as a result of the relativistic Doppler effect due to motion of the observer at speed $\vec{v}$ relative to the CMB rest frame, is calculable as [Peebles and Wilkinson (1968)]

$$\vec{D} = 2\frac{\vec{v}}{c}\vec{T}_0 + \mathcal{O}\left(\frac{\vec{v}^2}{c^2}\right). \tag{10.43}$$

Let us explain the above-sketched situation in more detail: On one hand, the velocity $\vec{v}_{LG,dir}$ generates by virtue of Eq. (10.43) a kinematic contribution to the CMB dipole, $\vec{D}_{LG,kin}$, in the rest frame of the Local Group. On the other hand, the CMB dipole $\vec{D}_{SS}$, as

---

[7]In [Nakamura *et al.* (2010)] a modulus of $|\vec{v}_{LG,dedu}| = (627 \pm 22)\,\mathrm{kms}^{-1}$ was obtained.

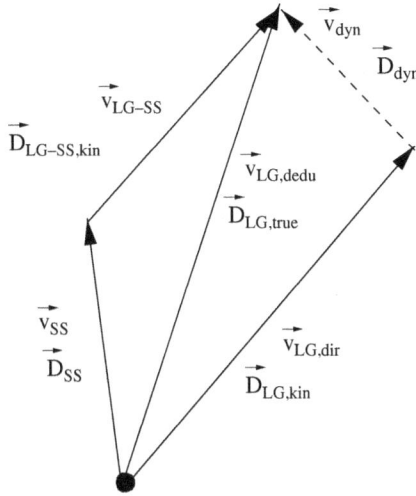

Figure 10.7. Diagram relating observed and deduced quantities concerned with CMB-dipole physics.

measured in the solar system rest frame, is in the rest frame of the Local Group supplemented by a purely kinematic contribution $\vec{D}_{LG-SS,kin}$ resulting from the relative velocity $\vec{v}_{LG-SS}$ between the solar system and the Local Group. The contribution $\vec{D}_{LG-SS,kin}$ is calculable by Eq. (10.43). Adding $\vec{D}_{SS}$ and $\vec{D}_{LG-SS,kin}$ yields the true CMB dipole $\vec{D}_{LG,true}$ in the rest frame of the Local Group. Thus we have for the dynamic contribution to the CMB dipole $\vec{D}_{dyn}$

$$\vec{D}_{dyn} = \vec{D}_{LG,true} - \vec{D}_{LG,kin}$$
$$= \vec{D}_{SS} + \vec{D}_{LG-SS,kin} - \vec{D}_{LG,kin}. \qquad (10.44)$$

In Fig. 10.7 this situation is sketched.

Let us now discuss the boundary conditions which Eq. (10.37) needs to be supplemented with. Equation (10.37) is an inhomogeneous, linear, second-order partial differential equation which can be solved using the numerical method of lines (see [Schiesser (1994)]). Four boundary conditions are required, two for the temporal and two for the spatial evolution. We assume the spatial distribution of the initial fluctuation at redshifts $z_i = 5$–30 to be of Gaussian shape with a height chosen such that $\frac{\delta T(z_i,\sigma=0)}{\bar{T}(z_i)} = 10^{-5}$ as is expected to be provided by primordial causes[8]:

$$\delta T(z_i, \sigma) = 10^{-5}\bar{T}(z_i)e^{-(\frac{\sigma}{w})^2}. \qquad (10.45)$$

Here the subscript i refers to "initial".

The width w of the Gaussian in Eq. (10.45) is varied to check for the robustness of the result against our ignorance concerning this boundary condition. Initially, it is assumed that the build-up of the fluctuation is slow since the source term $\frac{1}{2}\frac{d\delta\rho}{dT}\big|_{T=\bar{T}}$ in Eq. (10.37) driving this build-up is small for sufficiently large initial redshift z_i (see Fig. 10.8). That is, one prescribes

$$\partial_\tau \delta T(\tau,\sigma)|_{\tau=\tau_i} = 0 \qquad (10.46)$$

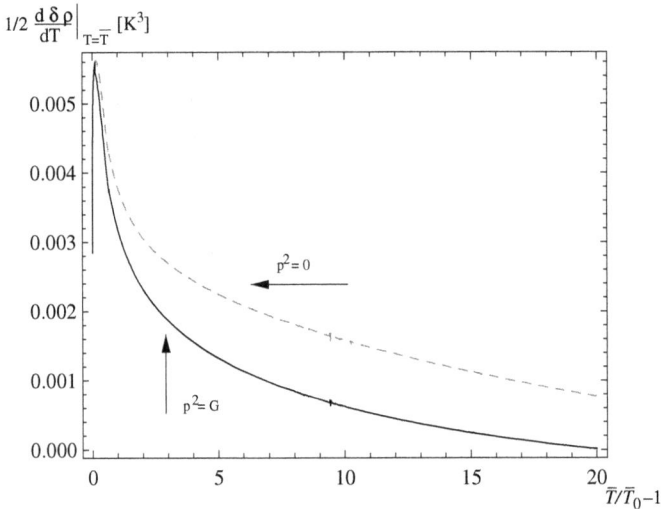

Figure 10.8. The "source term" $\frac{1}{2}\frac{d\delta\rho}{dT}\big|_{T=\bar{T}}$ in Eq. (10.37) as a function of $\bar{T}/\bar{T}_0 - 1$. The approximation $p^2 = 0$ [Szopa and Hofmann (2008)] (see Sec. 5.4.7.3) corresponds to the dashed line while the case of the selfconsistently determined screening function G (see Sec. 5.4.7.5) is depicted by the solid line.

---

[8] We also set $\frac{\delta T(z_i,\sigma=0)}{\bar{T}(z_i)} = 0$ at times.

and later checks for the independence of the result for the evolution on our choice of z_i. The profile δT, regardless of whether it is weakened or enforced by the evolution, remains extremal at $\sigma = 0$ where the initial inhomogeneity is located. So it should satisfy

$$\partial_\sigma \delta T(\tau, \sigma)|_{\sigma=0} = 0. \tag{10.47}$$

Finally, δT is zero for $\sigma \geq 1$ (horizon[9]) and for all times. Otherwise, the build-up of δT would be noncausal:

$$\delta T(\tau, \sigma \geq 1) = 0. \tag{10.48}$$

In order to be consistent with the boundary condition of Eq. (10.45), we approximate the boundary condition of Eq. (10.48) by simply prescribing the value of the profile δT at $z_i = 0$ and $\sigma = 1$ for all $z = 0$. This is in good agreement with Eq. (10.48) if $w \ll 1$.

10.2.3.2   Results

Here we present the results of the numerical computation of the solution to Eq. (10.37) subject to the background cosmology of Eq. (10.39) ($\Omega_m = 0.24$ and $\Omega_\Lambda = 1 - \Omega_m = 0.76$) and the boundary conditions (10.45)–(10.48).

In Fig. 10.8 a plot of $\frac{1}{2}\frac{d\delta\rho}{dT}\big|_{T=\bar{T}}$ as a function of $\bar{T}/\bar{T}_0 - 1$ obtained in the approximation $p^2 = 0$ [Szopa and Hofmann (2008)] (dashed line) and subject to a one-loop selfconsistent determination of the screening function G (solid line) are shown. Figure 10.9 is a plot of $\frac{\delta T}{T}$ as a function of $\bar{T}/\bar{T}_0 - 1$ for fixed distances $\sigma = 0.05; 0.5$, $k = 0.01868\bar{T}_0^2/H_0^2$, and a width $w = 10^{-2}$ of the initial Gaussian at $z_i = 20$. The value of k is determined such as to match the discrepancy $\bar{D}_{\text{dyn}}$ according to Eq. (10.42) (see also [Szopa and Hofmann (2008)]). The approximation $p^2 = 0$ [Szopa and Hofmann (2008)]

---

[9]This statement is only approximately valid because the employed relation between coordinates $\tilde{x}_i$ and x_i in Eq. (10.36) actually is only valid for their differentials.

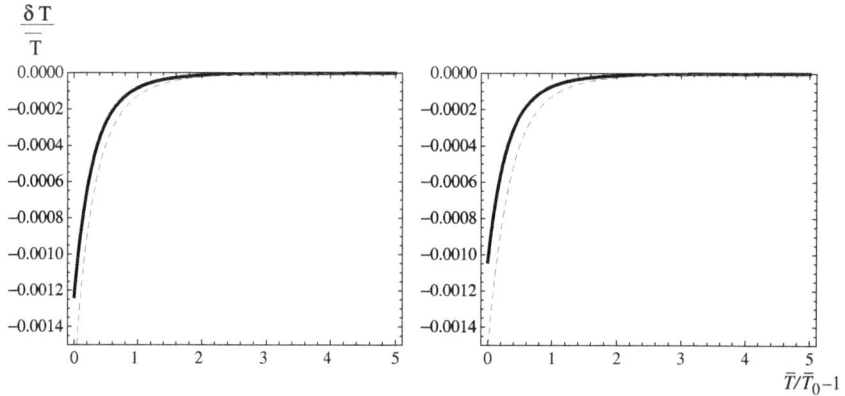

Figure 10.9. $\frac{\delta T}{\bar{T}}$ for two distances $\sigma = 0.05$ (left panel) and $\sigma = 0.5$ (right panel) as a function of $\bar{T}/\bar{T}_0 - 1$ for $k = 0.01868\bar{T}_0^2/H_0^2$. The width w of the initial Gaussian is $w = 10^{-2}$ at $z_i = 20$. The approximation $p^2 = 0$ corresponds to the dashed line [Szopa and Hofmann (2008)] while the case of the selfconsistently determined, transverse screening function G is depicted by the solid line.

corresponds to the dashed line while use of the selfconsistently determined, transverse screening function G yields the solid line.

In Fig. 10.10 plots of the profiles $\frac{\delta T}{\bar{T}}$ at $\bar{T}/\bar{T}_0 - 1 = 1$ and $z = \bar{T}/\bar{T}_0 - 1 = 0$ are shown when computed with the approximation $p^2 = 0$ and using the selfconsistent determined screening function G. In Fig. 10.11 plots of the profiles $\frac{\delta T}{\bar{T}}$ at $z = 0$ are shown. Notice that $k = 0.01868\,\bar{T}_0^2/H_0^2$ for the case $p^2 = 0$ while the k-value for the case $p^2 = G$ was adjusted such that the two profiles coincide at $\sigma = 0$. This is justifiable by the empirical uncertainty of k [Szopa and Hofmann (2008)]. Obviously, the difference between the two curves is small. Throughout our analysis we have varied z_i within the range $5 \le z_i \le 30$, and we have obtained identical results for the behavior of δT at $\bar{T}/\bar{T}_0 - 1 \le 1$.

10.2.4   *Alignment and suppression of low multipoles?*

Due to the black-body anomaly of SU(2)$_{\text{CMB}}$ primordial tempera-ture inhomogeneities are locally enhanced up to a temperature scale

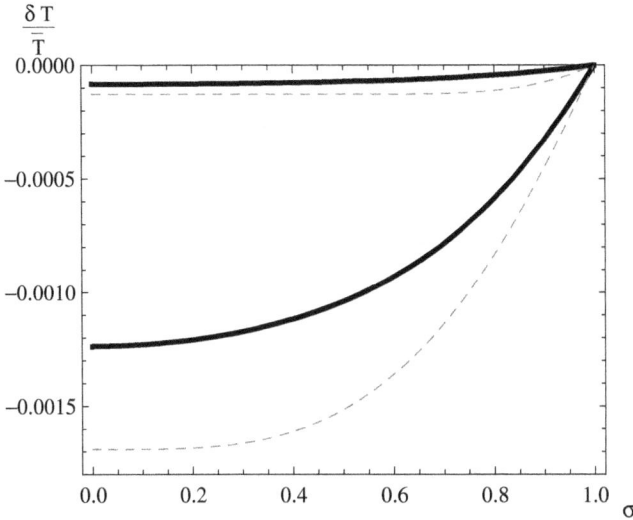

Figure 10.10.   The profile $\frac{\delta T}{T}$ at $\bar{T}/\bar{T}_0 - 1 = 1$ (upper curves) and $z = \bar{T}/\bar{T}_0 - 1 = 0$ (lower curves). The solid lines correspond the case of the selfconsistently determined, transverse screening function while the dashed lines are for the approximation $p^2 = 0$ [Szopa and Hofmann (2008)]. Again, we have used $k = 0.01868\bar{T}_0^2/H_0^2$.

$\frac{\delta T}{T} \sim 10^{-3}$. We have argued that this is related to the emergence of a sink-like spherical profile which is minimal at the center of the primordial inhomogeneity and which vanishes beyond the horizon of this center. The radial gradient of the profile is interpreted as a dynamic contribution to the dipole in the CMB temperature anisotropy thus accomodating the discrepancy between the directly observed and the inferred (by virtue of the relativistic Doppler effect) velocity of the Local Group with respect to the CMB rest frame.

Qualitatively seen, the rapid build-up at late times of such a spherical profile could be responsible for "inflating away" primordial large-scale anisotropies of temperature scale $\frac{\delta T}{T} \sim 10^{-5}$ thus explaining the suppression of the angular two-point correlation function of temperature for $\theta > 60°$ as observed by WMAP [de Oliveira-Costa *et al.* (2004);Copi, Huterer, and Starkman (2004);Copi *et al.* (2010)]. For a two-dimensional analog imagine a fluctuating

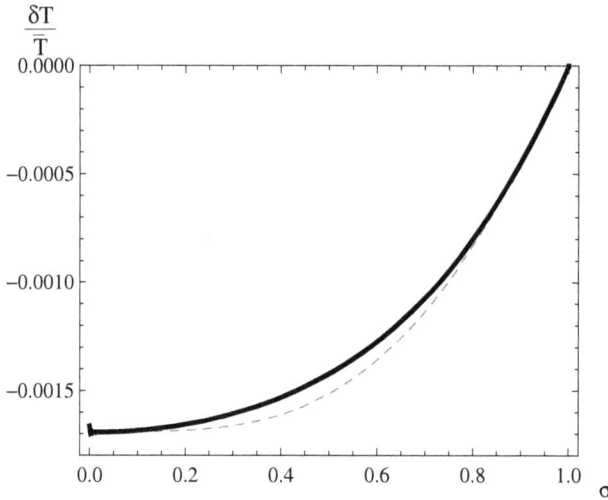

Figure 10.11. The profile $\frac{\delta T}{T}$ at $z = 0$. The dashed line is for the approximation $p^2 = 0$ [Szopa and Hofmann (2008)] and $k = 0.01868\bar{T}_0^2/H_0^2$. The solid line is obtained for the case of $p^2 = G$ and $k = 0.0136\bar{T}_0^2/H_0^2$. This k-value is generated by demanding the function $\frac{\delta T}{T}$ to coincide with the dashed line at $\sigma = 0$.

rubber scarf enframed by a spherical boundary. Large-scale fluctuations are smoothened out by the process of pinching the scarf centrally and quickly pulling it out of the plane into a conical shape. As a result, the large-scale fluctuations thus suppressed would necessarily be correlated due to the common cause for their suppression. To substantiate this scenario further a dedicated numerical investigation of small-amplitude modulations of the solution to Eq. (10.35) would be required.

Alternatively, to conceive the process of building up the sink-like profile in δT as a dynamical breaking of statistical isotropy, driven by the integrated black-body anomaly $\delta\rho \equiv \rho_{\mathrm{SU(2)_{CMB}}} - \rho_{\mathrm{U(1)}}$, one may proceed as follows [Hofmann (2013)]. When treated in the sense of a thermodynamical, canonical ensemble, $\mathrm{SU(2)_{CMB}}$: effects significantly bias a given fluctuation δT by a factor F which represents the ratio, $\mathrm{SU(2)_{CMB}}$ to U(1), of the *a priori* probability P for fluctuation δT to occur in a physical volume element ΔV at $\bar{T}/\bar{T}_0 - 1$.

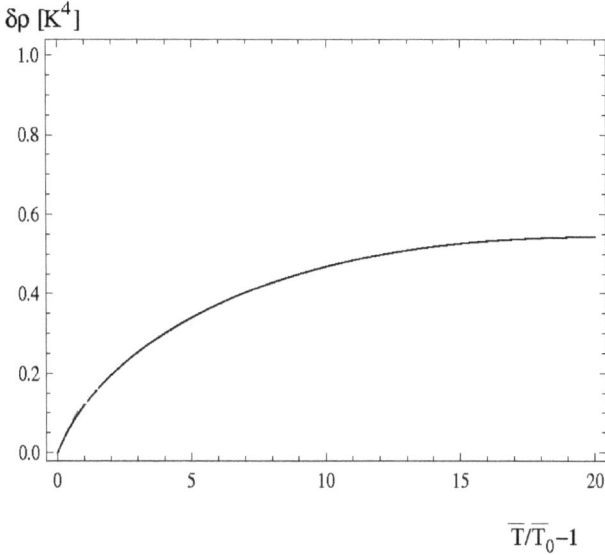

Figure 10.12.   The difference between photonic energy densities in SU(2)$_{CMB}$ and U(1)theories, $\delta\rho \equiv \rho_{SU(2)_{CMB}} - \rho_{U(1)}$, as a function of $\bar{T}/\bar{T}_0 - 1$.

One has

$$F(\bar{T}, \delta T) \equiv \frac{P_{SU(2)_{CMB}}}{P_{U(1)}} = N \exp\left(-\delta\rho\Delta V/\bar{T}\right), \qquad (10.49)$$

where

$$N \equiv \frac{\int_{\bar{T}_0}^{\infty} dT \exp\left(-\rho_{U(1)}\Delta V/\bar{T}\right)}{\int_{\bar{T}_0}^{\infty} dT \exp\left(-\rho_{SU(2)_{CMB}}\Delta V/\bar{T}\right)} \qquad (10.50)$$

and $\delta\rho$ is the positive function of positive slope shown in Fig. 10.12. Note that factor F is larger (smaller) for a negative (positive) fluctuation δT about a given CMB baseline temperature $\bar{T}$. Thus, a spatially local region, which by primordial chance is colder than its environment, becomes unstable against further cooling. Photons streaming out of this region start to cool a larger region, subjecting it, again, to SU(2)$_{CMB}$ biasing via factor F, and so forth. As a consequence, the spatial growth and deepening of a primordial, negative seed fluctuation apparently breaks statistical isotropy as judged by an observer who is located at the slope of the emergent temperature sink. Factor F should be easily implementable into CMB codes which evolve the spatial temperature distribution from one FRW time slice to the next. Qualitatively, the thermodynamical argument underlying the construction of factor F yields results identical to the ones obtained by evolution according to Eq. (10.35).

10.3   Planck-scale Axion and Dark Energy

In this section we would like to discuss a possible connection between dark energy — responsible for the universe's present accelerated expansion —, dynamical breaking of a global U(1) axial anomaly at the Planck scale $M_P \sim 10^{19}$ GeV, and the candidate theory SU(2)$_{CMB}$ for photon propagation.

10.3.1   *Model of the very early universe*

In [Giacosa, Hofmann, and Neubert (2008)] a Planckian physics model is investigated which, by a relaxation of Planckian vacuum energy, inaugurates a U(1)$_A$ Goldstone field A. In sub-Planckian epochs of the universe's evolution, that is, for times larger than $\sim 10^{-43}$ s, the (massless) field A would acquire a potential by its interaction with topological field configurations composing the ground states of non-Abelian gauge theories (for pioneering articles on the axial anomaly see [Adler (1969); Bell and Jackiw (1977); 't Hooft (1976); Fujikawa (1979, 1980)]). This generates dark energy if the spatially homogeneous field A is, by cosmological friction, constrained to roll down the slope of its potential slowly compared to the rate of cosmological expansion. As we will show below, this would be the case for an axion field emerging from Planckian physics which couples to SU(2)$_{CMB}$. Since a presentation of the details of the calculations in [Giacosa, Hofmann, and Neubert (2008)] would operate well beyond the scope of this book we only state here the main ideas and results.

In the model of [Giacosa, Hofmann, and Neubert (2008)] N species of massless, minimally coupled fermions are assumed to interact via gravitational torsion in a de Sitter spacetime. This inter-action causes a dynamical breaking of the associated $U_V(N) \times U_A(N)$ flavor symmetry down to its diagonal subgroup $U(N)$ rendering fermions, by virtue of a chiral condensate, massive and noninteract-ing in the limit $N \to \infty$. This chiral condensate is decomposed into its modulus ρ and N^2 phases (Goldstone directions) all of which must be regarded as quantum fields. In [Giacosa, Hofmann, and Neubert (2008)] only the flavor singlet pseudoscalar Goldstone field A, emerging as a result of the breakdown of the $U(1)_A$ factor in $U_V(N) \times U_A(N)$, and the field ρ are considered. In the following, we will refer to the field A as *Planck-scale axion*. A free fermion model in four-dimensional de Sitter spacetime is renormalizable [Candelas and Raine (1975)] in the sense that in integrating out the fermions divergences can be absorbed into renormalizations of two bare parameters of the theory: the cosmological constant Λ_b and the Hubble parameter H_b. The resulting effective potential appears to guarantee that in the limit $N \to \infty$ the physical Hubble parameter is driven to zero. This is equivalent to the statement that any pre-existing vacuum energy (of Planckian magnitude) is dynamically relaxed to zero.

The reason why out of N^2 Nambu–Goldstone fields only the flavor singlet field A is taken seriously in [Giacosa, Hofmann, and Neubert (2008)] is its capability of re-generating vacuum energy. This is an aspect of the so-called chiral anomaly [Adler (1969); Adler and Bardeen (1969); Bell and Jackiw (1977); 't Hooft (1976); Fujikawa (1979, 1980)] invoked by the coupling of A to the topological charge density of gauge-field configurations composing the ground states of Yang–Mills theories. On the microscopic level, that is, for a resolu-tion larger than $|\phi|$ of a given SU(2) or SU(3) Yang–Mills theory, the following interaction Lagrangian emerges due to quantum effects

$$\mathcal{L}_{\text{int}} \sim \frac{A}{M} \text{tr} \tilde{F}_{\mu\nu} F^{\mu\nu}. \tag{10.51}$$

Here M is the scale of chiral symmetry breaking. In our case M is approximately the Planck mass, $M \sim M_P \sim 10^{19}\, GeV$, and $F^{\mu\nu}$ ($\tilde{F}_{\mu\nu}$) denotes the fundamental field strength (dual field strength) of the Yang–Mills theory under consideration (see Sec. 1.4). Integrating out the ground-state portion of fundamental gauge field fluctuations, the field A acquires a potential. Roughly it is of the following form [Peccei and Quinn (1977); Preskill, Wise, and Wilczek (1983)]

$$\left(1 - \cos \frac{A}{M_P}\right) \Lambda_{YM}^4, \tag{10.52}$$

where Λ_{YM} refers to the Yang–Mills scale.

10.3.2 *Planck-scale axion: Cosmological evolution after CMB decoupling*

In this section we given an account on what the cosmological implications of the Planck-scale axion A and its interaction with the thermal ground state of SU(2)$_{CMB}$ are. SU(2)$_{CMB}$ is the only Yang–Mills theory among the chain of theories discussed in Sec. 9.1 which exhibit propagating gauge fields from the time of CMB decoupling at redshift $z = 1089$ ($T_{z=1089} \sim 2 \times 10^{-1}\, eV \sim 3000\, K$), see however Sec. 10.4, until present ($z = 0$). Our presentation closely follows the one of [Giacosa and Hofmann (2007)].

The possibility to interpret dark energy in terms of an ultralight Nambu–Goldstone boson field (Planck-scale axion A) is intensely discussed in the literature (see for example [Preskill, Wise, and Wilczek (1983); Frieman *et al.* (1992, 1995); Kim and Nilles (2003); Wilczek (2004); Kaloper and Sorbo (2005); Hall, Nomura, and Oliver (2005); Barbieri *et al.* (2005)]). A new aspect in this discussion is the concrete identification of the gauge dynamics, SU(2)$_{CMB}$, which provides for the axion potential by virtue of the axial anomaly [Giacosa and Hofmann (2007)].

If the associated Yang–Mills scale Λ_{YM} is far below the Planck mass M_P then A's slow-roll dynamics at late time can mimic a cosmological constant being in agreement with present observation [Perlmutter *et al.* (1998); Riess *et al.* (1998)]. In discussing the dark energy problem a connection between Planck-scale physics and strongly interacting gauge dynamics at present was already proposed in [Frieman *et al.* (1992)]. In that article the existence of a "hidden" gauge group to invoke the axion potential of Eq. (10.52) was proposed. To interpret this "hidden" gauge theory as the one underlying the propagation of electromagnetic waves and photons certainly is a bold step. As a physics model, such a step is, however, definitely falsifiable (see Chapter 9). Moreover, SU(2)$_{\text{CMB}}$ entails consequences for our expectations concerning the future evolution of the universe and for the nature of electroweak symmetry breaking in particle physics (see Sec. 9.1 and 9.6). The very light pseudoscalar bosons, associated with the field A, interact with ordinary matter feebly only and thus escape their detection in collider experiments. Because of a dynamically broken, global U(1)$_A$ symmetry associated with the very existence of the field A, the corresponding potential of Eq. (10.52) is radiatively protected.

Again, we wish to propose an axion-based scenario relating via SU(2)$_{\text{CMB}}$ the present temperature of the CMB, $T_{\text{CMB}} = 2.35 \times 10^{-4}$ eV, to the present scale of dark energy $\sim 10^{-3}$ eV. In other words, we intend to explore in connection with Planck-scale axion physics some cosmological consequences of the postulate that the gauge dynamics SU(2)$_{\text{CMB}}$ is responsible for photon propagation.

10.3.2.1   *Cosmological evolution from $z_{\text{dec}} = 1089$ to $z = 0$*

We consider a spatially flat universe whose expansion after CMB decoupling at redshift $z_{\text{dec}} = 1089$ is sourced by baryonic and dark, pressureless matter (M), a homogeneous Planck-scale axion field A and SU(2)$_{\text{CMB}}$ Yang–Mills thermodynamics. The evolution of the scale parameter $a = a(t)$ is determined by the Friedmann equation for a spatially flat universe [Eidelman *et al.* (2004); Spergel *et al.*

(2003)]

$$H(t)^2 = \left(\dot{a}/a\right)^2 = \frac{8\pi}{3}G\left(\rho_M + \rho_A + \rho_{\text{CMB}}\right), \qquad (10.53)$$

where $G \equiv \frac{1}{M_P^2}$ and $M_P \equiv 1.2209 \times 10^{19}$ GeV. Each of the contributions to the right-hand side of Eq. (10.53) is associated with a separately conserved cosmological fluid:

$$d\left(\rho_i a^3\right) = -p_i d(a^3) \quad (i = M, A, \text{CMB}). \qquad (10.54)$$

Since the matter-part is pressure-free, $p_M = 0$, we have $\rho_M(a) = \rho_M(a_0) \cdot (a_0/a)^3$ where t_0 is the present age of the universe (to be calculated) and $a_0 \equiv a(t_0)$. In terms of the critical density $\rho_c = 3H(t_0)^2/8\pi G = 4.08 \times 10^{-11}$ eV^4 the measured matter contribution reads

$$\Omega_M = \frac{\rho_M}{\rho_c}(a_0) = \Omega_{\text{Dark-Matter}} + \Omega_{\text{Baryon}} = 0.27 \pm 0.04. \qquad (10.55)$$

By virtue of Eq. (10.54) the dependence $\rho_{\text{CMB}} = \rho_{\text{CMB}}(a)$ is calculated numerically. Notice that at $t = t_{\text{dec}}$ the contribution of the excitations of SU(2)$_{\text{CMB}}$ to the critical energy density is about 10 % and decreases very rapidly for $t > t_{\text{dec}}$.

Let us rewrite the axion potential as follows

$$V(A) = (\kappa \cdot \Lambda_{\text{CMB}})^4 \left[1 - \cos\left(\frac{A}{F}\right)\right], \qquad (10.56)$$

where constraints on the scale F are discussed below. The dimensionless quantity κ parameterizes the uncertainty in the coupling of the topological defects of SU(2)$_{\text{CMB}}$ to the axion. The value of κ is expected to lie within $O(10^{-1})$ to $O(10^1)$ [Peccei and Quinn (1977)]. In our calculation we adjust κ such that the observed value of dark energy density is reproduced today [Eidelman *et al.*

(2004); Spergel *et al.* (2003)]:

$$\Omega_A = \frac{\rho_A}{\rho_c}(a_0) = 1 - \Omega_M = 0.73 \pm 0.04. \tag{10.57}$$

The axion energy density ρ_A and its pressure p_A are given as

$$\rho_A = \frac{1}{2}\dot{A}^2 + V(A), \quad p_A = \frac{1}{2}\dot{A}^2 - V(A). \tag{10.58}$$

From (10.54) and (10.58) the equation of motion for the homogeneous axion field A follows:

$$\ddot{A} + 3H\dot{A} + V'(A) = 0, \tag{10.59}$$

where $V' \equiv dV/dA$. The term $3H\dot{A}$ represents the friction exerted onto A by the expansion of the universe.

The authors of [Frieman *et al.* (1995)] conclude that the CMB-constraints on A-induced adiabatic density perturbations be such that the Hubble parameter during cosmological inflation is smaller than 10^{13} GeV. This entails that the scale F be larger than 10^{18} GeV$\simeq$ $0.1M_P$. Moreover, a quantum field-theoretic description in (3+1) dimensions, which underlies the axial anomaly, is likely meaningful only below the Planck mass. Thus it is natural to suppose that $F \sim M_P$.

The classical field A, representing a condensate of axion particles generated at $T \sim M_P$, is surely fixed to its starting value $A_\text{in} \sim F$ all the way down to CMB decoupling because of the large cosmological friction term in Eq. (10.59). This implies the following initial conditions at CMB decoupling:

$$A_\text{in} = A(t_\text{dec}) \sim F, \quad \dot{A}(t_\text{dec}) = 0. \tag{10.60}$$

We first consider $0 \leq A_\text{in} \leq \pi\frac{F}{2}$, i.e., a range for which the curvature of the potential is positive.

Let us now discuss the conditions for the axion field to behave like a cosmological constant at present. That is, we would like to prevent A from rolling down its potential until today. This happens if[10]

$$3H(t_0) \gg 2m_A, \qquad (10.61)$$

where $m_A \equiv (\kappa\Lambda_{CMB})^2/F$.

By using Eq. (10.53), $V(A) \simeq \frac{1}{2}m_A^2 A^2$, and by neglecting the small direct contribution of SU(2)$_{CMB}$ (excitations), we have

$$H(t_0)^2 = \frac{4}{3}\pi G m_A^2 A_{in}^2 \left(1 + \frac{\Omega_M}{\Omega_A}\right). \qquad (10.62)$$

Rewriting condition (10.61) by appealing to Eq. (10.62), we derive:

$$\frac{A_{in}}{M_P} \gg \frac{1}{\sqrt{3\pi(1 + \Omega_M/\Omega_A)}} \simeq 0.278. \qquad (10.63)$$

Even for $A_{in}/M_P \gtrsim 0.278$ slowly rolling solutions compatible with today's dark-energy density are numerically found (see discussion below). For $A_{in}/M_P \lesssim 0.278$ the parameter κ in Eq. (10.56) needs to assume unnaturally large values for the axion field A to generate today's value of dark energy density. Moreover, A would undergo many oscillations until present and thus would behave more like pressureless matter than dark energy. For $0 \le A_{in} \le \pi\frac{F}{2}$ to be meaningful when compared to the constraint (10.63) one needs to impose that $F/M_P > 0.177$. This is similar to the lower bound $F/M_P > 0.1$ arising from the consideration on adiabatic density perturbations in [Frieman et $al.$ (1995)].

In Fig. 10.13 admissible ranges for the initial conditions at $t = t_{dec}$ are shown. The triangular area α represents the allowed parameter range for a slowly rolling axion field at present. The horizontal line $A_{in}/M_P = 0.278$ indicates a rapid crossover from dark-energy-like (above) to oscillating (below) solutions. The allowed range is enlarged by including the trapezoidal area β corresponding to a negative curvature of the potential.[11] Notice that for $\pi F/2 \le A_{in} \le \pi F$

---

[10]$H(t)$ is a monotonic decreasing function, that is, if the condition $3H(t_0) \gg 2m_A$ is satisfied at t_0 then this holds for all t satisfying $t_{dec} \le t \le t_0$.

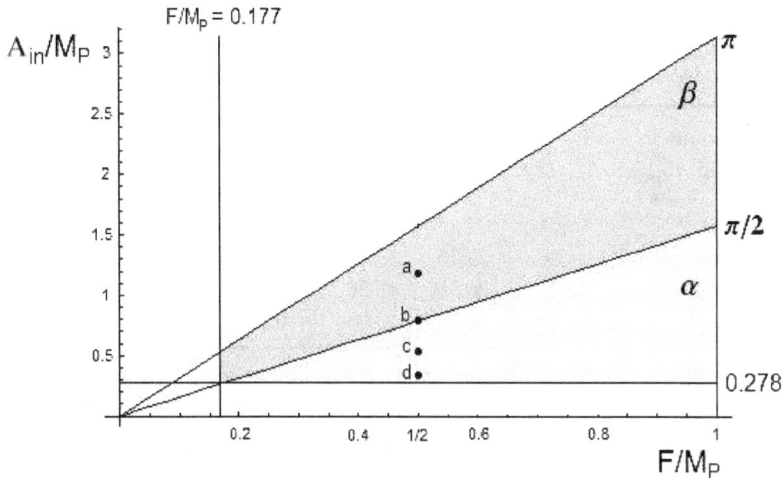

Figure 10.13.   Admissible ranges for values of the quantities A_{in}/M_P and F/M_P for dark-energy-like axion-field solutions today. The triangular area α (β) corresponds to A_{in}/M_P being below (above) the inflexion point of $V(A)$. The horizontal line $A_{in}/M_P = 0.278$ indicates a rapid cross-over from slowly rolling to oscillating solutions.

there are slowly rolling solutions with the required amount of present dark-energy density also for $F/M_P < 0.177$. However, for a decreasing value of F/M_P we observe that A_{in} needs to be closer to the maximum πF which is a somewhat fine-tuned situation [Kaloper and Sorbo (2005)]. We thus pick representative initial conditions as depicted by the points (a),(b),(c), and (d) in Fig. 10.13, keeping in mind that $\dot{A}(t_{dec}) = 0$, see Eq. (10.60).

In Table 1 numerical results are presented. The values of the following quantities are determined: κ such that $\Omega_A = 0.73$ at present, the present age t_0 of the universe, the present Zeldovich parameter for the axion fluid alone, $w_A(t_0) = (p_A/\rho_A)_{t=t_0}$ see Eq. (10.58), and for the entire universe, $w_{tot}(t_0) = (p_{tot}/\rho_{tot})_{t=t_0}$, and, finally, the value of redshift z_{acc} corresponding to the transition between decelerated

---

[11]If the field does not roll at $A_{in} = \pi F/2$ (inflexion point) then it also does not roll for $\pi F/2 \leq A_{in} \leq \pi F$.

Table 1. The values of selected cosmological parameters obtained for various initial values at CMB decoupling keeping $F/M_P = 0.5$ fixed.

Points in Fig. 10.13	A_{in}/M_P	κ	t_0(Gy)	$w_A(t_0)$	$w_{tot}(t_0)$	z_{acc}
(a)	$3\pi/4$	22.15	13.65	−0.97	−0.71	0.76
(b)	$\pi/4$	22.15	13.65	−0.97	−0.71	0.76
(c)	$\pi/6$	26.96	13.56	−0.91	−0.66	0.79
(d)	0.328	37.27	13.08	−0.61	−0.44	0.92

and accelerated cosmological expansion. For the set of initial values (a),(b), and (c) in Fig. 10.13 the axion field practically does not roll all the way until t_0. This is indicated by $w_A(t_0) \simeq -1$. For point (d) A_{in}/M_P is just above the bound in (10.63) causing the field A to roll at present: $w_\phi(t_0) = -0.61$. According to [Eidelman *et al.* (2004);Spergel *et al.* (2003)] $w_A(t_0) = -0.61$ is already inconsistent with observation [$w_A(t_0) < -0.78$ at 95% confidence level]. Decreasing A_{in}/M_P further, one rapidly runs into the regime $w_{tot}(t_0) > -1/3$ where the present universe does not accelerate. The values of z_{acc} obtained for (a), (b), and (c) are in approximate agreement with $z_{acc} = 0.75$ of a standard ΛCDM model. Moving $\frac{F}{M_P}$ within the allowed range $\alpha \cup \beta$ at fixed values of A_{in}/M_P (see Fig. 10.13), the values of the cosmological parameters in the table are almost unaffected.

10.3.2.2 *Massive CMB photons in the future*

Here we consider the future evolution up to the point where $SU(2)_{CMB}$ undergoes the transition to its preconfining phase. For simplicity we assume that the present age of the universe t_0 is given by the time when $T_c = T_0 = T_{CMB}$ is first reached. The system CMB + Planck-scale axion + cold dark matter + gravity remains in a supercooled state for $T < T_0$ so long as the energy density ρ_{CMB} of the deconfining phase is smaller than the energy density of the preconfining phase (see Sec. 6.2.3). At a certain value of (dimensionless) temperature, $\lambda_* = 2\pi T_*/\Lambda_{CMB} < \lambda_c = 2\pi T_c/\Lambda_{CMB}$, equality of the two energy densities takes place (see Fig. 6.4). At this point we may speak of a stable condensate of monopoles, and CMB photons can universally be considered massive. The numerical result for the scale factor at λ_* is $a(t_*)/a_0 = 1.17$.

According to Eq. (10.51) the anomaly-mediated decay width $\Gamma_{A \to 2\gamma}$ of the axion into two photons can be computed from a tree-level Feynman diagram. It is much smaller than the present Hubble parameter H_0:

$$\Gamma_{A \to 2\gamma} < \left(\frac{m_A}{F}\right)^2 m_A \sim 10^{-155} \text{ eV} \ll H_0 \sim 10^{-33} \text{ eV}. \quad (10.64)$$

Thus it is justified to treat the axion field as coherent for any practical purpose and to regard the axion fluid and the SU(2)$_{CMB}$ fluids as separately conserved.

The numerical value of the time interval $\Delta t_{m_y=0} = t_* - t_0$ follows from the solution to Eq. (10.53) for future cosmology. For the sets of initial values (a)–(d) at Table 1 of Sec. 10.3.2.1 we obtain the following numbers:

$$(a) \ \Delta t_{m_y=0} = 2.20 \text{ Gy}, \quad (b) \ \Delta t_{m_y=0} = 2.20 \text{Gy},$$
$$(c) \ \Delta t_{m_y=0} = 2.22 \text{ Gy}, \quad (d) \ \Delta t_{m_y=0} = 2.29 \text{ Gy}. \quad (10.65)$$

Thus the value of $\Delta t_{m_y=0}$ depends only weakly on the chosen parameter set. The error in determining the quantity $\Delta t_{m_y=0}$ is dominated by the observational uncertainty for the present Hubble parameter H_0. According to [Eidelman *et al.* (2004); Spergel *et al.* (2003)] we have $\frac{\delta H(t_0)}{H(t_0)} \sim \pm 0.056$. Running the simulations with the upper (lower) limit of the error range, this generates a decrease (increase) in $\Delta t_{m_y=0}$ of about 0.15 Gy.

10.3.2.3 *Slowly rolling Planck-scale axion: Dark energy and dark matter?*

A more unified but also more speculative picture arises if today's rolling axion field would be made responsible for both cosmological dark matter and dark energy (see [Padmanabhan (2003)]). On one hand, according to simulations performed with a canonical kinetic term in the axion Lagrangian density such a scenario would imply an age of the universe of about 20 Gy as opposed to 13.7 Gy with conventional cold dark matter. Also one would obtain $z_{\text{acc}} \sim 3$ as opposed to $z_{\text{acc}} \sim 0.75$, possibly endangering structure formation. On the other hand, structure formation and the flattening of the rotation curves of galaxies would require an explanation in terms of ripples and lumps of a coherent axion field [Wetterich (2001)], respectively. Moreover, the relation between luminosity distance and redshift as extracted from SNe Ia standard-candle observation [Perlmutter et al. (1998); Riess et al. (1998)] would have to be post-dicted with a pressureless contribution to the Hubble parameter that acquired nominal strength only very recently. The future will tell (gravitational lensing signatures for galaxies, theoretical results on the stability of the system axion-lump plus baryonic matter plus gravity) whether the possibility to dispose of a separately conserved cold dark matter component (in favor of a coherently oscillating axion field which is spatially homogeneous on cosmological length scales) in the mix of cosmological fluids is viable.

For completeness it was investigated in [Giacosa and Hofmann (2007)] how the scenario of a presently slowly rolling axion field affects the estimate $\Delta t_{m_y=0}$. By defining the quantity η through $\eta = -p_A$ and $\rho_A = \dot{A}^2 + \eta$ the axion fluid can be split into a component with $\rho_\Lambda = \eta$ ($w_\Lambda = -1$) and a component $\rho_{\text{DM}} = \dot{A}^2$ ($w_{\text{DM}} = 0$). Notice that the so-defined components are not separately conserved. The task then is to uniquely fix A_{in} and the parameter κ in Eq. (10.56) such that $\Omega_A = 0.96$ today ($\Omega_{\text{Baryon}} = 0.04$) and such that $\Omega_\Lambda = 0.73$ and $\Omega_{\text{DM}} = 0.23$. Using $\frac{F}{M_P} = 0.5$ we obtain $\frac{A_{\text{in}}}{M_P} = 0.53$ and $\kappa = 31.9$. This yields $\Delta t_{m_y=0} = 2.21$ Gy. Comparing with Eq. (10.65), one thus concludes that the estimate of $\Delta t_{m_y=0}$ is rather model-independent.

10.4   The Temperature-Redshift Relation of the CMB

This last section of the book addresses implications of the equation of state $P(\rho)$ of SU(2)$_{CMB}$ for the CMB temperature-redshift relation, conventionally based on the conformal scaling law $T = a^{-1}T_0$ (a the scale factor in the Friedmann-Lemaître-Robertson-Walker metric), and its consequences for a correct dating of cosmological late-time phenomena such as the re-ionization of the Universe. The latter occurs as a consequence of nonlinear structure growth, giving rise to the ignition of the first stars, with ultraviolet components in their light spectra ionizing interstellar hydrogen. We also speculate on a CMB induced mass-generation mechanism for cosmic neutrinos to reconcile the latest CMB data on the number of effective neutrino flavors, $N_{eff} \sim 3.36$ [Ade *et al.* I (2013)], with SU(2)$_{CMB}$. Finally, we speculate about changes implied by SU(2)$_{CMB}$ and a thus modified neutrino sector for the matter sector of the cosmological model at high redshift. The presentation in this section closely follows that in [Hofmann (2015)]. We remark that the treatment of the deconfining phase of SU(2)$_{CMB}$ here does not require any consideration of radiative corrections.

10.4.1   *Equation of state*

Let us first investigate how rapidly the parameter κ in a presumed equation of state of the form $P = \kappa\rho$ approaches that of a thermal gas of massless particles in deconfining SU(2) Yang-Mills thermodynamics as temperature increases away from T_c. Fig. 10.14 depicts κ as a function of[12] T/T_c: Already for $\frac{T}{T_c} \geq 4$ the equation of state is very close to that of a gas of free and massless particles ($P = \frac{1}{3}\rho$): $\frac{T}{T_c} = 4$ corresponds to $\kappa = \frac{P}{\rho} = 0.3201$ and $\frac{T}{T_c} = 6.2$ to $\kappa = \frac{P}{\rho} = 0.3297$.

---

[12]Note, however, that by definition, the actual equation $P = P(\rho)$ of state, $P(\rho)$ being a nonlinear function, exhibits no explicit T dependence.

P/ρ =κ

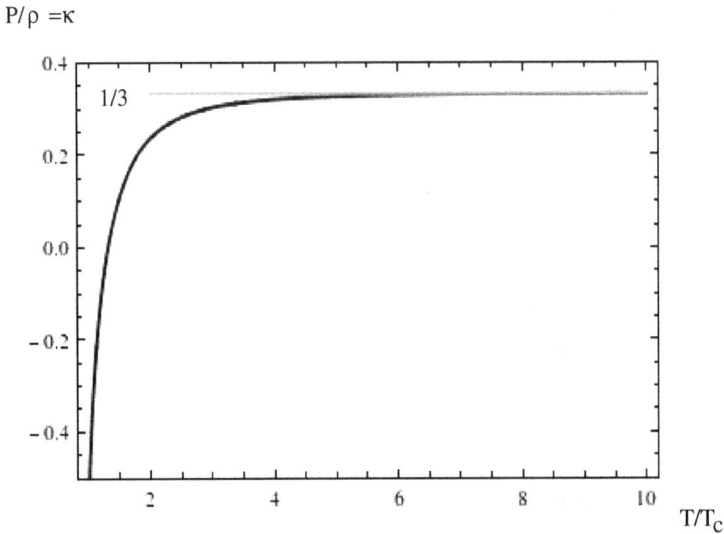

Figure 10.14.   Ratio of pressure and energy density, or equation-of-state parameter κ, as a function of T/T_C in the deconfining phase of SU(2) Yang-Mills thermodynamics. Note the rapid approach to the behavior of a thermal gas of free and massless particles ($P = \frac{1}{3}\rho$).

10.4.2   *Low-z and high-z scaling of CMB temperature in* $SU(2)_{CMB}$

Next, let us explore implications of $SU(2)_{CMB}$ for (i) the dependence of CMB temperature T on the cosmological scale factor, (ii) the cosmological evolution of energy density compared to that of a conventional photon gas, (iii) a resolution of tension between the redshift value z_{re} of instantaneous re-ionization as extracted from the CMB TT angular power spectrum and Baryon Acoustic Oscillations (BAO) ($z_{re} = 11.3 \pm 1$) on one hand [Ade *et al.* I (2013); Ade *et al.* II (2013)] and the detection/non-detection of the Gunn-Peterson trough in high-redshift quasar spectra ($z_{re} \sim 6$) on the other hand [Becker *et al.*(2001)], and (iv) the value of redshift z_{dec} at CMB decoupling and the present cosmological concordance model at high redshift.

10.4.2.1   *Energy conservation in an expanding Universe*

We start by assuming SU(2)$_{\text{CMB}}$ to be a separately conserved cosmic fluid, stretched by the expansion of a Friedmann-Lemaître-Robertson-Walker Universe. The latter is characterized by the scale factor a, normalized to unity at present: $a(T_0) \equiv a_0 = 1$. In discussing the temperature-redshift relation of the CMB no specific cosmological model is required. In particular, no assumption on the Universe's spatial curvature needs to be made. Separate conservation of the SU(2)$_{\text{CMB}}$ fluid predicts interesting consequences for re-ionization and CMB decoupling which are modified when this assumption is relaxed. Such a relaxation is proposed in Sec. 10.4.3.3 where we postulate the cosmic neutrino background to be only conserved together with SU(2)$_{\text{CMB}}$. Interestingly, however, there are no essential modifications under such an extension of SU(2)$_{\text{CMB}}$.

Let ρ and P denote energy density and pressure of SU(2)$_{\text{CMB}}$, respectively. The equation of energy conservation reads

$$\frac{d\rho}{da} = -\frac{3}{a}(P + \rho). \tag{10.66}$$

To solve Eq. (10.66) for $\rho(a)$, an equation of state $P = P(\rho)$ and a boundary condition $\rho^* = \rho(a^*)$ need to be prescribed. The former is obtained by solving $\rho = \rho(T)$ for $T = T(\rho)$ which then is substituted into $P = P(T)$. The choice of initial condition is explained in Sec. 10.4.2.2.

10.4.2.2   *Temperature vs. scale factor and energy density of SU(2)$_{\text{CMB}}$*

We now derive an SU(2)$_{\text{CMB}}$ prediction for $(T_0/T)(a)$. To do this, the initial temperature T^* (and hence energy density $\rho(T^*)$) is chosen such that, with a prescribed value $a^* < 1$ and using $T = T(\rho(a))$, the evolution $\rho(a)$ generates the value $(T_0/T) = 1$ at $a(T_0) = 1$ (today)[13]

---

[13]Again, the subscript '0' signals today's value of the associated quantity.

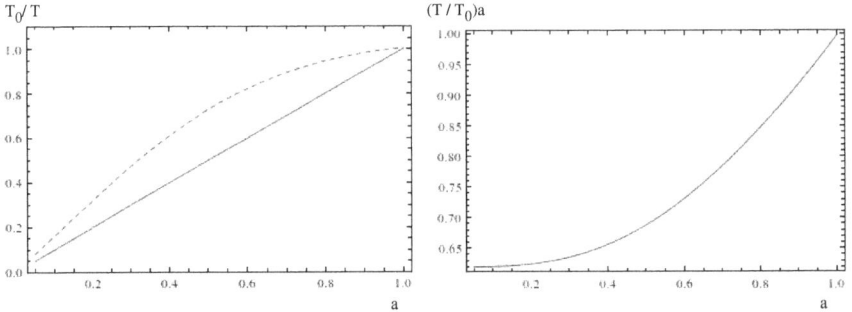

Figure 10.15. SU(2)$_{\text{CMB}}$ induced violation of (conformal) U(1) scaling of CMB temperature T with inverse scale factor a^{-1}. Left panel: $\frac{T_0}{T}$ as a function of scale factor a for SU(2)$_{\text{CMB}}$ (dashed line) and for the conventional U(1) theory (solid line). Right panel: $\frac{T}{T_0} \times a$ as a function of scale factor a for SU(2)$_{\text{CMB}}$. Note saturation of $\frac{T}{T_0} \times a$ to the value 0.62 for $a < 1/10$.

Prescribing $a^* = 1/10$, one obtains $T^* = 6.2\, T_0$, see right panel in Fig. 10.15.

Judging from the results of Sec. 10.4.1 and by saturation of $\frac{T}{T_0} \times a$ for $a < 1/10$ (right panel of Fig. 10.15), it is safe to (conformally) scale T with a^{-1} for $a < 1/10$. This is also expressed by the fact that the constant term 0.62 in fits of $\frac{T}{T_0} \times a$ to polynomials in a is stable under variations of the fit interval, contained in $\frac{1}{20} \le a \le 1$, and the polynomial degree. Thus we have

$$T = 0.62\, a^{-1} \times T_0, \quad (a \le \frac{1}{10}). \tag{10.67}$$

Eq. (10.67) and Fig. 10.15 state that for $a < 1$ the conformal scaling

$$T = a^{-1} \times T_0 \tag{10.68}$$

over-estimates the actual CMB temperature if it is SU(2)$_{\text{CMB}}$ that describes CMB photons correctly. Heuristically, this is a consequence of the fact that SU(2)$_{\text{CMB}}$ energy density ρ, chiefly residing in relativistic degress of freedom at high redshift, is not only reduced by spacetime growth, as in the conventional U(1) theory, but also by an investment into thermal ground-state structure and quasiparticle mass m. Recall that the former generates the latter by an increasingly efficient, adjoint Higgs mechanism as temperature drops towards T_c (or T_0), see Fig. 10.14.

Eq. (10.67) implies the ratio R_ρ of energy density ρ in SU(2)$_{\text{CMB}}$ to energy density $\rho_\gamma = \frac{\pi^2}{15} T^4$ of a conventional photon gas to be

$$R_\rho \equiv \frac{\rho}{\rho_\gamma} = 0.591 \quad (a \leq \frac{1}{10}). \tag{10.69}$$

Thus, although at high temperatures $(a \leq \frac{1}{10})$ SU(2)$_{\text{CMB}}$ possesses four times as many relativistic degrees of freedom as the conventional U(1) theory, its energy density ρ is considerably smaller than ρ_γ. Based on Eq. (10.67), we present in Sec. 10.4.2.3 observational evidence that, indeed, SU(2)$_{\text{CMB}}$ yields a better description of the CMB than conventional U(1) theory.

10.4.2.3 *The issue of early re-ionization*

Due to nonlinear structure growth at late times hydrogen gas is compacted under gravitational pull, and stars come into being. Due to their radiation, the Universe's intergalactic medium suffers re-ionization. A priori, there is no compelling reason why on cosmological time scales this process should not be considered instantaneous: The large-scale distribution of galaxies is homogeneous, and so, statistically speaking, each given and sufficiently large region of space experiences re-ionization of its hydrogen like any other region of similar extent does. This should preclude sizable retardation effects. Observationally, cosmological instantaneous re-ionization is supported by the rapid transition of the z dependence in quasar spectra. In [Becker et al.(2001)] a moderate resolution Keck spectroscopy of quasars at $z = 5.82, 5.99, 6.28$, discovered by the Sloan Digital Sky Survey (SDSS), was performed. While the two objects with $z = 5.82, 5.99$ do not exhibit the Gunn-Peterson trough, the object of highest redshift $z = 6.28$ cleary does so, suggesting that re-ionization indeed is a rapid transition occuring at $z_{re} \sim 6$. On the other hand, the latest value of a CMB based (WMAP and Planck) and Baryon Acoustic Oscillations (BAO) supported extraction of z_{re} for instantaneous re-ionization is $z_{re} \sim 11.3 \pm 1.1$, see Table 10 (last column) of [Ade *et al.* I (2013)] or Table 5 (last column) of [Ade *et al.* II (2013)]. Thus there is an obvious tension between the results for z_{re} from quasar spectra and those extracted from the CMB (plus BAO).

Let us now address this discrepancy. Quasar light propagates with energy densities orders of magnitude higher than that of the CMB. According to Sec. 9.62 this visible light should propagate in a wave-like fashion in SU(2)$_e$ and thus must be conformally red-shifted as in the conventional U(1) theory of electromagnetism. As a consequence, $z_{re} \sim 6$ of [Becker *et al.*(2001)] is to be trusted at face value as a physical redshift for instantaneous re-ionization: $z_{re} = 6$ $\Rightarrow a_{re}^{-1} = 7$.

We now test the validity of SU(2)$_{CMB}$ by appealing to the conventional assumption of conformal U(1) scaling as in Eq. (10.68) to deduce the hypothetical value of T_{re} associated with $z_{re} = 6$. This assumption underlies past and present CMB analysis, and in particular, the extraction of redshift for instantaneous re-ionization in [Ade *et al.* I (2013)]. According to Eq. (10.68) and accepting $z_{re} = 6$ as physical, we conclude that $T_{re} = 7 \times T_0$. But if nature indeed realizes SU(2)$_{CMB}$ then this (false) value of T_{re} would, according to Eq. (10.67), translate into

$$a_{re} = 0.62 \frac{T_0}{T_{re}} = 0.0886 \quad \Rightarrow \quad z_{re} = a_{re}^{-1} - 1 = 10.29 . \tag{10.70}$$

Within errors the thus determined value of $z_{re} = 10.29$ is consistent with $z_e \sim 11.3 \pm 1.1$ obtained from combined (conventional) CMB and BAO analysis (Planck plus WMAP, high l and BAO) [Ade *et al.* I (2013)]. Using the Planck data alone and invoking gravitational lensing, a somewhat lower value of z_e, subject to larger errors, was obtained by the Planck collaboration [Ade *et al.* I (2013)]: $z_{re} = 10.8_{-2.5}^{+3.1}$. Again, this is consistent with Eq. (10.70). Reasoning in this way, the discrepancy in the values of z_{re} is suggested to arise due to incorrect conformal U(1) scaling of CMB temperature when nature actually realizes SU(2)$_{CMB}$.

It is worth mentioning that the SU(2)$_{CMB}$ value of T_{re} is $T_{re} = 4.35 \times T_0 = 11.85$ K. In Sec. 10.4.3 we will see that corrections to Eq. (10.70) due to (unconventional) neutrino physics are not severe.

10.4.2.4   Redshift at CMB decoupling

The CMB decoupling temperature of about $T_{dec} = 3000$ K (for our purposes it is sufficient to assume that re-combination and CMB decoupling occur instantaneously and simultaneously) is unaffected[14] by SU(2)$_{CMB}$. According to Eq. (10.67) T_{dec} translates into a redshift z_{dec} at decoupling of

$$z_{dec} = \frac{1}{0.62} \frac{3000}{2.725} - 1 = 1775. \qquad (10.71)$$

This is substantially larger than the conventional value of $z_{dec} \sim 1100$ [Ade *et al.* I (2013)] and should have an impact on cosmological parameter values, notably matter density[15] and the Hubble parameter, H_0. A detailed investigation of this important point is beyond the scope of the present investigation.

10.4.3   *Massless neutrinos?*

10.4.3.1   *Adaption of standard treatment of neutrino temperature to SU(2)$_{CMB}$*

To start with, let us assume that neutrinos are massless and that there is no coupling between SU(2)$_{CMB}$ and the neutrino sector such that they represent separately conserved cosmic fluids. The standard

---

[14]This value of T_{dec} is a consequence of the ionisation energy of hydrogen, $E_{ion} = 13.6$ eV, and the Saha equation.

[15]Owing to SU(2)$_{CMB}$, matter density in the conventional concordance model would be $\left(\frac{1775}{1100}\right)^3 \sim 4.2$ times higher at CMB decoupling but would face the same (conventional) photon pressure.

argument of a conserved entropy density in the process of e^+e^- annihilation produces a ratio of neutrino temperature T_ν to CMB temperature T of

$$\frac{T_\nu}{T} = \left(\frac{g_1}{g_0}\right)^{1/3}, \tag{10.72}$$

where g_0 (g_1) denotes the number of relativistic degrees of freedom before (after) e^+e^- annihilation. In the conventional theory one has: $g_0 = 2 + \frac{7}{8}4$ and $g_1 = 2$ such that $\frac{T_\nu}{T} = \left(\frac{4}{11}\right)^{1/3}$. If we replace the conventional U(1) photon theory by SU(2)$_{\text{CMB}}$ then $g_0 = 8 + \frac{7}{8}4$ and $g_1 = 8$ which yields

$$\frac{T_\nu}{T} = \left(\frac{g_1}{g_0}\right)^{1/3} = \left(\frac{16}{23}\right)^{1/3}. \tag{10.73}$$

10.4.3.2 $N_{\text{eff}} \sim 3.36$ and separately conserved fluids of massless neutrinos

As we have seen in Sec. 10.4.2.2, it is safe to consider conventional, conformal scaling of T versus a^{-1} for $a \le 1/10$. Recall that $a = 1/10$ corresponds to $T = 6.2\,T_0$. Based on SU(2)$_{\text{CMB}}$ the effective number of neutrino flavours N_{eff} at $T = T_0$, as judged by the conventional theory in terms of the actual number of massless neutrino flavours N_ν, reads (total energy density in relativistic degrees of freedom minus energy density in photons divided by conventional energy density per massless neutrino flavour)

$$N_{\text{eff}} = \frac{\frac{7}{8}N_\nu(0.62)^4\left(\frac{16}{23}\right)^{4/3}}{\frac{7}{8}\left(\frac{4}{11}\right)^{4/3}}. \tag{10.74}$$

Note that SU(2)$_{\text{CMB}}$ effects creep into the numerator of Eq. (10.74) in terms of the factors $(0.62)^4$, related to the fact that T_ν of massless neutrinos always follows the conventional scaling law $T_\nu \propto a^{-1}$ while there are low-redshift violations thereof for T, and $\left(\frac{16}{23}\right)^{4/3}$, arising due to eight instead of two relativistic degrees of freedom in SU(2)$_{\text{CMB}}$ during e^+e^- annihilation (Sec. 10.4.3.1), deviating from unity and from $\left(\frac{4}{11}\right)^{4/3}$, respectively. For $N_\nu = 3$ we have $N_{\text{eff}} = 1.053$ which is far off[16] the observationally determined value of $N_{\text{eff}} \sim 3.36$ [Ade *et al.* I (2013)].

Interestingly, the value of N_{eff} depends on the redshift at which it is determined. For example, one obtains for $a < 1/10$

$$N_{\text{eff}} = \frac{\frac{7}{8} N_\nu \left(\frac{16}{23}\right)^{4/3} + 3}{\frac{7}{8} \left(\frac{4}{11}\right)^{4/3}}. \tag{10.75}$$

For $N_\nu = 3$ we would have $N_{\text{eff}} = 20.33$ instead of the value $N_{\text{eff}} = 1.053$ extracted at present. From now on we associate N_{eff} with its value today.

10.4.3.3   *CMB thermalized neutrinos*

The results of Sec. 10.4.3.2 suggest that SU(2)$_{\text{CMB}}$ and conventional neutrino physics are not compatible. Note that this extends to the case of neutrinos with fixed masses since the latter would reduce rather than enhance their contribution to the Universe's present energy density. Guided by results on how an SU(2) center-vortex responds to environmental conditions (putting forward an effective scale of resolution), see Sec. 7.3.1, a bold suggestion on how to circumvent these difficulties is to assume that a given neutrino

---

[16]If the ground-state energy density of SU(2)$_{\text{CMB}}$, $T_0^{-4}\rho_{\text{gs}}(T_0) = 32\pi^4\lambda_c^{-3}$ (scaled dimensionless by multiplication with T_0^{-4}), is added to the numerator of Eq. (10.74) then one obtains $N_{\text{eff}} = 6.2$. Clearly, this is also out of range. However, since (modulo small evanescence effects, see Sec. 9.4.3.2, CMB photons decouple from their ground state at T_0 the ground-state part of SU(2)$_{\text{CMB}}$ at present should be viewed as a (tiny) contribution to dark energy rather than dark radiation. Therefore, we do not in the following consider $\rho_{\text{gs}}(T_0)$ anymore when inferring N_{eff} from SU(2)$_{\text{CMB}}$ and neutrino physics at present.

flavor is represented by a single center vortex loop of a respective SU(2) Yang-Mills theory and that each of these theories underwent its preconfing-confining phase transition well before CMB decoupling. Due to its extendedness and (after an electric-magnetically dual interpretation) its unit of electric center flux such a center-vortex loop interacts with the CMB. As a consequence, neutrino temperature T_ν and CMB temperature T would coincide: $T_\nu \to T$ or $\left(\frac{16}{23}\right)^{1/3} \to 1$ in Eq. (10.72). Also, no additional split of T_ν and T at low z due to a violation of conformal scaling in the photonic sector would occur.

Because it is technically simpler let us first assume that neutrinos, due to their interactions with the CMB, exhibit the same temperature, $T_\nu = T$, but that they remain massless. In this idealization Eq. (10.74) modifies as

$$N_{\text{eff}} = \frac{\frac{7}{8} N_\nu}{\frac{7}{8}\left(\frac{4}{11}\right)^{4/3}} . \tag{10.76}$$

With $N_\nu = 3$ one obtains $N_{\text{eff}} = 11.56$ which is much too high. To reduce N_{eff} down to its physical value $N_{\text{eff}} = 3.36$ neutrinos need to acquire mass through interactions with the CMB.

Due to an SU(2) center vortex loop possessing electromagnetic properties only, it exclusively interacts with the photonic part of SU(2)$_{\text{CMB}}$. Modulo radiative corrections, which are inessential for the present discusssion, the thermal photon gas in SU(2)$_{\text{CMB}}$ is characterized by the single scale T. Thus, the response of the neutrino sector in terms of neutrino mass emergence must be such that $m_\nu = \xi T$ where ξ is a dimensionless constant of order unity. For cosmic neutrinos we assume m_ν to be universal, that is, flavor independent. We have given arguments in [Hofmann (2013)] on the viability of such a universal, temperature dependent neutrino mass in view of the overclosure bound of $\sim 15\,\text{eV}$, see [Lesgourges and Pastor (2014)]. Namely, for $0 \le \xi \le 10$ this bound is evaded up to $z \sim 10^4$ which is well before CMB decoupling. On the other hand, lower bounds on the sum of neutrino masses of the order 10^{-1} eV [Forero *et al.* (2012)], posed by the two scenari of mass hierarchy (normal and inverted) to explain neutrino oscillations, are invalid for the here proposed temperature dependent mass of cosmic neutrinos because they refer to neutrino environments which are largely disparate from the CMB (neutrino generation and propagation in long baseline reactor, atmospheric, and solar neutrino experiments).

With a universal, cosmic neutrino mass $m_\nu = \xi T$ the pressure P_ν and energy density ρ_ν is given as (2 spin orientations per flavor)

$$P_\nu = N_\nu T^4 \frac{1}{\pi^2} \int_0^\infty dx\, x^2 \log\left(1 + \exp\left(-\sqrt{x^2 + \xi^2}\right)\right) \equiv N_\nu T^4 \hat{P}_\nu(\xi),$$

$$\rho_\nu = N_\nu T^4 \frac{1}{\pi^2} \int_0^\infty dx\, \frac{x^2 \sqrt{x^2 + \xi^2}}{1 + \exp\sqrt{x^2 + \xi^2}} \equiv N_\nu T^4 \hat{\rho}_\nu(\xi). \qquad (10.77)$$

This model for the neutrino gas is not thermodynamically consistent by itself. (The CMB acts as a thermal background prescribing mass and inheriting its temperature to the neutrino gas.) Conservation of energy in an expanding Universe thus must be imposed onto the total energy density and pressure: $\rho \to \rho_{\text{tot}} = \rho + \rho_\nu$ and $P \to P_{\text{tot}} = P + P_\nu$ in Eq. (10.66). The model of Eqs. (10.77) exhibits energy density and pressure which both are proportional to T^4 like the CMB does (the photonic part of SU(2)$_{\text{CMB}}$ or the entire theory SU(2)$_{\text{CMB}}$ for $T \gg T_0$). That is, for $T \gg T_0$, the ratio $\frac{P_{\text{tot}}}{\rho_{\text{tot}}} \equiv \kappa$ is independent of T. It reads

$$\kappa = \frac{\frac{1}{3} + \frac{15}{4\pi^2} N_\nu \hat{P}_\nu(\xi)}{1 + \frac{15}{4\pi^2} N_\nu \hat{\rho}_\nu(\xi)}, \qquad (T \gg T_0). \qquad (10.78)$$

Fig. 10.16 depicts κ as a function of ξ for $T \gg T_0$ and $N_\nu = 3$. Because of the small deviation $\epsilon \equiv \frac{1}{3} - \kappa$, reaching its maximal value $\epsilon_{\text{max}} = 0.0315$ at $\xi = 2.77$, the high-temperature scaling of this combination of SU(2)$_{\text{CMB}}$ with $N_\nu = 3$ temperature dependent massive neutrino flavors exhibits only weak deviations from conformal scaling,

$$\frac{T}{T_p} = a^{-1 + \frac{3}{4}\epsilon}, \qquad (T \geq T_p; T_p \gg T_0; a(T_p) = 1), \qquad (10.79)$$

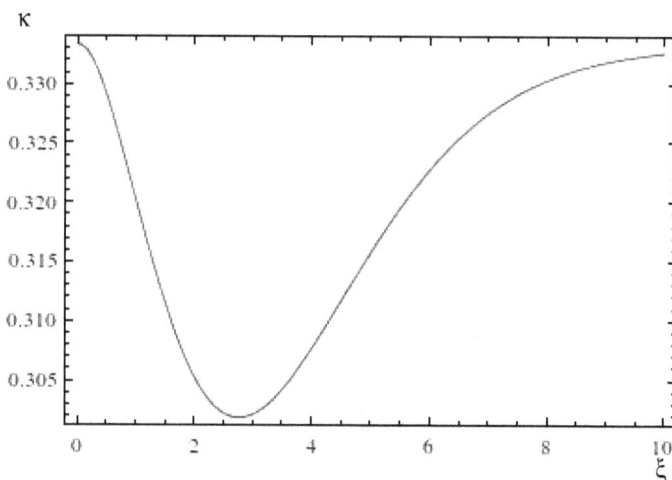

Figure 10.16.   The equation-of-state parameter κ for the model of SU(2)$_{\text{CMB}}$ combined with $N_\nu = 3$ temperature dependent massive neutrino flavors, see Eq. (10.77), as a function of ξ for $T \gg T_0$. Note the smallness of the deviation from $\kappa = \frac{1}{3}$, corresponding to the equation of state of a free gas of massless particles. The minimum of κ (or maximal deviation from $\kappa = \frac{1}{3}$) is at $\xi = 2.77$. For a comparison, the maximum of the black-body spectral energy density occurs at $\frac{\omega}{T} = 2.82$.

where T_p denotes a pivotal temperature within the high-temperature regime.

Let us now determine the value of ξ such that todays's value of $N_{\text{eff}}(\xi)$, defined as

$$N_{\text{eff}}(\xi) = \frac{N_\nu \hat{\rho}_\nu(\xi)}{\frac{7}{8}\frac{\pi^2}{15}\left(\frac{4}{11}\right)^{4/3}}, \tag{10.80}$$

matches $N_{\text{eff}} = 3.36$, the value observationally determined in [Ade et al. I (2013)]. We find

$$\xi = 3.973, \tag{10.81}$$

which is to the right of the minimum in Fig. 10.16. This corresponds to $\epsilon = 0.0263$. In Fig. 10.17 we show in analogy to Fig. 10.15 how scaling violation occurs in SU(2)$_{\text{CMB}}$ combined with $N_\nu = 3$ temperature dependent massive neutrino flavours ($\xi = 3.973$). In contrast

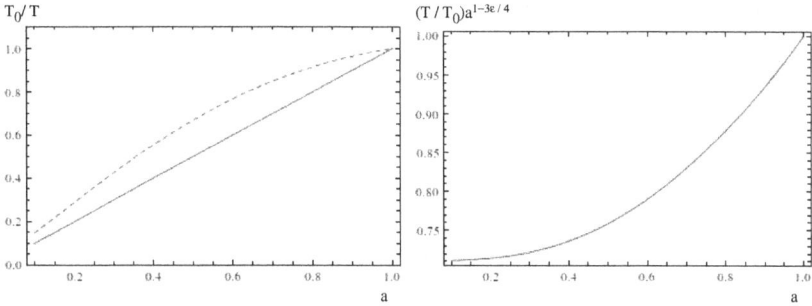

Figure 10.17. Violation of conformal scaling as induced by SU(2)$_{\text{CMB}}$ combined with $N_\nu = 3$ temperature dependent massive neutrino flavors. Left panel: $\frac{T_0}{T}$ as a function of scale factor a for this model (dashed line) and for a conventional U(1) photon theory (solid line). Right panel: $\frac{T}{T_0} \times a^{1-\frac{3}{4}\epsilon}$, compare with Eq. (10.79), as a function of scale factor a for this model. Saturation to $\frac{T}{T_0} \times a^{1-\frac{3}{4}\epsilon} = 0.71$ occurs for $a < 1/10$ which is consistent with $\frac{T}{T_0} = 6.8$ at $a = 1/10$ and $\epsilon = 0.0263$.

to the case of pure SU(2)$_{\text{CMB}}$ theory there is a violation of conformal U(1) scaling also for $T \gg T_0$. This is expressed by Eq. (10.79).

Employing Eq. (10.79) with $T_p = 6.8\,T_0$, we now estimate, in analogy to Sec. 10.4.2.4, the redshift z_{dec} at CMB decoupling as

$$z_{\text{dec}} = 10 \left(\frac{3000}{6.8 \times 2.725} \right)^{\frac{1}{1-\frac{3}{4}\epsilon}} - 1 = 1793.5\,. \tag{10.82}$$

Interestingly, this value of z_{dec} differs from $z_{\text{dec}} = 1775$, obtained for the case of pure SU(2)$_{\text{CMB}}$ (see Eq. (10.71)), by only 1%.

What about the redshift z_{re} for instantaneous re-ionization? The same reasoning as in Sec. 10.4.2.3 but now subject to the modified scaling of Eq. (10.79) yields

$$a_{\text{re}} = \frac{1}{10} \left(\frac{6.8\,T_0}{T_{\text{re}}} \right)^{\frac{1}{1-\frac{3}{4}\epsilon}} = 0.0970862 \quad \Rightarrow \quad z_{\text{re}} = a_{\text{re}}^{-1} - 1 = 9.3\,,$$

$$\tag{10.83}$$

where $T_{\text{re}} = 7 \times T_0$. This value of z_{re} remains compatible with the one extracted by the Planck collaboration using the CMB and gravitational lensing only, $z_{\text{re}} = 10.8\pm^{3.1}_{2.5}$ [Ade $et\,al.$ I (2013)], it differs from $z_{\text{re}} = 10.29$, obtained for the case of pure SU(2)$_{\text{CMB}}$ (see Eq. (10.70)), by -9.6%, and it is at slight tension with the value $z_{\text{re}} \sim 11.3 \pm 1.1$ obtained from combined (conventional) CMB and BAO analysis.

10.4.4   *Summary*

The physics addressed in the previous sections, which only requires the free quasiparticle level in describing thermodynamical quantities in the deconfining phase of SU(2)$_{\text{CMB}}$, is much more basic than SU(2)$_{\text{CMB}}$'s imprint on large-angle anisotropies due to particular radiative corrections as discussed in the early sections of this chapter. Essentially, due to its extra (massive) degrees of freedom and its thermal ground state, SU(2)$_{\text{CMB}}$ possesses a higher heat capacity than the conventional U(1) based photon theory. This implies that, as the Universe is compressed to a state of redshift $z = a^{-1} - 1$, CMB temperature T does not follow the conformal scaling law $T = a^{-1} T_0$ but increases with a slower rate as a function of a^{-1}. This could explain a discrepancy between the value of z_{re} for (instantaneous) re-ionization extracted from spectra of high-redshift quasars and the angular power spectrum of the CMB. We also have addressed how the SU(2)$_{\text{CMB}}$ scenario could affect cosmic neutrinos. An apparently viable scenario, where, by interacting with the CMB, cosmic neutrinos stream at CMB temperature T and acquire a mass $\propto T$, does not qualitatively change the prediction for z_{dec} (redshift at which CMB decouples) and for z_{re}. Essentially, the violation of the simple $T \propto a^{-1}$ scaling at low redshift moves z_{dec} from $z_{\text{dec}} \sim 1100$ to $z_{\text{dec}} \sim 1800$. To keep, at CMB decoupling, the ratio of thermal photon energy density to matter density unchanged, matter density would have to be rescaled by a factor $\left(\frac{1100}{1800}\right)^3 \sim 0.23$ yielding a new Ω_m of about 7 %. This, however, is close to the present baryonic density ($\sim 5.5\,\%$). At high redshifts the validity of the ΛCDM concordance model is thus questioned. In a cosmological model void of dark matter, the strong Ω_m component seen in cosmologically local signatures (luminosity distance - redshift curves, large-scale structure surveys) could be an indication of the onset of coherent oscillations of a homogeneous dark-energy field at late times, see Sec. 10.3.2, and conventional dark-matter halos, thought to be responsible for the flattening of the rotation curves of spiral galaxies, could be mimicked by (topologically stabilised) solitonic configurations of such a field [Wetterich (2001)].

To rule out or strengthen the here proposed scenario for CMB and cosmic neutrino physics, dedicated simulations of the implied cosmological model are required resting on data from both local cosmology and the CMB angular correlation functions.

References

Ade, P. A. R. *et al.* I (2013). Planck 2013 results. XVI. Cosmological parameters, arXiv:1303.5076v3.

Ade, P. A. R. *et al.* II (2013). Planck 2013 results. I. Overview of products and scientific results, arXiv:1303.5062v2 (2013).

Adler, S. L. (1969). Axial vector vertex in spinor electrodynamics, *Phys. Rev.*, **177**, 2426.

Adler, S. L. and Bardeen, W. A. (1969). Absence of higher order corrections in the anomalous axial vector divergence equation, *Phys. Rev.*, **182**, 1517.

Barbieri, R., *et al.* (2005). Dark energy and right-handed neutrinos, *Phys. Lett. B*, **625**, 189.

Becker, R. H. *et al.* Evidence for Reionization at $z \sim 6$: Detection of a Gunn-Peterson Trough in a $z = 6.28$ Quasar, *Astron. J.*, **122**, 2850.

Bell, J. S. and Jackiw, R. (1969). A PCAC puzzle: $\pi^0 \to \gamma\gamma$ in the sigma model, *Nuovo Cim. A*, **60**, 47.

Candelas, P. and Raine, D. J. (1975). General relativistic quantum field theory — An exactly soluble model, *Phys. Rev. D*, **12**, 965.

Copi, C. J., Huterer, D., and Starkman, G. D. (2004). Multipole vectors — A new representation of the CMB sky and evidence for statistical anisotropy or non-Gaussianity at $2 \leq l \leq 8$, *Phys. Rev. D*, **70**, 043515.

Copi, C. J., *et al.* (2010). Large angle anomalies in the CMB., *Adv. Astron.*, **2010**, 847541.

de Oliveira-Costa, A., *et al.* (2004). The Significance of the largest scale CMB fluctuations in WMAP, *Phys. Rev. D*, **69**, 063516.

Eidelman, S., *et al.* (2004). Review of particle physics. Particle data group, *Phys. Lett. B*, **592**, 1.

Erdogdu, P., *et al.* (2006). Reconstructed density and velocity fields from the 2MASS redshift survey, *Mon. Not. Roy. Astron. Soc.*, **373**, 45.

Erdogdu, P., *et al.* (2006). Talk given at 41st Rencontres de Moriond, Workshop on Cosmology: Contents and Structures of the Universe, La Thuile, Italy, 18–25 March 2006.

Frieman, J., Hill, C. T., and Watkins, R. (1992). Late time cosmological phase transitions. 1. Particle physics models and cosmic evolution, *Phys. Rev. D*, **46**, 1226.

Frieman, J. A., *et al.* (1995). Cosmology with ultralight pseudo Nambu–Goldstone bosons, *Phys. Rev. Lett.*, **75**, 2077.

Forero, D. V., Tortola, M., and Valle, J. W. F. Global status of neutrino oscillation parameters after Neutrino-2012, Phys. Rev. D, **86**, 073012.

Fujikawa, K. (1979). Path integral measure for gauge invariant fermion theories, *Phys. Rev. Lett.*, **42**, 1195.

Fujikawa, K. (1980). Path integral for gauge theories with fermions, *Phys. Rev. D*, **21**, 2848, Erratum-ibid. *Phys. Rev. D*, **22**, 1499.

Giacosa, F. and Hofmann, R. (2007). A Planck-scale axion and SU(2) Yang–Mills dynamics: Present acceleration and the fate of the photon, *Eur. Phys. J. C*, **50**, 635.

Giacosa, F., Hofmann, R., and Neubert, M. (2008). A model for the very early Universe, *JHEP*, **0802**, 077.

Goldsmith, P., Li, D., and Krčo, M. (2007). The transition from atomic to molecular hydrogen in interstellar clouds: 21 cm signature of the evolution of cold atomic hydrogen in dense clouds, *Astrophys. J.*, **654**, 273.

Gradshteyn, I. S. and Ryzhik, I. M. (1965). *Table of Integrals, Series, and Products* (Academic Press, New York and London).

Hall, J. H., Nomura, Y., and Oliver, S. J. (2005). Evolving dark energy with w deviating from -1, *Phys. Rev. Lett.*, **95**, 141302.

Hinshaw, G., *et al.* (2006). Three-year Wilkinson Microwave Anisotropy Probe (WMAP) observations: temperature analysis, *Astrophys. J. Suppl.*, **170**, 288 (2007).

Hofmann, R. (2005). Nonperturbative approach to Yang–Mills thermodynamics, *Int. J. Mod. Phys. A*, **20**, 4123, Erratum-ibid. A **21**, 6515, (2006).

Hofmann, R. (2013). The fate of statistical isotropy. Nature Phys., **9**, 686.

Hofmann, R. (2015). Relic photon temperature versus redshift and the cosmic neutrino background, *Annalen Phys.*, **527**, 254.

Kaloper, N. and Sorbo, L. (2005). Of PNGB quintessence, *JCAP*, **0604**, 007.

Kaviani, D. and Hofmann, R. (2007). Irreducible three-loop contributions to the pressure in Yang–Mills thermodynamics, *Mod. Phys. Lett. A*, **22**, 2343.

Keller, J., Hofmann, R., and Giacosa, F. (2008). Correlation of energy density in deconfining SU(2) Yang–Mills thermodynamics, *Int. J. Mod. Phys. A*, **23**, 5181.

Keller, J. (2008). Gauge-invariant two-point correlator of energy density in deconfining SU(2) Yang–Mills thermodynamics, Diploma thesis, Universität Heidelberg, arXiv:0801.3961v4 [hep-th].

Kim, J. E. and Nilles, H. P. (2003). A Quintessential axion, *Phys. Lett. B*, **553**, 1.

Knee, L. B. G. and Brunt, C. M. (2001). A massive cloud of cold atomic hydrogen in the outer Galaxy, *Nature*, **412**, 308.

Kogut, A., *et al.* (1993). Dipole anisotropy in the COBE DMR first year sky maps, *Astrophys. J.*, **419**, 1.

Lesgourges, J. and Pastor, S. (2014). Neutrino cosmology and Planck. New J. Phys., **16**, 065002.

Ludescher, J. and Hofmann, R. (2009). Thermal photon dispersion law and modified black-body spectra, *Ann. d. Phys.*, **18**, 271.

Ludescher, J. and Hofmann, R. (2009). CMB dipole revisited, arXiv:0902.3898 [hep-th].

Mather, J. C., *et al.* (1994). Measurement of the cosmic microwave background spectrum by the COBE FIRAS instrument, *Astrophys. J.*, **420**, 439.

Meyer, D. M. and Lauroesch, J. T. (2006). A cold nearby cloud inside the local bubble, *Astrophys. J.*, **650**, L67.

Nakamura, K., *et al.* (2010). Review of particle physics, *J. Phys. G*, **37**, 075021.

Padmanabhan, T. (2003). Cosmological constant: The weight of the vacuum, *Phys. Rept.*, **380**, 235.

Peccei, R. D. and Quinn, H. R. (1977). CP conservation in the presence of instantons, *Phys. Rev. Lett.*, **38**, 1440.

Peebles, P. J. E. and Wilkinson, D. T. (1968). Comment on the anisotropy of the primeval fireball, *Phys.Rev.*, **174**, 2168.

Perlmutter, S., *et al.* (1998). Measurements of omega and lambda from 42 high redshift supernovae, *Astrophys. J.*, **517**, 565.

Preskill, J, Wise, M. B., and Wilczek, F. (1983). Cosmology of the invisible axion, *Phys. Lett. B*, **120**, 127.

Redfield, S. and Linsky, J. L. (2007). The structure of the local interstellar medium IV: Dynamics, morphology, physical properties, and implications of cloud-cloud interactions, arXiv: 0709.4480 [astro-ph].

Riess, A. G., *et al.* (1998). Observational evidence from supernovae for an accelerating Universe and a cosmological constant, *Astronom. J.*, **116**, 1009.

Schiesser, E. (1994). Computational Mathematics in Engineering and Applied Science: ODEs, DAEs and PDEs (CR Press).

Schwarz, M., Hofmann, R., and Giacosa, F. (2007). Radiative corrections and the one-loop polarization tensor of the massless mode in SU(2) Yang–Mills thermodynamics, *Int. J. Mod. Phys. A*, **22**, 1213.

Spergel, D. N. (2003). First year Wilkinson Microwave Anisotropy Probe (WMAP) observations: Determination of cosmological parameters, *Astrophys. J. Suppl.*, **148**, 175.

Spergel, D. N., *et al.* (2007). Wilkinson Microwave Anisotropy Probe (WMAP) three year results: implications for cosmology, *Astrophys. J.Suppl.*, **170**, 377.

Szopa, M. and Hofmann, R. (2008). A model for CMB anisotropies on large angular scales, *JCAP*, **03**, 001.

Tegmark, M., de Oliveira-Costa, A., and Hamilton, A. J. S. (2003). A high resolution foreground cleaned CMB map from WMAP, *Phys. Rev. D*, **68**, 123523.

't Hooft, G. (1976). Symmetry breaking through Bell–Jackiw anomalies, *Phys. Rev. Lett.*, **37**, 8.

Wetterich, C. (2001). Are galaxies cosmon lumps?, *Phys. Lett. B*, **522**, 5.

Wilczek, F. (2004). A Model of anthropic reasoning, addressing the dark to ordinary matter coincidence, Solicited article for the book *Universe or Multiverse*, ed. B. Carr, hep-ph/0408167.

Acknowledgments

Much of the motivation for research in gauge theory as well as the shaping and consolidation of the results presented in this book was due to inspiring, conceptually and technically valuable, and, at times, highly critical discussions with colleagues. In particular, I would like to acknowledge eight productive and exhausting but often joyful years spent in collaborating with research students at the University of Heidelberg, the University of Karlsruhe and the Karlsruhe Institute of Technology: Carlos Falquez, Steffen Hahn, Ulrich Herbst, Dariush Kaviani, Jochen Keller, Niko Krasowski, Josef Ludescher, Julian Moosmann, Jochen Rohrer, Sebastian Scheffler, Markus Schwarz, and Michal Szopa. To Francesco Giacosa, whose excitement, common sense, critical judgement, high motivation, and exceptional devotion to joint work on Yang–Mills thermodynamics was instrumental in maintaining momentum after the year 2005, I am deeply grateful. I also would like to thank those colleagues whose help in presenting and discussing our ideas and results at seminars, conferences, workshops, or privately, by publishing them in journals, or by acquiring research funds was essential to keep the process alive: Tilo Baumbach, Robert Brandenberger, Stan Brodsky, Xavier Calmet, Carlo Contaldi, John Cornwall, Michael Creutz, the late Dima Diakonov, Gerald Dunne, Amand Faessler, Dale Fixsen, Herb Fried, Gregory Gabadadze, Thierry Grandou, Alan Guth, André Hoang, Gerard 't Hooft, Antal Jevicki, Bernd Kämpfer, Frans Klinkhamer, the late Lev Kofman, Alan Kogut, Nick Manton, John Moffat, Otto Nachtmann, Werner Nahm, John

Negele, Matthias Neubert, Joe Polchinski, Janos Polonyi, Tomislav Prokopec, Krishna Rajagopal, Hugo Reinhardt, Dirk Rischke, Martin Roček, Joao Rodriguez, Heinz Jürgen Rothe, Qaisar Shafi, Misha Shifman, Martin Speight, Nucu Stamatescu, Paul Sutcliffe, Chung-I Tan, Anthony Thomas, Gerhard Ulm, the late Pierre van Baal, Arkady Vainshtein, Jacobus Verbaarschot, Werner Wetzel, Frank Wilczek, Andreas Wipf, Mark Wise, and Toby Wiseman. For help with the literature concerning W. Pauli's role in the discovery of non-Abelian gauge theory and towards expositions of the ADHM construction and the Nahm transformation, I am indebted to the historian Helmut Rechenberg and the physicist Jan Pawlowski. To my editors at World Scientific, Mr. Yee Sern Tan, Mr. Mu Sen, and Mr. Low Lerh Feng go my special thanks for their professional help and patience during the process of manuscript preparation. Also, I would like to express my gratitude to Professor Phua Kok Khoo, Editor-in-Chief of World Scientific, for paving the way to publication and his continuous support. I owe a lot of moral and practical support to my family. In particular, I would like to thank Karin Thier and our children Cattleya and Emmy for their understanding of why I had to withdraw from social life for so many evenings to produce this book. Thank you, Karin, for braving the consequences! Markus Schwarz' and Heinz Jürgen Rothe's valuable comments on the manuscript and Julian Moosmann's overhaul of the layout are gratefully acknowledged. Last but not least, many thanks go to my father Rolf Hofmann for the creation of the water-color painting on the book jacket.

R. Hofmann

Author Index

Abrikosov, A. A., 22, 303, 364, 377
Ade, P. A. R. *et al.*, 493, 494, 498, 499, 501, 504, 505
Adler, S. L., 67, 87, 394, 482, 483
Altschuler, S. J., 331
Anderson, J. O., 19
Anderson, P. W., 6
Appelquist, T., 56
Arnold, P., 58
Atiyah, M. F., 67, 68, 78, 80, 88

Bali, G. S., 256, 379
Barbier, R., 484
Bardeen, W. A., 67, 394, 483
Barker, W. A., 5
Bassetto, A., 249
Baumbach, T., 218, 402, 416
Becchi, C. M., 31
Dell'Antonio, G., 31
Becherer, R. J., 416, 423
Becker *et al.*, 494, 497, 498
Belavin, A. A., 21, 67, 70
Bell, J. S., 67, 482, 483
Bender, C. M., 320, 329
Bethe, H., 386
Bjerrum-Bohr, N. E. J., 155
Bloch, F., 386
Boggess, N. W., 390
Bogomolnyi, E. B., 48
Borgs, C., 255, 256, 378
Bornyakov, V. G., 377
Bose, S. N., 30
Boyd, C., 372, 375, 376
Braam, P. J., 89
Braaten, E., 57, 58

Brodsky, S. J., 155
Broucke, M. E., 334, 403, 463
Brout, R., 6
Brown, F. R., 368, 371, 375
Bruckmann, F., 377
Brunt, C. M., 391, 403, 404, 447, 463
Busam, R., 325

Candelas, P., 483
Chern, S.-S, 70
Christ, N. H., 79
Collins, J. C., 151
Copi, C. J., 391
Corrigan, E., 79–81

D'Elia, M., 380
de Forcrand, P., 380
de Oliveira-Coasta, A., 465, 479
Del Debbio, L., 305
Dell'Antonio, G., 31
Deng, Y., 368, 374–376
Diakonov, D., 91, 96, 103, 105, 137, 243, 377, 378
Dirac, P. A. M., 62, 81, 85, 280, 282, 283, 319, 386, 388, 389
Dolan, L., 44
Donaldson, S. L., 67, 89
Donoghue, J. F., 155
Doroshkevich, A. G., 394, 395
Drinfeld, V. G., 68, 78
Duncan, A., 151
Dyson, F. J., 211, 386

Subject Index

www.ingramcontent.com/pod-product-compliance
Lightning Source LLC
Chambersburg PA
CBHW052116230326
41598CB00079B/3705